Darwin's Psychology

The Theatre of Agency

Darwin's Psychology

The Theatre of Agency

BEN BRADLEY

Professor Emeritus, Charles Sturt University, Australia

OXFORD
UNIVERSITY PRESS

OXFORD

UNIVERSITY PRESS

Great Clarendon Street, Oxford, OX2 6DP,
United Kingdom

Oxford University Press is a department of the University of Oxford.
It furthers the University's objective of excellence in research, scholarship,
and education by publishing worldwide. Oxford is a registered trade mark of
Oxford University Press in the UK and in certain other countries

First Edition published in 2020

Impression: 3

Published in the United States of America by Oxford University Press
198 Madison Avenue, New York, NY 10016, United States of America

British Library Cataloguing in Publication Data

Data available

Library of Congress Control Number: 2020944739

ISBN 978–0–19–870821–6

Printed and bound by
CPI Group (UK) Ltd, Croydon, CR0 4YY

I dedicate this book to:
Jane Selby, for her love, life-force, and intellectual reach
Peter Sylvester-Bradley, fossil-hunter and visionary evolutionist
Colwyn Trevarthen, mentor and inspiration

There is nothing new in the world except the history you do not know.[1]

[1] Merle Miller, *Plain Speaking: An Oral Biography of Harry S Truman* (London: Gollancz, 1974), p.26.

Preface

What discipline is more at issue in plumbing today's problems, in grounding human self-knowledge, in leading a happy life, or more fascinating for a curious naturalist, news buff, film watcher, novel reader, or flâneur, than the study of mind and behaviour? Yet modern psychology has faced crisis from birth. After several new dawns faded, 1913 finally saw publication of a manifesto that stuck: the launch of *behaviourism* (when italicized, specialized terms *may* be defined in the Glossary at the end of the book). 'Behaviourism' boosted itself by deriding the work of two men who had previously been the discipline's best-known exponents—Sigmund Freud and William James—along with their signature concepts, the unconscious and introspection.[2]

The proper focus for psychological studies has remained in dispute from that day to this. Should mind scientists research solely the visible, measurable movements that behaviourists call *responses*? Or should they target mental states, consciousness, the inner life, or even the repressed conflicts glimpsed through dreams? And, if so, how could psychology ever be truly scientific? By the 1920s, the word crisis was referring to the embarrassing proliferation of incommensurable schools within the discipline: structuralism; functionalism; behaviourism; psychoanalytical; Gestalt; cultural-historical; social; personality; developmental; biological; comparative.[3] More and more different starting-points were being invented and pursued at the same time in separate silos. Such proliferation has not slackened: humanist; ego; phenomenological; sociobiological; cognitivist; critical; discursive; dialectical; feminist; hermeneutic; postmodern; transpersonal; cognitive-behavioural; positive; and evolutionary psychologies continue to hatch and multiply.

Against this background, the need to present psychology as a coherent way of studying humans forced its adepts to bracket off discussion about starting-points and agree, instead, to agree on methods: the emblem of the laboratory; the ideal of experimentation; the importance of precisely defining variables; the careful aggregation of numerical data; and their statistical analysis.[4] Meanwhile the guild-like edifice of professionalized psychology has grown ever more muscular. Only

---

[2] John Watson, 'Psychology as the Behaviourist Views It,' *Psychological Review* 20, (1913); John Watson, *Behaviorism* (London: Kegan Paul & Co, 1925), pp.158–177.

[3] Csaba Pléh, 'Two Conceptions of the Crisis of Psychology: Vygotsky and Bühler,' in *Karl Bühler's Theory of Language*, ed. Achim Eschbach (Amsterdam: John Benjamins, 1988), pp.407–413.

[4] Hank Stam, 'Unifying Psychology: Epistemological Act or Disciplinary Maneuver?,' *Journal of Clinical Psychology* 60, (2004), pp.1259–1262.

members eligible to the guild today have the legal right to advertise themselves as psychologists. And eligibility depends on an increasingly long, costly, and intensive learning about psychology's methods of research, along with their most up-to-date fruit.[5]

But crises keep coming. From the 1960s on, the worm turned inwards, infecting trust in the validity of the experimental method itself. First came proof that the people upon whom psychologists experimented were far less passive than assumed, second-guessing the aims of the experiments in which they were enrolled, and then acting accordingly. This fed doubts about the relevance of research findings to anything beyond the little social drama of the particular study that gave rise to them.[6] Next came ethical questions about psychologists' habit of deceiving the people they studied.[7] Not far behind were political questions about psychologists' assumptions around normality, gender, society, and the uses of research to maintain the more powerful at the expense of the less.[8]

Lately, we have the replication crisis. Our discipline staggers under the discovery that many of the findings produced by our most prized methods of research cannot be replicated in follow-up studies.[9] Meanwhile, academic psychologists take pause before mounting evidence that the practical branch of the profession, counselling and psychotherapy, succeeds *despite*, not because of, the experiment-larded knowledge base provided by undergraduate courses—practitioners reporting they value reflection on their own and their colleagues' practical experience far more than statistical studies.[10] Practice-based evidence supplants evidence-based practice.

This book turns back from the never-ending crises of psychological modernism, with its venerable stresses on: experiments as the epitome of methodological rigour; 'narrowness' of analysis; precise categorization; and deduction over description (see Ch. 2).[11] My subject, Charles Darwin, developed an

---

[5] Ben Bradley, 'Pedagogy,' in *Encyclopedia of Critical Psychology*, ed. Thomas Teo (New York: Springer, 2014), https://doi.org/10.1007/978-1-4614-5583-7_213.

[6] Jerry Suls and Ralph Rosnow, 'Concerns About Artifacts in Psychological Experiments,' in *The Rise of Experimentation in American Psychology*, ed. Jill Morawski (New Haven CT: Yale University Press, 1988), pp.163–187; Kurt Danziger, *Constructing the Subject: Historical Origins of Psychological Research* (Cambridge: Cambridge University Press, 1990).

[7] Alan Elms, 'The Crisis of Confidence in Social Psychology,' *American Psychologist* 30, (1975), pp.967–976.

[8] Ian Parker, *The Crisis in Modern Psychology and How to End It* (London: Taylor & Francis, 2013).

[9] Scott Maxwell, Michael Lau, and George Howard, 'Is Psychology Suffering from a Replication Crisis?: What Does "Failure to Replicate" Really Mean?,' *American Psychologist* 70, (2015), pp.487–498; Marcus Munafo and John Sutton, 'There's This Conspiracy of Silence around How Science Really Works,' *Psychologist* 30, (2017), pp.46–49.

[10] Robyn Dawes, *House of Cards: Psychology and Psychotherapy Built on Myth* (New York: Free Press, 1996); Ben Bradley, 'Rethinking "Experience" in Professional Practice: Lessons from Clinical Psychology,' in *Understanding and Researching Professional Practice*, ed. Bill Green (Rotterdam: Sense Publishers, 2009), pp.65–82; Jodie Goldney, 'The Othering of a Profession,' *Psychreg Journal of Psychology* 2, (2018), pp.56–67.

[11] See James Angell, 'The Influence of Darwin on Psychology,' *Psychological Review* 16, (1909), pp.153, 158 who conspicuously named and extolled these virtues at the dawn of modern psychology.

approach to agency that looked askance at such stresses, preferring to regard psychological matters in the same broad and general light that he regarded all the other problems of nature.[12] Examining the main features of his approach thus allows me to reconnect the 'how' of studying our fellow creatures with common sense. In so doing, I exploit biology's growing awareness that evolution is organism-led—an awareness which brings further appreciation to the unsung breadth and insight of Darwin's understanding of the agency and interdependence of living things.

Modernist narrowness is a severe handicap when approaching a man of science as deep in his thinking and broad in his achievements as Darwin. Let alone psychology, and his fame as an evolutionist, Darwin was a pioneer in microscopy, taxonomy, ecology, geology, ethology, and plant physiology, besides inspiring new movements in theology and philosophy.[13] He has also long been the focus for enormous industry in the history of science.[14] Who on earth could comprehend such a wide domain of endeavour? Two types of 'who' need considering: readers and authors.

First, readers. I have done my best not to assume more than an introductory acquaintance with the fields of knowledge this book covers, because, given the variety of Darwin's domains of interest, few readers can assume they have ready-made expertise in them all. This hit me when I heard an eminent historian of science, who had devoted his career to mastering Darwin's manuscripts, dismiss the question of Darwin's influence on modern psychology by saying, 'Darwin influenced James. There's your proof.' His remark betrays an embarrassingly slight knowledge of modern psychology—which launched itself by declaring that James was 'as much out of touch with modern psychology as the stage coach would be with modern New York's Fifth Avenue'.[15]

Regarding the breadth required of an author attempting to examine Darwin's approach to psychology, I am acutely aware of my limitations in this regard. So it may be well to supplement the warning 'Reader beware!' with a glance at the path which has led me to write this book. I am son to a geography academic (later, a medical librarian) and a palaeontologist, founding professor of geology at Leicester University. From an early interest in animal behaviour, I enrolled at Oxford to take a then-new and controversial degree in Human Sciences, after a gap year including

[12] George Romanes, 'Charles Darwin, [Part] V. [Psychology],' *Nature* 26, (1882), p.169: 'in the same … nature.'

[13] Regarding philosophy and theology, see Bruno Latour, 'What Is Given in Experience? A Review of Isabelle Stengers, *Penser Avec Whitehead*,' *Boundary* 2, (2005), pp.222–237; Elizabeth Grosz, *Becoming Undone: Darwinian Reflections on Life, Politics and Art* (Durham N.C.: Duke University Press, 2011); Julian Huxley, *Religion without Revelation* (London: Parrish, 1957); John Haught, *God after Darwin: A Theology of Evolution* (Boulder: Westview, 2000).

[14] Timothy Lenoir, 'The Darwin Industry,' *Journal of the History of Biology* 20, (1987), pp.115–130; Gowan Dawson, 'Darwin Decentred,' *Studies in History and Philosophy of Biological and Biomedical Sciences* 46, (2014), pp.93–96.

[15] Watson, *Behaviorism*, p.105.

three months working with Uli Weidmann and Ken Simmons at Leicester University on the film analysis of duck courtship, and three months' assisting with a study of ducks' imprinting at Konrad Lorenz's Institute for Verhaltenphysiologie in Bavaria. The lure of Human Sciences was not just its breadth—it covered not only psychology but ethology, genetics, social geography, human ecology, sociology, social anthropology, demography, and human evolution (to which I added courses on literature and philosophy)—but its teaching staff. My boyhood hero Niko Tinbergen, the subtle and fascinating Edwin Ardener, Walter Bodmer FRS, A.H. Halsey, Jerome Bruner, and Michael Argyle all gave us memorable courses of lectures. Richard Dawkins taught us too, being then (1971–1974) in the throes of writing his *Selfish Gene*. Hair awry, he regaled us with provocative claims about lumbering robots and immortal coils, while puzzling us (hot from Tinbergen's and Bodmer's classes) by his blithe disregard for the place of whole organisms (phenotypes) in evolution's scheme.[16]

From Oxford I was lucky enough to land a Medical Research Council scholarship, advertised at Edinburgh by the brilliant psycho-biologist Colwyn Trevarthen, to conduct a doctoral video study of young infants as social beings. Colwyn is a great fan of Darwin, and delighted in Darwin's then-just-published Notebooks on 'man, morals, and metaphysics'.[17] 'Motives are units in the universe,' he would remind his little flock of supervisees, picnicking on the lawns of the Botanic Gardens.[18] Darwin pioneered research on infant development. Hence when I came to write *Visions of Infancy*, its first vision was Darwin's—a vision which I held to have tipped developmental psychology towards a form of Romanticism.[19]

After 40 years teaching, researching, and practicing psychology across the UK, USA, and Australia—including five years gathering the materials for this book—I have incurred many debts of gratitude. My first thanks go to my father, Peter Sylvester-Bradley, who, while a strong if unorthodox Christian, was equally an ardent evolutionist. He died two months before he could take up a position at the Open University (UK), alongside his friend and protégé Ian Gass, to design a course on evolution. This was to be accompanied by a new and long-premeditated magnum opus on the general theory of evolution—several proto-chapters of which exist in papers published between 1971 and 1978. He admired Julian Huxley's work on evolution, and followed Huxley in arguing to me that a truly evolutionary take

---

[16] Richard Dawkins, *The Selfish Gene: 30th Anniversary Edition* (Oxford: Oxford University Press, 2006).

[17] Howard Gruber and Paul Barrett, *Darwin on Man: A Psychological Study of Scientific Creativity, Together with Darwin's Early and Unpublished Notebooks* (London: Wildwood, 1974).

[18] Charles Darwin, 'Old & Useless Notes,' (Cambridge: Darwin Online, 1838–1840), p.25.

[19] Ben Bradley, *Visions of Infancy: A Critical Introduction to Child Psychology* (Cambridge: Polity Press, 1989). See also: Ben Bradley, 'Infancy as Paradise,' *Human Development* 34, (1991), pp.35–54; Ben Bradley, 'Darwin's Intertextual Baby: Erasmus Darwin as Precursor in Child Psychology,' *Human Development* 37, (1994), pp.86–102; Ben Bradley, 'Darwin's Sublime: The Contest between Reason and Imagination in *On the Origin of Species*,' *Journal of the History of Biology* 44, (2011), pp.205–232.

on human behaviour would build better on ethology than on experimental psychology. One of the books that he read aloud to his four children before our bedtime was Konrad Lorenz's *King Solomon's Ring*, a marvellous entrée to ethology.

I owe more than I can say to my wife, Jane Selby. Our research on babies in groups gave me the impetus to read more deeply in Darwin's work and so to realize that his approach to human behaviour was far more distinctive, extensive, and profound than I had known. In particular, the video-recordings we have been making of all-infant trios and quartets since 1999—which show that babies are capable of *group* interaction (not just one-to-one, 'dyadic' interaction) during the first months of life—chimed uncannily with Darwin's argument that the most human of human attributes result from the dynamics producing the group-life of our species' proto-human ancestors.[20] Jane has constantly lent her intellect, perceptiveness, and insight to this project: encouraging, questioning, illuminating, and extending the reach of my daily enthusiasms and discoveries, producing solutions to my puzzles, and constructively reading drafts.

Two colleagues from Charles Sturt University (CSU) Bill Green and Nick Drengenberg deserve special mention. It was Bill who, on the publication of a long-brewed paper on 'Darwin's Sublime' in 2011, suggested I consider writing a book on Darwin. At our weekly breakfasts he has fielded a hundred rants on Darwin, gently suggesting resonances and links to the works of (predominantly French) scholars. Over Thai lunches, Nick's breadth of scholarship and interest has extended my thinking in diverse directions, most notably to Latour and Tarde. I am also grateful to my colleague at CSU, Danielle Sulikowski, for raising several points relating to evolutionary psychology; and to Ross Chambers, for his intellectual engagement with, and early facilitation of, this project. I would particularly like to thank the librarians who have helped me around the world: notably in the Rare Books Room at the University Library at Cambridge and, particularly, Kirrarne Ianson at CSU—who has kept me superbly supplied with obscure tracts via 'interlibrary loan' in Bathurst. Also to Annette Goodwin for being my Endnote guru.

I owe the world of Darwin scholarship a great deal. Above all, the value of the support and friendship of Paul White, editor and research associate of the Darwin Correspondence Project in Cambridge, has been inestimable. I thank him for the unfailing and insightful help given to someone who has largely researched this book on the wrong side of the planet from the huge trove of Darwinalia, and Darwin expertise, in the University Library at Cambridge. Paul's original and graceful writings on Darwin never fail to nourish and inspire me, nor his conversation to inform and challenge my work. Beyond Paul, I have been most fortunate in the generosity

---

[20] Jane Selby and Ben Bradley, 'Infants in Groups: A Paradigm for the Study of Early Social Experience,' *Human Development* 46, (2003), pp.197–221; Jane Selby et al., 'Is Infant Belonging Observable? A Path through the Maze,' *Contemporary Issues in Early Childhood* 19, (2018), pp.404–416; Ben Bradley and Michael Smithson, 'Groupness in Preverbal Infants: Proof of Concept,' *Frontiers of Psychology* 8, (2017), https://doi.org/10.3389/fpsyg.2017.00385.

of many scholars and researchers who have helped me with this project, some-times in person, sometimes at a distance—though usually, when I first approached them, they did not know me from a bar of soap. In particular: Paul Stenner; Robert Richards; John Dupré; Daniel Nicholson; Kevin Laland; Katrina Falkenberg; Mary Jane West-Eberhard; Lynn Nyhart; Francis Neary; Gregory Radick; Sean Dyde; Jon Hodge; Evelleen Richards; Paul Griffiths; James Ounsely; Tobias Uller; Catherine Naum; Edwin Rose; Tiffany Mason; Matt Olsen; Stefanie Reichelt; David Siveter; Michael Summerfield; Michael Smithson; Matt Stapleton; John Forrester; Johanna Motzkau; Jill Morawski; Hank Stam; Marc Ereshevsky; Melanie Massaro; Jeremy Burman; Tim Lewens; Janet Browne; Anthony Campbell; Joel Krueger; Felicia Huppert; Fiona Green; Bill Blaikie; John and Karen Rennie; Martin Richards; and Mike Sheppard. Plus, I thank other friends, relations, and acquaintances who have kept asking after the book over the years—and sometimes opened up new vistas through their thoughtfulness. I particularly thank my son Peter Selby in this re-gard, for his gift of Tom Griffiths' *The Art of Time Travel*.[21]

Finally, I would like to thank four members of OUP's staff: Martin Baum, Matthias Buttler, Janine Fisher, and particularly, the gracious tending of the book's genesis from go to (almost) whoa by Charlotte Holloway. Thanks too to the an-onymous reader of the book's penultimate draft, found by Charlotte—and to that draft's three other readers: Jane Selby, Paul Stenner, and Paul White.

---

[21] Tom Griffiths, *The Art of Time Travel: Historians and Their Craft* (Carlton VIC: Black Inc., 2016).

Contents

1

Introduction

In psychology, everyone is an amateur. However many diplomas a psychologist
has, whatever our theoretical triumphs and the size of the laboratory in which we
work, we each must dust off our observational skills and pay attention like anyone
else when we get home. Or pay the price. Even one's own actions puzzle from time
to time, let alone those of one's familiars. Has the wife come in a little pale? Is her
quietness excitement? Has she some news? Or is she just tired? Off duty, whether
at home or out and about, theories and statistical generalizations may or may not
apply to any given occasion, and thus can mislead. One rather needs to know what
chain of events, what network of circumstances, a look or expression modifies. Did
my son just say, 'Fish slime'? Do our guppies ail? Is he being rude? Does that come
from school? From a dream? What undiscovered pattern do these words make
with the rest of his day?

The natural world is no different. I breakfast in a deck-chair on the veranda, faint
after a strenuous gym. A bird calls, its song a rusty gate. Hermit-thrush? Quick!
Look! Find! Follow! Alternatively, lie back and contemplate the lawn, this mingled
green slope of grass and cut weeds above which, today, stalks of wild garlic wave
white blossom, inviting the mower. Insects whirr in straight lines hither and yon,
beetles beetle, and worms crawl through the damp earth. Which is where Darwin
began, not with people, but with the art of paying attention to the interwoven histories of the meanest and most common among living things. Eventually he moved
on to people. But we will not make out what he made of humans unless we first take
time to examine how he saw the purposeful ways of humbler natives in the places
we inhabit.

Darwin recalled that his passion for birds, naming plants, trapping beetles,
and hoarding new minerals was manifest by age eight.[1] His fascination with all
manner of goings-on in the natural world stayed with him from boyhood through
old age. Here was not only his calling—'I was born a naturalist,' he said—but the
primary source of his eminence.[2] From 1853 to 1864, he won three of the highest
awards in Victorian science. Yet none of the citations for these medals mentioned
his work on evolution. Rather, they celebrated his 'minute attention,' his 'admirable

---

[1] Charles Darwin, *The Autobiography of Charles Darwin, 1809–1882: With Original Omissions
Restored*, ed. Nora Barlow (London: Collins, 1958), pp.44–45.
[2] Charles Darwin, 'Life. Written August — 1838,' (1838), http://darwin-online.org.uk/content/
frameset?itemID=CUL-DAR91.56-63&viewtype=text&pageseq=1, p.60.

Darwin's Psychology. Ben Bradley, Oxford University Press (2020). © Oxford University Press.
DOI: 10.1093/oso/9780198708216.001.0001

observations,' the data his inquiries had furnished, and the comprehensiveness of his research.[3]

Darwin saw the living world as a theatre of agency. This vision supplied the fulcrum both for his theory of evolution, and for his psychology. Natural selection was not for Darwin a causal process. Like gravity, it was a law that *resulted from* causal processes, namely: the transmission of heritable material and its individual development; the variability of structure and habit in animate beings—largely produced by what Darwin metaphorically called the struggle for existence: something which included not just combat, but relative reproductive success, interdependence, and even cooperation between members of the same species and of different species.[4] The same interdependencies of agency are, as we shall see, front and central to Darwin's take on psychological matters.

Straight off, we meet an obstacle. My claim that a theatre of agency produces evolution figures nowhere in the picture of Darwin modern science venerates. For eighty plus years, natural selection *itself* has been called the mechanism of evolution.[5] The mechanism which underpins the triumph of modern evolutionary biology: Darwin's one significant idea.[6]

Witness evolutionary psychology. Ever since the 1870s, psychology has been seeking some bedrock to ground its explanations—a material *something* that will provide the starting-point for any account of behaviour. Candidates have included: the brain; physiology; the causal nexus between stimulus and response; gene as blueprint; the model of mind as computer; or of action as text. Which makes recent proposals for a psychology based upon evolution a radical departure. Because evolution is not a here-and-now *thing* you can prod, parse, or measure. It is a set of *processes* which operate over vast lengths of time, embracing all living creatures,[7] body, and mind. Hence, an evolution-based psychology promises a fresh kind of coherence for the discipline, and indeed, for all walks of scholarship and research which deal with humans.

---

[3] These awards were: the Royal Medal (1853); the Wollaston Medal for geology (1859); and the Copley Medal (1864). In its citation for the Copley medal, the Royal Society explicitly excluded Darwin's evolutionary theory from the award. Cf. Frederick Burkhardt, 'England and Scotland: The Learned Societies,' in *The Comparative Reception of Darwinism*, ed. Thomas Glick (Chicago: University of Chicago Press, 1988), p.35; John Phillips, 'Award of the Wollaston Medal and Donation Fund,' *Quarterly Journal of the Geological Society of London 15*, (1859), p.xxiii.

[4] See Ch. 3 for further discussion.

[5] Denis Walsh, 'The Struggle for Life and the Conditions of Existence: Two Interpretations of Darwinian Evolution,' in *Evolution 2.0: Implications of Darwinism in Philosophy and the Social and Natural Sciences*, ed. Martin Brinkworth and Friedel Weinert (Berlin: Springer-Verlag, 2012), pp.191–209; e.g. Julian Huxley, *Evolution: The Modern Synthesis* (London: George Allen & Unwin, 1942), p.474; Richard Dawkins, *The Selfish Gene: 30th Anniversary Edition* (Oxford: Oxford University Press, 2006), p.18; Ernst Mayr, 'Darwin's Influence on Modern Thought,' *Scientific American 283*, (2000), p.80.

[6] Daniel Dennett, *Darwin's Dangerous Idea: Evolution and the Meanings of Life* (New York: Simon & Schuster, 1995), p.39 & passim.

[7] NB I apply the word 'creature' to *all* living organisms, plant, and animal.

So, what is Darwin's place in this new movement? He gets to be its deep-eyed figurehead, the prophet who, way back in 1859, foresaw the historic promise of the approach worked out in *On the Origin of Species*. Did not that masterpiece tell in its final pages of a distant future where fields would open on 'far more important researches,' those which would give psychology a new foundation, where every mental power and capacity was fruit of aeons-long development?[8] Unfortunately, lament psychology's new evolutionists, the social and behavioural sciences have remained largely untouched by the unique promise of Darwin's theory. Making it appear that it is *only now*, two centuries after his birth, that the distant future Darwin envisioned for psychology is upon us.[9]

All of which presumes that Darwin himself did nothing substantial to work out how evolution might bear on the complexities of agency. Darwin has only regained prominence in today's studies of mind and behaviour because it was he who long ago set science on the path towards a certain understanding of evolution—in which agency plays no part. It is this modern understanding in biology which is to revolutionize psychology, say evolutionary psychologists, not Darwin's own writings.[10]

Darwin's Psychology shows what happens when we switch this line of argument around and say: Look not at Darwin by the light of an evolutionary psychology remade in the image of twentieth-century biology. Examine at first hand Darwin's many, insightful, extensive, and largely unread investigations of agency, and then, through that prism, see how he envisaged the foundations for understanding human action. Do this, I argue, and one finds he resources a surprisingly socialized and agentic account of the weave of ways we and our fellow creatures live. Darwin was an unusually perceptive man of science, who advanced a sophisticated and vital vision of the earthly world. He extrapolated this vision to yield a not-just-evolutionary understanding of human agency, his complex understanding entailing interdependencies between species, bodies, actions, meanings, and cultures—benign and malignant.

But Darwin was an amateur, wasn't he? As such, twentieth-century moves to modernize science and psychology, to make them professions, give him an old-fashioned look.[11]

---

[8] Charles Darwin, *On the Origin of Species by Means of Natural Selection or the Preservation of Favoured Races in the Struggle for Life* (London: Murray, 1859), p.488.

[9] David Buss, 'The Great Struggles of Life: Darwin and the Emergence of Evolutionary Psychology,' *American Psychologist 64*, (2009), p.147: 'In 1859, Darwin provided a vision of a distant future in which psychology would be based on the new foundation. The distant future that Darwin envisioned is upon us.'

[10] E.g. this is the premise of Kevin Laland, *Darwin's Unfinished Symphony: How Culture Made the Human Mind* (Princeton: Princeton University Press, 2017); see Ben Bradley, Darwin 1.0: Is the EES Playing Catch-Up?, 2018, http://extendedevolutionarysynthesis.com/darwin-1-0-is-the-ees-playing-catch-up/.

[11] Sydney Ross, 'Scientist: The Story of a Word,' *Annals of Science 18*, (1962), pp.65–85; James Deese, *Psychology as Science and Art* (New York: Harcourt Brace Jovanovich, 1972), p.35. For an intriguing sidelight on psychology's professional stance, see Robert Joynson, *Psychology and Common Sense* (London: Routledge and Kegan Paul, 1974). I return to Darwin's so-called amateurism in Ch. 2.

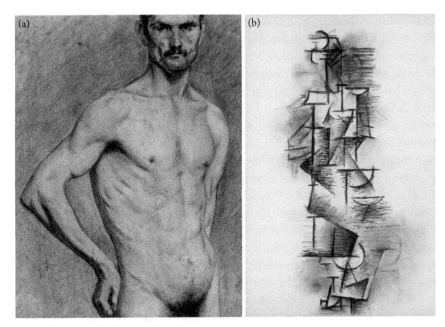

Figure 1.1 Two drawings by Picasso: (a) Naturalistic sketch of a matador, 1897; (b) Cubist portrait of a standing female nude, 1910 (right).

Sources: a) Pablo Picasso, *Matador Luis Miguel Dominguin*, 1897. © Succession Picasso/DACS, London 2019. Image © akg-images / Album / Kocinsky; b) Pablo Picasso, *Standing Female Nude*, 1910. Alfred Stieglitz Collection, 1949. © Succession Picasso/DACS, London 2019. Image © The Metropolitan Museum of Art/Art Resource/Scala, Florence.

1.1 Darwin and Modernism

Virginia Woolf once observed that in or about December 1910, the character of all human relations changed—whether between masters and servants, husbands and wives, or parents and children.[12] Far though it reverberated, Woolf's upheaval stood out most visibly in art galleries, where pictures representing recognizable places and people gave way to the abstractions of Cubism (Figure 1.1).

An image of science cradled this change. Thus, against what it called 'anecdote,' the Parisian manifesto *Cubism 1912* cast itself as a study of *the integration of the plastic consciousness*, something fuelled by various abstruse conceptualizations including theorems in geometry.[13] Such study required a development of mind that would tear painters away from any aim faithfully to portray with paints the natural

---

[12] Virginia Woolf, 'Mr. Bennett and Mrs. Brown,' in *The Captain's Deathbed and Other Essays*, ed. Leonard Woolf (San Diego: Harcourt Brace Jovanovich, 1978), p.96.

[13] See Edwin Ardener, 'Social Anthropology and the Decline of Modernism,' in *The Voice of Prophecy*, ed. Malcom Chapman (Oxford: Blackwell, 1989), pp.192ff, on modernism and manifestoes.

world of visible surfaces, as apprehended by common sense. Cubists inhabited 'a different kind of space,' structured by abstractions which anchored their pictures in a new ideal of organization, so as more deeply to harmonize 'with the totality of things, with the universe.'[14] When Cubists constructed a form, they did so according to their own theoretical notions, an achievement no one, 'save the man we call an artist,' could accomplish without external assistance. For example, a thoroughly modernized painter might magnify angles or features common sense made negligible. Conversely, characteristics salient in common sense could become infinitesimal or vanish: a woman's face rendered as three straight lines, two parallels for brows and mouth linked by a single nose-derived vertical (Figure 1.1b). The Cubist effect transformed 'quantity into quality,' selective alterations of size and angle seeding a brand new way of perceiving a vase or a naked girl.[15]

A like convulsion gave psychology its modern face. The manifesto for America's behaviourism appeared in 1913.[16] Paralleling *Cubism 1912*, psychologists were exhorted to abandon the natural world of ordinary observation and common sense. A new psychology should inhabit *a different kind of space* both physically and conceptually. Physically, it must migrate into the laboratory. Here psychologists' focus would shift away from what had hitherto seemed most essential to humanity (language, thought, moral action) to focus on the selection of simple, countable responses in—most often—a specialized breed of cage-reared rat.[17] In laboratories, quantities (analysis of numerical results) would forge qualities (e.g. conclusions about how best to raise a child).[18] By the Second World War, leading psychologists were contending that 'everything important in psychology' could be investigated through 'the continued experimental and theoretical analysis of the determiners of rat behaviour.'[19] Even in today's (largely) post-rat psychology, laboratory experiments remain the template for high-prestige research, and results meeting a *statistical* criterion are still taken to be results with *psychological* significance—quantities becoming qualities.[20]

---

[14] Albert Gleizes and Jean Metzinger, 'Cubism 1912,' in *Modern Artists on Art: Ten Unabridged Essays*, ed. Robert Herbert (Englewood-Cliffs, NJ: Prentice-Hall, 1964), pp.3–7.

[15] Gleizes and Metzinger, 'Cubism,' pp.6–7.

[16] John Watson, 'Psychology as the Behaviourist Views It,' *Psychological Review 20*, (1913), pp.158–177.

[17] Bonnie Clause, 'The Wistar Rat as a Right Choice: Establishing Mammalian Standards and the Ideal of a Standardized Mammal,' *Journal of the History of Biology 26*, (1993), pp.329–349.

[18] Gleizes and Metzinger, 'Cubism,' p.7: 'The painter ... changes quantity into quality.' Cf. Burrhus Skinner, *The Behaviour of Organisms: An Experimental Analysis* (New York: Appleton-Century-Crofts, 1938); Burrhus Skinner, *Beyond Freedom and Dignity* (London: Cape, 1972).

[19] Edward Tolman, 'The Determiners of Behaviour at a Choice Point,' *Psychological Review 45*, (1938), p.34. These were the last words of Tolman's presidential address to the American Psychological Association in 1937. See also Zing-Yang Kuo, 'A Psychology without Heredity,' *Psychological Review 31*, (1924), pp.427–448.

[20] Paul Meehl, 'Why Summaries of Research on Psychological Theories Are Often Uninterpretable,' *Psychological Reports 66*, (1990), pp.195–244; Michael Billig, 'Repopulating Social Psychology: A Revised Version of Events,' in *Reconstructing the Psychological Subject: Bodies, Practices and Technologies*, ed. Betty Bayer and John Shotter (London: Sage, 1998), pp.126–152; Michael Smithson, *Confidence Intervals* (Thousand Oaks CA: Sage, 2003). Doubts about this quantity→quality shift have

Having shifted into the laboratory, modern psychologists took to framing their work via a set of dichotomies: nurture against nature; biology *or* culture; subject-object; the individual and society. Take subject–object for example. In order to record the natural world *objectively*, psychologists had to become a specialized kind of *subject* before they could work scientifically. Hence the importance of their professional training. Just as rats were seen by behaviourists as machines, so researchers would need to become 'an automatic mechanism free from subjectivity', if they were to record others' behaviour without bias.[21]

To this end, psychologists have long been trained to base their records of behaviour on the *operations* they use to measure it. This means that, like Cubists, they should verify the forms they describe by means of *their own* previously worked-out ideas about the phenomena they study.[22] And, while psychologists become purified *subjects*, the creatures they study—rat or human—become their mirror-image: purely objects.[23] Laboratory studies stage behaviour as dependent mechanical response to whatever stimuli experimenters choose to expose targets to—electric shock plus tasty smell, reward plus scary photograph.[24] Hence the worry when evidence began to mount demonstrating that the 'objects' upon whom psychologists had thought they were experimenting proved far less passive than assumed: participants second-guessing the aims of the experiment in which they were enrolled, and then acting-out accordingly.[25] Even responses in rats were shaped by the expectations of their testers.[26]

For the purposes of this book, a most dubious trait of psychology's love affair with modernism has been its framing of the past. Modernizers—whether

---

contributed to the rise of language-based *qualitative* research in psychology: David Rennie, David Watson, and Michael Monteiro, 'The Rise of Qualitative Research in Psychology,' *Canadian Psychology* 43, (2002), pp.179–189.

[21] E.g. Watson, 'Psychology as Behaviourist,' p.158; John Watson, *Behaviorism* (London: Kegan Paul & Co., 1925), p.215; John Watson and Will Durant, 'Is Man a Machine? A Socratic Dialogue,' *The Forum* 82, (1929), pp.264–270; Jill Morawski, 'Organizing Knowledge and Behaviour at Yale's Institute of Human Relations,' *Isis* 77, (1986), p.235.

[22] Gleizes and Metzinger, 'Cubism,' p.6: 'To discern a form is to verify it by a pre-existing idea.' Cf. Sigmund Koch, 'Psychology's Bridgman Vs Bridgman's Bridgman,' *Theory & Psychology* 2, (1992), pp.261–290.

[23] Confusingly, these 'objects' were/are often called 'subjects' by psychologists—borrowing a tradition from medicine where the word *subject* had referred to a corpse employed for anatomical dissection: Kurt Danziger, *Constructing the Subject: Historical Origins of Psychological Research* (Cambridge: Cambridge University Press, 1990), p.54.

[24] Kerry Buckley, *Mechanical Man: John Broadus Watson and the Beginnings of Behaviourism* (New York: Guilford Press, 1989); Hank Stam and Tanya Kalmanovitch, 'E. L. Thorndike and the Origins of Animal Psychology: On the Nature of the Animal in Psychology' *American Psychologist* 53, (1998), pp.1135–1144.

[25] Jerry Suls and Ralph Rosnow, 'Concerns About Artifacts in Psychological Experiments,' in *The Rise of Experimentation in American Psychology*, ed. Jill Morawski (New Haven CT: Yale University Press, 1988), pp.163–187.

[26] Robert Rosenthal and Kermit Fode, 'The Effects of Experimenter Bias on the Performance of the Albino Rat,' *Behavioural Science* 8, (1963), pp.183–189.

in painting or psychology—underline their newness by contrast to a past constructed as archaic. Thus behaviourists cast Freud and the introspective methods of nineteenth-century psychologists William James and Wilhelm Wundt as laughably unscientific, as 'voodoo.'[27] So—what of Darwin?

Psychology's most esteemed first movers—Wundt and James among them—had widely acknowledged Darwin's impetus. So when the centenary of his birth came round in 1909, a leading journal commissioned the high-profile American psychologist, 40-year-old James Angell, to assess Darwin's contribution to the discipline.[28] Angell set out by saying that, for modern psychology, Darwinism had never been a vital issue. Evolution was by 1909 so widely accepted and broad a notion as to be beside the point. The modern psychologists of Angell's day preferred to focus, he said, on 'narrowly analytical problems of mind.'[29] Here he was aligning the new psychology with targeted experiments to define the meanings of mentalistic terms. This *analytic* task was best approached a priori or deductively, he said—opposing a *synthetic* approach, which built knowledge a posteriori from empirical facts previously gathered about the world.[30]

Darwin's work was of little help with a priori analysis, wrote Angell. Worse, many of the specific ideas in Darwin's writings—about the transmission of acquired habits, for example, or the shaping of instincts by natural selection—had either been proven false, or appeared so implausible as to invoke 'the miraculous.'[31] More up-to-date psychologists had better suggestions, Angell told his readers, which could be tested by physiologists and zoologists in laboratory experiments on animals like sea-urchins and jellyfish. Lit up by the burgeoning behaviour-focused experimentalism of modernist psychology, Darwin's observations invited ridicule. His studies of animals were naïve and anecdotal, 'highly archaic and scientifically anachronistic'—not least because they used common-sense categories without precise definition. So, his 'simple-minded' results could at best 'furnish amusement to the sophisticated animal psychologists of the present day.'[32]

And that, for 60 years, was that. Which meant, when sociobiology, and then evolutionary psychology, came on the scene in the 1970s and 1980s, their proponents could lament an astonishing 100 years of neglect of Darwin in the social sciences.[33] Yet, the Darwin these two movements undertook to re-introduce to psychology

---

[27] Watson, *Behaviorism*, p.18 & passim.

[28] James Angell, 'The Influence of Darwin on Psychology,' *Psychological Review 16*, (1909), pp.152–169. Angell (1869–1949) had been doctoral supervisor to John Watson (1879–1958) who launched the 1913 manifesto for behaviourism: Watson, 'Psychology as Behaviourist.'

[29] Angell, 'Darwin's Influence,' p.153.

[30] James Angell, 'Psychology at the St. Louis Congress,' *The Journal of Philosophy, Psychology and Scientific Methods 2*, (1905), pp.539–540

[31] Angell, 'Darwin's Influence,' p.155.

[32] Angell, 'Darwin's Influence,' pp.158–164.

[33] E.g. Dawkins, *Selfish Gene*, pp.xix, 1; John Tooby and Leda Cosmides, 'The Theoretical Foundations of Evolutionary Psychology,' in *The Handbook of Evolutionary Psychology*, ed. David Buss (Hoboken NJ: Wiley, 2016), p.3.

bore little resemblance to the Darwin accessible through his publications. Rather, sociobiologists and evolutionizers aimed to reform psychology by aligning it with an image of Darwin forged in a biology that celebrated *the gene*—an entity unknown to Darwin.

Mention of genes brings us to the second and most central prong of the modernist trident which threatens any reading of Darwin's work on psychological matters. For, the gene-centric view of evolution that emerged around 1940 still grounds the prevalent understanding of Darwin and his work, not just in evolutionary psychology, but in biology, the history of science, and general knowledge. Dubbing itself 'the Modern Synthesis', this initiative aimed to unify an early twentieth-century biology threatened by uncoordinated growth across a burgeoning array of specialisms: developmental physiology; embryology; ecology; comparative anatomy; systematics; geographical distribution; palaeontology; cytology; and mathematical analysis.[34]

Biology's mid-century modernizers found a way to weld a view of natural selection to the post-Darwin discoveries of Mendelian and population genetics. Mendel's interpretation of his studies on the breeding of peas invoked the *mutation* and *recombination* of heritable particles (christened 'genes' in 1909).[35] These two new processes filled what was held to be the biggest gulf in Darwin's understanding of evolution—his supposed ignorance of the origin of organic variability (but see Ch. 3).[36] Thanks to geneticists, biologists now knew that random mutations and chance chromosome changes provided the raw material of organic variation, from which the environment—more precisely, the mechanism of natural selection—winnowed out the most adaptive. This made any heritable characteristic of a living organism (any phenotype) a by-product of gene-replication.[37] Modernizing biologists thus denied any contribution to variation or adaptation from the agency of organisms. In Julian Huxley's words:

> It was one of the great merits of Darwin himself to show that the purposiveness of organic structure and function was apparent only. The teleology of adaptation is a pseudo-teleology, capable of being accounted for on good mechanistic principles, *without the intervention of purpose, conscious or subconscious,* either *on the part of the organism* or of any outside power.[38]

[34] Huxley, *The Modern Synthesis*; Betty Smocovitis, 'Unifying Biology: The Evolutionary Synthesis and Evolutionary Biology,' *Journal of the History of Biology 25*, (1992), pp.1–65.
[35] Wilhelm Johannsen, *Elemente Der Exakten Erblichkeitslehre Mit Grundzügen Der Biologischen Variationsstatistik* (Jena: Gustav Fischer, 1909).
[36] Huxley, *The Modern Synthesis*, p.125.
[37] Huxley, *The Modern Synthesis*, passim; Theodosius Dobzhansky, *Genetics and the Origin of Species*, 3rd ed. (New York: Columbia University Press, 1951), pp.16, 50, 21. Like 'gene,' 'phenotype' was also a term invented by Johannsen, Wilhelm Johannsen, 'The Genotype Conception of Heredity,' *American Naturalist 45*, (1911), pp.129–159.
[38] Huxley, *The Modern Synthesis*, p.412, my italics.

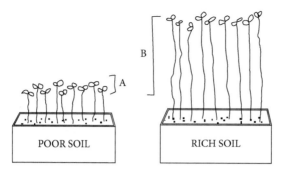

Figure 1.2 Samples from one set of pea-seeds sown in two different but uniform soils, one more fertile than the other. If only the poor soil is considered, it appears that all variations in height [A] result from genetic variation. But if both soils are considered, most of the variations in height [B] must result from differences between the soils.

The claim that animal agency had no relevance to evolution has long drawn criticism. 'Before an organism's environment can exert natural selection on it, the organism must select the environment to live in,' complained ethologist William Thorpe in 1965.[39] The denial that organisms contributed *as organisms* to the dynamics of evolution leant on biology's modernist migration, from observing plants and animals amid their natural habitats, into the laboratory. For decades, few commented that Mendel's methodology, like all modern biochemical, molecular, and developmental genetics, depended for progress on finding genetic differences that produced clear-cut, non-overlapping characters in an organism, under easily controlled conditions. Such characters are actually *very rare*. The vast majority of morphological, behavioural, and physiological differences between individuals do not, as biologists say, *Mendelize* (i.e. behave like they should according to Mendel).

Even the few differences that do Mendelize show considerable variation in extent, unless they develop in tightly controlled conditions—as in a laboratory. For example, variations in the heights of Mendel's pea plants can show almost 100% genetic control (or *heritability*) in a uniform environment. But, if tested in both dry and moist conditions, or in poor and rich soils, or at high and low altitudes—the heritability of their variations in stature comes close to 0% (Figure 1.2).[40] All of which must be forgotten whenever a modern synthesizer describes some characteristic of an organism as *determined by* its genes (see Ch. 3).[41]

[39] Thorpe quoted in Gregory Radick, 'Animal Agency in the Age of the Modern Synthesis: W.H. Thorpe's Example,' *British Journal for the History of Science: Themes 2*, (2017), p.36.

[40] Steven Rose, Leon Kamin, and Richard Lewontin, *Not in Our Genes: Biology, Ideology and Human Nature* (Harmondsworth: Penguin Books, 1984), pp.95–97.

[41] Richard Lewontin, 'Gene, Organism and Environment,' in *Evolution from Molecules to Men*, ed. Derek Bendall (Cambridge: Cambridge University Press, 1983), pp.276–277.

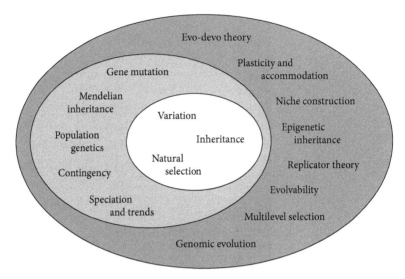

Figure 1.3 'The continuous expansion of evolutionary theory' as understood in modern biology: Darwin's views (centre field); the Modern Synthesis (intermediate field); and the Extended Synthesis (outer field).

Pigliucci, Massimo, and Gerd B. Müller, eds., *Evolution, the Extended Synthesis*, Figure 1.1 on page 11, © 2010 Massachusetts Institute of Technology, by permission of The MIT Press.

The story does not end here, of course. Over the last 20 years, more and more biologists have argued that the Modern Synthesis needs extending. And, amongst the many extensions proposed, we find a new emphasis on the organism's capacity to change or create its habitat through its own activities—a process some call *niche construction*.[42] This new *Extended* Evolutionary Synthesis represents an advance over, but *not* a refutation of the Modern Synthesis. Rather, just as the Modern Synthesis was bruited forth as an advance over Darwin, the Extended Synthesis represents yet further scientific progress (Figure 1.3).

In this picture of history, progress in evolutionary theory leaves Darwin's vision of the living world further and further behind. Accordingly, as Figure 1.3 implies, we are led to assume that Darwin had nothing worthwhile to say about the newly trumpeted phenomena of plasticity, niche construction, multilevel selection, the importance of individual development, evolvability—or, indeed, psychology. And if he did, these processes were not party to his theory of evolution.[43] Despite

[42] John Odling-Smee, Kevin Laland, and Marcus Feldman, *Niche Construction: The Neglected Process in Evolution* (Princeton NJ: Princeton University Press, 2003), p.419, where 'niche' is defined as *including* organisms' 'niche-constructing acts, and selection from [the] sources' available in their environment.

[43] Richard Dawkins, *The Blind Watchmaker: Why the Evidence of Evolution Reveals a Universe without Design* (Harlow: Longman, 1986), p.290.

Darwin's original treatment of evolution being enormously rich (as this book will document), scientific modernism makes him look like the purveyor of just one worthwhile idea: evolution by natural selection. In this way, modernist biologists suffer a severe constriction in their understanding of Darwin. Perhaps inevitably, therefore, as they increasingly recognize the need for a broader view of the natural world than the Modern Synthesis provides, they begin to rediscover—often unwittingly—much that has long been buried in Darwin's work.[44]

How to understand Darwin's writings brings in a third strand of modernist scholarship. Since 1960, Darwin has become a lodestone for historians of science, not just due to his prominence as a scientist, but because he left behind an enormous trove of unpublished writing. Aside from his millions of published words, the so-called Darwin industry[45] can mine: early drafts of his books; 15,000 letters; two unpublished autobiographies; private notebooks and diaries; hundreds of scribbled pages of detailed observations—of animals, plants, fossils, and rocks—made across 55 years; what he wrote about the expressions and agency of his infant children; lists of books he wanted to read and did read; plus more than a thousand books from his own library, with pencilled jottings in the margins. Beyond all this hums the intriguing universe of Victorian Britain.

Dear to the Darwin industry is a structure also grounding separation between the modern disciplines of psychology and sociology: separation between *the* (natural) *individual* and *society*.[46] One of modern historians' aims in focusing on Darwin has been to dispute a common-sense image among scientists: of Darwin's natural genius and achievements as transcending their social context. Did Darwin's works truly have no link to those of clay-footed contemporaries like Herbert Spencer, Robert Chambers, and Alfred Wallace? Did his theory of evolution show no imprint from the social, economic, and ideological forces affecting scientific authority in nineteenth-century Britain?[47] A heap of evidence has been swept up to refute such naïve ideas.[48] This context-stressing or 'contextualist' move against so-called 'Great Man' histories of Darwin has resulted in a denial that Darwin ever advanced a worthwhile psychology. This is for several reasons.

First, Great Man histories typically sanctify *On the Origin of Species* above everything else Darwin did, because, under the sway of the Modern Synthesis, that book

---

[44] Bradley, EES; Mary Jane West-Eberhard, 'Toward a Modern Revival of Darwin's Theory of Evolutionary Novelty,' *Philosophy of Science 75*, (2008), pp.899–908.

[45] Timothy Lenoir, 'The Darwin Industry,' *Journal of the History of Biology 20*, (1987), pp.115–130.

[46] Lisa Blackman, 'Reinventing Psychological Matters: The Importance of the Suggestive Realm of Tarde's Ontology,' *Economy and Society 36*, (2007), pp.577ff. For the centrality of this distinction to modernism in general, see Bruno Latour, *We Have Never Been Modern* (Cambridge MA: Harvard University Press, 1993), passim.

[47] Robert Young, 'Darwinism *Is* Social,' in *The Darwinian Heritage*, ed. David Kohn (Princeton: Princeton University Press, 1985), pp.609, 633; Gowan Dawson, 'Darwin Decentred,' *Studies in History and Philosophy of Biological and Biomedical Sciences 46*, (2014), pp.93–96.

[48] But see Martin Jay, 'Historical Explanation and the Event: Reflections on the Limits of Contextualization,' *New Literary History 42*, (2011), pp.557–571.

was held to contain Darwin's signal achievement, the 'best idea anyone has ever had' (i.e. his theory of natural selection).[49] The legend which needed challenging, historians said, was that Darwin's intentions and motives in writing *Origin* had produced a wholly original and unbiased picture of nature, innocent of social or ideological taint.[50] *Origin* therefore became the bull's-eye for contextualist revision. Yet, if Darwin's arguments about natural selection owed more to his contemporaries than had previously been admitted, this implied that what were deemed more peripheral parts of his work—such as his inquiries into agency—were even more second-hand.

Which raises a question. How should the fascinating contexts that professional historians unearth for Darwin inform a reading of his books? More precisely: how can a *context*, as historians put it, 'constitute' or 'construct' a scientific text?[51] By way of answer, a context is typically assumed to have impacted Darwin's mental life: a given set of circumstances, having shaped Darwin's way of seeing things, forges an *authorial cognition*—an intention, concept, belief, assumption, conviction, or 'unshakable faith'—which serves as a brand new key to unlock the meaning of a book like *Origin* or *Descent*, or even, sometimes, *all* his significant work.[52]

This approach risks serious oversimplification, both of Darwin's mental life and of his publications. No author's mentality is single or transparent, to themselves or anyone else. Moreover, motives other than truth-telling inform authors' self-presentations. This was notoriously the case with Darwin, who often described his own work in terms which prove disingenuous to the point of falsehood.[53]

For example, *The Descent of Man*—which arguably contains Darwin's most original psychological theorizing (see Chs 6–8)—opens by protesting that it presented 'hardly any original facts in regard to man.' Then it declares that, had Darwin received a recent book by Ernst Haeckel, 'before my essay had been written, I should probably never have completed it,'—because, 'Almost all the conclusions at which I have arrived I find confirmed by this naturalist, whose knowledge on many points is much fuller than mine.'[54] In private, however, Darwin was less flattering about

[49] Dennett, *Darwin's Dangerous Idea*, p.21.

[50] Steven Shapin and Barry Barnes, 'Darwin and Social Darwinism: Purity and History,' in *Natural Order: Historical Studies of Scientific Culture*, ed. Steven Shapin and Barry Barnes (Beverly Hills CA: Sage, 1979), p.127.

[51] Young, 'Darwinism,' p.635; David Kohn, 'Darwin's Ambiguity: The Secularization of Biological Meaning,' *British Journal for the History of Science* 22, (1989), p.218.

[52] Examples are legion, as will be seen in later chapters. They include: Jonathon Hodge, 'Darwin's Book: *On the Origin of Species,' Science & Education* 22, (2013), p.2267; Gregory Radick, 'Is the Theory of Natural Selection Independent of Its History?,' in *The Cambridge Companion to Darwin*, ed. Jonathon Hodge and Gregory Radick (Cambridge: Cambridge University Press, 2003), passim; Evelleen Richards, *Darwin and the Making of Sexual Selection* (Chicago: Chicago University Press, 2017), p.285.

[53] Sandra Herbert, 'The Place of Man in the Development of Darwin's Theory of Transmutation. Part 2,' *Journal of the History of Biology* 10, (1977), p.158.

[54] Charles Darwin, *The Descent of Man, and Selection in Relation to Sex* (London: Murray, 1874), pp.2–3, my italics; Ernst Haeckel, *Natürliche Schöpfungsgeschichte* (Berlin: Georg Reimer, 1868).

Haeckel's work[55]—whose popularized account of evolutionary theory actually has little overlap with *Descent*. Moreover, Darwin's letters show that he had received a copy of Haeckel's new book in November 1868, almost *two years* before he finished drafting *Descent*.[56] So his claim that the writing of *Descent* was too far-advanced to address Haeckel's work was untrue.[57] None of this has stopped historians quoting Darwin's phony self-belittlement as grounds to dismiss the psychological discussions in *Descent* as derivative, and hence uninteresting.[58]

Even more questionable than simplifying Darwin's mental life is minimizing the complexity of his publications. Not only do his books self-consciously address several different audiences, but any creative use of words evokes allusions and figures, the resonances for which an author cannot, and may not wish to, contain. So, as we shall see, such carefully crafted works as *Origin, Descent,* and *The Expression of the Emotions in Man and Animals* provide extraordinary examples of work which includes more than their maker at the time knew, despite all that he *did* know.[59] Each proves incorrigibly plural, and impossible to align with any single authorial cognition. Unsurprisingly, therefore, the following chapters show that each of these books has attracted a host of incompatible readings from context-first historians.

Useful though knowledge about a context may be, its relevance cannot be assessed without first undertaking independent analysis of the books upon which a context is claimed to throw light. This text-first approach stands in a long line of scholarship which makes step one in studying any writerly work an appreciation of the plurality that constitutes it.[60] The multivalence of each of Darwin's main books makes misleading its summing-up by means of a few quotations (see Chs 4 and 6).

---

[55] Charles Darwin, 'Letter to Huxley, 22nd December,' *Correspondence of Charles Darwin* (Cambridge: Darwin Correspondence Project, 1866), www.darwinproject.ac.uk/letter/?docId=letters/DCP-LETT-5315; Charles Darwin, 'Letter to Huxley, 14th October,' *Correspondence of Charles Darwin* (Cambridge: Cambridge University Press, 1869), www.darwinproject.ac.uk/letter/?docId=letters/DCP-LETT-6936.

[56] Charles Darwin, 'Letter to Haeckel, 7th November,' *Correspondence of Charles Darwin* (Cambridge: Darwin Correspondence Project, 1868), https://www.darwinproject.ac.uk/letter/?docId=letters/DCP-LETT-6450.

[57] And, when Darwin published a significantly revised edition of *Descent* in 1874, he did nothing further to revise his text so as to assimilate Haeckel's work.

[58] E.g. Robert Young, *Darwin's Metaphor: Nature's Place in Victorian Culture* (Cambridge: Cambridge University Press, 1985), pp.56–78; Greta Jones, 'The Social History of Darwin's *the Descent of Man*,' *Economy and Society 7,* (1978), pp.1, 10, 16; John Greene, 'Darwin as a Social Evolutionist,' *Journal of the History of Biology 10,* (1977), pp.1–17; Evelleen Richards, 'Darwin and the Descent of Woman,' in *The Wider Domain of Evolutionary Thought,* ed. David Oldroyd and Ian Langham (Dordrecht: D Reidel, 1983), p.76; Adrian Desmond and James Moore, *Darwin's Sacred Cause: Race, Slavery and the Quest for Human Origins* (London: Penguin, 2009), p.373.

[59] Cf. Gillian Beer, *Darwin's Plots: Evolutionary Narrative in Darwin, George Eliot and Nineteenth-Century Fiction,* 2nd ed. (Cambridge: Cambridge University Press, 2000), p.2.

[60] Roland Barthes, *S/Z* (Malden, MA: Blackwell, 1990), p.5. This overriding concern for the text-being-interpreted is a hallmark of phenomenological analysis, which requires that we 'dis-implicate that object [i.e. the text] from the various intentions of behaviour, discourse, and emotion' which might explain or 'reduce' it: see Paul Ricoeur, *Freud and Philosophy: An Essay on Interpretation* (New Haven CT: Yale University Press, 1970), pp.28ff. For arguments that, in the alchemy of creation, a written work may transcend, 'melt away,' textualize, or make irrelevant, its context, see Thelma Lavine, 'Reflections on the Genetic Fallacy,' *Social Research 29,* (1962), pp.321–336; Sigmund Freud, 'Dostoevsky and

Perhaps most obviously missing from context-driven accounts of Darwin's work is any role for his painstaking observations of living creatures in their worlds. What value for his science had his arduous collecting trips, his skills with the microscope, his geological apprenticeship, his maritime surveys, his dogged endeavours in taxonomy, his tireless experiments and observations on the physiology and behaviour of plants, of worms, of babies—if his texts were solely constructed by social forces? And does it matter that the findings produced by his long-honed practices of research retain weight, or even—as in the case of the return to a focus on live organisms in twenty-first century biology—*gain* weight, in more and more branches of contemporary scientific inquiry?

1.2 The Return to Living Organisms

An important context for this book is my reading of a growing body of work which encourages evolutionary biologists to rethink the role played by the doings of living creatures in constituting what Darwin called *the struggle for existence* (see Ch. 3).[61] The idea that individual organisms, or groups of organisms, *themselves* act to exploit, and so adapt to, the conditions in which they live has become crux to a contemporary movement in biology which explicitly harks back to Darwin. At its heart, this approach revalues what geneticists call phenotypes—the observable characteristics of a live creature.[62] This is important for many reasons, not least because description of the myriads of *phenotypes*, namely, the forms, structures, sizes, colours, and other characters of living organisms, was, before its eclipse by modernism, a chief aim of any naturalist's work—including Darwin's, of course.[63]

Modernist biology underpins the still-popular assumption that DNA (genes and genotypes) represents the blueprint for life, and hence, is the key to knowing,

Parricide', in *Standard Edition of the Complete Psychological Works of Sigmund Freud*, ed. James Strachey (London: Hogarth Press, 1961), p.173; Marcel Proust, *Against Sainte-Beuve and Other Essays* (Harmondsworth: Penguin Books, 1988), p.12; William Wimsatt and Monroe Beardsley, 'The Intentional Fallacy', *Sewanee Review 54*, (1946), p.480; Jacques Derrida, *Of Grammatology* (Baltimore, MA: Johns Hopkins University Press, 1976), p.158; Roland Barthes, 'The Death of the Author', in *Image-Music-Text* (Glasgow: Fontana, 1977); Sean Burke, *The Death and Return of the Author: Criticism and Subjectivity in Barthes, Foucault, and Derrida*, 3rd ed. (Edinburgh: Edinburgh University Press, 2010); Jay, 'Historical Explanation'; Martin Jay, 'The Textual Approach to Intellectual History', in *Force Fields: Between Intellectual History and Cultural Critique* (New York: Routledge, 1993), pp.158–166.

[61] Charles Darwin, *The Origin of Species by Means of Natural Selection, or the Preservation of Favoured Races in the Struggle for Life*, 6th ed. (London: Murray, 1876), pp.48–61.
[62] E.g. Steven Rose, *Lifelines: Biology, Freedom, Determinism* (Harmondsworth: Penguin Books, 1997).
[63] Johannsen, 'The Genotype', p.134. Mary Jane West-Eberhard, *Developmental Plasticity and Evolution* (New York: Oxford University Press, 2003), p.31 defines *phenotype* as 'all traits of an organism other than its genome', including 'enzyme products of genes, behaviours, metabolic pathways, morphologies, nervous tics, remembered phone numbers, and spots on the lung following a bout of flu.'

for example, what makes humans human.[64] It is thus typically said that DNA *pro-grammes* the characteristics of humanity, casting genes as the actors, whether selfish, cooperative, or otherwise. In this scenario, phenotypes become mere stooges, temporary vehicles for conveying genes from generation to generation.[65] However, a major conceptual about-face in twenty-first century biology makes phenotypes the leaders and genes the followers in evolutionary change.[66] It is this shift which gives Darwin's understanding of agency a threefold interest, simultan-eously being of previously unsuspected relevance to contemporary biology, while having both evolutionary and psychological consequences.

Phenotype-led biology exploits a crucial insight about living creatures. Life-forms defy the physical law that the level of organization in a sealed system crum-bles over time until reaching equilibrium at maximum chaos (or maximum *entropy*).[67] Clocks and cars obey this law. Plants, platypuses, and people do not. Living creatures are called *open* systems because, by constantly exchanging flows of matter and energy with their surroundings, they maintain themselves at a level of organization far above chaos.[68] Moreover, as both individual life-histories and species-level evolution affirm, open systems often grow *more organized* over time.

How do they do this? By transforming energy. The living gain energy by taking in and metabolizing food, air, water, and sunlight. They spend energy in the form of particular actions which keep them alive.[69] In fact, they must always be acting, both because—being open systems—they are always spending energy, and be-cause their very existence depends on a dynamic interdependency with their lo-cale. Thermodynamically, therefore: *all life-forms are inescapably agents*. It is their agency which gives them their capacity to adapt to challenges and, hence—as both Darwin and phenotype-led biology argue—to evolve.

On these grounds, *Darwin's Psychology* presents Darwin's view of the natural world as having two axes. These axes differ in terms of time-scale, in content, and in how they contribute to his discoveries. Obviously, therefore, we need to study

---

[64] E.g. Robert Plomin, *Blueprint: How DNA Makes Us Who We Are* (London: Penguin, 2018).

[65] Dawkins, *Selfish Gene*, p.25 & passim; Cf. Denis Walsh, 'Two Neo-Darwinisms,' *History and Philosophy of the Life Sciences 32*, (2010), pp.317–339; Walsh, 'Struggle for Life.'

[66] West-Eberhard, *Developmental Plasticity and Evolution*; West-Eberhard, 'Revival of Darwin's Theory'; Massimo Pigliucci, 'Phenotypic Plasticity,' ed. Massimo Pigliucci and Gerd Muller, *Evolution: The Extended Synthesis* (Cambridge MA: MIT Press, 2010), pp.370–371; Scott Gilbert and David Epel, *Ecological Developmental Biology: The Environmental Regulation of Development, Health, and Evolution*, 2nd ed. (Sunderland, MA: Sinauer Ass., 2015); Denis Walsh, *Organisms, Agency, and Evolution* (Cambridge: Cambridge University Press, 2015); Daniel Nicholson, 'The Return of the Organism as a Fundamental Explanatory Concept in Biology,' *Philosophy Compass 9*, (2014), pp.347–359.

[67] The second law of thermodynamics.

[68] Ilya Prigogine and Isabelle Stengers, *Order out of Chaos: Man's New Dialogue with Nature* (London: Fontana Paperbacks, 1984), pp.140–145; Peter Sylvester-Bradley, 'Evolutionary Oscillation in Prebiology: Igneous Activity and the Origins of Life,' *Origins of Life 7*, (1976), pp.10–11.

[69] Daniel Nicholson and John Dupré, eds., *Everything Flows: Towards a Processual Philosophy of Biology* (Oxford: Oxford University Press, 2018), pp.15–20; Nicholson, 'The Return of Organism,' p.355.

both dimensions before we can understand how Darwin construed what we call mental life. Axis one was the short-term, here-and-now world in which he could observe the life-forms he studied—his 'what.' Axis two explained how notable features of that world came about over immeasurably longer time-periods—the 'how come' of evolution.[70]

The idea that Darwin's had a two-ply vision of living things clashes with the modernism that still reigns over much biology and psychology. Hence, the next two chapters of *Darwin's Psychology* focus on the primary, yet less familiar, axis of his work—his view that living things constitute a theatre of agency. Chapter 2 focuses on his methods, and Chapter 3 on the understanding of agency that these methods helped Darwin construct when studying invertebrates and plants.

Several discords between Darwin's work and modernist science surround natural history. The most jarring relate to observation and description. Contemporary manuals of research in psychology give students scant guidance in how to observe or how to describe their fellow creatures. Rather, the term *observation* is used to refer to the reliable placing into pre-defined categories of a limited number of behaviours-of-interest, to be provoked in subjects by a well-designed array of experimental stimuli.[71] Nothing could contrast more with Darwin's approach to the problems of nature. I say 'Darwin's approach,' yet Chapter 2 shows that he did not invent his approach from scratch. Just the opposite. His descriptions of the intricate weave of relations within and between the members of different species, between their habits and their habitats—which, in his eyes, grounded the biology of any forest or field—built on centuries of collective endeavour to understand what he and his predecessors called *the economy of nature*.[72] Hence the need to examine Darwin's *practices* of science—not just because these may seem so archaic as to be incomprehensible through a modernist frame, but because, according to the logic of an organism-led biology, researchers are increasingly recognizing the need to return to a Darwin-like methodology so as to appreciate how evolution works at the level of real lives.[73]

Chapter 3 examines the findings about agency that Darwin's practices of observation and description produced when he targeted such relatively simple creatures as colonial invertebrates, climbing plants, and earthworms. Chapters 4 to 9 show that Darwin's understanding of agency grounds all his more-obviously psychological,

[70] Cf. Lavine, 'Genetic Fallacy.'

[71] E.g. Dennis Howitt and Duncan Cramer, *Introduction to Research Methods in Psychology*, 3rd ed. (Harlow UK: Pearson, 2011), p.11.

[72] Writers like Carl Linnaeus and Charles Lyell saw nature as a divinely constructed economy in which the division of labour between the interdependent activities and stations of its various inhabitants were 'fitted to produce general ends, and reciprocal uses,' and thus 'ensure the health and well-being of the natural world': Trevor Pearce, ' "A Great Complication of Circumstances"–Darwin and the Economy of Nature,' *Journal of the History of Biology 43*, (2010), p.498.

[73] Shirley Strum, 'Darwin's Monkey: Why Baboons Can't Become Human,' *Yearbook of Physical Anthropology 55*, (2012), p.7 & passim; Tim Clutton-Brock, 'Co-Operation between Non-Kin in Animal Societies,' *Nature 462*, (2009), p.54.

human-oriented, work: on emotional expression; self-consciousness; sex and desire; group-living; culture; and infancy. But, as Chapter 3 goes on to elaborate, Darwin's vision of the natural world as a theatre of agency also generated his understanding of evolution—the long-term axis of his vision of the natural world. In particular, it underpinned his formulation of natural selection, the prize exhibit of modernist evolutionary science, and hence, the site of the hottest debates about biology's re-emphasis on organisms.

Modernizers call themselves Darwinists because their synthesis, they say, remains true to the central tenets of Darwin's theory, while framing it in a more precise and up-to-date way. ('The selfish gene theory *is* Darwin's theory, expressed in a way that Darwin did not choose,' writes Richard Dawkins.)[74] Through this lens, natural selection becomes a mechanism which changes the genetic composition of populations, a mechanism whose investigation falls under the rubric of molecular biology and population genetics. Yet, *Origin* does not treat natural selection as a mechanism, but as *a law*. (Darwin compared it to the law of gravity.) As law, it results from the combined effect of three subsidiary processes: *inheritance*, which includes both the transmission and individual development of heritable particles; *variation*;[75] and the metaphorically-labelled *struggle for life*, the latter two of which *are generated through the interdependencies of agency.* Chapter 3 concludes that, for Darwin, as for today's phenotype-led biology, not 'genes' but *organisms are the leading actors of evolution.*[76]

1.3 The Theatre of Humanity

As Chapter 3 will relate, from his teens on, Darwin became actively engaged in the then-lively international efforts to discover distinctions between live creatures and inanimate things. This debate framed several strands of his research before and during his voyage on *HMS Beagle* (1831–1836), particularly, his studies of invertebrates and plants. He especially stressed the ways the interdependencies of agency and structure wove together the lives of conspecifics with those of their cohabitants and surroundings—thus challenging the central place given to a concept of 'the individual' in biology (and, later, psychology). Unsurprisingly, such early conclusions about the hallmarks of life in humble creatures shone through and structured much of what he subsequently had to say about life-forms of all kinds,

---

[74] Dawkins, *Selfish Gene*, p.xv, my italics; Jeremy Garwood, 'Darwin Reloaded,' *Lab Times 3*, (2017), p.20.
[75] NB the word 'variation' is ambiguous in *Origin*. In the singular and without an article (as here), 'variation' implies an act or process—meaning, for example, to try one thing after another. But as an instance or result of variation, *with* an article, *a* 'variation' implies an object or quality. In this second sense, *Origin* often glosses 'variations'—which it uses mainly in the plural—as 'individual differences.'
[76] Walsh, *Organisms*, p.241.

especially the agency of humans—which forms the central focus of the works by Darwin reviewed in Chapters 4 onwards.

Chapters 4 and 5 present an exposition of *The Expression of the Emotions in Man and Animals*. Chapter 4 underlines that *Expression* elaborates copious data to fill out *a theory of expression*, not a theory of emotion. Darwin's downplaying of emotion as a cause of expression raises a puzzle about how a person's facial and other gestures can be identified as having meaning—both in the normal run of social life, and in research. Through a focus on the methods of study described in *Expression*, I show that Darwin treated the significance of non-verbal behaviour as something relational rather than individual, best approached through its readings by others. This implies that expressions do not each have as their essence a unique, ready-made 'internal' upwelling of meaningful emotion. Sense is typically *read into* an expression by others, who interpret it according to the social situation of which it forms part.

Chapter 5 documents an extra layer of complexity added into Darwin's stress on situational reading, through his analyses of blushing. Blushing results from *reflexive* reading: I blush as a result of how I imagine you to be reading me. Darwin's concept of reflexive reading produces a nuanced understanding of differences between agency and self-consciousness, differences showing that humans have a dual self. *Expression*'s formulations on reflexive recognition prove fundamental to Darwin's treatment of animal desire (see Ch. 6) and human conscience (see Ch. 7).

Chapters 6, 7, and 8 discuss *The Descent of Man*. Chapter 6 disentangles *Descent*'s complex analysis of sexual agency, first in animals, then in humans—the theatre of action which Darwin deemed to generate *sexual selection*. This theatre is diverse, having four distinct plot-lines for animals, and five for humans. Each plot endows females and males with different characters. Unsurprisingly, therefore, when the book tries to generalize about men and women, it often seems to contradict itself—producing a series of muddles. Many of these confusions can only be disentangled by reference to the concerns of educated Victorian men about race and gender.

Chapter 7 examines *Descent*'s treatment of the most human aspects of human agency, amongst which conscience takes first place. Darwin's analysis of conscience sets out from the evolution of the moral sense, which gives each of us our feeling for right and wrong. He traces this—and other faculties, like speech—to the fact that human agency has largely been shaped by group living. In particular, group living gives the key to human self-criticism, something which *Descent* describes as a complex play in different characters. Darwin's argument that groupness laid the foundation for the formation of humanity throws down a challenge to more recent, individualistic psycho-biologies.

Chapter 8 uses phenotype-led evolutionary biology as a lens to survey the way that *Descent* places culture in the constitution of human action. Far from culture standing *beyond* his theory of biological evolution, Part I of Darwin's *Descent* shows that—at least in his treatment of European civilization—culture was dominant in specifying human action. This went equally for the details of: who reproduced with whom; aesthetic refinement; religious belief; languages; virtues; and differences between genders. According to Darwin's understanding, cultural investments in human social groups often override the processes that would otherwise give rise to natural selection.

Chapter 9 elaborates Darwin's proposal that infancy provides a test-case for evolutionary hypotheses about human agency. It first surveys Darwin's own research and publications on babies. It then reviews contemporary research bearing on the central plank of *Descent*'s account of the capacities that are most human in human agency: that they are based in group living. In particular, I describe several recent studies of all-infant groups. Contrary to theories which hold that a one-to-one infant-mother bond provides the prototype and template for human sociality, the new findings align better with Darwin's group-based theory of human action. For example, months before they can talk, babies prove capable of interacting with more than one other companion at the same moment.

1.4 Looking Ahead

While Darwin has long been hailed as a seminal influence in modern psychology, this is the first book ever to present a comprehensive exposition of his writings on psychological matters.[77] Growing disappointment with some fruit of twentieth-century modernization in biology and psychology gives firm ground for a thorough reconsideration of Darwin's work. Indeed, it is tempting to see the novelties championed by advocates of the 'extended' evolutionary synthesis (Figure 1.3) as, in large part, an unwitting 'back to Darwin' rediscovery of what got lost when the Modern Synthesis remade the history of evolutionary thinking in its own image.[78] A like dynamic animates recent calls by social theorists for a 'profound conceptual shift' in psychology: re-situating humans as responsive, embodied beings immersed within a reality of continuously intermingling, agential flows—an

---

[77] E.g. Edwin Boring, *A History of Experimental Psychology*, 2nd ed. (Englewood-Cliffs NJ: Prentice-Hall, 1950), p.506: 'By 1900 the characteristics of American psychology had become well defined. It had inherited its physical body from German experimentalism, but it had got its mind from Darwin.' The nearest thing this book has to a precursor is Robert Richards, *Darwin and the Emergence of Evolutionary Theories of Mind and Behaviour* (Chicago: University of Chicago Press, 1987). I have gained much from Richards' book, from his conversation, and other scholarly work. Yet, *Emergence* does not undertake to deal comprehensively with Darwin's writings about agency, focusing on those which influenced twentieth-century psychology, and thus largely bypassing Charles Darwin, *The Expression of the Emotions in Man and Animals*, 2nd ed. (London: Murray, 1890). Professor Richards (personal communication, 11 Jan 2017) told me that one reason for side-lining *Expression* 'had to do with the oddity of Darwin's own analysis, namely, that he understood no biological function for the expression of the emotions and had no role for natural selection'—which shows the strong sway the Modern Synthesis held over key historical treatments of Darwin in the late twentieth century.

[78] Bradley, EES.

immersion to which we owe significant aspects of our natures.[79] Should this book succeed, readers will hear in such rallying-cries the belated acknowledgment of a past that has long been overlooked, a past of great extent and unexpected contemporary relevance, a past well-captured by Darwin's writings on the lives and loves of plants and animals, full of insight and surprises.

Themes of embodiment and agency, unfolding and interdependence, of animate mingling with inanimate, of individual merging with group, and self with others—all are crucial to Darwin's vision of the living world. That said, there is a big difference between the clever quotation of extracts from a finished, archaic past and a systematic reacquaintance with a tradition that has always been available to anyone with eyes to read and ears to hear.[80] Such reacquaintance does not come easy. To perceive the scope of Darwin's take on agency, one's attention cannot be blinkered by any one discipline—whether genetics or psychology, sociology or history, cultural studies or philosophy. Nor can his work be perfunctorily condemned by anachronistic labelling of modernist sins—anecdotal, anthropomorphic, Lamarckian, sexist, racist (but see Chs 6–8). Its appreciation calls for a restructured field of vision regarding psychology.

American psychologist and philosopher, William James, noted that our brain's window on the present is slim—15 seconds at most. Even through this slit, our sight is partial. We catch only the facing portion of what reflects light to our eyes. So the here-and-now we intuit has roots in the invisible. James described the relation of what we eye to the meaning we give it as 'just like that of the fictitious space pictured on the flat back-scene of a theatre to the actual space of the stage.' The objects painted on the backdrop (trees, columns, houses in a receding street) give meaning to the objects standing on the stage, 'and we think we see things in a continuous perspective, when we really see thus only a few of them and imagine that we see the rest.'[81]

Darwin's writings flesh out a way of seeing and conceiving living beings, both high and humble, which contrasts dramatically with the stagings of mental life giving sense to most twentieth- and some twenty-first-century psychologies, and the biologies they invoke—individuals as asocial; behaviour as mechanically stimulated; cognition as computer-output; genes as blueprints for behaviour; and

[79] E.g. John Shotter, 'Agential Realism, Social Constructionism, and Our Living Relations to Our Surroundings: Sensing Similarities Rather Than Seeing Patterns,' *Theory & Psychology 24*, (2014), p.306 Cf. Paul Stenner, 'Psychology in the Key of Life: Deep Empiricism and Process Ontology,' in *Theoretical Psychology: Global Transformations and Challenges*, ed. Paul Stenner, et al. (Ontario: Captus, 2011), pp.48–58.
[80] Latour, *Never Been Modern*, p.74.
[81] James's commentary in William James, *The Principles of Psychology* (New York: Holt, 1890), pp.570–605 on the 'synthetic' or 'specious' present was neither the first nor the last. See e.g. Merlin Donald, *A Mind So Rare: The Evolution of Human Consciousness* (New York: Norton, 2001), p.25. Cf. Ben Bradley, *Psychology and Experience* (Cambridge: Cambridge University Press, 2005), pp.48–55.

we humans as 'lumbering robots.'[82] For Darwin, whether observing bindweed, babies, or baboons, the backcloth he had embroidered defined an 'infinitely complex' theatre of agency, in which relatedness was all.[83] It is to his weaving of that backcloth that I now turn.

1.5 Overview of the Argument

A bare bones summary of the book's argument goes like this:

- *Darwin's Psychology*, the first book comprehensively to examine Darwin's writings on psychological matters, is grounded by the premise that ongoing revisions to evolutionary theory in biology and psychology increasingly broaden the relevance of his work to today's science.
- These revisions include:
  - findings garnered by extending the descriptive basis for evolutionary generalization (as in ecology and ethology) beyond that provided by laboratory research (Ch. 2);
  - the crucial role played in evolutionary change by whole organisms;
  - the prevalence of symbiosis and cooperation across species and levels of organic organization;
  - the view that the agency and plasticity of phenotypes lead evolutionary change (genes being followers);
  - a resulting recognition of the key role played by individual development in the production of individual differences;
  - increased attention to niche construction and the place of what Darwin called 'habits' and 'culture' in evolutionary change.
- The starting-point for my exegesis of Darwin's writings is the recognition that natural selection is *a law* which embraces a perceived *sequence of events* or phenomena (he compares it to the law of gravity). It is *not* a causal mechanism that *explains* those events (as it is when viewed through the lens of the Modern Synthesis; see Ch. 3);
  - For Darwin, the causal mechanisms are inheritance, variation, and the struggle for life;

---

[82] E.g. Dawkins, *Selfish Gene*, pp.19, 46–87, 270–271. As Bruno Latour, *What Is the Style of Matters of Concern?* (Assen: Van Gorcum, 2008), p.32, remarks: 'there is nothing more amazingly artificial, more carefully staged, more historically coded than meeting a matter of fact.' To understand any such scientific staging as Darwin's, he advises us to shift our attention 'from the stage to the whole machinery of [the] theatre' which frames the matters with which he concerns us (p.38)—as I do in Ch. 2.

[83] Darwin, *Origin 1876*, pp.49, 62, 92.

- Inheritance, according to Darwin, includes both *transmission* (of his 'gem-mules,' our 'genes' or 'DNA') and individual *development* (of what he called 'habits,' 'culture'; what we call epigenetics and learning);
- What he called his *metaphor* of 'the struggle for existence' covered the phe-nomena of collaboration and interdependence, *not just those of competition.*
- Darwin's distinction between law and causal mechanism (or '*vera causa*') opens the question: What processes did he hold to produce the sequences of event covered by the law of natural selection—and the law of sexual selection?
- I paraphrase his answer as: 'The theatre of agency produces them' (Ch. 3).
- In this theatre, a*gency* or organic purposiveness:
  - (a) is ubiquitous in the living world (see his research on the ova of *Flustra*, movement in climbing and insectivorous plants, problem-solving in earthworms);
  - (b) has many levels;
  - (c) produces individual differences or *variations* through individual development;
  - (d) means different actions *rebound* upon their agents in different ways— producing winners and losers:
    - agency thus *leads* evolutionary change (in Darwin's 'struggle for life'), while genes/gemmules follow—as in what Darwin called 'transitional habits,' or what we now call 'the Baldwin effect.'
  - (e) agentic rebounding, or reagency, also produces *interdependency*: eco-logical complexity (his 'web of complex relations'); co-adaptations (see his monograph on orchids and insects);[84] and, most importantly here, *social animals.*
- Social animals *evolve differently* from the non-social, via: mutual aid; sympathy; group coherence;
- As chapters comprising the second half of the book document in detail, re-agency gives the key to Darwin's take on human psychology. To wit:
  - An inborn capacity for reading (or 'recognizing') *others'* emotional expres-sions *in situ* gives those expressions their meaning (rather than the meaning being endowed by the hidden mental states of the expresser) (Ch. 4);
  - A more complex *doubled* reagency—in which my reading of your reading of my appearance or actions—produces blushing (Ch. 5);
  - Such a doubled dynamic also structures sexual display and desire, in both animals and humans (Ch. 6);
  - A like doubled reagency produces conscience and group cohesiveness (Ch. 7);

---

[84] Charles Darwin, *The Various Contrivances by Which Orchids Are Fertilised by Insects*, 2nd revised ed. (London: John Murray, 1882).

- And so creates cultures (Ch. 8).
- Darwin's psychology can thus provide a coherent framework for understanding a wide array of otherwise-disparate contemporary research findings across biology and psychology—as I illustrate from recent research on the agency of human infants (Ch. 9).

2

The Art and Craft of Describing

'*Respect* good describers,' Darwin urged.[1] Why? Because science sets out from the question, *what?* Which means science must begin by describing. Only when we have described what we want to explain will we be in a position to find out how and why it is the way it is. And the better the description, the more likely it is our explanation will stick. So, how do we best set about describing? For Darwin, and the long line of naturalists to which he belonged, this was not primarily a question about how to use language. It was a question of how best to look. A practical question. For a notable assumption grounding the rise of science rendered finding and making, our discovery of the world and our crafting of it, a single process. Attentive looking, transcribed by the hand—what might be called the observational craft—was the best way to record the multitude of things that make up the visible world.[2]

The seventeenth-century tradition captured by the mantra, 'to see is to picture,' had, by Darwin's day, spawned a long list of instrumental crafts deemed essential to any serious scientific enterprise.[3] Even at the birth of the Royal Society in 1660—sometimes taken to mark the dawn of contemporary science—*picturing* did not simply mean drawing. Giving images epistemic status aligned them with truth, and thus, with the opposite of truth, deception. The eye itself was a picture-making instrument. And, as with any instrument, errors in seeing could arise from the form and function of the eyes. Vision could be trusted, but only advisedly, when the optics of seeing were properly understood. Hence the need for a robust, self-critical discipline of right observation, backed by a sophisticated craft of truthful representation.[4] Central to this discipline were instruments that could reduce the fallibility of the eye, most notably, the microscope. Three years into its life, the Royal Society asked Robert Hooke (1635–1703), pioneer of the microscope, to bring one of his

---

[1] Charles Darwin, "E," (1838–1839), p.52, his emphasis.

[2] Svetlana Alpers, *The Art of Describing: Dutch Art in the Seventeenth Century* (London: Penguin, 1989), pp. xxvii, 273.

[3] Alpers, *Art of Describing*, p.xix contrasts Kepler's '*ut pictura, ita visio*' [to see is to picture] with 'the ubiquitous doctrine *ut pictura poesis* [which] was invoked in order to explain and legitimize [Italian, Renaissance] images through their relationship to prior and hallowed texts'—most notably, biblical texts. '*Ut pictura poesis*' [as is painting so is poetry] makes both painting and poetry text-based practices, and the human the focus and the measure in painting. Dutch art—like Baconian science—made text subordinate to vision and its foibles. 'Man' was dethroned: all things in nature were to be represented 'exactly and unselectively.'

[4] Alpers, *Art of Describing*, pp.50, 33, 36, 34.

Darwin's Psychology. Ben Bradley, Oxford University Press (2020). © Oxford University Press.
DOI: 10.1093/oso/9780198708216.001.0001

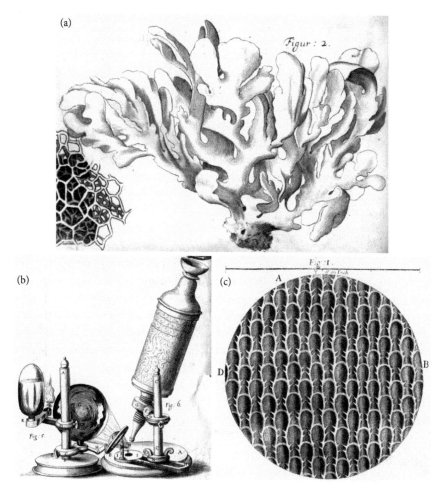

Figure 2.1 Images from Hooke's *Micrographia*: (a) illustration of the sea-dwelling plant-like animal (or 'zoophyte') *Flustra*; (b) Hooke's microscope and light source; (c) the magnified 'leaf' of *Flustra*.

Hooke R. (1665). *Micrographia: Some Physiological Descriptions of Minute Bodies Made by Magnifying Glasses with Observations and Inquiries Thereupon*, London : Printed by Jo. Martyn and Ja. Allestry, printers to the Royal Society.

p.92, Figure 2, Schem. XIV, Figure 1 & Observation 23. Hooke wrote: 'this Sea-weed … is cover'd over with a most curious kind of carv'd work, which consists of a texture much resembling a Honey-comb; for the whole surface on both sides is cover'd over with a multitude of very small holes, being no bigger then so many holes made with the point of a small Pinn, and rang'd in the neatest and most delicate order imaginable.'

observations to their every weekly meeting. A year later, the Royal Society put out Hooke's beautiful trail-blazing *Micrographia: or Some Physiological Descriptions of Minute Bodies Made by Magnifying Glasses* (see Figure 2.1).[5]

---

[5] Robert Hooke, *Micrographia: Or Some Physiological Descriptions of Minute Bodies Made by Magnifying Glasses. With Observations and Inquiries Thereupon* (London: Royal Society, 1664).

To understand how Darwin saw living things—my task in this chapter—we need to respect the long-standing traditions of practical craft through which he accessed the natural world. The chapter's challenge results in part from the modernist abandonment of these traditions in favour of what has sometimes been called a *nominalist* approach to observation.[6] As someone who has spent forty years on teams of lecturers in psychology, I know it makes sense to say: 'Scientific discoveries are not based on reality; they are based on *descriptions* of reality'—for the reason that observations and facts only count as scientific when fixed by words.[7] Stop there, and the relation between *observing* a thing and *describing* it amounts to noticing the minimum required to name a given object in conformity with the rules of the delimited language devised for that purpose by a researcher.[8] Hence, psychologists train students to *operationalize* their variables.[9]

The procedure of *operationalization* makes the meaning of a concept synonymous with the operations pre-assigned to measure it. Imagine, for example, you are a psychologist who wants to conduct a study of love. A viable operational definition might comprise just three measures, namely, that a lover: seeks their loved one's company above anyone else's; gives the loved one gifts; and dates the loved one.[10] Only these three categories of behaviour would get included in your study. Moreover, no behaviour would actually need to be observed at first hand, should you choose to collect your data by administering a questionnaire to elicit self-reports from the people you study—a very common practice.

The descriptive craft of a psychology that set out from natural history, as Darwin's did, could not contrast more with the modernist subordination of description to definition by operation.

2.1 Vocation

Let me begin by reconsidering the Darwin whom modernist psychologies cast as a gifted amateur.[11] Today, the word amateur has two connotations. On one

---

[6] Michel Foucault, *The Order of Things: An Archaeology of the Human Sciences* (London: Routledge, 2002), p.322.

[7] An argument invoking Karl Popper, *Objective Knowledge: An Evolutionary Approach* (Oxford: Clarendon Press, 1972). This view invites social constructionism in psychology: e.g. Kenneth Gergen, 'The Social Constructionist Movement in Modern Psychology', *American Psychologist* 40, (1985), p.268, who, with odd logic, argues that, because description (partly) depends on social processes of signifying, 'observation of persons . . . is questionable as a corrective or guide to descriptions of persons.'

[8] Foucault, *Order of Things*, passim.

[9] E.g. Dennis Howitt and Duncan Cramer, *Introduction to Research Methods in Psychology*, 3rd ed. (Harlow UK: Pearson, 2011), pp.41, 52–54.

[10] For this example, see: Richard Kasschau, *Psychology: Exploring Behaviour* (Englewood Cliffs, NJ: Prentice-Hall., 1980), pp.27–29; also Sigmund Koch, 'Psychology's Bridgman vs Bridgman's Bridgman', *Theory & Psychology* 2, (1992), pp.261–290. Obviously, much would escape such a definition, not least Pierre Bezuhof's love for Natasha Rostova throughout all but the last few chapters in Leo Tolstoy, *War and Peace* (London: The Reprint Society, 1960).

[11] E.g. James Deese, *Psychology as Science and Art* (New York: Harcourt Brace Jovanovich, 1972), p.35. Cf. Soraya De Chadarevian, 'Laboratory Science Versus Country-House Experiments. The

hand it relates to unpaid work: a layperson or hobbyist. On the other, it dispar-
ages the ineptitude of dabblers lacking in technical expertise. Both senses defer to
the professional. Yet Britain had no professional scientists in 1831, when Darwin
first dedicated himself to full-time investigation of the natural world. Even the
word 'scientist' did not exist. Leaving aside the special case of medicine, scientific
training had no place in university education until the 1870s. So science could
not become a career in the modern sense for much of the nineteenth century.[12]
This made nearly all scientific discoveries prior to the 1890s the fruit of unsalaried
work—not just Darwin's. Even Einstein's special theory of relativity, announced in
1905, we owe to his spare time.[13]

To call Einstein scientifically incompetent sounds ridiculous. But, as seen in
Chapter 1, the rise of modernism in psychology has made the opposite ring true
of Darwin. Where the laboratory is deemed the preeminent venue for scientific
discovery, natural histories of the kind conducted by Darwin come to look so archaic as to appear scientifically anachronistic.[14] The tradition of natural history is
best recognized today through beautiful, high-production TV films featuring the
hidden worlds and antics of lesser-known plants and animals. Go back two centuries, however, and natural history gave foundation to every branch of empirical
science (but see the distinction of the *empirical* from the *mathematical* sciences in
2.4 Empirical Orientation).[15]

For today's professionals, a gap separates life from work—the people scientists
are from their 'day-job'—such that, where job swamps identity, the term *work-
aholic* applies. Three centuries ago, natural history invoked a way of life akin to
religious vocation and the mediaeval monastic practice which tied observation to
observance, as a 'process whereby reflection on the physical objects of Creation led
to knowledge of spiritual truths.'[16] By the time Darwin embarked on *HMS Beagle*
in October 1831, the naturalist's way of life was deep-grooved. The best of the breed
undertook extensive travels, especially when young. A century before Darwin,
the famous Swedish taxonomist Carl Linnaeus (1707–1778) had undertaken
to prepare his acolytes to venture into unknown lands, ensuring they were able
quickly to orient themselves, to recruit native populations as both informants and

Controversy between Julius Sachs and Charles Darwin,' *British Journal for the History of Science* 29, (1996), pp.17–41.

[12] Paul White, 'The Man of Science,' in *A Companion to the History of Science*, ed. Bernard Lightman (London: Wiley, 2016).
[13] Albert Einstein, 'On the Electrodynamics of Moving Bodies,' in *The Principle of Relativity* (New York: Dover, 1952).
[14] James Angell, 'The Influence of Darwin on Psychology,' *Psychological Review* 16, (1909), p.159; see too De Chadarevian, 'Laboratory Science.'
[15] Phillip Sloan, 'Natural History,' in *The Cambridge History of Eighteenth Century Philosophy*, ed. Knud Haakonssen (Cambridge: Cambridge University Press, 2006), pp.903–937; Lorraine Daston, 'The Empire of Observation, 1600–1800,' in *Histories of Scientific Observation*, ed. Lorraine Daston and Elizabeth Lunbeck (Chicago: University of Chicago Press, 2011), pp.81–113.
[16] Katherine Park, 'Observation in the Margins, 500–1500,' in *Histories of Scientific Observation*, ed. Lorraine Daston and Elizabeth Lunbeck (Chicago: Chicago University Press, 2011), pp.22–23.

spokespersons, thus to return with 'useful knowledge.'[17] Before Queen Victoria had long settled on her throne, not just Darwin, but many of his peers had made extensive scientific voyages: Joseph Hooker, Thomas Huxley, Henry Bates, Robert Brown, and Alfred Wallace among them.[18]

Travel had long been a staple of scientific advance. In 1620, well before the Royal Society opened its doors, its luminary, Francis Bacon (1561–1626), had advised that science would benefit from picking the brains of travellers and collating the resultant knowledge. By 1652, Samuel Hartlib (c.1600–1662) had inaugurated the distribution of lists of queries to peripatetic observers. For example, Gerald Boate (1604–1650), a Hartlib associate, had contributed 'An Interrogatory Relating more particularly to the Husbandry and Natural History of Ireland.' This had 25 sections.[19]

Such broadcast use of interrogatories (surveys) underlines that natural history, while conducted by individuals, had collective aspirations. To observe in detail the entire natural world went beyond the capacity of any single naturalist. Yet, when put together, the discoveries of individuals might pave the way for a more universal and reliable natural history.[20] A main means of sharing and checking observations was letter-writing. The publication of Darwin's enormous correspondence has so far taken four decades of editorial work and already fills twenty-five heavy volumes—to which several more are yet to be added. A large proportion of these letters were written to fellow naturalists, underlining Darwin's indebtedness to the tradition of collegial observation, both scientifically and socially.

Interrogatories underline too the extraordinary breadth of attention expected of a naturalist: weather; myths; tides; terrain; human and animal habits; human languages; geology; geography; flora; fauna; stars and planets—nothing was alien. Thomas Huxley wrote that, although naturalists might set out from 'the investigation of habits,' they were 'insensibly led into Physiology, Psychology, Geographical and Geological distribution.' Equally, the investigation of form necessarily led them into 'systematic Zoology and Botany, into Anatomy, Development, and Morphology.' Whilst each such science, if pursued into all its details, might fill a lifetime, it was 'in their aggregate only,' that they amounted to the science of natural

[17] Michael Dettlebach, '"A Kind of Linnaean Being": Forster and Eighteenth-Century Natural History,' in *Observations Made During a Voyage around the World by Johann Reinhold Forster*, ed. Nicholas Thomas, Harriet Guest, and Michael Dettlebach (Honolulu: University of Hawai'i Press, 1996), p.lix.

[18] See e.g. Iain McCalman, *Darwin's Armada: Four Voyages and the Battle for the Theory of Evolution* (New York: Norton, 2009).

[19] Daniel Carey, 'Compiling Nature's History: Travellers and Travel Narratives in the Early Royal Society,' *Annals of Science* 54, (1997), p.272.

[20] E.g. from 1789, Gilbert White, *The Natural History of Selborne* (Oxford: Oxford University Press, 2013), p.71: 'as no man can alone investigate all the works of nature, these partial writers may, each in their department, be more accurate in their discoveries, and freer from errors, than more general writers; and so by degrees may pave the way to an universal correct natural history.'

history. The title of 'naturalist' was deserved only by one who had 'mastered the principles of all.'[21]

For Darwin, the ties between the study of forms and habits were far tighter than Huxley's words suggest—as we shall now see.

2.2 The Economy of Nature

The phrase *economy of nature* today sounds odd. Yet, it was a central focus for natural history in the 1700s and early 1800s, and so too for Darwin.[22] As the phrase implies, writers like Carl Linnaeus and Charles Lyell saw nature as a divinely constructed economy in which the division of labour between the interdependent activities and stations of its various inhabitants were 'fitted to produce general ends, and reciprocal uses,' and thus 'ensure the health and well-being of the natural world.'[23]

According to this idea, nature regulated itself through relationships between different types of organism, both destructive and constructive. The balance might change, but balance would persist. In *The Natural History of Selborne*, Darwin's boyhood hero Gilbert White (1720–1793) took earthworms to illustrate the minute and complex interdependencies between the different species making up the economy of nature. The loss of so small and despicable a creature from the 'chain of Nature,' White said, would do more than compromise 'half the birds, and some quadrupeds which are almost entirely supported by them.' For worms were also 'great promoters of vegetation, which would proceed but lamely without them.'[24]

Given this understanding, naturalists necessarily observed and recorded animals *in vivo* and *in situ*, because every organism's character was constituted by its diverse relations with the other organisms and the physical conditions which made up its particular habitat. The dwelling-place of a plant or an animal was always relationally performed—beings did not pre-exist their relatings.[25] Thus White defined natural history as inquiry 'into the life and conversation of animals'—where *conversation* had the now-archaic sense of 'manner of conducting oneself in a place,' and there, 'having dealings with others.' Hence the need to unravel the heterogeneous amalgam through which an organism constituted—and was constituted

[21] Thomas Huxley, 'On Natural History, as Knowledge, Discipline, and Power,' *Proceedings of the Royal Institution* 6, (1856), http://aleph0.clarku.edu/huxley/SM1/NatHis.html, p.306; After his death, Darwin's son Francis observed that his father had been: 'a Naturalist in the old sense of the word, that is, a man who works at many branches of science, not merely a specialist in one': Francis Darwin, 'Reminiscences of My Father's Everyday Life,' in *The Life and Letters of Charles Darwin*, ed. Francis Darwin (London: John Murray, 1887), pp.155–156.
[22] Trevor Pearce, ' "A Great Complication of Circumstances"–Darwin and the Economy of Nature,' *Journal of the History of Biology* 43, (2010), pp.493–528.
[23] Pearce, 'Complication,' p.498.
[24] White, *Natural History of Selborne*, pp.172–173. Another of Darwin's influences who used this concept was Charles Lyell. See Pearce, 'Complication'; Denis Walsh, 'Two Neo-Darwinisms,' *History and Philosophy of the Life Sciences* 32, (2010), pp.317–339.
[25] Matthew Watson, 'Performing Place in Nature Reserves,' p.51, (2003), p.145; Donna Haraway, *The Companion Species Manifesto: Dogs, People, and Significant Otherness* (Chicago: Prickly Paradigm, 2003), p.6.

by—its conditions of life. A fox's territorial marker, the stone a thrush picks as anvil to smash a snail, a bee's discovery of nectar-rich flowers three fields away from its hive, gain meaning for the animal concerned solely through intersections between its own embodied habits, needs, and characters, and those of the other existents that inhabit the same material site. Unlike bench-classifiers, therefore, naturalists needed strong boots if they were to describe facts, 'which cannot be seen in specimens in spirits.'[26]

Interdependencies with the things and creatures in an organism's surrounds were mediated by habits (see 2.6.3 Habitat versus environment). The word *habit* did not figure in White's natural history, but became central to Darwin's. (White talked of *manner* and *instinct*.) Nineteenth-century science gave *habit* a wider reach than 'a repeated action' (today's main usage). Darwin quickly learnt that, to understand the taxonomy and anatomy of an organism, he needed to study how it *used* its 'conditions of country.'[27] His *Beagle* voyage proved that even the subtlest difference in habit could serve as his most reliable clue to identity.[28]

Take a hunch Darwin had, during his voyage down the east coast of South America, that two populations of similar-looking mocking-birds, living adjacently in Argentina, comprised different species. The northerners acted tame and bold, stealing meat from country houses. Their southern neighbours proved shyer, had a slightly different tone of voice, and frequented plains and valleys 'thinly scattered with stunted and thorn-bearing trees. It does not appear to move its tail so much.' However, once he had shot members of both groups and inspected them close up, he concluded they were identical. Back in London, he passed them on to the ornithologist, John Gould, as being from the same species. Gould confirmed Darwin's first impression, however. Meticulous anatomical examination proved the two tribes comprised separate species: 'a conclusion in conformity with the trifling difference of habit'—of which, however, Gould was not aware.[29]

---

[26] Charles Darwin, 'Letter to Henslow, 24th July to 7th November,' (Cambridge: Darwin Correspondence Project, 1834), https://www.darwinproject.ac.uk/letter/?docId=letters/DCP-LETT-251.xml. Regarding historians' need for 'strong boots,' see Tom Griffiths, *The Art of Time Travel: Historians and Their Craft* (Carlton VIC: Black Inc., 2016), pp.55–56. Cf. Darwin's letter to his sister from Uruguay, requesting a further '4 pair of very strong walking shoes': Charles Darwin, 'Letter to Catherine Darwin, 22nd May to July 14th,' (Cambridge: Darwin Correspondence Project, 1833), https://www.darwinproject.ac.uk/letter/?docId=letters/DCP-LETT-206.xml.

[27] Darwin, 'Notebook E,' pp.51–52: 'Thinking of effects of my theory ... Ascertainment of closest species (& naming them) with relation to habits, ranges & external conditions of country, most important & will be done to all countries, — but naming mere single specimens in skins worse than useless. — I may say all this, having myself aided in such sins.'

[28] 'Darwin was one of the first to use observations of behaviour in species diagnosis.' See Nora Barlow, 'Introduction,' [pp.201–207] to Charles Darwin, 'Darwin's Ornithological Notes [1837]. Edited with an Introduction, Notes, and Appendix by Nora Barlow,' *Bulletin of the British Museum (Natural History) Historical Series* 2, (1963), p.206.

[29] Charles Darwin, *Journal of Researches into the Natural History and Geology of the Countries Visited During the Voyage of H.M.S. Beagle Round the World, under the Command of Capt. Fitz Roy R.N.* (London: John Murray, 1890), p.54–55; John Gould, *Birds Part 3 of the Zoology of the Voyage of H.M.S.*

HEAD OF SCISSOR-BEAK.

RHYNCHOPS NIGRA, OR SCISSOR-BEAK.

Figure 2.2 The Scissor-Beak (*Rhynchops nigra*).

Darwin, C. (1890). *Journal of Researches into the Natural History and Geology of the Countries Visited During the Voyage of H.M.S. Beagle Round the World, under the Command of Capt. Fitz Roy* R.N., p.145.

Habits were important to Darwin because they showed how a creature integrated into the economy of nature. To account for the anatomical peculiarities of a single specimen 'in skins' without reference to the live animal's relationships to the beings and things that made up its world would be to play blind man's buff. For example, on 15th October 1833 in Uruguay, Darwin shot a bird he later identified as a scissor-beak (*Rhynchops nigra*; see Figure 2.2). It had short legs, webbed feet, extremely long-pointed wings, and was about the size of a tern. The beak was flattened vertically, as if twisted 'at right angles to that of a spoonbill,' and was as elastic 'as an ivory paper-cutter.' The lower mandible, uniquely among birds, was an inch and a half longer than the upper.[30]

Beagle (London: Smith Elder and Co., 2006), http://darwin-online.org.uk/content/frameset?pageseq=1&itemID=F9.3&viewtype=text. pp.60–61.

[30] Darwin, *Journal of Researches* p.138.

How to explain this extraordinary anatomy? Detective Darwin donned his boots to go look. As dusk fell, he saw small groups of the birds come out and skim rapidly to and fro across a still lake. Bills wide, their long lower mandibles ploughed the water, leaving narrow wakes on the mirror-like surface. The scissor-beaks were hunting tiny fish. In flight, they would twist and turn with extreme speed, dexterously managing to secure prey, once scooped up, by snapping down their upper mandible onto its longer twin. Such a habit made sense of the asymmetrical bills, *and* the value of the long primary feathers of their wings in keeping the birds' bodies dry, *and* their forked tails which were 'much used in steering their irregular course.' Darwin hypothesized that scissor-beaks generally fished by night, because that was the time at which their aquatic quarry rose most abundantly to the surface. The whole structure of the bird, concluded Darwin, was 'adapted for such habits.'[31]

2.3 Instrumental Aids to Discovery

Biology is often said to have undergone a transformation as the twentieth century dawned, that led it to abandon natural history.[32] The new biology, we hear, would no more inspire budding botanists and zoologist to embark on heroic voyages, thus to uncover large-scale patterns in the economy of living nature. Twentieth-century biologists would train to work in laboratories, and aim to penetrate the internal workings of prostrate or deceased organisms, the better to discover 'fundamental causes.'[33] No longer producible by amateurs or comprehensible to general readers, typical research findings would henceforth rely on expensive, highly technical, machines of analysis used in rigorously choreographed experiments, to be conducted by a gated guild of properly-qualified professionals.

Modern psychology came to life hot on the heels—and in the image—of this new indoor biology. As first-year undergraduate textbooks still tell, psychology was born in 1879 when the 'father' of psychology, Wilhelm Wundt, set up the world's first laboratory to study mental life.[34] Wundt acted as midwife, students hear, because he made 'psychology an independent discipline, separate from

---

[31] Darwin, *Journal of Researches*, pp.136–139. See also Charles Darwin, 'Zoology Notes,' in *Charles Darwin's Zoology Notes & Specimen Lists from H.M.S. Beagle*, ed. Richard Keynes (Cambridge: Cambridge University Press, 1832–1836), p.233; Darwin, 'Darwin's Ornithological Notes [1837]. Edited with an Introduction, Notes, and Appendix by Nora Barlow,' p.233.

[32] But see Lynn Nyhart, 'Natural History and the 'New' Biology,' in *Cultures of Natural History*, ed. Nick Jardine, James Secord, and Emma Spary (Cambridge: Cambridge University Press, 1996), p.426.

[33] Nyhart, ''New' Biology,' p.426.

[34] William James had earlier (1875) set up a less-prepossessing laboratory in a basement at Harvard University. Alternatively, Stanley Hall is held to have set up the first (because more prepossessing) psychological laboratory in the USA, at Johns Hopkins University (in 1882–1883). In short, *the first psychological laboratory* has become a founding trope in the mythology of modernist psychology.

philosophy ... by insisting on systematic observation and measurement.'[35] By 1900, American psychologists had brought Wundt's lesson home, founding a score of laboratories where they worked amid prominent arrays of apparatus, glinting with the fittings which invited the name *brass-instrument* psychology.[36]

All of which might imply that nineteenth-century naturalists went bare-handed to seek their discoveries. Not so. By the time Darwin joined the *Beagle* at the age of twenty-two, the needs of naturalists for brass and other instruments had become extensive. In the two months before sailing, Darwin assembled such volumes of equipment for his work as ship's naturalist that he felt he should promise his captain, Robert Fitzroy, 'two big cases' could remain behind 'without very material injury' if required.[37] He had consulted numerous experts about what he should take. Geological tools topped the list: 'a heavy hammer, with its two ends wedge-formed and truncated'; 'a light hammer for trimming specimens'; plus 'some chisels and a pickaxe for fossils.'[38] He purchased a set of aneroid or 'mountain' barometers, along with tables for their use, to measure the altitudes at which he collected specimens.[39] Darwin always carried a pocket lens when in the field,[40] and a compass, which, with his clinometer, could be used to measure the gradient and bearing of the slopes from which he collected his samples (see Figure 2.3).

Once Darwin had carried his specimens back to his cabin, he submitted them to further ordeals, all scrupulously recorded in notebooks. He had a blowpipe, along with borax and charcoal, to test rocks' responses to heat. He had supplies of three acids to help analyse the chemical make-up of his samples. He had a reflective goniometer to measure the angle of crystal faces in rock, which, by using reference books, helped him identify his finds (Figure 2.3). For animals, Darwin had various tools for dissection. And, when recording or roughly diagramming descriptions of his samples in his specimen notebooks—once tagged, measured, and analysed—he could draw on Patrick Syme's book of plates of different hues, to

---

[35] Dennis Coon and John Mitterer, *Introduction to Psychology: Gateways to Mind and Behaviour*, 2nd ed. (Belmont CA: Wadsworth, 2010), p.22; Wayne Weiten, *Psychology: Themes and Variations*, 9th ed. (Belmont CA: Wadsworth, 2013), p.3. Systematic observation, measurement—and indeed experiments—that we would now retrospectively call psychological had been done for decades prior to 1879. But Helmholtz, Fechner, and their colleagues called their work *physiology* or *psychophysics*, not 'psychology.' Wundt overtly championed *psychology*.

[36] Robert Davis, 'The Brass Age of Psychology,' *Technology and Culture* 11, (1970), pp.604–612.

[37] Charles Darwin, 'Letter to Fitzroy, 4th or 11th October 1831,' (Cambridge: Darwin Correspondence Project, 1831), https://www.darwinproject.ac.uk/letter/?docId=letters/DCP-LETT-139.xml.

[38] Charles Darwin's 'Advice to the Admiralty for Travelling Geologists,' quoted in Sandra Herbert, *Charles Darwin, Geologist* (Ithaca, NY: Cornell University Press, 2005), pp.103–105.

[39] Charles Darwin, 'Letter to Robert Fitzroy, 10th October,' (Cambridge: Darwin Correspondence Project, 1831), https://www.darwinproject.ac.uk/letter/?docId=letters/DCP-LETT-139.xml: 'Have you a good set of mountain barometers—Several great guns in the Scientific World have told me some points in geology to ascertain which entirely depend on their relative height.' The note to that letter tells us he also 'took with him a set of aneroid barometers. The tables for their use, with CD's [marginal] notes, are in [a copy of] *Jones's companion to the mountain barometer & tables.*'

[40] Darwin, 'Zoology Notes,' p.320.

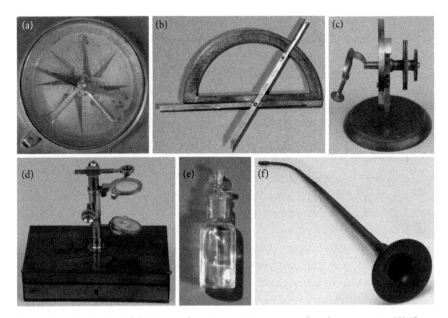

Figure 2.3 Examples of the types of equipment Darwin used on his voyage in *HMS Beagle*. (a) Compass clinometer, to measure the orientation and slope of rock features; (b) Contact goniometer, to identify mineral crystals; (c) Reflecting goniometer, for more accurate mineral identification; (d) Microscope; (e) Acid dropper; (f) Blowpipe.

All images come from the webpage for the Sedgwick Museum: http://www.sedgwickmuseum.org/index.php?page=charles-darwin.

give precision when describing the tints of whatever he had collected.[41] Then came the need for careful systems of preserving organic specimens prior to boxing them up for shipment back to England and expert examination by others. Darwin had undertaken lessons in taxidermy when a student, from the freed Guyanan slave John Edmonstone.[42] His choice preservative for more perishable organisms was gin.[43]

---

[41] Patrick Syme, *Werner's Nomenclature of Colours: With Additions, Arranged So as to Render It Highly Useful to the Arts and Sciences, Particularly Zoology, Botany, Chemistry, Mineralogy, and Morbid Anatomy. Annexed to Which Are Examples Selected from Well-Known Objects in the Animal, Vegetable, and Mineral Kingdoms*, 2nd ed. (Edinburgh: Blackwood, 1821);See Charles Darwin, 'Letter to Henslow, 9th September,' (Cambridge: Darwin Correspondence Project, 1831), https://www.darwinproject.ac.uk/letter/?docId=letters/DCP-LETT-123 Darwin's use of Syme's colour chart gives the lie to Foucault's odd claim—apparently based on his reading of a few sentences from Linnaeus—that the 'proper objects' of natural history were 'cleansed even of their colours': Foucault, *Order of Things*, p.146.

[42] Charles Darwin, *The Autobiography of Charles Darwin, 1809–1882: With Original Omissions Restored*, ed. Nora Barlow (London: Collins, 1958), p.51.

[43] Charles Darwin, 'Letter to Charles Whitley, September 9th,' *Correspondence of Charles Darwin* (Cambridge, 1831), https://www.darwinproject.ac.uk/letter/?docId=letters/DCP-LETT-121.xml.

Half way around the world, Darwin found his field notes becoming bulky: 'about 600 small quarto pages full; about half of this is Geology, the other imperfect descriptions of animals.'[44] Although Darwin had equipped himself with a telescope for observing animals (as well as rock exposures), guns proved his main implement for collecting terrestrial specimens. He had six, amongst which pride of place went to an expensive new rifle.[45] At sea, he invented a device to drag behind the ship for catching small marine animals.[46] And, as with rocks, he would sometimes test the reactions of the organisms he had caught to novel stimuli. 'Neither Spirits of Wine or fresh water had any perceptible effect,' he noted of a red coralline algae.[47]

Every one of Darwin's instruments required practice in a set of skills peculiar to itself. Already, while a schoolboy, he had gained plenty of hands-on experience in chemistry in the company of older brother Erasmus (1804–1881) and younger sister Catherine (1810–1866). Together they had colonized a room in their father's house, christened it 'The Laboratory,' and equipped it for the analysis of minerals and crystals. In 1825, aged sixteen, Darwin followed Erasmus to Edinburgh University to study medicine. Here he completed top-notch courses in geology and natural history, alongside those on more obviously medical topics like anatomy and surgery. But he preferred fieldwork to lectures—some fruits of which we will meet in Chapter 3. He undertook extensive collecting trips along the shores of the Firth of Forth, his inquiries in marine zoology abetted by long hours peering at his captures down a simple microscope.

Darwin continued to pursue natural history after, at age eighteen, he squeamishly switched from medicine at Edinburgh to read Arts at Cambridge—both by attending lectures on geology and botany, and in practice, learning to use geological instruments from his mentor, Professor John Henslow (1796–1861), and avidly collecting beetles in high-spirited competition with his friends.[48] Degree done, his credit was such that he was invited to accompany Cambridge's Professor of Geology, Adam Sedgwick (1785–1873), on a ten-day field-trip to north Wales in August

---

[44] Darwin, 'Letter to Henslow, 24th July to 7th November.' Here he neglected to mention that he had also collected and made notes on plants: Duncan Porter, 'Charles Darwin's Notes on Plants of the Beagle Voyage,' *Taxon* 31, (1982), pp.503–506.

[45] See Charles Darwin, 'Letter to Charles Whitney, 9th September,' (Cambridge: Cambridge Correspondence Project, 1831), http://www.darwinproject.ac.uk/entry-121 NB Although Darwin was an excellent shot, much of the shooting of animals while on the *Beagle* was done for him by servants and sailors, most notably Syms Covington (1816–1861): pp.100–109, 228–229 in Janet Browne, *Charles Darwin: Voyaging*, 2 vols., vol. 1 (London: Pimlico, 1996); Roger McDonald, *Mr Darwin's Shooter* (Sydney: Random House, 1988).

[46] Charles Darwin, *Diary of the Voyage of H.M.S. Beagle*, ed. Paul Barrett and Richard Freeman, vol. I, The Works of Charles Darwin (London: William Pickering, 1986), p.56 (Jan 10th & 11th, 1832): Not that this was the end of his hydrographic craft; see e.g. Alistair Sponsel, 'An Amphibious Being: How Maritime Surveying Reshaped Darwin's Approach to Natural History,' *Isis* 107, (2016), pp.254–281.

[47] Darwin, 'Zoology Notes,' p.13.

[48] Browne, *Darwin, Vol.1*, 1; pp.28–31, 69–72, 99–100, 136.

1831. This turned into an intensive practical training in fossil-identification, rock-recognition, dip-measuring, geological mapping, and stratigraphic surveying—an apprenticeship which set him up 'wonderfully' for his later inquiries.[49]

Over the years, Darwin became a celebrated exponent of microscopy. Secreted within the twenty-two-year-old recruit's 'frightfully bulky' luggage for the *Beagle*, had been two state-of-the-art microscopes, with a variety of lenses[50]—and an additional double convex lens on a manoeuvrable stand to focus illumination on opaque objects.[51] Eighteen months into the voyage, he wrote to a bug-hunting cousin back in Britain that the pleasure of working with his microscope ranked second only to geology.[52]

Darwin's expertise in microscopy continued to deepen into middle age, in tandem with progress in the design of microscopes, and others' practical discoveries.[53] But it was the demands of inquiry which mainly pushed him to improve his skills, for, in 1846, the thirty-seven year-old Darwin began an enormous eight-year-long project designed to unravel the taxonomy of barnacles, in which he aimed to identify different living forms by their miniscule *internal* organs. This was a pioneering move. In April 1847, Darwin was forced to buy a newfangled compound microscope, with achromatically corrected lenses, to aid him develop this angle of his barnacle project—the best microscope then available, worth about £3,400 (U\$4,300) in 2019 money.

In 1848, the top British instrument-maker began to market a microscope of Darwin's own design. In 1849, Darwin's growing renown as a microscopist meant he was asked to write on microscopes for John Herschel's *Manual of Scientific Enquiry; Prepared for the Use of Her Majesty's Navy: and Adapted for Travellers in General* (1849). Darwin's dense pages of instruction cover: the type of microscopes that work best; the types of lenses useful for different purposes; how best to condense light onto the focal object; the design of arm supports best to steady the hands; the tools needed for manipulating live or dead tissue under the microscope; and the tools required for the preparation of organisms and tissues for microscopic examination. Darwin then appended a list of one hundred and fifteen 'heads of

[49] Browne, *Darwin, Vol.1*, 1, pp.136–142. Earlier that year, Darwin had accompanied Sedgwick 'in a survey of the geology around the Cambridge area. In July, Darwin made his own geological survey of the region around [his home town of] Shrewsbury': p.25 in Phillip Sloan, 'The Making of a Philosophical Naturalist,' in *The Cambridge Companion to Darwin*, ed. Jonathan Hodge and Gregory Radick (Cambridge: Cambridge University Press, 2003), pp.17–38.
[50] Six different focal distances for lenses are mentioned in Darwin, 'Zoology Notes,': 1/3″, ¼″, 1/5″, 1/10″, 1/20″ & 1/30″. Regarding Darwin's two microscopes, see Phillip Sloan, 'Darwin's Invertebrate Program, 1826–1836: Preconditions for Transformism,' in *The Darwinian Heritage*, ed. David Kohn (Princeton: Princeton University Press, 1985), pp.92ff & 116.
[51] For a discussion of how he used the illuminating lens see Darwin, 'Letter to Catherine Darwin, 22nd May to July 14th.'
[52] Charles Darwin, 'Letter to Fox, May 23rd,' *Correspondence of Charles Darwin* (Cambridge: Darwin Correspondence Project, 1833), https://www.darwinproject.ac.uk/letter/?docId=letters/DCP-LETT-207.
[53] Boris Jardine, 'Between the Beagle and the Barnacle: Darwin's Microscopy, 1837–1854,' *Studies in History and Philosophy of Science* 40, (2009), pp.382–395.

observation [which] will guide the dissector to the facts that it is most desirable to determine and note down ... when time and opportunity concur for the anatomical examination of an animal.'[54] In 1853, Darwin won the Royal Society's 'Royal Medal,' partly in recognition of his microscopic analyses of barnacles.[55]

Besides their mutual reliance on instruments, both Victorian naturalists and laboratory scientists emphasized draughtsmanship. 'What you have not drawn, you have not seen,' went the maxim of pioneering experimental botanist Julius Sachs (1832–1897).[56] Likewise Darwin, who deplored his own incapacity to draw, complimented his friend Huxley's graphic skills by commenting: 'What an advantage to be able to sketch easily! No one has a right to attempt to be a naturalist who cannot.'[57]

Darwin's doubts about his abilities did not stop him drawing. As we will see, his teenage notes on the ova of *Flustra* (see Ch. 3) were accompanied by three sketches, the last of which displays what the eggs looked like ordinarily versus under magnification (Figure 3.1).

Decades later, Darwin was still sketching in order to see, whether while examining the ornamentation on the feather of an Argus pheasant; or vascular tissue in the leaf of a sundew plant (Figure 2.4).[58] His sketches show many corrections—underlining the dialectic between drawing, error, and better seeing, which lent precision to his observations.[59]

Darwin also drew to check his own hypotheses: 'The explanation given is wholly erroneous, as I have discovered by working out an illustration in figures,' he apologized, in a last-minute addition to *Descent*.[60] Or he might sketch conceptual diagrams to help him think through the intersections between evolution and taxonomy, for example (e.g. a branching, genealogical 'tree' of life: Figure 2.5; see too Figure 3.9).[61]

---

[54] Charles Darwin, 'On the Use of the Microscope on Board Ship,' in *A Manual of Scientific Enquiry; Prepared for the Use of Her Majesty's Navy: And Adapted for Travellers in General* ed. John Herschel (London: Murray, 1849). Dissection of the stomach was the best guide to what an animal ate, said Darwin.

[55] Anon, 'Royal Medal,' in *Wikipedia* (2018).

[56] Sydney Vines and Dukinfield Scott, 'Reminiscences of German Botanical Laboratories in the 'Seventies and 'Eighties of the Last Century,' *New Phytologist* 24, (1925), p.10: 'Was man nicht gezeichnet hat, hat man nicht gesehen.'

[57] Darwin, *Autobiography*, p.47; Charles Darwin, 'Letter to Huxley, 3rd September,' (1855), https://www.darwinproject.ac.uk/letter/?docId=letters/DCP-LETT-1759.xml.

[58] Phillip Prodger, *Darwin's Camera: Art and Photography in the Theory of Evolution* (Oxford: Oxford University Press, 2009), pp.45–46.

[59] Describing Darwin's drawings of the pheasant's feather which provided templates for Fig.2.4 right, Julia Voss, *Darwin's Pictures: Views of Evolutionary Theory, 1837–1874* (New Haven CT: Yale University Press, 2010), p.131, notes that 'All four sheets contain areas that have been pasted over with slips of paper and redrawn—apparently to fix mistakes.'

[60] Charles Darwin, *The Descent of Man and Selection in Relation to Sex Vol.1* (London: Murray, 1871), p.ix.

[61] This was a common practice among predecessor naturalists: see Pascal Tassy, 'Trees before and after Darwin,' *Journal of Zoological Systematics and Evolutionary Research* 49, (2011), pp.89–101; Nathalie Gontier, 'Depicting the Tree of Life: The Philosophical and Historical Roots of Evolutionary Tree Diagrams,' *Evolution* 5, (2011).

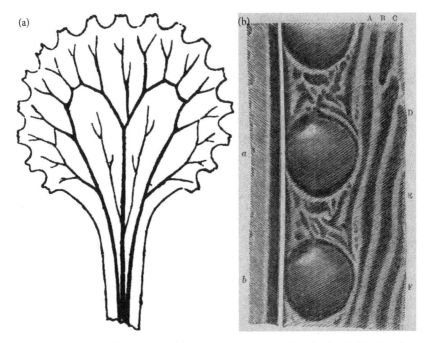

Figure 2.4 (a) Darwin's drawing of the vascular tissue in a Sundew leaf, tidied up for publication by his son George; (b) an engraver's woodcut of the eye-like 'ocelli' in the wing-feather from an Argus pheasant from *Descent*.

Source: (a) Darwin, C. (1875). *Insectivorous Plants*, (London: John Murray, Albemarle Street).
(b) Darwin, C. (1871). *The Descent of Man and Selection in Relation to Sex*. (London: John Murray, Albemarle Street).

Charles Darwin, *Insectivorous Plants*, 2nd ed. (London: John Murray, 1888), p.201; Charles Darwin, *The Descent of Man, and Selection in Relation to Sex* (London: Murray, 1874), p.435—henceforth abbreviated as *Descent*—where the legend reads: 'Part of secondary wing-feather of Argus pheasant, shewing two perfect ocelli, a and b. A, B, C, D, &c., are dark stripes running obliquely down, each to an ocellus.' The woodcut may have been by T.W. Wood.

Beyond this, illustration played a crucial role in the *publications* of naturalists. Trail-blazed by Hooke's *Micrographia*, the success of the burgeoning Victorian industry in publishing books on natural history largely depended upon pictorial beauty—especially in books displaying dramatic images of exotic birds, by the likes of Audubon and Gould.[62] Copious drawings and etchings by engravers and other craftsmen illustrated Darwin's books. Darwin never himself drew these public pictures. But he exerted close control over their accuracy, arrangement, and content.[63]

---

[62] Voss, *Darwin's Pictures*, pp.22–45.
[63] Prodger, *Darwin's Camera*, pp.26–32.

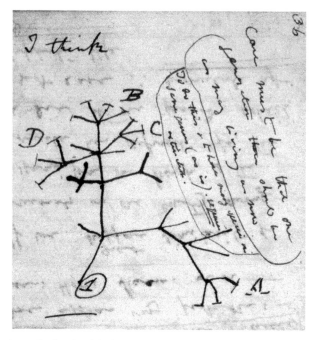

Figure 2.5 An early sketch of the 'tree' or 'coral' of life in Darwin's Notebook B (1837).

Source: Detail from Darwin, C. *Notebook B*, c.1837–38. Reproduced by kind permission of the Syndics of Cambridge University Library, Classmark: DAR 121, p.36.

Charles Darwin, 'Notebook B,' (1837–1838), pp.25, 36.

Of particular importance for his psychological work was Darwin's imaginative, pioneering, scientific usage of the camera (see Ch. 4). Perhaps most importantly, and just like drawing, photography aided accurate observation: 'It is easy to observe infants whilst screaming,' Darwin wrote, 'but I have found photographs made by the instantaneous process the best means for observation, as allowing more deliberation.'[64] Without the distancing effect made possible by photography, Darwin's son Francis later added, 'his sympathy with the grief spoiled his observation.'[65]

Darwin also used photographs experimentally, to test the meanings people read into different facial expressions—making the first of what psychologists now call 'judgement tests' (see Ch. 4). Beyond this, Darwin innovatively constructed displays of photographs to *illustrate* his scientific books—most notably, in *The Expression of the Emotions in Man and Animals* (e.g. Figure 2.6; see Chs 4 and 5).[66]

[64] Charles Darwin, *The Expression of the Emotions in Man and Animals*, 2nd ed. (London: Murray, 1890), p.155—henceforth abbreviated as *Expression*.

[65] Darwin, 'Reminiscences,' p.132.

[66] Prodger, *Darwin's Camera*; Paul White, 'Darwin Wept: Science and the Sentimental Subject,' *Journal of Victorian Culture* 16, (2011), pp.195–213.

Figure 2.6 'Infants whilst screaming'—one of the six images published under this heading in *Expression*.

Source: Darwin, C. (1872). *The Expression of the Emotions in Man and Animals*, Plate I, (London: John Murray, Albemarle Street).

Darwin, *Expression*, p.154.

2.4 Empirical Orientation

By the time Darwin had mastered it, the practice of natural history was a sophisticated and technique-dependent craft—much at odds with its modern image as the hobby of dilettantes. Furthermore, it employed criteria which focused directly on the issue central to this chapter: how can science best render living things in language? As we will shortly see, these criteria strongly shaped the distinction Darwin drew between his own approach to the study of agency and that of Victorians more renowned as psychologists—both now and in their own day: two of the most prominent being Herbert Spencer (1820–1903) and Alexander Bain (1818–1903).[67]

---

[67] See Robert Young, *Mind, Brain, and Adaptation in the Nineteenth Century: Cerebral Localization and Its Biological Context from Gall to Ferrier*, History of Neuroscience (New York: Oxford University Press, 1990); Rick Rylance, *Victorian Psychology and British Culture, 1850–1880* (Oxford: Oxford University Press, 2000); Roger Smith, *Free Will and the Human Sciences in Britain 1870–1910* (Pittsburgh PA: University of Pittsburgh Press, 2016), pp.35–56, 88–95.

For centuries, philosophers and taxonomists had debated whether their classifi-
cations of minerals, fauna, and flora reflected the true and natural arrangement of
different types and species. This hung on answering two questions. First, whether
a *natural system* or non-haphazard organization of living things really existed.[68]
And, secondly, if it did, how best to gain knowledge of that natural system so as to
arrange the labels and levels of scientific taxonomy (or *systematics*) to match the organization intrinsic to the natural world. Over the time Darwin jousted with these
issues, he moved towards two seemingly contradictory positions. On one hand,
he came to the view that the term *species* was one 'arbitrarily given for the sake of
convenience' to a set of individuals closely resembling each other.[69] On the other,
like many fellow naturalists, he foresaw that evidence would emerge proving the
existence of a natural system. Indeed, this was the Holy Grail which the *Origin* later
argued his evolutionary theory to have delivered: the 'hidden bond of connection'
between different kinds of living creature proved *genealogical*, equating to their degree of common descent.[70]

The taxonomic argument about how best to access the natural system traditionally revolved around, at one extreme, whether the anatomy of a single feature or
system (e.g. the reproductive system) should provide the source of decisive clas-
sificatory characters, or, at the other, whether to rely for classification on resemblance in *all* characters between two groups of organisms, without giving greater
weight to any particular resemblance. Darwin's thinking about this issue was complex. However, one proposition from which he never resiled was that discrimi-
nating between species is, above all, an *empirical* task.[71] If two groups of animals
were hypothesized to belong to different species, they should be called A and B
until data had been collected: no taxonomic decisions could be made until both
groups' ranges and habits had been ascertained and compared. Comparing range
and habits would point to what further should be observed, for example, what
physiological or anatomical features might prove taxonomically most relevant.[72]
Even then, the distinction between calling two closely related but distinguishable
groups of organisms *species* rather than *sub-species* was somewhat arbitrary—and
therefore, as he put it, a matter of convenience (see previous paragraph).[73]

While Newtonian science drew on prior observations—some of them commonplace (apples falling), some less so (a prism splits white light into a spectrum of

---

[68] Phillip Sloan, 'John Locke, John Ray, and the Problem of the Natural System', *Journal of the History of Biology* 5, (1972), pp.1–53; Sloan, 'Natural History.'

[69] Charles Darwin, *The Origin of Species by Means of Natural Selection, or the Preservation of Favoured Races in the Struggle for Life*, 6th ed. (London: Murray, 1876), p.42.

[70] Darwin, *Origin 1876*, p.369.

[71] Charles Darwin, 'Notebook C,' (1838), p.70.

[72] Darwin, 'Notebook C,' pp.122ff.

[73] Kevin De Queiroz, 'Branches in the Lines of Descent: Charles Darwin and the Evolution of the Species Concept,' *Biological Journal of the Linnean Society* 103, (2011), p.23.

colours; planets have elliptical orbits)—Phillip Sloan has argued that the growth of Linnaeus' taxonomic programme constituted natural history as an alternative to Newton's approach.[74] Newton had sought to deduce the underlying *mathematical* structure of reality, favouring mechanistic models and reductive explanations. By contrast, naturalists like the influential French Count Buffon (1707–1788) disparaged deductive, mathematical, knowledge, because it yielded only truths 'of definition,' grounded in the relations between ideas, not on observable things.

Physical truth, in contrast to Newton's mathematical truth, was less arbitrary because it did not depend primarily on human deduction. Instead of drawing from human suppositions, physical truths were grounded upon *facts*: 'a frequent repetition and uninterrupted succession of the same events constitutes the essence of physical truth ... One goes from definition to definition in the abstract sciences, but one proceeds from observation to observation in the sciences of the real,' wrote Buffon.[75] Knowledge gained through a careful observation of nature was firmer than the abstractions of mathematical physics, and other definition-based sciences.[76]

What naturalists called observation included what we would call experiments—and often, *repeated* experiments.[77] In natural history, experiments would test the fidelity of initial observation: 'the nature of things betrays itself more readily under the vexations of art [i.e. experiments] than in its natural freedom,' wrote Francis Bacon in 1620.[78] Not that there was a hard line between the two. Observations were often contrived, and experiments might be natural (e.g. effects of lightning strikes; cf. the test of Einstein's theory by observations during an eclipse).[79] So when a naturalist wrote up a history, they not only reported what they saw, but, like modern psychologists, regaled their readers with tales of 'the procedures, tricks, and techniques' which had made their observations possible.[80] Equally, in Darwin's

---

[74] Sloan, 'Natural History,' pp.909ff, where he refers to James Larson, 'An Alternative Science: Linnaean Natural History in Germany, 1770–1790,' *Janus* 66, (1979), pp.267–283.

[75] Buffon quoted by Sloan, 'Natural History,' pp.918–919.

[76] Réaumur (1683–1757) expressed a similar sentiment: 'The spirit of observation, the kind of spirit essential to naturalists, and commonly assigned to them, is equally necessary to progress in every other science. It is the spirit of observation that causes us to perceive what has escaped others, that allows us to grasp the relations among things that appear different, or that causes us to find the differences among those that seem similar': Mary Terrall, *Catching Nature in the Act: Réaumur and the Practice of Natural History in the Eighteenth Century* (Chicago: Chicago University Press, 2014), p.6.

[77] Darwin's botanist-son Francis noted his father, 'would patiently go on repeating experiments where there was any good to be gained.': Darwin, 'Reminiscences,' p.145; Cf. Daston, 'Empire of Observation,' pp.93–95.

[78] Bacon quoted in Alpers, *Art of Describing*, p.103.

[79] Frank Dyson, Arthur Eddington, and Charles Davidson, 'A Determination of the Deflection of Light by the Sun's Gravitational Field, from Observations Made at the Total Eclipse of May 29, 1919,' *Philosophical Transactions of the Royal Society of London* 579, (1920), pp.291–333.

[80] Terrall, *Catching Nature*, p.7; cf. Anne Larsen, 'Equipment for the Field,' in *Cultures of Natural History*, ed. Nick Jardine, James Secord, and Emma Spary (Cambridge: Cambridge University Press, 1996), p.358: '... natural history in this period [1700–1900] was a science based on *specimens* ... Specimens were not, however, natural objects: they were artificial things, designed and constructed by naturalists to answer various scientific needs.'

geological fieldwork, finding and collecting specimens made only a first step. As we have seen, long hours on the *Beagle* followed, vexing his samples with acids—Darwin used nitric, sulphuric, and 'muriatic' (hydrochloric)—and with flame (see Figure 2.3).[81]

In his later years, Darwin conducted thousands of experiments, particularly botanical experiments.[82] Some of his psychological experiments are now being recognized as trail-blazing contributions to scientific knowledge (see Ch. 4).[83] Nor was Darwin a naïve experimentalist. Latterly, his experiments frequently involved comparisons between different conditions, under careful control. His most sophisticated studies systematically compared the results, across many generations, of self- and cross-fertilization in the hedgerow climber Morning Glory (*Ipomoea*). Seeds generated by self- versus cross-fertilization were separated by glass but planted in the same pot filled with the same sandy soil, exposed to the same heat and equal light, germinated at the same speed, equally watered at the same time, and all protected from insects and extraneous matter by a finely meshed, white, cotton, net. As Darwin wrote: 'I do not believe it possible that two sets of plants could have been subjected to more closely similar conditions, than were my crossed and self-fertilised seedlings.'[84] The results of these experiments were then mathematically analysed by the founder of today's statistical methods in psychology, Francis Galton, showing, with a 'very good' degree of mathematical confidence that cross-fertilization was of significant benefit to the vigour of the experimental plants, when compared to self-fertilization.[85]

Natural history saw observations and experiments cycling in tandem: 'Observation, by the curiosity it inspires and the gaps that it leaves, leads to experiment; experiment returns to observation by the same curiosity that seeks to fill and close the gaps still more; thus one can regard experiment and observation as in some fashion the consequence and complement of one another.'[86] So it was with Darwin. For example, in old age, he commented that he had spent eleven years making the 'numerous experiments' recorded in his book on floral crosses, prompted 'by a mere accidental observation ... [of] the remarkable fact

[81] Charles Darwin's 'Advice to the Admiralty for Travelling Geologists,' quoted in Herbert, *Charles Darwin, Geologist*, pp.103–105.
[82] Mea Allan, *Darwin and His Flowers* (New York: Tablinger, 1977), p.180.
[83] E.g. Peter Snyder, Rebecca Kaufman, John Harrison, and Paul Maruff, 'Charles Darwin's Emotional Expression 'Experiment' and His Contribution to Modern Neuropharmacology,' *Journal of the History of the Neurosciences* 15, (2010), pp.158–170.
[84] Allan, *Darwin's Flowers*, pp.249ff; Charles Darwin, *The Effects of Cross and Self Fertilisation in the Vegetable Kingdom*, 2nd ed. (London: John Murray, 1878), p.13.
[85] Darwin, *Cross Fertilisation*, pp.18, 146, 235.
[86] Jean Le Rond d'Alembert, 'Expérimental,' in Diderot and d'Alembert, eds., *Encyclopédie* (1751–1780), quoted by Daston, 'Empire of Observation,' p.86.

that seedlings of self-fertilised parentage are inferior, even in the first generation, in height and vigour to seedlings of cross-fertilised parentage.'[87]

2.5 Darwin's Approach to the Study of Humans

Darwin's round-the-world trip on *HMS Beagle* ended on Sunday 2nd October 1836. In July 1837, he began to fill a sequence of six fascinating notebooks in which he developed his thinking—begun while sailing back to England—about what he called the *transmutation* of species, that is, evolution. They are now called his 'Transmutation Notebooks.' The first two of these he labelled 'B' and 'C' (a Notebook A was begun at the same time as 'B,' devoted to 'Geology'). On finishing Notebook C in July 1838, Darwin began *two* new volumes simultaneously, labelled 'D' and 'M.' Notebook D continued the trains of thought that had filled Notebook C, excepting the implications of his theory for human beings, which he reserved for Notebook M. Both 'D' and 'M' had successor volumes: labelled 'E' and 'N' respectively. 'E' and 'N' were both completed during the summer of 1839. Darwin continued to fill one further collection of notes into 1840. These also concerned humans, later being thrust into a box labelled, 'Old and useless Notes about the moral sense & some metaphysical points.'

The two transmutation notebooks dealing with humans—'M' and 'N'—reveal the beginnings of a refinement of approach to the study of psychological problems which culminated in Darwin's books on *The Descent of Man* (1871) and *The Expression of the Emotions* (1872). These two late publications capitalize on a careful consideration—and then rejection—of what the twenty-nine-year-old Darwin had called 'metaphysics.' *Metaphysics* was among the words that, in 1856, aged forty-seven, Darwin had retrospectively scrawled as summaries inside the covers of both 'M' and 'N' (see Ch. 4). Today, some scholars suppose that Darwin and his peers used the word 'metaphysics' to refer to what we would call psychology. Hence, on the supposition that the transmutation notebooks represent the unsurpassed peak of Darwin's brainwork, 'M' and 'N' are said to cover 'almost all the main theories' about psychological matters which later filled *Descent* and *Expression*.[88] I demur. Because, when we step into the library of the 1830s, the meanings of the word *metaphysics* prove cloudy and contentious.

The entry on 'Metaphysics' in Brewster's *Edinburgh Encyclopaedia* of 1830 comprises an essay of 30,000 words which nowhere discusses psychology. Brewster redirected readers to an entry on 'Pneumatology' (which subsumed psychology),

---

[87] Darwin, *Autobiography*, p.133. Cf. Darwin, *Cross Fertilisation*.

[88] Michael Ghiselin, 'Darwin and Evolutionary Psychology,' *Science* 179, (1973), p.964; Jonathon Hodge, 'The Notebook Programmes and Projects of Darwin's London Years,' in *The Cambridge Companion to Darwin*, ed. Jonathon Hodge and Gregory Radick (Cambridge: Cambridge University Press, 2003), pp.40–41; Tim Lewens, *Darwin* (London: Routledge, 2007), p.2.

with the enticing if confusing comment that Pneumatology was 'the most interesting branch of metaphysics.' The entry on Metaphysics in the then-current edition of the *Encyclopaedia Britannica* did not mention psychology either, though, amongst a swathe of ontological matters (time, space, motion, God), it did debate some proto-psychological topics (e.g. consciousness).[89]

Darwin's earliest philosophical mentor, family friend Sir James Mackintosh (1765–1832), disparaged metaphysics as 'a specimen of all the faults which the name of a science can combine.' Even the make-up of the word itself, *meta*-physics, encouraged 'the pernicious error of believing that it seeks something more than the interpretation of nature.' Worse, it varied in meaning from century to century and tongue to tongue. It was pretentious too. No wonder then that it had 'become a name of reproach and derision among those who altogether decry it.'[90]

If we trace Darwin's use of the term metaphysics over time, we find that he initially used it descriptively, to point to an area of enquiry which he would in future need to explore. Its first mention was in a note from early 1838, where metaphysics figured in a list of topics to which 'my theory would give zest.'[91] A few months later, we find Darwin quoting a posthumous biography of Mackintosh which underlined a great difficulty in setting out to reason about something so complex as human nature: 'there is really no natural starting place, because there is nothing more elementary than that complex nature itself with which our speculations must end as well as begin.' To which Darwin added: 'The centre is everywhere & the circumference nowhere as long as this is so —!! Metaphysics!!!'[92] Perhaps his thought was that, with no obvious place from which to start inquiries into so complex a topic as human nature, he would need to study metaphysics to work out a defensible approach. Another possibility is that he saw this quote as an outright refutation of metaphysics: if 'complex nature' is where reasoning must both start and finish, then, as per Buffon, don't bother about metaphysics! Holidaying in August, we find him noting that he had read 'a good deal of various amusing books, & paid some attention to Metaphysical subjects.'[93]

---

[89] Anon, 'Metaphysics,' in *The Edinburgh Encyclopaedia*, ed. David Brewster (Edinburgh: Blackwood, 1830).; Anon, 'Metaphysics,' in *Encyclopaedia Britannica or, a Dictionary of Arts, Sciences and Miscellaneous Literature*, ed. David Millar (Edinburgh: Archibald Constable & co., 1810).

[90] James Mackintosh, *Dissertation on the Progress of Ethical Philosophy, Chiefly During the Seventeenth and Eighteenth Centuries* (Edinburgh: Adam and Charles Black, 1862): p.4. Mackintosh here used the word 'science' to include what we would now exclude from science as philosophy. Darwin was closely reading Mackintosh's *Dissertation* in August 1838, and again in April–May 1839: Charles Darwin, 'Notebook M,' (Cambridge: Darwin Online, 1838), pp.75, 132e; Charles Darwin, 'Notebook N,' ed. Paul Barrett (1838–1839), pp.89, 92; Charles Darwin, 'Old & Useless Notes,' (Cambridge: Darwin Online, 1838–1840), p.42.

[91] Darwin, 'Notebook B,' p.228.

[92] Darwin, 'Notebook C,' p.218.

[93] Charles Darwin, 'Darwin's Journal,' *Bulletin of the British Museum (Natural History). Historical Series 2*, (1959), p.27.

From this point on, Darwin's attitude to metaphysics grew more informed, and began to alter. His reading swiftly embraced Hume's *Enquiry Concerning Human Understanding*, the Scottish common-sense philosopher Dugald Stewart (1753–1828), Brewster's 1838 review of Comte, Abercrombie's *Inquiries Concerning the Intellectual Powers and the Investigation of Truth*, and an essay on 'the philosophy of consciousness,' newly published by James Ferrier (see Ch. 4).[94] All of these authors belittled or derided metaphysics. For instance, Hume's *Enquiry* famously ended with the words:

> If we take in our hand any volume; of divinity or school metaphysics, for instance; let us ask, *Does it contain any abstract reasoning concerning quantity or number?* No. *Does it contain any experimental reasoning concerning matter of fact and existence?* No. Commit it then to the flames: for it can contain nothing but sophistry and illusion.[95]

Unsurprisingly therefore, we find Darwin's remarks on metaphysics growing more critical as his reading progresses. He became fond of quoting Auguste Comte's (1798–1857) three-step schema for decoding the history of knowledge—which cast metaphysics as pre-scientific, because non-observational. First and lowest in Comte's hierarchy came theological-cum-fictitious explanations for phenomena such as thunder or dreams, as in 'savage' societies. Next came the metaphysical-cum-abstract state of knowledge, typified by Western philosophy. And finally, humans would attain to truly scientific or *positive* knowledge, systematically built up from observation. Take for example Darwin's exuberant note: 'Origin of man now proved.—Metaphysic must flourish.—He who understands baboon would do more towards metaphysics than Locke.' This appeared relatively early in his reading on the topic.[96] Yet, even at this point, Darwin was just as critical of Locke's metaphysics as Hume's *Enquiry* had been, rejecting it in favour of a scientific (observational) understanding of primates. By 3rd October, his view had firmed:

---

[94] Anon. [David Brewster], 'Review of Cours De Philosophie Positive, by Auguste Comte,' *Edinburgh Review* 67, (1838), pp.271–308; John Abercrombie, *Inquiries Concerning the Intellectual Powers and the Investigation of Truth*, 8th ed. (London: John Murray, 1838); James Ferrier, 'An Introduction to the Philosophy of Consciousness, Part I,' *Blackwood's Edinburgh Magazine* 43, (1838), pp.186–201.

[95] David Hume, *An Enquiry Concerning Human Understanding* (Chicago: Open Court, 1921), p.276, italics his—a book in which Hume (p.22) targeted Locke's philosophy as one 'betrayed ... by the schoolmen' into an 'ambiguity and circumlocution' which seemed to 'run through that philosopher's reasonings ... on most subjects.' What Hume called the 'enmity' between empirical science and metaphysics had been deepening for a long time—at least from the era of Robert Boyle (1627–1691). Cf. Ferrier, 'Consciousness'; [Brewster], 'Review of Comte'; Abercrombie, *Inquiries Concerning the Intellectual Powers and the Investigation of Truth*; Steven Shapin and Simon Schaffer, *Leviathan and the Air-Pump: Hobbes, Boyle, and the Experimental Life* (Princeton NJ: Princeton University Press, 1985): pp.80ff & passim.

[96] August 16th, 1838.

To study Metaphysics, as they have always been studied appears to me to be like puzzling at astronomy without mechanics.—Experience shows the problem of the mind cannot be solved by attacking the citadel itself.—the mind is function of body.—we must bring some *stable* foundation to argue from.[97]

In later years the word *metaphysical* served in Darwin's correspondence as an insult. A book or article that was 'metaphysical,' was 'mere verbiage,' 'barely intelligible,' dealt in 'far-fetched analogies,' was 'rubbish,' being the work of a 'wind-bag,' with 'muddled . . . brains,' and 'an entire want of common sense.'[98] The twenty-nine-year-old Darwin's remark that 'we must bring some *stable* foundation to argue from' shows an inkling of the distinctive approach to human and animal agency he elaborated three decades later in *Expression* and *Descent* (see Chs 4–9). Unlike other writers on the topics he discussed, he would not treat them as problems in metaphysics or philosophy. He would approach them exclusively 'from the side of natural history.'[99]

Given all this, we can understand why Darwin criticized as *unscientific* any non-empirical or deductive approach to the observable world. This was most obviously true when he judged his fellow-psychologists. For example, here is what Darwin

[97] Darwin, 'Notebook M,' pp.84e, 136; Darwin, 'Notebook N,' p.5, Darwin's emphasis.

[98] Darwin, *Descent*: p.78; Charles Darwin, 'Letter to Lyell, 8th October,' (Cambridge 1845), https://www.darwinproject.ac.uk/letter/?docId=letters/DCP-LETT-919.xml; Charles Darwin, 'Letter to Hooker, 25th December,' (Cambridge: Darwin Correspondence Project, 1857), https://www.darwinproject.ac.uk/letter/?docId=letters/DCP-LETT-2194.xml; Charles Darwin, 'Letter to Gray, 26th November,' (Cambridge: Darwin Correspondence Project, 1860), https://www.darwinproject.ac.uk/letter/?docId=letters/DCP-LETT-2998.xml; Charles Darwin, 'Letter to Gray, 11th April,' (Cambridge: Cambridge Correspondence Project, 1861), https://www.darwinproject.ac.uk/letter/?docId=letters/DCP-LETT-3115.xml; Charles Darwin, 'Letter to Gray, 13th September,' (Cambridge: Cambridge Correspondence Project, 1864), https://www.darwinproject.ac.uk/letter/?docId=letters/DCP-LETT-4611.xml; Charles Darwin, 'Letter to Murray, 13th April,' (Cambridge: Darwin Correspondence Project, 1871), https://www.darwinproject.ac.uk/letter/?docId=letters/DCP-LETT-7680.xml. Note too Darwin's retrospective labelling of his 'old metaphysical' notes from the 1830s as 'Useless.' It is also worth noting that, according to Lord Alfred Tennyson, the Metaphysical Society (1869–1880)—which included 'sixty-two of the most eminent thinkers, churchmen, editors, men of affairs, writers, and scientists of the period'—broke up 'because, after years of strenuous effort, no one had succeeded even in defining the term "metaphysics"'; see Robert Young, *Darwin's Metaphor: Nature's Place in Victorian Culture* (Cambridge: Cambridge University Press, 1985): pp.151–152.

[99] Darwin, *Descent*, p.97; George Romanes, 'Charles Darwin, [Part] V. [Psychology],' *Nature* 26, (1882), p.169. Darwin's refusal to have any truck with philosophy was a defining trait (see Ch. 10), because the lure of philosophy has been constant in psychology, from the 1870s up to the present day—as can be seen by comparing the works of Bain and Spencer—and their historians—with today's vanguard, the Gergens, Shotter, and Burman: see e.g. Smith, *Free Will and the Human Sciences*; Thomas Dixon, *From Passions to Emotions: The Creation of a Secular Psychological Category* (Cambridge: Cambridge University Press, 2003); Kenneth Gergen, *Realities and Relationships: Soundings in Social Construction* (Cambridge MA: Harvard University Press, 2009); John Shotter, 'Agential Realism, Social Constructionism, and Our Living Relations to Our Surroundings: Sensing Similarities Rather Than Seeing Patterns,' *Theory & Psychology* 24, (2014), pp.30–325; Erica Burman, 'Limits of Deconstruction, Deconstructing Limits,' *Feminism & Psychology* 25, (2015), pp.408–422.

had to say in his late *Autobiography* about Herbert Spencer, perhaps the most famous Victorian psychologist:

> Herbert Spencer's conversation seemed to me very interesting ... Nevertheless I am not conscious of having profited in my own work by Spencer's writings. His deductive manner of treating every subject is wholly opposed to my frame of mind. His conclusions never convince me: and over and over again I have said to myself, after reading one of his discussions,—'Here would be a fine subject for half-a-dozen years' work.' His fundamental generalisations (which have been compared in importance by some persons with Newton's laws!)—which I daresay may be very valuable under a philosophical point of view, are of such a nature that they do not seem to me to be of any strictly scientific use. They partake more of the nature of definitions than of laws of nature. They do not aid one in predicting what will happen in any particular case. Anyhow they have not been of any use to me.[100]

Spencer had no training as a naturalist. 'If he had trained himself to observe more,' judged Darwin, he would have been 'a wonderful man.'[101] For example, Spencer's remarks on babies smacked more of clairvoyance than observational rigour. 'To the incipient intelligence of an infant,' he airily wrote, 'noise does not involve any conception of body. In an oft-recurring echo, the sound has come to have an existence separate from the original concussion—[it] continues after the vibrating body which caused it has become still.'[102] Darwin's path-breaking study of infancy, in contrast, drew on a notebook filled over many years with brilliant and detailed first-hand observations of the Darwins' own children—as we shall see in Chapter 9.[103]

2.6   Resonances

Darwin's Psychology aims to describe and analyse the approaches, themes, and details that inform Darwin's writings about psychological matters. I would not have taken up this task, if I had not seen valuable synergies between Darwin's work and wider debates across the years in the biological and human sciences, right up to

---

[100] Darwin, *Autobiography*, pp.108–109.

[101] Charles Darwin, 'Letter to Hooker, 10th December,' *Darwin Correspondence Project* (Cambridge1866), https://www.darwinproject.ac.uk/letter/?docId=letters/DCP-LETT-5300.xml.

[102] Herbert Spencer, *The Principles of Psychology, Vol.2*, 2nd ed. (London: Williams & Norgate, 1872), p.141.

[103] Charles Darwin and Emma Darwin, *Notebook of Observations on the Darwin Children*, (Cambridge: Darwin Online, 1839–1856), http://darwin-online.org.uk/content/frameset?itemID=CUL-DAR210.11.37.; Charles Darwin, 'A Biographical Sketch of an Infant,' *Mind: A Quarterly Review of Psychology and Philosophy* 2, (1877), pp.285–294.

today. Thus every chapter ends with a brief guide to major harmonies and discords with the work of other scientists and scholars.

2.6.1 Natural histories of man

The idea that humans are animal and belong to a continuous scale of nature (sometimes called the *scala natura*; or 'great chain of being'), proves of ancient vintage. Pre-Darwinian taxonomies drawn up by Linnaeus and Buffon included human beings.[104] So: was there a single human species? Or several? Both the Bible, and early versions of the 'human science,' or anthropology, advanced by philosophers like Immanuel Kant (1724–1804) and David Hume (1711–1776), answered that there was just one human nature. Hence, much though different circumstances might shape this essential nature, there was only one human species—a view later dubbed *monogenism*.

Throughout the second half of the eighteenth and the early nineteenth century, however, taxonomists were increasingly drenched in exotic materials about humanoid creatures from globe-sailing naturalists. Scientific travellers brought home, not just reports and specimens of novel plants and animals, but of mannish apes, aborigines, feral children, and curious person-like creatures with tails living in woods or underground caves. Such a mass of new information about the geographical variations, the far-flung island inhabitants, the odd customs, the diets, ornaments, skin colours, and diverse mores of human-like beings, challenged assumptions about the uniformity of human nature. The new natural histories of man pushed towards a different conclusion—called *polygenism*: if novel animals and plants formed distinct species confined to bounded regions, surely human beings should also be considered to comprise several species.[105]

As we shall see in later chapters, it was the continuing clash between monogenists and polygenists that gave Darwin a rationale for his first extended publication on humans—*The Descent of Man*, and arguably its sequel, *Expression*.[106] Inflamed by the derogation of slaves during the American civil war (1861–1865), Britain had seen a fierce institutional struggle between its two principal fora for debates about the ethnographic evidence furnishing natural histories of man: the Ethnological Society, founded in 1843; and the Anthropological Society of London, which split

---

[104] Phillip Sloan, 'The Gaze of Natural History,' in *Inventing Human Science: Eighteenth Century Domains*, ed. Christopher Fox, Roy Porter, and Robert Wokler (Berkeley CA: University of California Press, 1995).

[105] Sloan, 'Natural History,' pp.930–932; Sloan, 'Gaze of Natural History.'

[106] Adrian Desmond and James Moore, *Darwin's Sacred Cause: Race, Slavery and the Quest for Human Origins* (London: Penguin, 2009); Gregory Radick, 'Darwin's Puzzling *Expression*,' *Comptes Rendus Biologies* 333, (2010), pp.181–187.

from it in 1863. The older Ethnological Society favoured monogenism. The new Anthropological Society favoured polygenism.

A prominent view among prominent Victorian monogenists in the Ethnological Society, the so-called *ethnologicals,* made humans originate in the Garden of Eden, whence a diffusion of different nations or races had taken place, over a brief, biblical, time-period. Cultural change went in the direction of degeneration, not progress: making so-called *savages* (e.g. Native Americans; Australian Aborigines) 'degraded' humans. *Anthropologicals,* on the other hand, typically took Europeans to have evolved *progressively* from more primitive humans, and over a much longer period—anything from 100,000 to ten million years.[107] Among polygenists in the Anthropological Society, the focus was less on explaining human diversity as represented *spatially,* than on accounting *temporally* for the origin of (their own) civilization. Anthropologicals' methods were comparative, focusing on beliefs, artefacts, and customs. The ethnologicals favoured language-centred methods.[108]

From early on, Darwin hated slavery—descending on both his mother's and father's sides from outspoken abolitionists. Hence, both *Descent* and *Expression* weighed in on the side of monogenism (traditionally, the view of ethnologicals). In them, he aimed to show that slavery could not be justified by any scientific argument that Negroes belonged to a different species from Europeans. Yet, as we shall see, Darwin argued for monogenism on progressive, *evolutionary* grounds—a perspective formerly used by anthropologicals to justify polygenism.

2.6.2 Three takes on observation

Modernist psychology invites two takes on observation. The first deems transparency of mind and action in the experimenter most productive of truth about the world. Such *subjective* clarity entails both commissions and omissions. The modernist researcher must describe exactly what they *commit* to do in their investigation, and then do just that—as epitomized in the principle of so-called operational definition. Equally, researchers must *omit* to do anything else: influence the people they study in unspoken ways; import personal or cultural bias into their research; hide negative results; or jettison data that might disprove their conclusions. Their reward for such subjective rigour is *objective* knowledge.[109]

A second form of modernist psychology—sometimes called deconstruction, social construction, or *post*-modernism—provocatively denies the possibility of the subjective purification required for operational definition, and so concludes that

---

[107] Alfred Wallace, 'The Origin of Human Races and the Antiquity of Man Deduced from the Theory of "Natural Selection"', *Journal of the Anthropological Society of London* 2, (1864), pp.clviii–clxxxvii.

[108] George Stocking Jr, *Victorian Anthropology* (New York: Free Press, 1987), pp.144–185.

[109] Jill Morawski, 'Organizing Knowledge and Behaviour at Yale's Institute of Human Relations,' *Isis* 77, (1986), pp.219–242.

experiment and observation are futile as truth-producing exercises.[110] Proponents of this approach contend that 'the "reality" of the world, or the "facts" of any matter, are generated through social discourse, and not through observation'—and so conclude that observers' own theoretical, cultural, and personal preconceptions *create* the facts and the knowledge that they claim to gain through careful watching and listening.[111] Such arguments take their main warrant from epistemologies (philosophies of knowledge) which make language the shibboleth of knowing, and so deny both our capacity to grasp non-human natures and animals' capacity to know.[112]

Underlying both these takes on observation is the bifurcation of the world into a subjective part, which colours our realities with qualities like feelings, and an objective part (things in themselves)—the subject–object dualism that we met in Chapter 1 ('Darwin and Modernism').[113] For Darwin, this split could not have held water, because agency, for him, melded embodiment with purpose (mentality)— as Chapter 3 will elaborate. Likewise, observation *inevitably* entangled theoretical interest with visible, tangible, tasteable events.[114] Hence its difficulty. And hence the arduous labour spent in prosecuting the many crafts required of any naturalist hoping to access the 'natural system.'[115] Natural history perforce proved a slow science,[116] as Darwin's maxim implied: 'It's dogged as does it.'[117]

Perhaps unsurprisingly, therefore, Darwin's notebooks and practice, his private comments and his public statements, reveal a complex, sometimes confusing, commentary on the tortuous difficulties of handling the tensions between theory and observation in pursuing the art and craft of natural history. To wit: the opening sentences of the *Origin* imply, and Darwin's *Autobiography* asserts outright, that, for five years after quitting the *Beagle:* 'I worked on true Baconian principles, and

---

[110] See Bruno Latour, *We Have Never Been Modern* (Cambridge MA: Harvard University Press, 1993) for the argument that post-modernism lies within the ambit of modernism.

[111] Mary Gergen, 'Induction and Construction: Teetering between Two Worlds,' *European Journal of Social Psychology* 19, (1989), pp.431–437; Benjamin Bradley, 'Deconstruction Reassigned? 'The Child,' Antipsychology and the Fate of the Empirical,' *Feminism & Psychology* 25, (2015), pp.284–304.

[112] E.g. Gergen, 'Induction and Construction'; Gergen, *Realities and Relationships*; Burman, 'Deconstructing Limits,' p.413; Jacques Derrida, *The Animal That Therefore I Am* (New York: Fordham University Press, 2008).

[113] For further discussion, see Benjamin Bradley and Colwyn Trevarthen, 'Babytalk as an Adaptation to the Infant's Communication,' in *The Development of Communication*, ed. Natalie Waterson and Catherine Snow (London: Wiley, 1978), pp.75–92; Paul Stenner, *Liminality and Experience: A Transdisciplinary Approach to the Psychosocial* (London: Palgrave Macmllan, 2017).

[114] Darwin was not above smelling and tasting new finds; e.g. when mouthed, a native guava eaten by the giant Galapagos tortoises, had 'an acid astringent taste': Darwin, 'Zoology Notes,' p.296.

[115] Darwin, *Origin 1876*, p.369.

[116] Isabelle Stengers, *Another Science Is Possible: A Manifesto for Slow Science* (Cambridge: Polity Press, 2018).

[117] Darwin, 'Reminiscences,' p.149; a maxim he took from Anthony Trollope, *The Last Chronicle of Barset* (London: Smith & Elder, 1867), http://www.gutenberg.org/files/3045/3045-h/3045-h.htm. Ch. 61. The full quote is: 'It's dogged as does it. It's not thinking about it.'

without any theory collected facts on a wholesale scale.'[118] Later, to his son, he said: 'No one could be a good observer unless he was an active theoriser.'[119] A sympathetic correspondent read: 'How odd it is that everyone should not see that all observation must be for or against some view, if it is to be of any service.' Theory, Darwin went on, comprised 'the very soul of observation'[120]—but with a crucial nuance. Emma Darwin used to quote her husband's saying: 'It is a fatal fault to reason whilst observing, though so necessary beforehand and so useful afterwards.'[121] If an observer stuck too strongly to their theory *whilst observing*, they became blind to unforeseen aspects of what they saw. Hence the premium he put on 'never letting exceptions pass unnoticed,' which his son Francis called: the 'one quality of mind which seemed to be of special and extreme advantage in leading him to make discoveries.'[122]

Here we meet a third take on observation as a venerable set of collaborative, doggedly self-undeceiving practices aimed at true seeing: a take hall-marking Darwin's works, as the following chapters prove; a take extolled by ethologists; a take side-lined or scorned by modernist psychologies.[123]

Regarding which, I remember my surprise in 1970 upon discovering that the Experimental Psychology degree at Oxford University invited no input from Nobel prize-winning ethologist Niko Tinbergen—who worked in the same building that housed Oxford's psychologists. Only by choosing to study the new Human Sciences degree (as I did) could undergraduates benefit from the fruits of Tinbergen's extraordinary craft.[124] Perhaps this stand-off was mutual. Frowning at psychology, Tinbergen wrote that 'contempt for simple observation' was 'a lethal trait in any science.'[125] In particular, he held that ethology trumped behaviourism, which erred by focusing on 'a few phenomena observed in a handful of species which were kept in impoverished environments, to formulate theories claimed to be general.' From Tinbergen's standpoint, modern psychology's haste 'to step into the twentieth century' and become a 'respectable' science, meant that it had 'skipped the preliminary

---

[118] Darwin, *Origin 1876*, p.1; Darwin, *Autobiography*, p.119.

[119] Darwin, 'Reminiscences,' p.149.

[120] Charles Darwin, 'Letter to Fawcett, 18th September,' *Darwin's Correspondence* (Cambridge: Darwin Correspondence Project, 1861), https://www.darwinproject.ac.uk/letter/?docId=letters/DCP-LETT-3257.xml.

[121] Darwin, *Autobiography*, p.159.

[122] Darwin, *Autobiography*, pp.119 & 159; Darwin, 'Reminiscences,'; Darwin, 'Letter to Fawcett, 18th September.'

[123] E.g. the scornful treatment in Gergen, 'Induction and Construction' of Moscovici's argument that social psychology should follow the observational practices of ethology and ethnography so as to gain rich descriptions of the facts that it aims to explain: Serge Moscovici, 'Notes toward a Description of Social Representations,' *European Journal of Social Psychology* 18, (1988), pp.211–250; Serge Moscovici, 'Preconditions for Explanation in Social Psychology,' *European Journal of Social Psychology* 19, (1989), pp.407–430.

[124] Or, even better, by accompanying Tinbergen on his annual vacation to study black-headed gulls on Walney Island.

[125] Niko Tinbergen, 'On Aims and Methods of Ethology,' *Animal Biology* 55, (2005), p.300.

descriptive stage that other natural sciences had gone through,' and so was soon 'losing touch with the natural phenomena.'[126] This was not a criticism that he could have levelled at Darwin's psychology, however, as the following chapters will prove.

2.6.3 Habitat versus environment

It is easy to miss crucial differences between Darwin's day and ours. Take the current terms *behaviour* and *environment*: words which have no obvious etymological tie. Darwin hardly used the word environment, and only late in life.[127] He rarely used 'behaviour' in the sense we do. What words did he use? *Habit* came closest to *behaviour*. And the nearest to *environment*? Perhaps *habitat, habitation,* or, most frequently, *conditions of life.*

Note the common root of the words habit, inhabitant, habitation, and habitat. All four come from the Latin *habere*, meaning a manner of 'having' or 'holding oneself.'[128] Further, a strong allusive chime obtains between habit and habitat: one's habitat means the place one *in-habits*—the place one's habits make one's home. This chime well conveys Darwin's vision of interdependency between an organism's life pattern and its place of habitation. Likewise with *conditions of life*, a phrase which posits no separation between organism and externalities. As with the 'if→then' of logical conditionality, any organism's existence *necessarily* implies specific conditions—as George Lewes (1817–1878) emphasized. In Darwin's writings 'conditions of life' meant the relations which afforded the existence of *a particular* organism, 'external and internal, physical, organic, and social.'[129]

Environment was for Victorians a far newer word than *habitat*, and had different connotations. It entered English in the 1830s,[130] ('habitat' dated from around 1760), first being used in an evolutionary context by Spencer, in his *Principles of Psychology* (1855). This gave it a particular colouring. Etymologically, environment implied something encircling or surrounding an entity. No chime here of indwelling or interfusion between an organism and its place in the world. *The* environment swiftly became a thing in itself: free-standing, without implication of a particular inhabitant. Thus, Spencer gave a specific, causative function to what he

---

[126] Tinbergen, 'Aims and Methods,' p.299.

[127] He first publicly used the word 'environment' in 1875, according to Angelique Richardson, 'Darwin and Reductionisms: Victorian, Neo-Darwinian and Post-Genomic Biologies,' *19: Interdisciplinary Studies in the Long Nineteenth Century* (2010). The first private use was in 1868, so far as I can find.

[128] See 'Habit, n.' and 'Habitat, n.' in Anon, *Oxford English Dictionary Online* (Oxford: Oxford University Press, 2018), http://www.oed.com.ezproxy.csu.edu.au/. Cf. Edwin Ardener, "Behaviour"—a Social Anthropological Criticism,' in *The Voice of Prophecy and Other Essays*, ed. Michael Chapman (Oxford: Blackwell, 1989), pp.105–108.

[129] George Lewes, 'Mr Darwin's Hypotheses, I,' *Fortnightly Review* 3, (1868), p.367. Cf. Tim Ingold, 'Bindings against Boundaries: Entanglements of Life in an Open World,' *Environment and Planning A* 40, (2008), pp.1796–1810.

[130] Courtesy of Thomas Carlyle (1795–1881).

called 'the environment': evolution resulted from 'adjustments of inner relations to outer relations,' and animal action was 'caused by the immediate contact [with] *something in* the environment.'[131]

As early as 1868, Lewes pointed out that Spencer's conceptions of evolution (and of animal action)—when compared to Darwin's—made 'the common mistake of supposing the Organism to be passive under the influence of external conditions,' that is, of the environment. Spencer's theory, like his definition of life as 'inner actions so adjusted as to balance outer actions,' implied that action can 'come before the agent.' But this was absurd, said Lewes, because an action 'is the agent in act.'[132]

Spencer's language—which is also our language—implies an environmentally shaped evolution which hollows agency and creativity out of organisms. The vacuum such language creates attracts the idea that heredity played the determinative role in shaping the fates of individuals (soon taken up by Darwin's half-cousin, Francis Galton: 1822–1911). The organism itself thus becomes little more than an empty space in which the dictates of the past (heredity) face off against the dictates of free-standing, external conditions. This conceptualization sets up the endless— some would say fruitless—debates about *genes versus environment* in popular and academic psychology. The defining terms of such debate provide no way to conceptualize interdependency. Hence the need to coin a new term to catch the interactions between different species and their environments, namely, *ecology*.[133] As its inventor, Ernst Haeckel, explained: 'by ecology we mean the body of knowledge concerning the economy of nature ... in a word, ecology is the study of all those complex interrelationships referred to by Darwin as the conditions of the struggle for existence.'[134] Arguably, however, the terms *ecology* and *interaction* still fail to capture the implications of mutuality, conditionality, and agency in Darwin's understanding of organic interdependence (see Ch. 3).[135]

2.7 Conclusion

Wise ones say that, if you want to understand what a science is, you should look first, 'not at its theories or at its findings, and certainly not at what its apologists

---

[131] Herbert Spencer, *The Principles of Psychology*, 2 vols., vol. 1 (London: Williams & Norgate, 1870), pp.140, 294, 298–299, my emphases.

[132] George Lewes, 'Mr. Darwin's Hypotheses, Part III,' *Fortnightly Review (New Series)* 4, (1868), pp.66, 73.

[133] Hans-Jorg Rheinberger and Peter McLaughlin, 'Darwin's Experimental Natural History,' *Journal of the History of Biology* 17, (1984); Ferdinando Boero, 'From Darwin's *Origin of Species* toward a Theory of Natural History,' *F1000 Prime Reports* 7, (2015), https://f1000.com/prime/reports/b/7/49.

[134] Haeckel quoted in Robert Stauffer, 'Ecology in the Long Manuscript Version of Darwin's *Origin of Species* and Linnaeus' *Oeconomy of Nature*,' *Proceedings of the American Philosophical Society* 104, (1960), p.235.

[135] As argued by, e.g. John Odling-Smee, Kevin Laland, and Marcus Feldman, *Niche Construction: The Neglected Process in Evolution* (Princeton NJ: Princeton University Press, 2003), pp.419 & passim.

say about it; you should look at what the practitioners of it do.'[136] This chapter has surveyed the practical craft that inspired and grounded Darwin's adventures in evolutionary theory and the study of agency. I have taken pains to underline the technical and practical proficiency manifest in Darwin's work, because that work has often been belittled by modern science for lack of these qualities. Nor did Darwin inherit solely practical skills from the tradition of natural history his work extended and challenged. Such supposedly amateurish quirks as anthropomorphic description (but see Ch. 6), as well as key constructs like the economy of nature, reflect his uptake of pre-Darwinian traditions.

Most of all, natural history was a *vocation*—not a nine-to-five job. Should one doubt the depth or durability of Darwin's calling as naturalist, one need only inspect his post-*Beagle* career. Disembarking in October 1836, he spent the next five years in London, networking with members of several august scientific societies and serving as secretary to perhaps the most active of these, the Geological Society.[137] He thereby built up a lifelong circle of well-positioned friends and scientific colleagues, some of whom he first recruited as specialized experts to help him analyse the collections he had made on his five-year circumnavigation in *HMS Beagle*. In September 1842, thirty-three-year-old Darwin and his wife Emma, aged thirty-four, resettled with their two children in a village called Down, seventeen miles south of London.[138] This became the lasting home for the Darwins' large family—eight more children were born at Down House—plus their many servants. It also became a veritable factory for scientific research and publication.

Far from being a retreat from earnest scientific labour into dilettantism, or from men of science themselves, Darwin at Down House 'embraced the work discipline of industrial capitalism' seven days a week—as witnessed by his time-keeping, his regularity, his precise accounting of production and expenditure—whilst tirelessly pursuing a thousand scientific questions.[139] Here he eventually led, or bred, a workforce which included dedicated and accomplished research assistants, technicians, translators, and editorial staff (predominately made up of family members).[140] Here he constructed or bought both Heath Robinson and state-of-the-art

---

[136] Clifford Geertz, *The Interpretation of Cultures: Selected Essays* (New York: Basic Books, 1973), p.5. Cf. Albert Einstein, 'On the Method of Theoretical Physics,' *Philosophy of Science* 1, (1934), p.163: 'If you want to find out anything from the theoretical physicists about the methods they use, I advise you to stick closely to one principle: don't listen to their words, fix your attention on their deeds'; Andrew Pickering, 'From Science as Knowledge to Science as Practice,' in *Science as Practice and Culture*, ed. Andrew Pickering (Chicago: University of Chicago Press, 1992), pp.1–26.

[137] Martin Rudwick, 'Charles Darwin in London: The Integration of Public and Private Science,' *Isis* 73, (1982), pp.186–206; Sandra Herbert, 'The Place of Man in the Development of Darwin's Theory of Transmutation. Part 2,' *Journal of the History of Biology* 10, (1977), pp.155–227.

[138] The village is now called Downe.

[139] Paul White, 'Darwin's Home of Science and the Nature of Domesticity,' in *Domesticity in the Making of Modern Science*, ed. Donald Opitz, Staffan Bergwik, and Brigitte Van Tiggelen, (Basingstoke, UK: Palgrave Macmillan, 2016), p.69.

[140] Darwin drew on the skills of three sons: Francis, an accomplished botanist; George, a mathematician who became Professor of Astronomy and Experimental Physics at Cambridge; Horace, a skilled technician who later founded the Cambridge Scientific Instrument Company. Darwin also drew on the

Figure 2.7 Darwin's study at Down House, as pictured in *The Century Magazine*, January 1883. Here, books would systematically be moved from one shelf to another, after they had been read, and before they had been catalogued, prior to a third, longer-term, shelving. Meanwhile Darwin's systematic noting of ideas, quotations, references from books, and observations, would be accumulating as appropriate in his publication-oriented portfolios (visible in the racks on the right-hand side of the fireplace).

Source: The Study at Down from *The Century Illustrated Monthly Magazine*, v.25 Nov 1882–April 1883, p. 420, Indiana University Library. Illustration from a painting by Alfred Parsons. Engraved by James Tynan.

Image credit: Wellcome Collection. Distributed under the terms of the Creative Commons Attribution 4.0 International (CC BY 4.0). https://creativecommons.org/licenses/by/4.0/#_ga=2.241103562.87728 4273.1565857924-514625952.1510241134.

Darwin, *Autobiography*, pp.137–138.

scientific equipment; installed specially designed experimental hot-houses; and hosted many meetings of elite scientists. In his study, 'the only comfortable chair was converted as if for industrial use, with metal legs and wheels for ease of movement about the room. It looked like it belonged in a workshop or factory, and lent a sense of ruthless efficiency to the space' (behind the table in Figure 2.7).[141] Here

valued editorial skills of daughter Henrietta. Wife Emma was a gifted linguist, and when translation from German was required, a governess might be drawn in. Darwin's gardeners also assisted.

[141] White, 'Darwin's Home of Science,' pp.68–69.

he read voraciously across many fields, systematically collating his notes. And here he wrote his books.

Before we get to his books, however, we need to understand the vision of the natural world underpinning all his work, which his practice as a naturalist produced. This appears most clearly in his observations and hypotheses about such relatively simple creatures as colonial invertebrates, climbing plants, and earthworms—as the next chapter documents.

3

Agency and Its Effects

We each have two eyes—and sense the solidity of the world by combining their different lines of sight. Darwin's vision of the living world likewise fuses two contrasting perspectives. First come the events comprising the economy of nature, their time-scale ranging from moments to lifespans—the theatre of agency. At right angles comes the resultant of that theatre: the slow modification of species, effecting common descent. This second line of sight has a far vaster time-scale than the first, ranging from millennia to periods Darwin likened to eternity, which 'mind cannot grasp.'[1]

Chapter 3 examines the ways Darwin constructed these two axes. I begin with the shorter-term, showing that, because, for Darwin, agency embodies purpose—it implies mentality. Which makes the theatre of agency intrinsically psychological. Moreover, because Darwin's understanding of agency (axis 1) underpins his formulation of evolution (axis 2), his approach has results which differ markedly from those psychologies which set out by presuming that the immense antiquity of human descent has always already shaped present actions. The book's remaining chapters will elaborate these results in detail.

3.1 Contours of Agency

3.1.1 Agency itself

If the living world is to make sense as a theatre of agency, the word *agency* needs attention. Today's dictionaries call agency *a capacity to act* (among other things). Darwin proves profligate and idiosyncratic in his use of the term. Profligate, in that he applied the words agent and agency to all sorts of thing and process: inanimate, animate, vegetable, animal, human, spiritual, and hypothetical. Idiosyncratic, if we believe that his Victorian peers essentially saw agency as a human attribute: exercise of a person's rational will.[2]

---

[1] Charles Darwin, *On the Origin of Species by Means of Natural Selection or the Preservation of Favoured Races in the Struggle for Life* (London: Murray, 1859), pp.87, 285.

[2] As argued by Roger Smith, *Free Will and the Human Sciences in Britain 1870–1910* (Pittsburgh PA: University of Pittsburgh Press, 2016), p.104. Smith concludes that, because Victorian culture equated agency with exercise of the will, it had, for them, a *moral* quality. To mount this argument, Smith has to systematize the typically uncorseted Victorian usage of the words *agent* and *agency*, so

Darwin's Psychology. Ben Bradley, Oxford University Press (2020). © Oxford University Press.
DOI: 10.1093/oso/9780198708216.001.0001

In more detail: Darwin's frequent use of the words agent and agency largely applied to generic powers, both of inanimate and animate entities. In this generic sense, he wrote of the agency of worms, of insects, and of 'man'.[3] And, while living birds were 'highly effective agents' in the transportation of seeds, *inanimate*, or what he called 'external' agencies, also affected the growth of plants: the daily alternations of light and dark; light itself; changes in temperature; or the pull of gravity.[4] In other contexts he discussed 'chemical agents', and the agencies of wind (in fertilizing plants), of dust (in thickening good soil), of rain-soaked ground (which swells), and of breaking waves (as affecting coral).[5] 'Internal' agencies included such actions as the self-burying movement of flower-heads in the subterranean clover (*Trifolium subterraneum*—an agency derived from the vegetable power of *circumnutation*; discussed later in this chapter).[6] With regard to humans, Darwin described as agencies the effects of habit, of reasoning powers, of instruction, of religion, modesty, and the will.[7] With regard to the kinds of taste that led to the domestication of plants and animals, and to the production of new breeds and cultivars, human agency frequently proved 'unconscious'.[8] Too, Darwin discussed human belief in unseen or 'spiritual' agencies.[9] Further uses of the words *agents* and *agencies* applied to *hypothetical* processes: the proposed evolutionary effects of the disuse of organs (in rendering them rudimentary); and selection, whether natural or sexual.[10]

Finally, *agent* occasionally occurs where Darwin described one individual's self-caused movements—or at least a perception of such. Thus, an open, breeze-tweaked parasol twitching 'by itself' on Darwin's lawn spooked his dog into barking: 'He must, I think, have reasoned to himself in a rapid and unconscious

---

projecting onto Victorian debates a human-only, meaning for agency—which clashes with its several roles in Darwin's (and e.g. Huxley's) work.

[3] Charles Darwin, *The Formation of Vegetable Mould, through the Action of Worms, with Observations on Their Habits* (London: John Murray, 1882), p.177; Charles Darwin, *The Various Contrivances by Which Orchids Are Fertilised by Insects*, 2nd revised ed. (London: John Murray, 1882), p.247; Charles Darwin, *The Descent of Man, and Selection in Relation to Sex* (London: Murray, 1874), pp.341, 356.

[4] Charles Darwin, *The Power of Movement in Plants* (London: John Murray, 1880), p.560; Charles Darwin, *The Origin of Species by Means of Natural Selection, or the Preservation of Favoured Races in the Struggle for Life*, 6th ed. (London: Murray, 1876), p.326.

[5] Charles Darwin, *Insectivorous Plants*, 2nd ed. (London: John Murray, 1888), p.80; Darwin, *Worms*, pp.119, 237; Charles Darwin, *The Structure and Distribution of Coral Reefs* (London: Smith, Elder & Co, 1842), pp.44, 130; Darwin, *Origin 1876*, p.161.

[6] Darwin, *Movement in Plants*, p.517.

[7] Darwin, *Descent*, p.618; Darwin, *Origin 1876*, p.108; Charles Darwin, *The Expression of the Emotions in Man and Animals*, 2nd ed. (London: Murray, 1890), pp.253, 374. NB while Darwin twice referred to the will as an agency—this is not the same thing as equating agency *as such* with exercise of the will. Darwin's main discussion of the will is in *Expression* (see Chs 4 & 5).

[8] Darwin, *Origin 1876*, p.411: the 'unconscious form of selection by man ... follows from the preservation of the most valued individuals of each breed, without any intention on his part to modify the characters of the breed'; p.615.

[9] Darwin, *Descent*, pp.65ff.

[10] Darwin, *Origin 1876*, pp.400–401; Darwin, *Descent*, pp.61, 198.

manner, that movement without any apparent cause indicated the presence of some strange living agent, and that no stranger had a right to be on his territory.'[11]

In what follows, I identify agency with a capacity to act. This narrows its purchase in comparison to Darwin's texts. I do this to reflect findings from Darwin's lifelong research on purposive movement in plants and animals. As I now show, this narrower definition captures a sense of agency as applying to occasions when some entity moves of its own accord (viz. the parasol described previously), which implies, *moving as means to an end.*[12] Such a definition would only make cars and trains agents if we failed to mark off the purposive, flexible, adaptiveness of living forms from the automatic routines and invariant stimulus–response couplets, typical of machines. As William James observed: with inorganic entities, you alter the initial conditions and bring forth each time a different result. 'But with intelligent agents,[13] altering the conditions changes the activity displayed, but not the end reached.' James' link of agency to intelligence was also crucial to Darwin's understanding of life, both in plants and animals, and so to his whole psychology. As James underlines: '*The pursuance of future ends and the choice of means for their attainment are thus the mark and criterion of the presence of mentality* in a phenomenon. We all use this test to discriminate between an intelligent and a mechanical performance.'[14]

I will soon return to the link between agency and intelligence. But to show how Darwin first came to understand that agency implied mentality, we must consider his investigations into two key questions: Is there a distinction between life and matter? and How do the kingdoms of plants and animals relate? These two questions touch because, if different laws applied to organic than to inorganic nature, then the organic laws should apply equally to *all* living beings, both animal and vegetable.[15] It is on this assumption that Darwin wrote: 'The grand question which every naturalist ought to have before him when dissecting a whale or classifying a mite, a fungus, or an infusorian is, "What are the Laws of Life." '[16]

---

[11] Darwin, *Descent*, p.95.
[12] Hence, philosophers say, agency implies 'normativity': the linking of ends to their appropriate means. Barham, 'Normativity, Agency, and Life,' *Studies in History and Philosophy of Biological and Biomedical Sciences* 43, (2012), p.93.
[13] I.e. agents having a modicum of understanding.
[14] William James, *The Principles of Psychology* (New York: Holt, 1890), pp.20–21, his italics.
[15] This was also germane to Darwin's evolutionism. Theoretical unification of plants with animals would remove a barrier in the way of transformism, 'since this would allow that all life, and not simply the separate animal and plant series, could be unified in a common source': Phillip Sloan, 'Darwin's Invertebrate Program, 1826–1836: Preconditions for Transformism,' in *The Darwinian Heritage*, ed. David Kohn (Princeton: Princeton University Press, 1985), p.120.
[16] Charles Darwin, 'Notebook B,' (1837–1838), p.229.

3.1.2 What is life?

Almost as soon as he arrived at Edinburgh University in 1825, sixteen-year-old Darwin began collecting the plant-like animals known as *zoophytes* during walks along the margin of the Firth of Forth. Edinburgh was the leading venue for invertebrate zoology in Europe during the 1820s—and marine invertebrates became an enduring fascination for Darwin. Ten years later, on his voyage around the world, he continued to delight in marine invertebrates, especially sea pens and coral reefs. This delight later culminated in, and was perhaps killed off by, the eight years of labour he put into his magisterial, four-volume, taxonomy of living and fossil barnacles (1846–1854).[17]

Darwin's enthusiasm for sea-life was undoubtedly piqued by a finding he made in 1827. Looking down a microscope at the seaweed-like animal called *Flustra*, he saw eggs using hair-like oars (or 'cilia') to swim. This discovery of the tiny eggs' 'spontaneous motion' was new to the science of his day, and the topic of Darwin's first serious scientific presentation. His discovery was quickly taken up as his own by a prominent professor at Edinburgh, Robert Grant.[18]

Underwater, *Flustra* looks like a hand-sized limp-armed tree (Figure 2.1, top). Colonies of polyps stud its branches, each apparently unconnected to the rest. Stimulation of one polyp does not cause reactions in others, and the polyps are able to carry out their own autonomous reproduction 'almost as if independent creatures.'[19] How colonial invertebrates bred remained of abiding interest to Darwin while at university, throughout his *Beagle* voyage, and beyond. His interest was not unique, zoophytes being a popular topic for observation and fantasy among Victorians.[20] And a topic of scientific controversy. Questions particularly surrounded the so-termed 'granular matter,' or 'gemmules,' found within the ovaries of zoophytes like *Flustra*, and presumed responsible for their reproduction (Figure 3.1).

---

[17] 'I have now for a long time been at work at the fossil cirripedes, which take up more time even than the recent;—confound & exterminate the whole tribe; I can see no end to my work.' Charles Darwin, 'Letter to Hooker, 3rd Feb,' (1850).

[18] Infuriating Darwin. See Charles Darwin, 'On the Ova of Flustra, or, Early Notebook, Containing Observations Made by Cd. When He Was at Edinburgh, March 1827,' in *The Collected Papers of Charles Darwin.*, ed. Paul Barrett (Chicago: Chicago University Press, 1977).

[19] Sloan, 'Darwin's Invertebrate Program,' p.81.

[20] Isabelle Stengers and Ilya Prigogine, 'The Reenchantment of the World,' in *Power and Invention: Situating Science* (Minneapolis: University of Minnesota Press, 1997), pp.33–60; Danielle Coriale, 'When Zoophytes Speak: Polyps and Naturalist Fantasy in the Age of Liberalism,' *Nineteenth-Century Contexts* 34, (2012), p.21: 'The unusual appearance, habits, and behaviours of these zoophytes captured the Victorian imagination, shocking observers and inviting them to take a closer look with microscopes, magnifying glasses, and other ocular technologies. As the popular natural history craze swung into full effect in the mid-nineteenth century, knowledge of the polyp and its curiosities became a badge of honour, a sign of careful attention and willingness to engage the living world.'

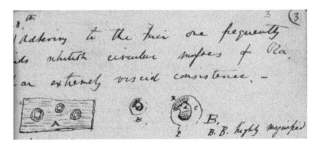

Figure 3.1 Darwin's drawings of the ova of *Flustra*: 'one frequently finds whitish circular masses of Ova, of an extremely viscid consistence,—& have the appearance represented at A [A is the left-hand image], when magnified however, it appears to be a mass [of] capsules containing animals united together by a transparent gelatinous matter [right-hand image].'

Source: Detail from Darwin, C. *Natural History Notebook*, 1827–1842. Reproduced by kind permission of the Syndics of Cambridge University Library, Classmark: DAR 118, p.7.

Darwin, 'Ova of Flustra,' pp.285–286.

Under magnification, granular matter showed similarities to the vital particles in pollen. Did such particles furnish the common denominator of life? Most microscopists agreed that granular particles moved or rotated, whether taken from plant or animal. But, what made them move? Did their motion result from chemical reaction—or physical force, like heat or collision? Or, if they were held to be alive, was their motility conferred from outside onto otherwise-passive matter? Alternatively, had the atoms inherent life?[21]

Ten years Darwin delved these questions, eye to the microscope, collecting specimens and reading books, changing his views as the debate widened to include arguments and observations from France and then Germany. By 1837 he had settled on a view of the granules (or gemmules) as 'living atoms'—that is, as *immanently dynamic*—common to animals and plants.[22] Their dynamism propelled and organized each newly budded creature's development. This tie between the life that animates granular matter and its development into a particular kind of organism implied an agency which operated according to a plan—the plan realized in the organization and activities of the fully formed life into which the gemmule grew.[23]

---

[21] My chief source for what follows is Phillip Sloan, 'Darwin, Vital Matter, and the Transformism of Species,' *Journal of the History of Biology* 19, (1986), pp.369–445.

[22] Phillip Sloan, 'The Making of a Philosophical Naturalist,' in *The Cambridge Companion to Darwin*, ed. Jonathon Hodge and Gregory Radick (Cambridge: Cambridge University Press, 2003), p.30.

[23] Richard Owen, 1837, quoted in Sloan, 'Vital Matter' p.414: 'the germ is amorphous matter, vivified by an organizing principle, which, operating upon the surrounding amorphous nutrient substance, arranges and forms it into the organs by whose harmonious action, Life is afterwards to be maintained.' During the late 1830s, Owen was significantly influencing many of Darwin's ideas, Sloan argues.

3.1.3 Purposive movement in plants

Linkage between agency and the type-specific outcomes of growth—both of anatomy and of habit—was refined and advanced in Darwin's intensively researched, post-*Origin* monographs on the movements of plants. These concerned: orchids, which move so as to attach their pollen to insects, thus effecting beneficial cross-fertilization (1862); climbing plants like vines, honeysuckle, and mallow, which attach themselves to supports which they then ascend (1865); and, plants like sundew (*Drosera*), in which purpose-grown tissues enable the plant's leaves to move and catch insects, curling inwards to digest insects in a temporary stomach awash with secreted acids (1875).[24] A late masterpiece, *The Power of Movement in Plants* (1880), elaborated on these studies.[25]

Still a landmark in botany and plant physiology, this last book distinguishes two kinds of vegetable action: reversible and irreversible. Well-known types of reversible movement include the many plants which regularly fold their leaves at night, and unfold them by day (Figure 3.2 top). Or the reversible spring-like movements resulting from local changes in cell-pressure or 'turgescence' which plants like sundew and the Venus fly-trap use to capture insects—movements which Darwin showed experimentally could be triggered in many artificial ways (Figure 3.2 bottom).

Darwin found a second kind of *irreversible* movement or 'habit' in all young plants, based on a fundamental tendency to twist, or, as he termed it, to *circumnutate*: 'Apparently every growing part of every plant is continually circumnutating,' he wrote: most obviously, a twining, climbing plant, 'continually ... bends successively to all points of the compass, so that the tip revolves' (Figure 3.3).[26] Even non-climbing plants incessantly twisted as they grew, albeit with a movement less fleet and on a smaller-scale than in what Darwin called twiners. Radicles—the first root to sprout from a seed—excelled at this dance, but you could see it almost ubiquitously: in any growing root-tip; in young leaves; in runners; rhizomes; tendrils; and at any apex of shoot or stem. Modifications of circumnutation created all the myriad 'habits,' 'tropisms,' or 'agencies'[27] that idiosyncratically adapted different species of plants, and different parts of the same plant, to the circumstances in which they lived—their growth away from or

---

[24] Darwin, *Insectivorous Plants*, pp.15, 18, 86, 106.

[25] Charles Darwin, *The Movements and Habits of Climbing Plants* (London: John Murray, 1882); Darwin, *Orchids*; Darwin, *Movement in Plants*; Darwin, *Insectivorous Plants*; Craig Whippo and Roger Hangarter, 'The "Sensational" Power of Movement in Plants: A Darwinian System for Studying the Evolution of Behaviour,' *American Journal of Botany* 96, (2009), pp.2115–2127; Ulrich Kutschera and Karl Niklas, 'Evolutionary Plant Physiology: Charles Darwin's Forgotten Synthesis,' *Naturwissenschaften* 96, (2009), pp.1339–1354.

[26] Darwin, *Climbing Plants*, p.95; Darwin, *Movement in Plants*, p.3.

[27] Darwin, *Movement in Plants*, p.199.

(a)

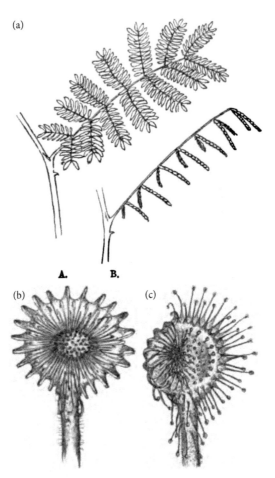

A.   **B.**

(b)   (c)

Figure 3.2 (a) the leaves of a sweet acacia plant (*Acacia farnesiana*) A, by day, and B, at night. (b) Darwin put a tiny piece of meat on the left side of the leaf pictured at left to illustrate the swift *reversible* movement of the sundew, *Drosera*, by which they catch insects (c).

Source: (a) Darwin, C. (1880). *The Power of Movements in Plants*; (b) Darwin, C. (1888). *Insectivorous Plants*; (c) Darwin, C. (1888). *Insectivorous Plants*.

towards: gravity; physical contact; sun; chemical gradients; moisture; injury; and varying textures of soil.

Given the wide range of references Darwin gave to the word agency, can we conclude from his calling circumnutation an 'innate' agency that he held plants to have had a kind of *mentality*?[28] Yes we can. His numerous, minute, experiments

---

[28] Darwin, *Movement in Plants*, pp.165, 263ff.

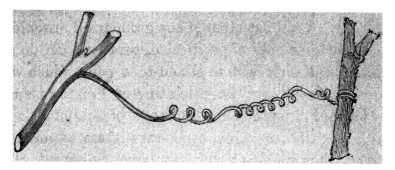

Figure 3.3 Circumnutation: The spiral movements of the tendril of a climbing plant, like this bryony, are *irreversible*.

Source: Darwin, C. (1882). *The Movements and Habits of Climbing Plants*.

Image credit: Wellcome Collection. Distributed under the terms of the Creative Commons Attribution 4.0 International (CC BY 4.0). https://creativecommons.org/licenses/by/4.0/#_ga=2.241103562.87728 4273.1565857924-514625952.1510241134.

on plants gave him grounds to conclude that plant movement manifested both irritability and sensibility—if not some sort of intelligence, though of a kind alien to humans.[29]

In the first place, plants showed an exquisite sensitivity to stimulation: they had 'senses.' The swift response of mimosa leaves to touch furnished a widely discussed example at Victorian dinner-tables. But Darwin's paragon—the fly-trapping sundew—far out-performed mimosa. He would brush the palm of a sundew's leaf with a droplet of, say, diluted camphor. Within minutes, the tentacles edging the leaf began to close. Alternatively, he would place a tiny bit of hair (8/1000ths of an inch long, weighing 0.000822 milligrams) on the palm of a sensitive leaf, and observe the same effect. He also demonstrated the extraordinarily weak chemical dilutions of phosphate of ammonia to which sundew would respond: 'the absorption by a gland of only the 1/19,760,000 of a grain (0.00000328 mg), that is, a little more than the one-twenty-millionth of a grain, is sufficient to cause the tentacle bearing this gland to bend to the centre of the leaf.' Darwin's awe at this extreme sensitiveness—'exceeding that of the most delicate part of the human body'—ballooned when he considered that plants 'transmitted various impulses from one part of the leaf to another ... without the intervention of any nervous system.'[30] Which brings us to a key distinction.

---

[29] 'Why do we find it so much more difficult to understand plants than animals? It's because of the history of evolution, which split us off from vegetation very early on. All our senses developed differently, and so we have to use our imaginations to get even the slightest idea of what is going on inside trees': Peter Wholleben, *The Hidden Life of Trees: What They Feel, How They Communicate* (Carlton, Vic.: Black Inc., 2016), p.227.

[30] Darwin, *Insectivorous Plants*, pp.135, 213, 219.

All Darwin's experimental results described in the previous paragraph, impressive though they were, could be subsumed under the long-known category of *irritability*. According to eighteenth century medical scholars like Albrecht von Haller (1708–1777), irritability was an instantaneous, contractive reactivity to a direct stimulus—a mechanical quality of animal muscles. In contrast, *sensibility*—the capacity to feel—was the quality transmitted from one place to another by nerves, and pertained to the mind or soul.[31] Whilst, by Darwin's day, few doubted that plants showed irritability to direct stimulation (e.g. mimosa leaves close when touched), sensibility was still thought absent from the vegetable kingdom.[32]

Hence it was significant that *The Power of Movement in Plants* dwelt at length on plants' sensibility. The upshot? The experiments the book describes proved that an influence or 'impulse' from light or gravity could travel 7–10 mm to the site of response. When Darwin cut off the tips of roots or shoots, the plants became insensitive to changed conditions. Alternatively, once exposed to relevant stimulation, immediate removal or injury of the organ of sensation (the growing tip) did not eliminate the plant's subsequent response. Hence, though plants lacked what in an animal would be called nerves, Darwin's observations showed that they acted in ways which resembled 'many of the actions performed unconsciously by the lower animals.' He wrote of the supremely sensitive tip of the radicle as 'perceiving' environmental changes, concluding that 'having the power of directing the movements of the adjoining parts,' the tip 'acts like the brain of one of the lower animals; the brain being seated within the anterior end of the body, receiving impressions from the sense-organs, and directing the several movements.'[33]

Comparison of the radicle to animal brains brought derision from some of Darwin's peers (especially the eminent German plant physiologist, Julius Sachs).[34] Yet Darwin's observations of growing plants' purposive response to obstacles rebutted such criticism. One experiment began by Darwin placing sheets of glass to block the downward growth of the radicles of broad beans. Bean rootlets would flatten against the glass and then turn at right-angles to move across the hard surface. Next, Darwin placed small wooden bars on the glass, so as to obstruct the radicles' path again. Each would hesitate, and then turn to creep along the slip of wood, 'until it came to the end of it, round which it bent rectangularly. Soon afterwards when coming to the edge of the plate of glass, it was again bent at a large angle, and descended perpendicularly into the damp sand.'[35] Darwin's study neatly aligned baby beans with James' definition of agency: 'with intelligent agents, altering the

---

[31] Ildiko Csengei, *Sympathy, Sensibility and the Literature of Feeling in the Eighteenth Century* (Basingstoke: Palgrave Macmillan, 2012), pp.81ff.

[32] Soraya De Chadarevian, 'Laboratory Science Versus Country-House Experiments. The Controversy between Julius Sachs and Charles Darwin,' *British Journal for the History of Science* 29, (1996), p.27.

[33] Darwin, *Movement in Plants*, pp.571–573.

[34] De Chadarevian, 'Laboratory Science.'

[35] Darwin, *Movement in Plants*, p.130.

conditions changes the activity displayed, but not the end reached.'[36] Whether or not Darwin's beans met obstacles, they eventually found different ways to achieve the same end, plunging straight down into the earth. Did this mean plants were, in James' word, *intelligent*?

3.1.4 Intelligence in simple creatures

While Darwin's books never overtly call plants intelligent, they often apply this word to less-lauded animals, like spiders and fiddler crabs.[37] His last book contained his most notable study of intelligence. Its first two chapters report twenty-eight experiments that systematically explored 'the agency of worms.'[38] Most of these experiments investigated the senses of ordinary earthworms. He found them completely deaf: 'indifferent to shouts' and responding neither to a shrill metal whistle nor to 'the deepest and loudest tones of a bassoon.' Yet they proved acutely sensitive to touch, had a limited sense of taste, and responded crudely to light, providing it shone on the front portion of their bodies, 'where the cerebral ganglia lie.' The observant Darwin took pause over worms' reactions to light. A worm, when suddenly illuminated, 'dashes like a rabbit into its burrow,' he wrote. Did they dash *reflexively*, he wondered. No, because a worm did not always flee in this way, 'as if it were an automaton ... even when the light was concentrated on them through a large lens.' Earthworms only fled light when not otherwise occupied—in eating, pulling leaves into their burrows, or 'whilst they are paired.' This showed that, far from behaving mechanically, worms had a power of attention, 'and attention implies the presence of a mind.'[39]

Darwin directed many experiments to testing whether worms 'exhibit some degree of intelligence instead of a mere blind instinctive impulse.' Here he exploited an obviously adaptive habit: 'their manner of plugging up the mouths of their burrows,' to insulate themselves against the cold. Typically worms plugged their holes with leaves (Figure 3.4). Darwin found that they generally dragged leaves tip-first into the hole: which he deemed the most efficient way of using leaves as

---

[36] James, *Principles*, pp.20–21.

[37] Darwin, *Descent*, pp.271–272: Spiders possess 'acute senses, and exhibit much intelligence,' while fiddler crabs are 'highly intelligent.'

[38] Darwin, *Worms*, p.177. The word 'experiment' is only used once in Darwin's monograph on worms. Darwin's word was 'observation,' even when the worms he was studying were kept in artificial conditions (flower-pots with glass lids, indoors), and being tested with artificial stimuli (leaf-like triangles of paper)—see Ed Reed, 'Darwin's Earthworms: A Case Study in Evolutionary Psychology,' *Behaviourism* 10, (1982), pp.165–185. Cf. Ch.2.

[39] Darwin, *Worms*, pp.23–25. NB the physiologist William Carpenter (1813–1885) saw attention as regulated by the will. However, I can find no evidence that Darwin followed suit. Cf. William Carpenter, *Principles of Mental Physiology, with Their Applications to the Training and Discipline of the Mind, and the Study of Its Morbid Conditions*, 2nd ed. (London: Henry S. King, 1875), pp.20–26; Darwin, *Descent*, pp.73–74.

Figure 3.4 Earthworm (left) dragging a leaf into its burrow.
Source: Denis Crawford / Alamy Stock Photo.

draught-excluders. He demonstrated that worms plugged their holes less carefully when the ambient temperature rose. Then he showed that, if it proved impractical to tug a leaf into a burrow by its tip—as with forked pine needles—worms would grab the base instead, even if they had never before encountered a forked pine-needle. What would they do when confronted with specially-tailored leaf-size triangles, made of 'moderately stiff writing-paper, which was rubbed with raw fat on both sides, so as to prevent their becoming excessively limp when exposed at night to rain and dew'? Hundreds of trials showed that worms largely acted in the most appropriate and efficacious way, even when faced by entirely novel pseudo-leaves: they act in 'nearly the same manner as would a man, who had to close a cylindrical tube with different kinds of leaves, petioles, triangles of paper, &c, for they commonly seize such objects by their pointed ends. But with thin objects a certain number are drawn in by their broader ends. They do not act in the same unvarying manner in all cases ...' For instance, they did not drag in leaves 'by their foot-stalks, unless the basal part of the blade is as narrow as the apex, or narrower than it.'[40]

---

[40] Darwin, *Worms*, pp.315–316.

Furthermore, worms did not solve the problems Darwin set by randomly try-try-trying until they accidentally succeeded. Their first grab with their mouth typically hit the place most likely to accomplish burrow-blockage (in 68/89, or 76%, of the trials). So, given that worms' leaf-dragging adjusts appropriately to a host of varied conditions, and does not proceed by trial and error: 'we can hardly escape from the conclusion that worms show some degree of intelligence in their manner of plugging up their burrows.'[41]

Represented by his experiments on earthworms, agency (for Darwin) entailed refined sensory capacities *and* intelligence—understood as the capacities both to respond flexibly (rather than mechanically, 'as if it were an automaton') to changing circumstances, and thus, to overcome a variety of obstacles in achieving an advantageous goal. Such a criterion would render plants intelligent too.

3.2 Agency Implies Interdependence

3.2.1 Every act rebounds (reagency)

Agency requires a world upon which to act. And, as in the laws of physics, the world that is acted upon will oppose to every action a *reaction*.[42] Everything that moves 'rebounds and resonates.'[43] Self-evidently, therefore, agency has effects. (Why else act?) More precisely, agency or purposive movement implies certain *preferred* effects: the intrinsic tie between agency and a given animated being's distinctive physical form setting criteria for what that creature 'should' or needs to do.[44] But not all an act's effects will answer an agent's needs. Accidental, 'knock-on' effects may occur. Mistletoe grows on trees. But, added Darwin, 'if too many of these parasites grow on the same tree, it languishes and dies,' killing the mistletoe.[45]

Because agency entails intelligence, the unintended—or, more precisely, unpurposed—ways in which the world rebounds will feed back to and

[41] Darwin, *Worms*, pp.85–95. See also recent research on intelligence in bees ('Resonances').

[42] For references to 'action and reaction,' see e.g. Darwin, *Origin 1859*, pp.74–75, 350, 408. The reference to 'physics' is to Newton's 3rd law of motion, Isaac Newton, *Newton's Principia: The Mathematical Principles of Newton's Philosophy* (New York: Adee, 1846), p.83: 'To every action there is always opposed an equal reaction: or the mutual actions of two bodies upon each other are always equal, and directed to contrary parts ... If you press a stone with your finger, the finger is also pressed by the stone.'

[43] Jacques Lecoq, *The Moving Body: Teaching Creative Theatre* (London: Methuen, 2002), p.187.

[44] Daniel Nicholson, 'The Return of the Organism as a Fundamental Explanatory Concept in Biology,' *Philosophy Compass* 9, (2014), p.355.

[45] Darwin, *Origin 1876*, p.50. NB one of the knock-on effects of action is, loosely speaking, experience. Darwin's take on psychological events such as 'experiences' was as co-produced *between* people and circumstances (Chs 4–5). Germane discussions of this approach to the psychological field are to be found around p.31 in Alfred Whitehead, *Nature and Life* (Cambridge: Cambridge University Press, 1934); Steven Brown and Paul Stenner, *Psychology without Foundations: History, Philosophy and Psychosocial Theory* (London: Sage, 2009), pp.28ff.

Figure 3.5 Hexagonal cells built by honey-bees.
Source: iStock.com / Druzhinina.

(potentially) *inform* agents in unexpected ways about their changing circumstances. They can then adjust their subsequent actions to the feedback, honing the effectiveness of their manoeuvres. In this way, agency constantly weaves dependencies between a creature and its counterpart conditions of life. Furthermore, when these counterparts are themselves animate—members of the same or of other species—the reflexive effects of agency can accelerate, and their complexities multiply.

Some unpurposed side-effects prove fatal. Trial may solely produce error, and no lucky hit. The worm which tugs the leaf may draw the blackbird's eye. Other side-effects may contribute accidentally to supplying a creature's wants. The hexagonal shape of the cells in bee-hives represents an ideal and mathematically recondite 'perfection in architecture' for the storage of honey, with minimal use of labour and wax (Figure 3.5).[46] Darwin's experiments showed that, provided bees started building equidistant from one another, cells built initially as spheres would change shape into hexagons when a building ran up against its neighbours' cells, each bee compromising on sphericality by flattening adjoining walls—their hexagonal design originating as a lucky accident.[47]

---

[46] Darwin, *Origin 1876*, p.227.
[47] Darwin, *Origin 1876*, pp.220–227.

Similarly, as earthworms colonize their patch over many generations, its soil becomes enriched and deepened to form humus or 'vegetable mould.' Worms 'mingle the whole intimately together,' like a gardener who prepares fine soil for his choicest plants. The more it is mingled, the better soil is fitted to retain moisture and to absorb all soluble substances, as well as for the process of 'nitrification'—which renders it hospitable to nitrogen-needy plants.[48] Worms secrete digestive acids which break down stones and rocks; they pull leaves into their burrows which eventually rot and so act as compost; they ingest and grind down rough particles of earth due to the action of small stones in their muscular gizzards; and they aerate soil by their burrowing, making it more able to retain water, meaning that plants can more easily grow in it. All these actions have a knock-on effect for worms and their descendants, because worms live more easily and numerously in the fine, thick, worm-processed, humus-rich earth than in unwormed, stony ground. Darwin made this point by calculating, from the weight of extruded castings, that more than twice as many worms lived in a square yard of earth at humus-rich locations than in a square yard that was humus-poor.[49]

Regarding more complex side-effects, Darwin liked to refer to the exquisite modifications of vegetable colour, form, and movement that tempt insects to sip at nectar-laden flowers and thus be showered with pollen, meaning that, on their next trip to a similar flower, they cross-fertilize it, gifting the concerned plants an advantage over their self-fertilized neighbours. So, dependence becomes *inter*dependence—a crux for Darwin, both aesthetically and scientifically. And a central theme in the remainder of this book.[50]

3.2.2 Mutual relations

Aged twenty-three, sailor Darwin had found the forests of Brazil sublime . . . a bewitching and bewildering tapestry of interwoven life-forms: 'Twiners entwining twiners. Tresses like hair. Beautiful Lepidoptera . . . If the eye attempts to follow the flight of a gaudy butter-fly, it is arrested by some strange tree or fruit; if watching an insect one forgets in the strange flower it is crawling over,' he exclaimed.[51] Forty

[48] Darwin, *Worms*, p.313.
[49] Darwin, *Worms*, pp.161–173. See also p.316, where Darwin likens earthworms' formation of vegetable mould to the agency of corals, which have 'constructed innumerable reefs and island[s] in the great oceans.' Coral-built reefs and islands underpin the survival of all successor coral generations, because the outcrops, that originally hosted the corals' ancestors, are sinking. Hence new polyps must build on top of the sinking remains of their dead ancestors to keep their heads (almost) above water: Darwin, *Coral Reefs*.
[50] This phenomenon is discussed at length on pp.69–76 in Darwin, *Origin 1876* and in Darwin, *Orchids*; Charles Darwin, *The Effects of Cross and Self Fertilisation in the Vegetable Kingdom*, 2nd ed. (London: John Murray, 1878). *Cross and Self Fertilisation* experimentally proves the advantages in vigour and fertility for plants that breed by cross- versus self-fertilization.
[51] See p.27b [punctuation somewhat normalized] in Charles Darwin, 'Rio Notebook,' (Darwin-Online1832); Charles Darwin, *Diary of the Voyage of H.M.S. Beagle*, ed. Paul Barrett and Richard Freeman, vol. I, The Works of Charles Darwin (London: William Pickering, 1986), p.42. Cf. Ben

years later, the profuse interdependencies of the 'tangled bank, clothed with many plants of many kinds, with birds singing on the bushes, with various insects flitting about, and with worms crawling through the damp earth', remained the focus of his vision, and what he taxed his theory to explain (see 3.3 Deriving Evolution).[52]

Time and again the *Origin* describes the reciprocal relations that grow up between 'the many beings which live around us', relations so 'infinitely complex and close-fitting' as far to exceed human comprehension.[53] In writing thus, Darwin particularly emphasized interdependencies arising through the effects of rebounding actions of organism upon organism.[54] The *Origin's* first illustration cites mistletoe. Later, the book elaborates several scenarios to illustrate 'the ever-increasing circles of complexity' that constitute the mutual relations conditioning the life of any given animated being, circles so finely balanced that 'the merest trifle' may gift it advancement, or spell doom.[55]

Take relations between English clover, bees, mice, and cats. Among insects that ply the meadows, only bumblebees both weigh enough to open the clover-plant's nectaries, and have tongues long enough to sip their syrup. Bumblebees breed in underground nests, often built in disused mouse-tunnels. If mice find these nests, they eat them. Cats destroy mice. So, says Darwin, allow cats to roam in green fields, and the fields will purple with clover, proving how, even 'plants and animals, remote in the scale of nature, are bound together by a web of complex relations.'[56]

How much more intimately, then, might the rebounding effects of the actions and reactions of neighbours closest in the scale of nature bind them together, that is, of conspecifics? Here we meet a crucial move in Darwin's take on the natural world—particularly for psychology: his splitting off of *social* organisms from all others. Many organisms fare well in isolation, or in family groups. But social plants (e.g. certain grasses; thistles), like social animals, thrive only in groups, even on the edge of their range, as with trees in the Arctic (see Ch. 7).[57]

3.2.3 What is an individual?

Darwin's apprenticeship as a naturalist plunged him into debates about *individuality*. Both his teenage interest in the autonomous movements of granular matter

Bradley, 'Darwin's Sublime: The Contest between Reason and Imagination in *on the Origin of Species*', *Journal of the History of Biology* 44, (2011).

[52] Darwin, *Origin 1876*, p.429.
[53] Darwin, *Origin 1876*, pp.4, 49, 427. Interdependencies between organism and organism were, for Darwin, the main driver of evolution: 'the most important of all causes of organic change' (see 3.3.2 Struggles for existence).
[54] Dov Ospovat, *The Development of Darwin's Theory: Natural History, Natural Theology, and Natural Selection, 1838–1859* (Cambridge: Cambridge University Press, 1981), pp.205–207.
[55] Darwin, *Origin 1876*, pp.2, 57.
[56] Darwin, *Origin 1876*, p.57.
[57] Charles Darwin, *Charles Darwin's Natural Selection: Being the Second Part of His Big Species Book Written from 1856 to 1858* (Cambridge: Cambridge University Press, 1975), p.203.

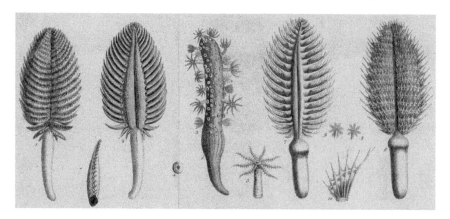

Figure 3.6 Various sea pens, pictured in 1763.

Source: Ellis, John. LIII. An account of the sea pen, or *Pennatula phosphorea* of Linnaeus; likewise a description of a new species of sea pen, found on the coast of South-Carolina, with observations on sea-pens in general. In a letter to the Honourable Coote Molesworth, Esq; M. D. and F. R. S. from John Ellis, Esq; F. R. S. and Member of the Royal Academy at Upsal. *Phil. Trans. R. Soc.*, 53: 419–35. Scanned images © 2017, Royal Society. http://doi.org/10.1098/rstl.1763.0054.

Image credit: From Susannah Gibson, 'On Being an Animal, or, the Eighteenth Century Zoophyte Controversy in Britain,' *History of Science* 50, (2012) p.471.

and his first scientific presentation on reproduction in the colonial invertebrate *Flustra* speak to this issue. Across Europe in the 1830s, established men of science were disputing the same phenomena as Darwin, particularly: how colonial invertebrates bred; the division of labour in social insects like bees; grafting and asexual reproduction in plants; the then-new theory of cells; cutting a single flat-worm in half, which 'healed' to become *two* viable worms; and alternating and metamorphic life-cycles, as found in marine life (e.g. sea pens, barnacles) and insects (e.g. caterpillar–chrysalis–butterfly).[58]

Enthusiastic observations on all these topics stud Darwin's notes and publications—especially those made aboard *HMS Beagle*. For example, on 17th October 1832, exploring nearby ocean-shallows after anchoring in Bahia Blanca (Argentina), Darwin spotted a submerged colony of sea pens, *Virgularia patagonica*, two feet tall. Under water, they looked like big quills with their nibs stuck in the mud, occasionally giving out coordinated flashes of light (Figure 3.6). Around the stem on each sea pen's plume wound numerous folds carrying tiny polyps, 'many thousands' to a plume (Figure 3.7). The folds reached out from the

[58] For Darwin's own flat-worm experiments, see pp.312ff (February–March 1835, Hobart, Australia), Charles Darwin, 'Zoology Notes,' in *Charles Darwin's Zoology Notes & Specimen Lists from H.M.S. Beagle*, ed. Richard Keynes (Cambridge: Cambridge University Press, 1832–1836). For the scientific context of Darwin's writings, see Lynn Nyhart and Scott Lidgard, 'Individuals at the Center of Biology: Rudolf Leuckart's 'Polymorphismus Der Individuen' and the Ongoing Narrative of Parts and Wholes. With an Annotated Translation,' *Journal of the History of Biology* 44, (2011), pp.373–443.

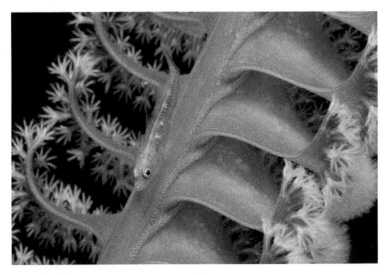

Figure 3.7 Magnification of a sea pen (*Virgularia*), showing the individual polyps along its branches (and a visiting fish).
Source: Daniela Dirscherl / Getty Images.

shaft into polyp-laden 'arms ... like spokes of a wheel.' [59] Dug up, and carried back to his cabin, Darwin's microscope showed that, as in coral, every polyp looked like a separate animal: each, 'though closely united to his brethren,' had a distinct mouth, body, and tentacles. 'Yet,' said Darwin, 'we see they act by one movement; they have also one central axis connected with a system of obscure circulation, and the ova are produced in an organ distinct from the separate individuals.' Not only were each pen's flashes of light synchronized across the whole organism, but a single injury to its stem would cause all the polyps to collapse. 'Well may one be allowed to ask,' Darwin concluded, '*What is an individual?*' [60]

While on the *Beagle*, Darwin also took note of a Chilean apple tree which showed a proclivity for asexual breeding. Its lower branches had 'conical, brown, wrinkled points' that were prone to dive into the earth and become roots, as soon as mud was splashed against the tree—producing new trees, or, were they extensions of the old tree? Reflecting on this and like cases, Darwin commented that it was 'very doubtful whether the flowers and leaf-buds, annually produced from the same bulb, root, or tree, can properly be considered as parts of the same individual, though in some respects they certainly seem to be so.' He later reported that he

---

[59] Darwin, 'Zoology Notes,' p.401.
[60] Darwin, *Voyage*, pp.98–99 (my italics).

enjoyed looking at the trees in Kensington Gardens as 'great compound animals'—a phrase he would subsequently apply to corals and hydra.[61]

So: *do independent, self-contained individuals make up the biological universe?* Darwin's challenge to this idea never flagged. He ultimately concluded that humans were united with all vertebrates in having descended from the larvae of colonial invertebrates (sea squirts or 'Ascidians').[62] Accordingly, he grounded his understanding of moral life—for him, the highest feature of humanity—in the psycho-social mergence of individual with group. As we will see, this grounding has profound consequences, not just for his psychology, but for its framing of moral sense and action (Chs 4–7).

3.3 Deriving Evolution

3.3.1 Law versus causal mechanism

How did Darwin derive from the theatre of agency the second axis of his vision of the natural world, the modification of species?[63] Comparative anatomy, the nesting of smaller taxonomic groups within larger, the geographical distribution of flora and fauna, and geological series of fossils—all furnished evidence for *the fact* of 'change of species by descent.' But, he also sought details of phenomena that might illuminate *how* such change had come about, including: natural selection; correlated growth; sexual selection; use and disuse of parts; and direct action of the physical conditions of life. Today, *evolution by natural selection* is often counted as Darwin's signal contribution to knowledge. Yet the conflation of natural selection with evolution is one Darwin warned against. *How* evolution came about was a subsidiary problem—signifying 'extremely little'—compared to proving 'that *species have descended from other species* and have not been created immutable.'[64]

Darwin did not invent the idea of evolution.[65] So, without his formulation of natural selection, *On the Origin of Species* might have drawn little scientific notice on its

---

[61] Jonathon Hodge, 'Darwin as a Lifelong Generation Theorist,' in *The Darwinian Heritage*, ed. David Kohn (Princeton, NJ: Princeton University Press, 1985), pp.209, 236 & passim; Charles Darwin, 'Notebook M,' (Cambridge: Darwin Online, 1838), p.41; Charles Darwin, *The Variation of Animals and Plants under Domestication*, 2nd ed., 2 vols., vol. 1 (London: John Murray, 1875), pp.397–398.

[62] Darwin, *Descent*, pp.159–160.

[63] Darwin mainly used the terms 'the modification of species,' 'community of descent,' and, especially in his letters, 'common descent'—to refer to what we call 'evolution.'

[64] Charles Darwin, 'Origin of Species,' *Athenaeum* 1854, (1863), my italics. Darwin also stressed this point in Charles Darwin, 'Letter to Gray, 11th May,' *Correspondence of Charles Darwin* (Cambridge: Darwin Correspondence Project, 1863), www.darwinproject.ac.uk/letter/?docId=letters/DCP-LETT-4153.

[65] Peter Bowler, *Evolution: The History of an Idea*, 2nd ed. (Berkeley: University of California Press, 1989).

debut in 1859. Nowadays, we often hear natural selection framed as a cause, process, or mechanism. *Origin*'s rich metaphorical language did indeed sometimes cast natural selection as an all-powerful agency, incessantly working to improve the living world—as Darwin soon came to regret.[66] But *the argument* of the book framed things differently, making natural selection *a law* that resulted from what it called the struggle for life, *not a causal mechanism or process*. After *Origin* came out, Darwin repeatedly rebutted critics who argued that his book had failed because it had not shown natural selection to be a *true cause* ('vera causa') of species change.[67] *Origin* called natural selection a *general law*, 'leading to the advancement of all organic beings,—namely, multiply, vary, let the strongest live and the weakest die.'[68] By *law*, Darwin meant merely, a 'sequence of events as ascertained by us.'[69] In this, natural selection was like Newton's law of gravity, he said.[70] The law of gravity gave a new coherence to various observable sequences of events, such as falling apples, tidal flows, and orbiting planets. But the law of gravity did *not* specify what mechanisms caused these events—a conundrum which taxes physicists to this day.[71] Likewise, certain processes *caused* the sequences of events brought together under the law of natural selection. In this sense, natural selection did not refer to a causal power.[72] It described effects.[73]

---

[66] Leading to changes in later editions. Cf. Charles Darwin, 'Letter to Hooker, 29th March,' *The Correspondence of Charles Darwin* (Cambridge: Darwin Correspondence Project, 1863); Darwin, *Origin 1859*, p.84; Darwin, *Origin 1876*, p.65; Gillian Beer, *Darwin's Plots: Evolutionary Narrative in Darwin, George Eliot and Nineteenth-Century Fiction*, 2nd ed. (Cambridge: Cambridge University Press, 2000); Robert Young, *Darwin's Metaphor: Nature's Place in Victorian Culture* (Cambridge: Cambridge University Press, 1985).

[67] E.g. of Huxley, Darwin told Hooker: 'he rates higher than I do the necessity of Natural Selection being shown to be a vera causa always in action.' Charles Darwin, 'Letter to Hooker, 14th Feb,' *The Correspondence of Charles Darwin* (Cambridge: Darwin Correspondence Project, 1860), www.darwinproject.ac.uk/letter/?docId=letters/DCP-LETT-2696.

[68] Darwin, *Origin 1876*, p.234.

[69] Darwin, *Origin 1876*, p.63.

[70] Darwin, *Origin 1876*, p.421; Charles Darwin, 'Letter to Gray, February 24th,' *Correspondence of Charles Darwin* (Cambridge: Darwin Correspondence Project, 1860), https://www.darwinproject.ac.uk/letter/?docId=letters/DCP-LETT-2713; Charles Darwin, 'Letter to Lyell, February 23rd,' *Correspondence of Charles Darwin* (Cambridge: Darwin Correspondence Project, 1860), https://www.darwinproject.ac.uk/letter/?docId=letters/DCP-LETT-2707.

[71] The best approximation to its solution was worked out by Albert Einstein, but there are many competing post-Einstein theories.

[72] Even historians have claimed Darwin designed *Origin* to show that natural selection qualifies as a *vera causa*: most notably, Jonathon Hodge, 'The Structure and Strategy of Darwin's 'Long Argument',' *British Journal for the History of Science* 10, (1977), p.242. Hodge ignores *Origin*'s explicit framing of natural selection as a law—and Darwin's rejection of natural selection as a *vera causa* by reference to gravity—both in *Origin* and in several letters. Hence Hodge is forced to argue both that Darwin's book was structured to prove natural selection a *vera causa, and* that *Origin* 'violates' that structure. The 'three general evidential considerations' Darwin *should* have observed when establishing a *vera causa*, 'do not map onto the *Origin*'s three clusters of chapters,' writes Hodge—an observation which would make better sense if Darwin were not following the *vera causa* principle in the first place: Hodge, 'Structure and Strategy,' p.244; Jonathon Hodge, 'Darwin's Book: *On the Origin of Species*,' *Science & Education* 22, (2013), p.2274.

[73] Darwin, *Origin 1859*, p.433: '... natural selection, *which results from the struggle for existence ...*' (my italics). 'Strictly speaking, Natural Selection is not a cause at all, but is the mode of operation of a certain quite limited class of causes': Chauncey Wright, 'Review [of *Contributions to the Theory of*

The first sentence of *Origin*'s fourth chapter, on 'Natural Selection,' names the two main processes which, according to Darwin, produce the phenomena described by the law of natural selection: 'How will *the struggle for existence*, briefly discussed in the last [*Origin*'s third] chapter, act in regard to *variation* [discussed in *Origin*'s first and second chapters]?' *Origin* then reminds us: 'how infinitely complex and close-fitting are the mutual relations of all organic beings to each other and to their physical conditions of life.' Grinding exceeding slow and exceeding small, this mill of mutual relations—the often-agentic struggle for existence—would not only induce variation, but winnow better variations from the worse, making the better adapted more apt to survive and reproduce than the worse. The result?—the '*preservation of favourable variations and rejection of injurious variations*,' which, Darwin wrote, 'I call Natural Selection.'[74]

This distinction between law and causal mechanism is crucial to the reappraisal of the theory of evolution—and of Darwin's writings—currently under way in biology, and carried forward by *Darwin's Psychology*.[75]

3.3.2 The struggle for existence

All previous evolutionary theories had most glaringly failed, Darwin believed, because they had focused on explaining change over time in *individual* species, viewed in isolation, not 'the numerous and beautiful co-adaptations which we see throughout nature.'[76] His theory would derive such *mutual* transformation directly from his vision of the economy of nature as an interdependent, co-adapted whole, bound together by a web of complex relations. It is this vision which underpins his concept of the struggle for existence—and the achievement which his book's last paragraph resoundingly proclaims:

> It is interesting to contemplate a tangled bank, clothed with many plants of many kinds, with birds singing on the bushes, with various insects flitting about, and with worms crawling through the damp earth, and to reflect that these elaborately constructed forms, so different from each other, and dependent on each other in

---

Natural Selection. A Series of Essays by Alfred Russell Wallace],' *North American Review* 111, (1870), p.293. Cf. Denis Walsh, 'Two Neo-Darwinisms,' *History and Philosophy of the Life Sciences* 32, (2010), pp.317–340.

[74] Darwin, *Origin 1859*, pp.80–81, my italics.
[75] E.g. Denis Walsh, 'The Struggle for Life and the Conditions of Existence: Two Interpretations of Darwinian Evolution,' in *Evolution 2.0: Implications of Darwinism in Philosophy and the Social and Natural Sciences*, ed. Martin Brinkworth and Friedel Weinert (Berlin: Springer-Verlag, 2012), pp.191–208.
[76] Darwin, *Origin 1876*, p.xvii.

so complex a manner, have all been produced by laws acting around us. These laws, taken in the largest sense, being Growth [individual development] with Reproduction; Inheritance which is almost implied by reproduction; Variability from the indirect and direct action of the external conditions of life, and from use and disuse: a Ratio of [populational] Increase so high as to lead to a Struggle for Life, and *as a consequence to Natural Selection*, entailing Divergence of Character and the Extinction of less-improved forms.[77]

'Struggle for life' was used synonymously in *Origin* with *struggle for existence*, a phrase which dated from before Darwin, being most notably used by Thomas Malthus (1766–1834) and Darwin's geological mentor, Charles Lyell (1797–1875).[78] Malthus used it to invoke gladiatorial combat—induced by invasion or population pressure—a view still popular today.[79] *Origin's* chapter on natural selection duly dwells on the geometrical rate of increase of animals and plants where no destruction befalls their progeny, such that they would 'soon' cover the earth, if unchecked.[80] However, while Darwin did move on to illustrate natural checks to unbridled increase, he warned that he only used the term struggle for existence in 'a large, and metaphorical sense.' This included, not just competition, but relative reproductive success and, the 'dependence of one being on another.'[81]

Origin's chapter on the struggle for existence labours long over this last point. It first points to the woodpecker, with its feet, tail, beak, and tongue, 'so admirably adapted to catch insects under the bark of trees.'[82] Then to the parasitic mistletoe, which depends on: its host tree for support and sustenance; on attracting insects (like the mistletoe moth) to cross-pollinate its flowers; and birds (like the mistletoebird and the mistle thrush), to eat its berries, and so disperse its seeds (Figure 3.8). All of which made it 'preposterous' to account for the structure of mistletoe,

---

[77] Darwin, *Origin 1876*, p.429, my italics—which highlight the logical dependency of Natural Selection on the Struggle for Life.

[78] Thomas Malthus, *An Essay on the Principle of Population, or a View of Its Past and Present Effects on Human Happiness; with an Inquiry into Our Prospects Respecting the Future Removal or Mitigation of the Evils Which It Occasions*, 6 ed. (London: Murray, 1826): '... the frequent contests [of European conquistadors] with tribes in the same circumstances with themselves, would be so many struggles for existence, and would be fought with a desperate courage, inspired by the reflection, that death would be the punishment of defeat, and life the prize of victory,' Book I, 6; Charles Lyell, *Principles of Geology, or the Modern Changes of the Earth and Its Inhabitants Considered as Illustrative of Geology*, 9th ed. (New York: Appleton, 1854), pp.605, 688, 700, 796; Alistair Sponsel, *Darwin's Evolving Identity: Adventure, Ambition, and the Sin of Speculation* (Chicago: University of Chicago Press, 2018)

[79] Conway Zirkle, 'Natural Selection before the *Origin of Species*,' *Proceedings of the American Philosophical Society* 84, (1941), p.74; Richard Dawkins, *The Selfish Gene: 30th Anniversary Edition* (Oxford: Oxford University Press, 2006), p.2. Witness too David Attenborough's fatalistic commentary on a pod of orcas relentlessly hounding the calf of a gray whale in the opening episode of *Blue Planet*: Alastair Fothergill, 'Ocean World,' in *The Blue Planet* (London: BBC, 2001).

[80] Darwin, *Origin 1859*, p.64.

[81] Darwin, *Origin 1876*, p.50, my italics.

[82] Darwin, *Origin 1876*, p.2.

Figure 3.8 From Australia: (a) Mistletoe moth (*Comocrus behri*); (b) Mistletoe bird (*Dicaeum hirundinaceum*).

Source: (a) Denis Crawford / Alamy Stock Photo; (b) Pete Evans / Shutterstock.com.

with its relations to 'several distinct organic beings,' by the effects of 'external conditions, or of habit, or of the volition of the plant itself.'[83]

Origin's longest examples are ecological. In the lordly estates of a rich uncle, Darwin had compared the flora and fauna of a barren heath of several hundred acres, part of which had been fenced off twenty-five years previously, and planted with Scotch fir. Twelve new species of plant now inhabited the plantation—not counting sedges and grasses. Insect-life abounded among the trees, sustaining six species of insectivorous birds lacking in the unfenced land—though the heath had two or three species of bug-eating bird not found among the firs. This observation was given to underline the potent knock-on effects for many different kinds of creature of 'the introduction of a single tree, nothing whatever else having been done, with the exception of the land having been enclosed, so that cattle could not enter.'[84]

Like examples follow. All tell of extensive knock-on effects for numerous apparently unrelated species from the removal or addition of a single species to a given habitat. All indicate how *sympatric* species—those inhabiting the same locale—'which for long ages have there struggled together, [will] have become mutually co-adapted.'[85]

[83] Darwin, *Origin 1876*, p.2.
[84] Darwin, *Origin 1876*, p.56. Obviously he meant, a single *type of* tree. NB too his 'one exception': the fencing of the firred area, 'so that cattle could not enter'—a not inconsiderable exception!
[85] Darwin, *Origin 1876*, pp.348–349.

Species that struggle together, in Darwin's metaphorical sense, are not necessarily warring all against all. Their struggles interweave. So different individuals will exploit any advantage offered by fellow beings, and with all the more effect should their counterparts also benefit from the arrangement.[86] *Origin* gives the example of social plants, instancing wheat, corn, and canola (charlock). Here large stocks of individuals from the same species—relative to the numbers of its 'enemies'—must stand together for any to survive.[87] Even parasitic mistletoe may benefit its host— as the birds attracted to the mistletoe improve distribution of the seeds of the host tree, not just the mistletoe's own seeds, lending tracts of land that support mistletoe far greater biodiversity than like tracts that lack the kissing bough.[88]

Finally, Darwin drives home his point, asserting: 'the structure of every organic being is related, in the most essential yet often hidden manner, to that of all other organic beings, with which it comes into competition for food or residence, or from which it has to escape, or on which it preys.' He instances: the structure of 'the teeth and talons of the tiger,' alongside 'the legs and claws of the parasite which clings to the hair on the tiger's body'; the 'beautifully plumed seed' of the dandelion; and 'the flattened and fringed legs of the water-beetle.' At first sight, dandelion seed and the water-beetle legs might seem to be solely adapted to drifting on air or swimming underwater. Yet the advantage of plumed seeds also derives from the land 'being already thickly clothed by other plants; so that the seeds may be widely distributed and fall on unoccupied ground.' And the legs of water-beetles allow them to compete 'with other aquatic insects, to hunt for its own prey, and to escape serving as prey to other animals.'[89]

3.3.3 Variability and divergence

Taxonomic trees branch (Figure 3.9). But why do they branch? For almost twenty years, Darwin would have answered in terms of organisms' reactions to intermittent changes in their external conditions: climate change; water-supply; accidents of transport; geographical isolation; availability of foodstuffs; the arrival or extinction of allies or competitors. But in the 1850s he came up with a more proactive theory of evolutionary divergence. This explained how a single widely distributed species would itself agentically diversify into many species, *without any change in*

[86] E.g. Tim Clutton-Brock, 'Co-Operation between Non-Kin in Animal Societies,' *Nature* 462, (2009), pp.51–57.

[87] Darwin, *Origin 1859*, p.70.

[88] Making mistletoe a 'keystone species': David Watson, 'Mistletoe: A Keystone Resource in Forests and Woodlands Worldwide,' *Annual Review of Ecology and Systematics* 32, (2001), pp.219–249; Ron Van Ommeren and Thomas Whitham, 'Changes in Interactions between Juniper and Mistletoe Mediated by Shared Avian Frugivores: Parasitism to Potential Mutualism,' *Oecologia* 130, (2002), pp.281–288.

[89] Darwin, *Origin 1859*, p.77.

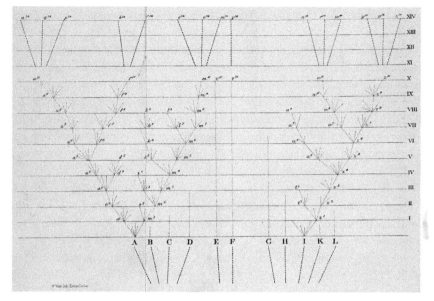

Figure 3.9 'Let A to L represent the species of a[n imaginary] genus large in its own country,' Darwin wrote, explaining *Origin*'s only drawing. Each horizontal line represented one thousand generations, or 'better ... ten thousand generations.'

Source: Darwin, C. (1859). *On the Origin of Species by Means of Natural Selection or the Preservation of Favoured Races in the Struggle for Life.*

Image credit: Wellcome Collection. Distributed under the terms of the Creative Commons Attribution 4.0 International (CC BY 4.0). https://creativecommons.org/licenses/by/4.0/#_ga=2.241103562.87728 4273.1565857924-514625952.1510241134.

Darwin, *Origin 1859*, pp.116–117.

external conditions. These newly formed species would, in due course, cohere into higher taxonomic categories (first, genera, then families, and so on).[90] It was partly through the specialization wrought by such diversification that *Origin* explained the evolution of the human brain.

Darwin reasoned as follows. Through arithmetic study of the potency of various genera of plants, he had found that large genera (having more species) were more 'potent' than small, in that they supported more individuals per unit of space, and had more varieties (species and sub-species) per genus.[91] Darwin construed this finding in terms of a division of labour. More forms of life could be supported in any one area if they all performed more specialized functions in the overall 'drama of existence.' Such division of labour allowed many more varieties of creature, and many more individual organisms, to live together—provided individuals of each

---

[90] Ospovat, *Darwin's Theory*, pp.170ff.

[91] Janet Browne, 'Darwin's Botanical Arithmetic and the 'Principle of Divergence', 1854–1858,' *Journal of the History of Biology* 13, (1980), pp.63, 86.

new variety deviated enough from individuals of its parent species, to permit their coexistence.[92]

Inspired by this botanical arithmetic—and also by the idea that increasing division of labour lends greater efficiency—Darwin formulated a 'principle of divergence': 'the more diversified the descendants from any one species become in structure, constitution, and habits, by so much will they be better enabled to seize on many and widely diversified places in the polity of nature, and so be enabled to increase in numbers.'[93] Speciation resulted from organisms *constantly and actively* striving to find new and better ways of living, as compared to their competitors. To vary here meant, principally, to improvise new habits.

Origin illustrates this point by imagining a population of four-legged carnivores which has attained the maximum size supportable by their habitat. In stable conditions, this population could only grow further by its descendants 'seizing on places at present occupied by other animals': finding ways to feed on new kinds of prey; or 'inhabiting new stations, climbing trees, frequenting water, and some perhaps becoming less carnivorous. *The more diversified in habits and structure the descendants of our carnivorous animals become, the more places they will be enabled to occupy.*'[94] Such evolutionary divergence depended on habits in several ways: as bars to interbreeding; as means of colonizing new places currently unfilled in the economy of nature; and/or as means of more effectively filling occupied places than did their current inhabitants.[95]

Divergence endowed evolution with *progressiveness* too. More potent genera would reach 'a higher stage of perfection' than the less potent, because they would have had actively to out-compete more varieties of cousin.[96] This drew in Darwin's ideas about division of labour. New places were opened up in a given habitat by means of increased specialization, of physiological organization, and *of action*: 'The enormous *number* of animals in the world depends [on] their varied structure & complexity. – hence as the forms become complicated, they opened fresh means of adding to their complexity.'[97] While Darwin proposed no law of *inevitable* evolutionary progress,[98] his principle of divergence meant that, as species diverged, they would become more specialized and more complex in what they did. Among

---

[92] Browne, 'Darwin's Botanical Arithmetic,' p.69. NB Browne's phrase 'the drama of existence' is genuinely Darwinian. Darwin likened the incompleteness of the fossil record to 'an occasional scene, taken almost at hazard, in an ever slowly changing drama': Darwin, *Origin 1876*, p.292.

[93] Darwin, *Origin 1876*, p.87.

[94] Darwin, *Origin 1876*, pp.87–88, my italics.

[95] Darwin, *Natural Selection*, p.582.

[96] Darwin, *Origin 1876*, p.340.

[97] Charles Darwin, "E," (1838–1839), p.95.

[98] Sometimes a species might achieve success by becoming simpler, as in the case of *Lernaea*—a parasitic freshwater Anchor Worm—that had become reduced to 'a mere reproductive sack': Darwin, 1845, quoted in Ospovat, *Darwin's Theory*, p.217. Darwin was contradicting Lamarck here, who, Darwin thought, had contended that there was a universal 'will to progress' in nature. NB Characteristically,

other things, this would have resulted in 'advancement of the brain for intellectual purposes.'[99]

3.3.4 Agency directs selection

According to the principle of divergence, new habits lead to changes in structure (cf. the scissor-beak; see 2.2 The Economy of Nature). How come? Because change of habit, if repeated over generations, was likely to produce an inherited effect.[100] Which meant habits could pilot variation in evolution. Today, this sounds like Lamarckism, because the most popular and widespread take on evolution in the twentieth century made it a process driven by *random changes at the molecular level* of genes.[101] But *Origin* dismisses the idea of organisms varying 'due to chance' as 'wholly incorrect', the word 'chance' serving solely as a coverall for our failure to comprehend the enormous complexity of laws and causes that govern organic variation and other natural phenomena.[102] Instead, Darwin argued that habit-led evolution was consistent with natural selection—an argument which today's biologists nowadays file under 'the Baldwin effect,'[103] but *Origin* discusses under *transitional habits*.

Cue the flying squirrel. *Origin* tempts us to imagine that a squirrel has gained the habit of launching itself, not just from branch to branch, but from tree-tops. This would create a new selection pressure that would advantage any squirrels aided by an anatomical or physiological character that helped them jump greater distances (stronger spring at take-off, better depth vision, lighter body-weight, more aerodynamic tail, broader flanges of skin between front and back legs). Tree-surfing would put a new premium on glide-friendly changes to the squirrel's physique, such that any 'chance' heritable variation that fitted it better to its novel habit would likely increase its reproductive success compared to its unchanged conspecifics.

Darwin sometimes more glibly stated his law of natural selection in a way that contradicts this point, e.g. as 'leading to the advancement of all organic beings': Darwin, *Origin 1876*, p.234.

[99] Darwin, *Origin 1876*, p.98.

[100] Darwin, *Origin 1876*, p.8. NB Darwin here states that the *effects* of the habit are inherited, not the habit itself. Elsewhere, he sometimes wrote as if the habit itself were inherited (see 4.2 A Drama of Surfaces).

[101] Richard Dawkins, *The Blind Watchmaker: Why the Evidence of Evolution Reveals a Universe without Design* (Harlow: Longman, 1986), p.290.

[102] Darwin, *Origin 1876*, pp.58, 106. Cf. the extended discussion of the illusions of 'chance' (which Darwin likens to the illusions of 'free will') in Darwin, 'Notebook M,' pp.26–27, 30–31, 72–74, etc. NB, random genetic variation figures in phenotype-led biology, because it helps *to stabilize* non-random phenotypic changes. The problem with the gene-led vision of evolution is that it side-lines the non-random phenotypic production of variations. E.g. Dawkins, *Selfish Gene*, pp.31, 248.

[103] Mary Jane West-Eberhard, *Developmental Plasticity and Evolution* (New York: Oxford University Press, 2003), pp.24–25, 151–153.; Jacy Young, 'The Baldwin Effect and the Persistent Problem of Preformation Versus Epigenesis,' *New Ideas in Psychology* 31, (2013), pp.355–362.

Hence, 'it would be easy for natural selection to fit the animal, by some modification of its structure, for its changed habits.'[104] Thus, while the production of what we now call genetic variations—which stabilized the bodily changes that make tree-surfing easier for squirrels—might seem random, the *direction* of adaptation would conform to the *non-random* agentic innovations of the flying squirrel.

In this way, while admitting difficulty in explaining how any specific variation has come about, *Origin* undertakes to show that developmental processes not only express but can *produce* heritable variation. Habit and 'use and disuse'—which amount to the same thing[105]—figure prominently here, though habit shades into *development*. The atrophy of wings in flightless birds, like penguins or ostriches, has resulted from behavioural change, but was presumably effected by alterations to the developmental processes that produce wings. Likewise, the blindness of moles bespeaks their burrowing, and enlarged wings in flower-feeding beetles and butterflies must have followed a choice of diet. Indeed, observations showing that 'changed habits produce an inherited effect' occur throughout *Origin*.[106] And this was because the organization of living things was '*plastic*,' said Darwin, or 'readily capable of change.'[107]

Plasticity affected plants' and animals' reproductive systems in particular, reproduction proving to be the generative process most vulnerable to creating variants in consequence of direct impact from the environment.[108] Viewed in this way, what Darwin called reproduction merged with development. For example, as we now know, the embryos in unhatched turtle eggs change sex from male to female when the ambient temperature rises by one or two degrees.[109] Such temperature change has no effect on hatched turtles. Likewise, radiation has far greater effects on the form of very young mammalian embryos than it does later in pregnancy, or postnatally.[110]

Changes occurring earlier in the production of offspring were the more likely to have knock-on effects, because of the imperative to maintain the anatomical and functional integrity of the whole animal. Darwin called such collateral effects *correlation of growth* or *correlated variation*: 'the whole organisation [of the organism] is so tied together during its growth and development, that when slight variations in any one part occur ... other parts become modified.'[111] A favourite example was

---

[104] Darwin, *Origin 1876*, p.141.

[105] Darwin, *Origin 1876*, p.xiv.

[106] Darwin, *Origin 1876*, pp.8, 138–143.

[107] Darwin, *Origin 1876*, pp.112–115, 438.

[108] Darwin, *Descent*, pp.189–190. NB I use the term 'generative' here in the sense developed by Hodge, 'Generation Theorist.'

[109] James Bull, 'Sex Determination in Reptiles,' *Quarterly Review of Biology* 55, (1980), pp.3–21.

[110] Marco De Santis et al., 'Radiation Effects on Development,' *Birth Defects Research (Part C)* 81, (2007), pp.177–182.

[111] Darwin, *Origin 1876*, p.114.

the relation in cats between complete whiteness of fur, blue eyes, and deafness.[112] Another instance was correlation of hair colour and hair texture in some Native American tribes.[113]

3.3.5 Transmission and development

Given all this, how would Darwin have countered the modernist bromide that he was 'ignorant' of the origins of organic variability—because he did not know that 'gene mutations and chromosome changes are the source of variations'?[114] *Origin* is characteristically humble about what its author knew. Yet its first five chapters all deal with the topic of variation, also called *individual differences* (a phrase taken up with alacrity by psychological pioneers like Francis Galton, 1822–1911, the founder of *differential psychology*). In 1868, these five chapters were capped by a two-volume behemoth: *The Variation of Animals and Plants under Domestication*.[115]

With growing criticism of the gene-centric view of evolution known as the Modern Synthesis, Darwin's extensive studies of the sources of evolutionary variation have regained the attention of twenty-first century biologists.[116] Regarding inheritance, it is true that Darwin's understanding of the transmission of heritable characters has largely been superseded by twentieth-century genetics. But a great deal of what he wrote about inheritance has to do with individuals' development of specific characters—a topic of rising importance in today's biology.[117] Darwin stressed that two distinct processes contribute to inheritance—the *transmission*, and the *development* of characters; 'but as these generally go together, the distinction is often overlooked.'[118] To make this point, he instanced characters which are transmitted to all members of a species, but develop only at maturity, or during old age, or in one sex—like breasts and beards, 'which are transmitted through both sexes, though developed in one alone.'[119] As we shall see, Darwin's recognition of the place of development in the production of evolutionarily significant variation

[112] Darwin, *Origin 1876*, p.115.
[113] Darwin, *Descent*, p.197.
[114] E.g. Theodosius Dobzhansky, *Genetics and the Origin of Species*, 3rd ed. (New York: Columbia University Press, 1951), pp.16, 50, 21; Brian Charlesworth and Deborah Charlesworth, 'Darwin and Genetics,' *Genetics* 183, (2009), pp.757–766.
[115] Darwin, *Variation*, Vol.1, 1; Charles Darwin, *The Variation of Animals and Plants under Domestication*, 2 vols., vol. 2 (London: John Murray, 1875).
[116] Mary Jane West-Eberhard, 'Toward a Modern Revival of Darwin's Theory of Evolutionary Novelty,' *Philosophy of Science* 75, (2008), pp.899–908.
[117] West-Eberhard, *Developmental Plasticity and Evolution*, pp.188ff; Susan Oyama, Griffiths, Paul and Gray, Russell, *Cycles of Contingency: Developmental Systems and Evolution* (Cambridge, MA: MIT Press, 2001); Gerd Muller, 'Evo–Devo: Extending the Evolutionary Synthesis,' *Nature Reviews: Genetics* 8, (2007), pp.943–949.
[118] Darwin, *Descent*, p.227.
[119] Darwin, *Descent*, pp.227–228.

was also critical to his understanding of the place of cultural differences in evolution (cf. Chs 8–9).

3.4   Resonances

Origin names the two engines which drive natural selection as *variation* and *the struggle for life*. Both derive from the theatre of agency. Hence the early chapters of *Origin* reveal how the first, quick-tempo axis of Darwin's vision of the natural world spawns a second slow-tempo axis: the modification of species. These two axes have evoked two types of resonance (and discord) in successor literature: one concerning agency, and its link to mentality; the other concerning community of descent.

3.4.1   Living as agency

Darwin's early research on zoophytes and other simple organisms led him to distinguish laws governing life from the laws governing inanimate matter. As he confided to a notebook a year or two after he disembarked from the *Beagle*: living matter was governed by certain laws 'different from those that govern in the inorganic world.' Life itself was: 'the capability of such matter obeying a certain & peculiar system of movements, *different from inorganic movements*.'[120] Everything moves, but the living move in 'a certain and peculiar' way.

   Darwin's distinction of animate movement—or agency—from inanimate movement, proved crucial to the early uptake of his work in psychology, especially by William James (1842–1910). James was steeped in Darwin's work, all the way from the 'Metaphysical Club' that he frequented in his twenties—which included one of Darwin's most esteemed American allies, Chauncey Wright (1830–1875)—through James's careful reading of Darwin's books (including *Expression* and *Descent*), his interests in physiology, and his long friendship with the prominent French evolutionist Henri Bergson.[121] James had personal links to Darwin's family through friends and relatives, being acquainted with Darwin's eldest son William (1839–1914). Both his first two publications (reviews of Huxley and Wallace), and his first two psychological essays, debated Darwin's bearing on psychology. In fact, William Darwin, after hearing James lecture 'Are we automata?' at Harvard in

---

[120]   Charles Darwin, 'Old & Useless Notes,' (Cambridge: Darwin Online, 1838–1840), p.34, my italics. Cf. Sloan, 'Vital Matter.'

[121]   Phillip Wiener, 'Peirce's Metaphysical Club and the Genesis of Pragmatism,' *Journal of the History of Ideas* 7, (1946), pp.218–233.; Henri Bergson, *Creative Evolution* (New York: Henry Holt, 1911).

1877, asked James if he could copy the lecture to send to his father, to help Darwin senior 'confound Huxley.'[122]

The harmonies between Darwin and James were twofold. The first emerged strongly in James's essays, 'Brute and Human Intellect,' and 'Are We Automata?,' which refuted the reductive physicalism of Spencer and Huxley, by differentiating life from matter. This echoed Darwin. Life differed from matter and so escaped mechanical laws. And, in a second echo, Darwin's principle of continuity in the living world ('*natura non facit saltum*': nature does not make jumps)[123] meant that life was coterminous with mind. Hence, if humans had mental capacities, so would oysters or worms—albeit to a far tinier degree. Plants also had forms of sentience.[124]

This second insight was key for James. He had initially identified life with 'feeling,' which he then equated to streams of 'consciousness' (in humans). But he became unhappy about identifying (the psychological side of) life with consciousness. This was partly because lower animals and plants could not easily be thought of as conscious, even though they were very much alive. So James slowly came to embrace a broader, more pluralistic notion of *experience* as the 'aboriginal stuff or quality of being.' Consciousness was relegated to a second-order concept, a 'function' by which the primary stuff of pure experience was augmented or doubled over on itself to become knowledge.[125] James' critique of consciousness gained scant traction in twentieth-century psychology—especially with the rise of a cognitive psychology focused primarily on phenomena of consciousness like attention, problem-solving, language-use, recollection, perception, creativity, and thinking.[126]

Darwin's principle of continuity has fared no better. Far from agency being coterminous with life wherever organisms are found, twentieth-century theorists commonly equated agency with the capacity *reflexively* to monitor one's acts, that is, *account for them in words*. This meant only humans could be agents—non-human creatures being dismissed, for example, as 'outside the realm of subjective experience, and alien to it,' inhabiting a 'material world, governed by impersonal relations of cause and effect.'[127] Often, writers like Ludwig Wittgenstein (1889–1951)

---

[122] William James, 'Are We Automata?,' in *Essays in Psychology* (Cambridge MA: Harvard University Press, 1984), pp.38–61; Thomas Huxley, 'On the Hypothesis That Animals Are Automata, and Its History,' *Fortnightly Review* 16, (1874), pp.555–580; Eugene Taylor, 'William James on Darwin: An Evolutionary Theory of Consciousness,' *Annals of the New York Academy of Sciences* 602, (1990), p.10.

[123] Darwin, *Origin 1876*, p.166 & passim.

[124] James, 'Brute and Human Intellect,'; James, 'Are We Automata?'; Charles Darwin, 'Notebook N,' ed. Paul Barrett (1838–1839), p.49.

[125] William James, 'Does Consciousness Exist?,' in *Essays in Radical Empiricism* (New York: Longmans, Green, & Co., 1912), p.3.

[126] Howard Gardner, *The Mind's New Science: A History of the Cognitive Revolution* (New York: Basic Books, 1987); Ben Bradley, *Psychology and Experience* (Cambridge: Cambridge University Press, 2005).

[127] E.g. Anthony Giddens, *The Constitution of Society: Outline of the Theory of Structuration* (Cambridge: Polity Press, 1984), pp.2–3: 'To be human is to be a purposive agent, who both has reasons

were taken to establish a social-scientific discontinuity between *the laws* that governed the behaviour of non-human creatures and *the rules* that should be invoked to understand the lives of language-users.[128] All of which testifies to an unresolved theoretical disquiet at Darwin's placement of humans among animals.[129]

Furthermore, Darwin's stress on the distinctiveness of the animate from the inanimate world, and continuity among living creatures, invokes a vision of life inimical to a split between body and mind. Unlike the more explicitly psychological work of his contemporaries Spencer and Bain, in which *the mind* itself attains an independent ontological status, no such emphasis occurs in Darwin (see Ch. 5).[130] Rather than there being two systems of reality—material and mental—Darwin's treatment of agency implies an always-already *embodied* mentality. Hence, the aboriginal stuff of life has no 'inner duplicity'—as William James put it—or 'bifurcation,' to quote Alfred North Whitehead (1861–1947).[131]

This does not equate to Darwin *being a materialist*—as some conclude.[132] A closer parallel would be with the renewed recognition in biology that living organisms are open systems which, by constantly exchanging flows of matter and energy with their surroundings, maintain themselves at a level of organization far above chaos, and hence escape the second law of thermodynamics that applies

for his or her activities and is able, if asked, to elaborate discursively upon those reasons (including lying about them).'

[128] E.g. Peter Winch, *The Idea of a Social Science and Its Relation to Philosophy* (London: Routeldge and Kegan Paul, 1958); Kenneth Gergen, 'The Social Constructionist Movement in Modern Psychology,' *American Psychologist* 40, (1985), pp.266–275.

[129] Jacques Derrida, *The Animal That Therefore I Am* (New York: Fordham University Press, 2008), p.136—see Ch.10.2 The Cost of Suspicion; Paul Stenner, 'On the Actualities and Possibilities of Constructionism: Towards Deep Empiricism,' *Human Affairs* 19, (2009), pp.194–210.

[130] Herbert Spencer, *The Principles of Psychology*, 2 vols., vol. 1 (London: Williams & Norgate, 1870); Herbert Spencer, *The Principles of Psychology, Vol.2*, 2nd ed. (London: Williams & Norgate, 1872); Alexander Bain, *The Emotions and the Will*, 2nd ed. (London: Longmans Green, 1865); Alexander Bain, *The Senses and the Intellect*, 3 ed. (London: Longmans Green, 1868).

[131] James, 'Does Consciousness Exist?,' p.9; Paul Stenner, 'A.N. Whitehead and Subjectivity,' *Subjectivity* 22, (2008), pp.90–109.

[132] E.g. Howard Gruber, *Darwin on Man: A Psychological Study of Scientific Creativity* (Chicago: University of Chicago Press, 1981), p.30. Several commentators on Darwin's Notebooks echo Gruber, e.g. Jonathan Hodge, 'Chance and Chances in Darwin's Early Theorizing and in Darwinian Theory Today,' in *Chance in Evolution*, ed. Grant Ramsey and Charles Pence (Chicago: Chicago University Press, 2016), p.50: By 1838, says Hodge, Darwin was 'explicit ... in his materialism about the mind as the workings of the brain.' Hodge implies Darwin's views on materialism were firm and unequivocal 'by 1838,' not changing in later years. That they *did* change can be seen from Charles Darwin, 'Old & Useless Notes,' (Cambridge 1837–1840), p.41v, where, by 1839, Darwin had concluded that 'all that can be said' was that thought and brain-process/organization 'run in a parallel series'—which is not materialism, in the sense of identifying mind with matter: see Sandra Herbert and Paul Barrett, 'Introduction: Old & Useless Notes,' in *Charles Darwin's Notebooks, 1836–1844*, ed. Paul Barrett et al (Ithaca NY: Cornell University Press, 1987), p.597. What labelling Darwin 'a materialist' might mean is unclear, given the many brands of Victorian materialism, whether invoking: atheism; a vaguer denial of spirit; 'vital materialism'; 'physical materialism'; or merely a claim that all observable phenomena could be explained by natural laws: see Edward Manier, *The Young Darwin and His Cultural Circle. A Study of Influences Which Helped Shape the Language and Logic of the First Drafts of the Theory of Natural Selection* (Dordrecht: Reidel, 1978), Ch.4.

to non-living 'closed' systems (as discussed in Chapter One).[133] Life-forms gain energy by metabolizing food, air, water and sunlight. They spend energy by performing actions to keep them alive.[134] This means all organisms are *inherently* agents.

Finally,150 years of research emphasize Darwin's evidence that a long list of 'lower' animals exhibit intelligence. Recent research on bees, for example, shows them capable of: intelligently solving problems using imagination rather than trial and error; observational or social learning; tool use; and cultural transmission.[135]

3.4.2 Interdependency

As seen in Chapter Two, Darwin's stress on interdependency between different species in a particular habitat swiftly led to the invention of the term *ecology*. 'Everywhere we see organic action & reaction,' wrote Darwin. Consequently, 'all nature is bound together by an inextricable web of relations.'[136] Twenty-first century biology has taken up such webbedness with gusto.[137]

John Dupré puts particular emphasis on the consortia of microbes which inhabit any large organism. It has long been known that nitrogen-fixing bacteria help plants like lupins survive in inhospitable soil. But more than half of the weight of life today comprises microbes. Microbes form ninety percent of the cells in a human body, and 99% of the genes in you or me belong to bacteria. Just like the lupin, we depend on these microbes—to help digest our food, for example. And they depend on us. And they often depend on each other. Biofilms for example. Made up of many different species of microscopic organism, biofilms can act cooperatively. Dental plaque spells decay for human teeth, but the species involved act collaboratively—both in their initial colonization of a mouth and in forming a biofilm strong enough to prevent a swallow-dive into the acidic churn of the stomach.[138]

---

[133] Ilya Prigogine and Isabelle Stengers, *Order out of Chaos: Man's New Dialogue with Nature* (London: Fontana Paperbacks, 1984), pp.140–145; Peter Sylvester-Bradley, 'Evolutionary Oscillation in Prebiology: Igneous Activity and the Origins of Life,' *Origins of Life* 7, (1976), pp.10–11.

[134] Daniel Nicholson and John Dupré, eds., *Everything Flows: Towards a Processual Philosophy of Biology* (Oxford: Oxford University Press, 2018), pp.15–20; Nicholson, 'The Return of Organism,' p.355.

[135] Sylvain Alem et al., 'Associative Mechanisms Allow for Social Learning and Cultural Transmission of String Pulling in an Insect,' *PLoS Biology* 14, (2016), https://doi.org/10.1371/journal.pbio.1002564.; Olli Loukola et al., 'Bumblebees Show Cognitive Flexibility by Improving on an Observed Complex Behaviour,' *Science* 355, (2017), pp.833–836.

[136] Darwin, *Natural Selection*, p.272.

[137] Scott Gilbert and David Epel, *Ecological Developmental Biology: The Environmental Regulation of Development, Health, and Evolution*, 2nd ed. (Sunderland, MA: Sinauer Ass., 2015), p.xiii; John Dupré, *Processes of Life: Essays in the Philosophy of Biology* (Oxford: Oxford University Press, 2012).

[138] Dupré, *Processes of Life*, pp.75, 85ff, 146; Paul Kolenbrander et al., 'Communication among Oral Bacteria,' *Microbiology and Molecular Biology Reviews* 66, (2002), p.486: 'The bacteria on human teeth and oral mucosa have developed the means by which to communicate and thereby form successful organizations.'

Or consider the wood-wide-web. In 1997, Suzanne Simard and her colleagues showed that different plants in a wood actively connect with each other through thin strands of fungus called *hyphae*. Hyphae pass information, nutrients and chemicals from one plant to another—leading to resource-sharing between trees and a greater capacity to withstand environmental threats such as drought or disease. The cooperativeness made possible through the operation of this symbiotic web improves both the longevity of the forest and the longevity of the fungi.[139]

From 'wrinkly spreader' bacteria (which must collaborate to float) to ants and olive baboons, mutual aid is a widespread and beneficial feature of life, even when involving unrelated animals.[140] Indeed, nests and hives are now seen as *superorganisms*, 'collections of agents which can act in concert to produce phenomena governed by the collective.'[141] Martin Nowak has tagged this focus in today's evolutionary research 'the *snuggle* for survival.'[142]

3.4.3 Psycho-sociality

The idea that social organisms have evolved as such, and gain many advantages through their sociality, shaped Darwin's approach to human agency (see Chs 4–9).[143] Yet an asocial 'individual' has traditionally been the focus of scientific psychology: an abstract, average cipher, whose mental life could be specified, studied, and discussed without mentioning the actual circumstances in which any real man or woman lived.[144] Thus students' first textbook typically leads them through

---

[139] Suzanne Simard et al., 'Net Transfer of Carbon between Ectomycorrhizal Tree Species in the Field,' *Nature* 388, (1997), pp.579–582.

[140] E.g. West-Eberhard, *Developmental Plasticity and Evolution*; David Wilson and Edward Wilson, 'Rethinking the Foundation of Sociobiology,' *Quarterly Review of Biology* 82, (2007), pp.327–348; Clutton-Brock, 'Co-Operation in Animals'; Shirley Strum, 'Darwin's Monkey: Why Baboons Can't Become Human,' *Yearbook of Physical Anthropology* 55, (2012), pp.S3–S23.

[141] Kevin Kelly, *Out of Control: The New Biology of Machines, Social Systems and the Economic World* (Boston: Addison-Wesley, 1994), p.98; e.g. Thomas O'Shea-Wheller, Ana Sendova-Franks, and Nigel Franks, 'Differentiated Anti-Predation Responses in a Superorganism,' *PLoS ONE* 10, (2015), e0141012. doi:10.1371/journal.pone.0141012.

[142] Martin Nowak, 'Why We Help,' *Scientific American* 307, (2012), p.34.

[143] Cf. William Gaunt, *The Aesthetic Adventure* (London: Jonathan Cape, 1945), p.121, who shows 'mutual help' was a part of Pre-Raphaelite thinking (a source for Darwin?). See John Ruskin, *Modern Painters*, vol. five (Boston: Estes and Lauriat, 1894), Part VIII, Ch. 1, 'The Law of Help,' e.g., 'A pure or holy state of anything … is that in which all its parts are helpful or consistent. They may or may not be homogeneous. The highest or organic purities are composed of many elements in an entirely helpful way. The highest and first law of the universe – and the other name of life is, therefore, 'help'. The other name of death is 'separation'. Government and co-operation are in all things and eternally the Laws of Life. Anarchy and competition, eternally and in all things, the Laws of Death' (p.175).

[144] Michael Billig, 'Repopulating Social Psychology: A Revised Version of Events,' in *Reconstructing the Psychological Subject: Bodies, Practices and Technologies*, ed. Betty Bayer and John Shotter (London: Sage, 1998), pp.126–151; Mostafa Rad, Alison Martingano, and Jeremy Ginges, 'Toward a Psychology of *Homo Sapiens*: Making Psychological Science More Representative of the Human Population,' *Proceedings of the National Academy of Sciences of the United States of America* 115, (2018), pp.1401–1405. Modern psychology began as 'the science of finite individual minds'; p.vi in

a series of topics—from biological foundations, through sensation, perception, consciousness, reasoning, learning, motivation, and memory. Then comes a chapter on individual development, followed by a series of overviews of 'individual differences', highlighting perhaps intelligence, personality, and mental disorders. Finally, tacked on at the end, comes a chapter on social psychology. This layout implies that all the preceding chapters of the textbook have been dealing with an *asocial*, generic human being.

Such 'methodological individualism' has become a hallmark of psychology.[145] Thus, for most of last century, child and developmental psychology operated under a rubric of 'socialization'—which implies that, at birth humans are essentially *asocial*, however cute, labile, or perceptive.[146] Likewise, several voices have sketched the problems which arise when aggregated measures of a few predefined behaviours of nameless subjects—most often young college students—desert-islanded from their social worlds and familiar surroundings, and tested one by one in antiseptic university laboratories, became the gold standard for the evidence which supported twentieth-century psychology's vision of humanity.[147] The recent rise of social constructionism, critical psychology, and qualitative research are partly responses to these problems.[148]

An individualistic vision of human beings tallies neither with Darwin's understanding, nor with twenty-first century biology.[149] Nowadays, biological doubts about individuality go from debates about the unexpected prevalence of hybrids between what were formerly thought to be discrete species, all the way down the organic scale.[150] As we have seen, colonies and social groups may act as individuals,

James, *Principles* Cf. Charles Taylor, *Sources of the Self: The Making of the Modern Identity* (Cambridge, MA: Harvard University Press, 1989).

[145] Steven Lukes, *Individualism* (Colchester: ECPR, 2006).
[146] This assumption is explicitly stated as the first sentence in Rudolph Schaffer, *The Growth of Sociability* (Harmondsworth: Penguin, 1971), p.1.
[147] Denise Riley, *War in the Nursery: Theories of the Child and Mother* (London: Virago, 1983); Kurt Danziger, *Constructing the Subject: Historical Origins of Psychological Research* (Cambridge: Cambridge University Press, 1990); Julian Henriques et al., *Changing the Subject: Psychology, Social Regulation and Subjectivity* (London: Routledge, 1998); Ben Bradley, 'Language and the Dissolution of Individuality,' *MOSAIC Monographs* 3, (1988).
[148] E.g. Gergen, 'Social Constructionist Movement'; David Rennie, David Watson, and Michael Monteiro, 'The Rise of Qualitative Research in Psychology,' *Canadian Psychology* 43, (2002), pp.179–189; Ian Parker, *The Crisis in Modern Psychology and How to End It* (London: Taylor & Francis, 2013).
[149] Peter Godfrey-Smith, 'Darwinian Individuals,' in *From Groups to Individuals: Evolution and Emerging Individuality*, ed. Frédéric Bouchard and Philippe Huneman (Cambridge MA: MIT Press, 2013), pp.17–35; see also p.126 in Gilbert and Epel, *Ecological Developmental Biology*; Scott Gilbert, Jan Sapp, and Alfred Tauber, 'A Symbiotic View of Life: We Have Never Been Individuals,' *Quarterly Review of Biology* 87, (2012), pp.325–341; Scott Lidgard and Lynn Nyhart, eds., *Biological Individuality: Integrating Scientific, Philosophical, and Historical Perspectives* (Chicago: Chicago University Press, 2017).
[150] James Mallett, 'Hybrid Speciation,' *Nature* 446, (2007), pp.279–283; Karen Barad, *Meeting the Universe Halfway: Quantum Physics and the Entanglement of Matter and Meaning* (Durham, North Carolina: Duke University Press, 2007); Robert Wilson and Matthew Barker, 'The Biological Notion of Individual,' ed. Edward Zalta, *Stanford Encyclopedia of Philosophy* (2017), <https://plato.stanford.edu/

that is, as superorganisms—the group subsuming its members. Likewise, a single living organism subsumes many lower levels of agency. 'An organic being is a microcosm,' wrote Darwin, 'a little universe, formed of a host of self-propagating organisms, inconceivably minute and numerous as the stars in heaven.'[151] Today's biology affirms that my organs and tissues have their own moderate autonomy, as do my cells, as do the mitochondria within my cells. Even the food I eat does not wholly convert into 'me' when I digest it, continuing to exert an independent metabolic and genetic agency after my breakfast.[152] The discreteness of 'the gene' also proves a chimera. Overlapping sections of DNA can code different proteins; and 'the same' gene, when read in opposite directions, may code different chemicals. Conversely, the same chemical can be coded by different sequences of DNA.[153]

Chapters Four to Nine examine how the amorphous boundaries between individual and group play out in Darwin's understandings of human agency, especially in embarrassment, desire, and conscience. This aspect of Darwin's work resonates with recent psycho-social treatments of supra-individual phenomena like suggestibility, affect, and groupness in humans.[154] It also chimes with Lev Vygotsky's (1896–1934) basing of social-developmental psychology in the proximal, 'inter-mental' zone established *between* collaborating individuals (see Ch. 9). Accordingly, Vygotsky framed the dynamics that drove development as being akin to those of drama or live performance—an ongoing, improvised affair in which events unfold on a moment-to-moment basis, 'where competing forces come into play and often collide, where everything emerges out of live relations among players participating within an ensemble, in an intricate balance of oppositions and reconciliations.'[155]

archives/spr2017/entries/biology-individual/>; Frédéric Bouchard and Philippe Huneman, eds., *From Groups to Individuals: Evolution and Emerging Individuality* (Cambridge MA: MIT Press, 2013).

[151] Darwin, *Variation, Vol.2*, 2, pp.364–366, 398–399.

[152] Hannah   Landecker,   'Metabolism,   Autonomy,   and   Individuality,'   in   *Biological Individuality: Integrating Scientific, Philosophical, and Historical Perspectives*, ed. Scott Lidgard and Lynn Nyhart (Chicago: Chicago University Press, 2017), pp.225–247; John Dupré and Maureen O'Malley, 'Varieties of Living Things: Life at the Intersection of Lineage and Metabolism,' *Philosophy Theory and Practice in Biology* 1, (2009), http://dx.doi.org/10.3998/ptb.6959004.0001.003.

[153] Dupré, *Processes of Life*, pp.135ff.

[154] Lisa Blackman, 'Affect, Relationality and the Problem of Personality,' *Theory, Culture and Society* 25, (2008), pp.23–47; Johanna Motzkau, 'Exploring the Transdisciplinary Trajectory of Suggestibility,' *Subjectivity* 27, (2009), pp.172–194; Benjamin Sylvester Bradley and Michael Smithson, 'Groupness in Preverbal Infants: Proof of Concept,' *Frontiers in Psychology* 8, (2017), https://doi.org/10.3389/fpsyg.2017.00385; Paul Stenner, *Liminality and Experience: A Transdisciplinary Approach to the Psychosocial* (London: Palgrave Macmllan, 2017); It also strikes a chord in recent philosophy; e.g. Barad, *Meeting the Universe*, p.ix.

[155] Lev Vygotsky, *Mind in Society: The Development of Higher Psychological Processes* (Cambridge MA: Harvard University Press, 1978); Anna Stetsenko, 'Darwin and Vygotsky on Development: An Exegesis on Human Nature,' in *Children, Development and Education: Cultural, Historical, Anthropological Perspectives*, ed. Michalis Kontopodis, Christoph Wulf, and Bernd Fichtner (Dordrecht: Springer, 2011), p.32.

3.4.4 Psychology and the modern synthesis

As discussed in Chapter One, the gene-centred view of evolution first promulgated by Theodosius Dobzhansky, Ernst Mayr, Julian Huxley and others in the 1930s and 1940s has cast a long shadow over understandings of Darwin's work, both in biology and in psychology. For as long as this self-styled Modern Synthesis holds sway, formulations pertaining to agency in *Origin*, and its successor volumes, can be dismissed as scientifically anachronistic, because everything in them—bar the idea of natural selection—has apparently been superseded by a molecular understanding of evolution. Today, however, the Modern Synthesis is challenged by a more comprehensive or 'extended', phenotype-led, understanding of evolution. This invites a rehabilitation of many of Darwin's original hypotheses and data (see 3.4.5 Phenotype-led evolution). As yet, this 'quiet revolution' has had little impact in psychology, a state of affairs I will now briefly consider (and which *Darwin's Psychology* aims to remedy).[156] Most obviously, the Modern Synthesis has shaped the three forms of Darwinism best-known to today's behavioural science: sociobiology; Evolutionary Psychology (among its several variants, I will only capitalize the one associated with Santa Barbara, upon which I mostly focus); and theories of cultural evolution.

Sociobiology, born in the 1970s, invests deeply in the gene's eye view of evolution. The gene is its measure of all things—the organism is the measure of none. The gene furnishes not only the unit of heredity, but the unit of selection, *and* the unit of adaptation: 'if adaptations are to be treated as "for the good of" something,' writes Dawkins, 'that something is the gene.'[157] Natural selection follows pat as a causal process operating at the level of genes.[158] Darwin's theatre of agency is entirely erased by this vision or, more precisely, it is relocated. For Dawkins, *genes* are the agents: having personality traits (selfishness); 'choosing to group themselves together'; 'programming' brains and bodies; and 'playing out tournaments of manipulative skill.'[159] This makes organisms puppets, 'vehicles', 'lumbering robots', 'machines created by our genes.' Meanwhile, evolution becomes 'the external and visible manifestation' of the survival of *replicators*.[160] Genes are replicators.

---

[156] Gilbert and Epel, *Ecological Developmental Biology*, p.xiii; but see Karola Stotz, 'Extended Evolutionary Psychology: The Importance of Transgenerational Developmental Plasticity', *Frontiers in Psychology* 5, (2014), https://www.frontiersin.org/articles/10.3389/fpsyg.2014.00908/full.

[157] Richard Dawkins, *The Extended Phenotype: The Long Reach of the Gene* (Oxford: Oxford University Press, 1982), p.vii. The sociobiology of E.O. Wilson was equally invested in the Modern Synthesis. See Edward Wilson, *Sociobiology: The New Synthesis* (Cambridge MA: Harvard University Press, 1975). Unlike Dawkins, however, Wilson's view of evolution has altered significantly since the 1970s. See e.g. Wilson and Wilson, 'Rethinking Sociobiology'; Martin Nowak, Corina Tarnita, and Edward Wilson, 'The Evolution of Eusociality', *Nature* 466, (2010), pp.1057–1062.

[158] John Tooby and Leda Cosmides, 'The Theoretical Foundations of Evolutionary Psychology', in *The Handbook of Evolutionary Psychology*, ed. David Buss (Hoboken NJ: Wiley, 2016), p.53; Dawkins, *Selfish Gene*, p.xix.

[159] Dawkins, *Selfish Gene*, p.55 & passim; Dawkins, *Extended Phenotype*, p.7 & passim.

[160] Dawkins, *Selfish Gene*, pp.2, 19, 126.

Figure 3.10 In 2007, New Yorker Wesley Autrey (a) jumped down onto a subway track in front of an oncoming train and saved a student who had had a seizure and fallen onto the lines. Both survived. French philosopher Anne Dufourmantelle (b) died in 2017 at a beach near St Tropez after plunging into the Mediterranean to save two children caught in turbulent seas. They both survived.

Source: (a) Mariela Lombar / ZUMA Press, Inc. / Alamy Stock Photo; (b) Photo by Jean-Marc ZAORSKI/Gamma-Rapho via Getty Images.

https://www.nytimes.com/2007/01/03/nyregion/03life.html.

https://www.nytimes.com/2017/12/18/world/heroism-altruism-courage-compassion-2017.html
https://www.tripadvisor.com.au/Attraction_Review-g186225-d212978-Reviews-Cambridge_
American_Cemetery_and_Memorial-Cambridge_Cambridgeshire_England.html.

Organisms and groups of organisms are 'best not regarded as replicators; they are *vehicles* in which replicators travel about.'[161]

Of most relevance to psychology are claims like Dawkins' that genes have an external and visible manifestation. Not only does this assume genotypes *cause* phenotypes. It imputes a direct correspondence between genes and observable characters. And it implies a form of deductivism: the *kinds* of things organisms can do are *pre-defined by theory*. Most notably, and in direct contradiction to Darwin (see Ch. 7), no individual can have 'group-related' adaptations.[162] The way natural selection works precludes such adaptations. Thus even if you (think you) see a group-related adaptation, you err—genetic theory trumps the art and craft of describing. Most famously, under the aegis of the gene, altruism does not exist. Or, more precisely, altruism no longer entails acting with selfless concern for the well-being of others (cf. Figure 3.10). It occurs when 'the effect of an act is to lower' the 'survival prospects' of the genes of the 'presumed altruist' while raising those of

---

[161] Dawkins, *Extended Phenotype*, p.82.
[162] George Williams, *Adaptation and Natural Selection* (Princeton NJ: Princeton University Press, 1966), p.93; Dawkins, *Selfish Gene*.

the genes of 'the presumed beneficiary'.[163] Which is an unknown. For, as Dawkins concedes: 'It is a very complicated business to demonstrate the effects of behaviour on long-term survival prospects,'—especially when *long-term* means millions of years. So altruistic acts cease to be observable phenomena. Altruism is a surmise from best-guess calculations of the long-term pros and cons of a given act for its participants' genes—best left to clairvoyants, one imagines.

Evolutionary Psychology, the more-sophisticated but self-confessed heir of sociobiology, dates from around 1990. While several 'evolutionary psychologies' exist, the label most often attaches to the so-called Santa Barbara School.[164] This most differs from sociobiology in proposing that—rather than programming behaviour without intermediary—genes programme 'mental modules,' 'psychological mechanisms,' or 'neuro-computational programs,' which, *in turn*, generate behaviour. Requiring millions of years to evolve, these modules supposedly have a design that fits humans to the long-ago conditions of the Stone Age.[165] So, evolutionary psychologists devote themselves to 'reverse engineering' current human behaviour to discover the Palaeolithic mechanisms generating it.

A critical assumption grounds this approach: that natural selection is 'the only known natural physical process' capable of building 'highly ordered functional organization (adaptations) into the designs of species.' Thus—in this view—although not everything is functional, 'whenever functional organization is found in the architecture of species, its existence and form can be traced back to a previous history of selection.'[166]

Three things need noting here. Most obviously, there are forms of highly-ordered, functional, organization which are culturally induced on a time-frame that excludes explanation by reference to the Stone Age—like literacy.[167] This underlines that, secondly, Evolutionary Psychology has no place for the relative autonomy of culturally-shaped variations which may, incidentally, feed the mill of natural selection (see Ch. 8). Thirdly, like sociobiology and the Modern Synthesis, Evolutionary Psychology puts great weight on the *causal power* of natural selection: 'natural

[163] Dawkins, *Selfish Gene*, pp.3ff.

[164] Tim Lewens, *Darwin* (London: Routledge, 2007), pp.146–156; Darren Burke, 'Why Isn't Everyone an Evolutionary Psychologist?,' *Frontiers in Psychology* 5, (2014), https://www.frontiersin.org/articles/10.3389/fpsyg.2014.00910/full; Danielle Sulikowski, 'Evolutionary Psychology: Fringe or Central to Psychological Science?,' *Frontiers in Psychology* 7, (2016), https://www.frontiersin.org/articles/10.3389/fpsyg.2016.00777/full.

[165] Tooby and Cosmides, 'Theoretical Foundations,' p.10: 'Complex adaptations necessarily reflect the functional demands of the cross-generationally long-enduring structure of the organism's ancestral world, rather than the modern, local, transient, or individual conditions.'

[166] Tooby and Cosmides, 'Theoretical Foundations,' p.10.

[167] A flaw Tooby and Cosmides implicitly acknowledge. Reading shows high organization, is functional in today's world, but has *not* evolved as such: Tooby and Cosmides, 'Theoretical Foundations,' pp.28–30. Cf. Peter Robinson, 'Literacy, Numeracy, and Economic Performance,' *New Political Economy* 3, (1998), pp.143–149; Richard Hoggart, *The Uses of Literacy* (London: Chatto & Windus, 1957).

selection is an engineer that designs organic machines.'[168] Further back lies the assumption that: 'Genes are the means by which functional design features replicate themselves from parent to offspring.'[169] This cements both the Modern Synthesis' gene-level vision of heredity as transmission *without development*, and the assumption that genes *cause* so-called *design features*, such as the 'evolved programs that underlie sexual behaviour, mate choice, attractiveness, intrasexual competition, intersexual conflict, and mateship maintenance'—once again denying any efficacy to phenotypic agency.[170] All this runs counter to the formulations on natural selection in *Origin*.

Finally, the Santa Barbara School has long focused on explaining the psychological foundations of culture.[171] Foundations do not a whole building make, however. Hence the rise of a set of theories complementary to Evolutionary Psychology, positing *cultural evolution*.

The idea of *cultural evolution* entails several debts to the Modern Synthesis (see also 8.4.3 Theories of cultural evolution). To wit, the Modern Synthesis holds everything within the *biological* sphere to result from *gene-level* processes operating over millions of years. Which puts fast-changing cultural content beyond the reach of biological explanation. So the Modern Synthesis necessitates a dichotomy between genes and culture.[172] Hence, if, for some reason, one believes that *all aspects of humanity require an evolutionary explanation*—which Darwin did not (see Ch. 8)—the Modern Synthesis will make it seem that a theory of cultural evolution is required to complement one's theory of biological evolution in order to complete one's explanation of human agency.

Which presages another debt. As the perceived problem of cultural evolution only exists thanks to the Modern Synthesis, the same source is held to solve it. Witness Dawkins proposing that, in cultural evolution there is a unit of information that faithfully replicates itself, mutates, and recombines *just like a gene*—albeit over a much shorter time-scale—which he christens *the meme*.[173] Alternatively, we find authors like Peter Richerson and Robin Boyd constructing their so-called

---

[168] Tooby and Cosmides, 'Theoretical Foundations,' p.22; cf. Daniel Nicholson, 'Organisms ≠ Machines,' *Studies in History and Philosophy of Biological and Biomedical Sciences* 44, (2013).

[169] Tooby and Cosmides, 'Theoretical Foundations,' p.23.

[170] Tooby and Cosmides, 'Theoretical Foundations,' p.5; cf. Karola Stotz, 'Extended Evolutionary Psychology.'

[171] E.g. John Tooby and Leda Cosmides, 'The Psychological Foundations of Culture,' in *The Adapted Mind: Evolutionary Psychology and the Generation of Culture*, ed. Jerome Barkow, Leda Cosmides, and John Tooby (New York: Oxford University Press, 1992), pp.19–135.

[172] E.g. The split between 'psychosocial' and 'genetic' phases of evolution in Julian Huxley, 'Introduction to the Second Edition,' in *Evolution: The Modern Synthesis* (London: George Allen & Unwin, 1963), pp.xlivff; Peter Richerson and Robert Boyd, *Not by Genes Alone: How Culture Transformed Human Evolution* (Chicago: University of Chicago Press, 2005), p.6: 'we are largely what our genes and our culture make us ...'

[173] Dawkins, *Selfish Gene*, pp.189ff.

kinetic theory of cultural evolution in the image of the *population thinking* championed by Modern Synthesizer Ernst Mayr.[174]

3.4.5 Phenotype-led evolution

This century has heard a swelling chorus of arguments for a revision of the Modern Synthesis. A keynote has been that the Modern Synthesis neglects development. Thus the most prominent alternatives to the Modern Synthesis get grouped as: 'evolutionary developmental biology' (or 'evo-devo'); 'developmental systems theory'; and 'developmental plasticity'. However, acknowledging the need for a renewed focus on *development* in evolutionary thinking raises a question: the development *of what*?[175] Most of the approaches that fall under the three headings I have just listed blur or beg this question.[176] I concentrate on one that does not.

To the question—what develops?—Mary Jane West-Eberhard answers: *phenotypes*. The first eight chapters of her book *Developmental Plasticity and Evolution* outline an integrated theory of the phenotype, the lack of which she argues to have cramped both the Modern Synthesis, and post-Synthesis understandings of evolution. Only when we have understood the phenotype and its responsiveness to the environment (or *plasticity*), she says, can we understand its development, evolution, and selection—or the place of genes therein: 'Why start with the phenotype and its development? Because that is where evolution starts.'

---

[174] Richerson and Boyd, *Not by Genes Alone*.

[175] The extent to which this question remains unanswered by an approach, measures the extent to which that approach risks committing the *genetic fallacy*, discussed in the opening pages of this chapter: Thelma Lavine, 'Reflections on the Genetic Fallacy', *Social Research* 29, (1962), pp.321–336; Bradley, *Psychology and Experience*, pp.113–132.

[176] **Evo-devo**'s promise of integration falls on the concept of development: failure to consider development in the Modern Synthesis supposedly underpinning all its inadequacies. Yet, in evo-devo, *development* remains an abstraction, representing (irreversible) directional change in *any* time-series of observations of any biological entity: West-Eberhard, *Developmental Plasticity and Evolution*, p.28; Muller, 'Evo-Devo'; Scott Gilbert, Thomas Bosch, and Cristina Ledón-Rettig, 'Eco-Evo-Devo: Developmental Symbiosis and Developmental Plasticity as Evolutionary Agents', *Nature Reviews: Genetics* 16, (2015), pp.611–622; Kevin Laland, John Odling-Smee, and Gilbert Scott, 'Evodevo and Niche Construction: Building Bridges', *Journal of Experimental Zoology (Mol Dev Evol)* 310B, (2008), pp.549–566; Massimo Pigliucci, 'Do We Need an Extended Evolutionary Synthesis?', *Evolution* 61, (2007), pp.2743–2749. **Developmental Systems Theory** (DST) implies that 'systems' are 'what develops.' Yet its advocates disagree about what a 'system' is. After all candle flames are as much open systems as cheetahs. Hence DST gets aimed at contradictory targets: organisms; the entire life-cycle of organisms; and parts of organisms, like teeth. But mostly, the 'system' DST theorizes is the organism *plus its environment*—'the organism/niche dyad.' To the extent that DST does not strongly identify organisms as its explanatory target, agency plays little part in determining outcomes. It makes scant sense to say an organism–niche dyad 'acts.' So, behaviour and agency typically go missing in DST accounts, as standing 'beyond the organism': Jason Robert, Brian Hall, and Wendy Olson, 'Bridging the Gap between Developmental Systems Theory and Evolutionary Developmental Biology', *BioEssays* 23, (2001), pp.954–955; Thomas Pradeu, 'The Organism in Developmental Systems Theory', *Biological Theory* 5, (2010), pp.216–222; Dupré, *Processes of Life*, p.255; Muller, 'Evo-Devo.'

Why treat the phenotype *before* development? Because, development depends at every step 'on the pre-existent structure of the phenotype.' Why prioritize phenotypes over genotypes? Because, 'genes are usually followers, not leaders, in evolutionary change.'[177] In elucidating her answers to these questions, West-Eberhard's book fattens with argument and illustration from the lives of beetles, butterflies, buttercups, birds, and myriad other organisms to illuminate the plasticity of phenotypes, how they develop, and how they evolve. Moreover, as she documents throughout her book, her approach harmonizes with Darwin's take on evolution as driven by the struggle for existence in a theatre of agency, being afforded by each creature's 'power of action.'[178] So West-Eberhard's theory paves the way for the arguments I go on to make in the later chapters of *Darwin's Psychology*.

West-Eberhard's stress on responsiveness or plasticity in the phenotype implies a central role for the agency of both plants and animals in the process of adaptation and the genesis of variation. Not least because, while calling phenotypic plasticity a developmental process, West-Eberhard accepts that development incorporates agency. Moreover, as plasticity includes both irreversible *and reversible* phenotypic changes during an individual's lifetime, even a momentary expression counts as a 'development.'[179]

Through her eyes, the focus of evolutionary thinking shifts from genes towards how 'the activities of organisms affect the structure of populations.'[180] This sounds nebulous when stated so baldly. To make it plausible, West-Eberhard shows that the 'activities of organisms' go far beyond what modern psychologists call *behaviour*. She describes and illustrates several organismic processes that contribute to plasticity, instancing: exploratory growth in plants; the trial and error learning of song dialects in young sparrows; hyper-variability and so-called *somatic selection* (the induction of pattern through reinforcement) as shown in orientation among single-cell paramecia; the mediating power of hormones to affect both form and behaviour; homeostatic mechanisms, including movements of the stomata that control transpiration in plants; and the defensive release of toxins by trees when under attack by beetles.[181]

Amongst these processes, her theory gives top billing to *phenotypic accommodation*, where organisms develop functional phenotypes through 'adaptive mutual adjustment among variable parts during development, without genetic change.'[182]

---

[177] West-Eberhard, *Developmental Plasticity and Evolution*, pp.28–29.

[178] Charles Darwin, *The Autobiography of Charles Darwin, 1809–1882: With Original Omissions Restored*, ed. Nora Barlow (London: Collins, 1958), p.89.

[179] West-Eberhard, *Developmental Plasticity and Evolution*, p.32.

[180] Walsh, 'Two Neo-Darwinisms,' p.335.

[181] West-Eberhard, *Developmental Plasticity and Evolution*, p.37: somatic selection is where 'large numbers of random, or possibly chaotic variants, modifications, movements, or positions are produced, and then some are selectively preserved or reinforced, while the remainder are unoccupied or eliminated.'

[182] West-Eberhard, *Developmental Plasticity and Evolution*, p.51.

These new adjustments could, over long time periods, become genetically preserved or genetically 'accommodated'—producing agentically-led or, in her terms, *plasticity-led* evolution. West-Eberhard's first example features an unlucky if resourceful baby goat, which emerged from the womb with only two hind legs. Lacking forelegs, the plucky kid adopted a 'semi-upright posture and bipedal locomotion from the time of its birth.' By the time it died due to an accident at the age of one, 'it had developed several behavioural and morphological specializations similar to those of kangaroos and other bipedal mammals, including the ability to hop rapidly when disturbed, enlarged hind legs, a curved spine, and an unusually large neck.'[183]

A central role in her theory is given to the *modularity* of phenotypic traits, as in the different leaf-forms a plant pops out in successive stages of its growth. West-Eberhard defines *modular traits* as: 'subunits of the phenotype that are determined by the switches or decision points that organize development, whether of morphology, physiology, or behavior.' The plasticity of development results from the many alternatives opened by a branching sequence of developmental 'switching points.'[184] Critically, any switch can as easily be thrown by environmental change (e.g. the presence or absence of predators at laying affects the time to hatching of the eggs of tree-frogs) as by genetic change (e.g. mutation).[185] For example, the larvae of some tropical wasps can become either queens or workers—two castes which are very different in morphology and in behaviour—depending on *how they are fed*. Genes help fix what queens and workers look like, and what they do. Yet *all* the wasps' larvae have these genes. It is diet which throws the switch that makes a larva queen or worker.[186] Similarly, ambient temperature throws the switch that makes turtle hatchlings male or female (see 3.3.4 Agency directs selection).

One immediate pay-off from West-Eberhard's theory needs noting. Once the active phenotype is seen as fulcrum to evolution, any need for a stand-alone theory of 'cultural evolution' vanishes (see Ch. 8). West-Eberhard defines the most general and essential property of evolution as 'cross-generational change in phenotypic frequencies or dimensions *involving change in gene frequencies*.'[187] As her book amply documents, cross-generational change in phenotypes often takes place *without* involving change in gene frequencies (e.g. by phenotypic accommodation). Hence, her theory of the phenotype undertakes: (a) to analyse and explain *all* phenotypic change and, *only then*; (b) to determine whether and which of such changes involve changes in gene frequencies.

Whilst plasticity itself may have evolved, not all manifestations of organic flexibility are coded by genes. Thus, Darwin observed that all the growing parts of all plants circumnutate, but added that, although this movement appeared in some

[183] West-Eberhard, *Developmental Plasticity and Evolution*, p.51.
[184] West-Eberhard, *Developmental Plasticity and Evolution*, p.56.
[185] West-Eberhard, *Developmental Plasticity and Evolution*, pp.249–250.
[186] Mary Jane West-Eberhard, 'Mary Jane West-Eberhard,' *Evolution & Development* 11, (2009).
[187] West-Eberhard, *Developmental Plasticity and Evolution*, p.31.

cases to help plants adapt, it was not always helpful—circumnutation being so universal a phenomenon, that we should not 'suppose it to have been gained for any special purpose.'[188] This shows that not everything phenotypes do is relevant to—and even less, is a direct, adaptive *consequence* of—evolution. There is a phenotypic *excess*, or what West-Eberhard calls 'non-adaptive' plasticity.[189] Such excess is a crux of Darwin's psychology, because, while agency inheres in all phenotypes, simple or complex, its 'frontier' (see Ch. 4) is human *culture*—something which not infrequently has *mal*adaptive consequences, according to Darwin (see Ch. 8).

The concept of excess underlines that phenotypic plasticity *goes beyond* what is explained by theories of evolution regarding organic process, whether the theory be Darwin's, post-Darwinian, or neo-Darwinian. [190] Darwin often referred to the performance of an instinctual action as pleasurable—tying such pleasure to evolved, *functional* behaviour.[191] As *Origin* adds, however, evolutionary theory does not require every living creature always and for ever to be struggling for existence, or behaving 'functionally,' but only *at intervals*:

> each organic being ... at some period of its life, during some season of the year, during each generation or *at intervals,* has to struggle for life and to suffer great destruction. When we reflect on this struggle, we may console ourselves with the full belief, that *the war of nature is not incessant* ... and that the vigorous, the healthy, and the happy survive and multiply.[192]

Descriptively, *phenotypic excess* includes those times when a plant or animal is *not* struggling: this well-fed pelican lazing among companions on a sunlit beach; that eagle soaring on a thermal rising from woods far below. *Theoretically*, it stands for those aspects of phenotypic accommodation unallied to genetic change, the existence of which abolishes the need for a stand-alone 'non-biological' theory of cultural evolution, because it closes the ostensibly 'unbridgeable gap' between the cognitive capabilities and achievements of humanity and those of other animals—the gap which creates the need for theories of cultural evolution.[193] Culture proves susceptible to explanation by just the same biological processes that underlie other

---

[188] Darwin, *Movement in Plants*, p.263. He went on: 'We must believe that it follows in some unknown way from the manner in which vegetable tissues grow.'

[189] West-Eberhard, *Developmental Plasticity and Evolution*, pp.43ff & passim.

[190] *Excess* is sometimes used to translate the French word *jouissance* which refers to the excessive kinds of pleasure which *go beyond* the pleasure principle in psychoanalytic theory: Kelly Ives, *Cixous, Irigaray, Kristeva: The Jouissance of French Feminism* (Kidderminster: Crescent Moon, 1996).

[191] E.g. Darwin, *Descent*, pp.104, 107.

[192] Darwin, *Origin 1876*, p.61. Cf. Erasmus Darwin, *The Temple of Nature, or, the Origin of Society: A Poem, with Philosophical Notes* (Baltimore: Butler Bonsal & Niles, 1804), p.194 which talks of mountains as 'mighty monuments of past delight.'

[193] Kevin Laland, *Darwin's Unfinished Symphony: How Culture Made the Human Mind* (Princeton: Princeton University Press, 2017), p.2.

kinds of excessive action, once we have in our possession an adequate theory of the phenotype.

3.4.6 Ancestry

The shorter-term axis of Darwin's vision of the living world—the theatre of agency—grew out of the venerable traditions of natural history. How might that axis best be put together with the deep time of his second axis, common descent? Straight after *Origin* came out, Darwin took off on a tangent to indulge in a *jeu d'esprit*, 'a little book on flowers.'[194] Or so it might seem. In fact, as he told his publisher John Murray, *On the Various Contrivances by which British and Foreign Orchids are Fertilised by Insects* had a precise aim: 'to illustrate how Natural History may be worked under the belief of the modification of species.'[195] For, as the movement of his prose in *Origin* suggests, the *web of complex relations* that gave its hallmark to his theatre of agency now had a new orthogonal dimension that underpinned the here-and-now: a genealogical *web of affinities*.[196] How should these ancestral affinities alter observation and description of the natural world?

The fertilization of orchids by insects furnished a brilliant, topical illustration of *Origin*'s central strength: its explanation of co-adaptation.[197] Previously, botanists had assumed that the often tiny variations between different kinds of orchid had no functional significance. Orchids' often flamboyant blooms sported a host of consistent differences from species to species, not just in colouring, scent and pattern, but in miniscule details affecting pistils and stamens, ovaries and nectaries, shapes and numbers of hairs and pouches, the size and form of discs, lips, and hoods. To Bible-followers, including many of Darwin's scientific contemporaries, the wondrous profusion of orchid flowers was testament to their divine origin. But, in Darwin's eye, every tint, ridge, and wrinkle of an orchid's anatomy had a function that its evolutionary past had helped adapt to its world. Nothing had been created from scratch. Everything had been 'modified.' Each feature had a history of serving diverse previous purposes, which, by comparison with related species,

---

[194] Janet Browne, *Charles Darwin: The Power of Place* (London: Jonathan Cape, 2002), p.166.

[195] Darwin, *Orchids*; Charles Darwin, 'Letter to Murray, 24th September,' *Correspondence of Charles Darwin* (Cambridge: Cambridge University Press, 1861), www.darwinproject.ac.uk/letter/?docId=letters/DCP-LETT-3264 More covertly, Darwin confessed to the Harvard botanist Asa Gray, *Orchids* aimed to perform 'a "flank movement" on the enemy,' the enemy being anyone who believed that there was 'an intelligent designer anywhere in Nature': Charles Darwin, 'Letter to Gray, 23rd–24th July,' *The Correspondence of Charles Darwin* (Cambridge: Darwin Correspondence Project, 1862), www.darwinproject.ac.uk/letter/?docId=letters/DCP-LETT-3662; Asa Gray, 'Letter to Darwin, 2nd–3rd July,' *The Correspondence of Charles Darwin* (Cambridge: Darwin Correspondence Project, 1862), www.darwinproject.ac.uk/letter/?docId=letters/DCP-LETT-3637.

[196] Darwin, *Origin 1859*, p.434.

[197] Tovah Martin, *Once Upon a Windowsill: A History of Indoor Plants* (Portland OR: Timber Press, 1988).

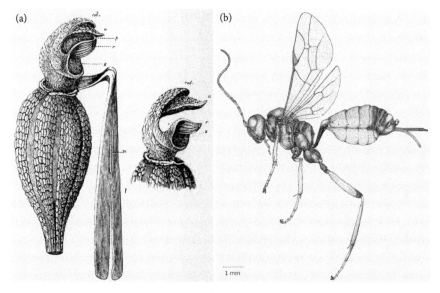

Figure 3.11 The flowers of twayblade orchids (*Listera ovata*; a) attract scorpion-wasps (*Ichneumonidae*; b). Each flower has a tiny explosive structure that spits out a quick-set glue to stick grains of pollen to their visitor's probing head. So a hungry wasp cannot help but fertilize the next nectar-laden twayblade it calls on—becoming the unwitting medium of what Darwin quaintly called the orchids' 'marriage ceremony.'

Source: (a) Darwin, C. (1904), *The Various Contrivances by Which Orchids Are Fertilised by Insects*, 2nd edition.; (b) Copyright © 1996 An-Ly Yao.

'Scorpion-wasp' is the common name for Ichneumonids. Darwin discusses their role as marriage-celebrants of twayblades at length in Darwin, *Orchids*: pp.115–128, and in Charles Darwin, 'Letter,' *Gardeners' Chronicle*, (1861). For outraged comments on these wasps' 'abhorrent' method of reproduction (they inject their eggs into living caterpillars which are then eaten alive from inside out by the grubs), and its bearing on theological explanations for natural form, see *Darwin, Origin 1859*: 199–200, 244, 472 and a letter to his friend, the American botanist, Asa Gray: Charles Darwin, 'Letter to Gray, 22nd May,' (Cambridge: Darwin Correspondence Project, 1860), https://www.darwinproject.ac.uk/letter/?docId=letters/DCP-LETT-2814.xml.

he set out to reconstruct. (It was these *organismic purposes* which, as Modern Synthesizers would say, had been stabilized by lucky if aimless genetic mutations and recombination.)

The fact most central to this vortex of vegetable variation was the gain in vigour and fecundity conferred by plants' cross-fertilization or 'inter-marriage,' when compared to the effects of self-fertilization.[198] Hence the drive for plants to lure insects to sip their syrup: whether wasp, bee, fly, or moth—and thereafter to take away pollen to fertilize their neighbours. Each orchid needed to delight a specific kind of bug to maximize the probability it would commute between conspecific orchids (Figure 3.11). Hence the enormous diversity in the particulars of orchid

[198] Darwin, *Cross Fertilisation*.

structure and movement. By the same token, the habits, body and mouth-parts of the attracted insect would need to have become adapted, both to probe between the petals of the orchid to suck its nectar while, at the same time, being positioned in just the right way to pick up and drop off pollen, before it flew onto feed at (and fertilize) the next such plant. This bespoke tailoring of the forms and behaviours of two completely different species, animal and vegetable, is the first well-worked-out example of what we now call 'co-evolution.'[199]

The critical chapter of *Orchids* came at the end. It focused on genealogical affinities—*homologies* in the flowers of orchids.[200] Without that finale, a Christian naturalist might at a pinch have explained all the exquisite co-adaptations documented in Darwin's book by the creative fiat of an ultra-whimsical Divine Designer. The chapter on homologies showed that an evolutionary natural history needed to advance in *two* directions. It could not solely describe the current utility of a character—the enormously elongated floral tube of *Angaecum sesquipedale,* for example.[201] It had to address 'propinquity of descent.'[202]

An organism's adaptations were doubly relative. They were relative, first—*synchronously*—to its physical conditions of life and the adaptations of members of its own and other species, with which the organism was currently struggling to survive. But secondly, if an adaptation had evolved through natural selection, it must also have given the organism's *ancestors* relative advantages in the past. To test whether this was the case requires that adaptations be mapped back onto a hypothesized tree of common descent—the species' *phylogeny*—to construct a comparative test (between taxa) as to whether and how the proposed adaptation might have been generated *historically* for its current role—something that is sometimes missing from contemporary uptakes of evolutionary theory in psychology.[203]

For example, some sociobiologists test only the *current utility* of, say, a courting Bluebird cock's threats to its own mate when an intruders appears. Once such utility has been shown, the sociobiologist flits immediately to construct a just-so

---

[199] Regarding orchids 'behaving,' one of Darwin's 'curious' observations was that some orchids responded to an insect's arrival by moving their pollen-masses on their stalks so as better to adhere to their visitor's body: Darwin, *Orchids*, pp.24, 154–155 & passim.

[200] Darwin, *Orchids*, pp.226ff.

[201] Darwin was so confident of his theory explaining the diverse forms of orchid bloom that, when he received a white orchid (*Angaecum sesquipedale*), originating from Madagascar, with a nectar tube eleven and a half inches long, he concluded, without further evidence, that 'in Madagascar there must be moths with proboscides capable of extension to a length of between ten and eleven inches!': Darwin, *Orchids*, p.198. During his lifetime he was ridiculed for this conjecture: e.g. by the Duke of Argyll: Browne, *Darwin II*, p.195. Yet just such a Madagascan moth was discovered in 1903: Claire Micheneau, Steven Johnson, and Michael Fay, 'Orchid Pollination: From Darwin to the Present Day,' *Botanical Journal of the Linnean Society* 161, (2009), pp.1–19.

[202] Darwin, *Origin 1859*, p.413.

[203] Allan Larson, 'Review of John Alcock's *'the Triumph of Sociobiology',* Isis 93, (2002), pp.348–349; David Baum and Allan Larson, 'Adaptation Reviewed: A Phylogenetic Methodology for Studying Character Macroevolution,' *Systematic Zoology* 40, (1991), pp.1–18; Elliott Sober, *Evidence and Evolution: The Logic Behind the Science* (Cambridge: Cambridge University Press, 2008); Gregory Radick, 'Evidence-Based Evolutionism,' *Biological Theory* 5, (2010), pp.289–291.

story of its phylogeny. However, if a scientist truly wished to prove the Bluebird's action is an evolved adaptation, they should undertake a second, *phylogenetic* analysis of the Bluebird's closest relatives to identify the comparisons appropriate for testing their hypothesis, regarding antecedent conditions, associated characters, environmental contexts, and selective regimes across the whole group of Bluebird-related species.[204]

3.5 Conclusion

Darwin's vision of the living world did not just underpin his psychology. It grounded his theory of evolution too. Nor did he invent his 'theatre of agency' from scratch—as seen from the many connections to the work of his peers and predecessors noted throughout this and the previous chapter. Assumptions that evolution solely entailed the operation of natural selection, acting at the level of the gene, dominated the twentieth century. This vision invited a narrowed and simplified view of Darwin's contribution to science. However, with the advent of genuine alternatives to last century's Modern Synthesis—most notably, West-Eberhard's recent theory of the phenotype—a great deal of Darwin's work regains scientific relevance. In particular, Darwin's theatre of agency regains its status as the generative domain of organic variation, and of the struggles which, over time, will apportion fates in the living world. Dramas observable in the present gave him his key to the past.[205]

The following chapters show that the readmission to biology of Darwin's two-sided take on evolutionary change has repercussions for any account of agency or mentality in psychology. So let me stress one more time the crucial importance of the twin axes which frame Darwin's vision of living things. Both in psychology and evolutionary biology, the twentieth century gloried in an *in vitro* theatre of mechanism.[206] This prioritized linear conceptions of temporal succession (diachrony, causation) which side-lined observing, describing, and understanding the *in situ*, as-it-happens, pell-mell, blooming, buzzing, synchronous pluriverse of the here-and-now. Yet our brains and minds deal first in here-and-nows, whatever such

---

[204] E.g. Larson, 'Review of Alcock.'

[205] 'The present is the key to the past' is a credo sometimes associated with *uniformitarianism*, a geological purview traceable to Darwin's mentor Chares Lyell, among others. But see Martin Rudwick, 'The Strategy of Lyell's Principles of Geology,' *Isis* 61, (1970), pp.4–33. Cf. John Tooby and Leda Cosmides, 'The Past Explains the Present,' *Ethology and Sociobiology* 11, (1990), pp.375–424.

[206] E.g. Steven Pinker, 'The False Allure of Group Selection,' in *Handbook of Evolutionary Psychology*, ed. David Buss (Hoboken NJ: Wiley, 2016), p.869: 'What's satisfying about the [gene-centric] theory [of natural selection] is that it's so mechanistic.' For early critiques of mechanism in twentieth-century psychology, see John Shotter, *Images of Man in Psychological Research* (London: Methuen, 1975); Rom Harré and Paul Secord, *The Explanation of Social Behaviour* (Oxford: Basil Blackwell, 1972).

immediacy comes to look like after reconstruction.[207] The here-and-now turns out to have a like primacy in Darwin's understanding of the natural domain—for it is not recondite genes but the manifest theatre of agency that drives evolution. The main mystery of Darwin's world is the visible, not the invisible.[208]

It is dusk in the estuary and the chorus of a thousand cicadas comes to a strident crescendo, then ebbs with the light. Each cicada has its motives, its pitch, its patch, its particular orientation to nearby intimates and antagonists. Enacted moment by moment, hour by hour, day by day, the resultant dramas propel each cicada's unique temporal fate, and, over immeasurably longer time-spans, will construct its descendants' evolutionary history. As with cicadas so with all creatures, including the human: the seeming successiveness when we dig 'vertically' into the evolutionary past only makes sense when juxtaposed with the 'horizontal' simultaneities of presence. 'History, to be worthy of the name, should bring us a stereoscopic view of man's life. Without that extra dimension, strangely poignant as well as vivid, it is flat, and because it is flat it is false.'[209]

---

[207] Bradley, *Psychology and Experience*, pp.46ff.

[208] Oscar Wilde, *The Picture of Dorian Gray* (London: Simpkin Marshall Harrison, 1920), p.29.

[209] John Priestley, 'The Linden Tree: A Play in Two Acts,' in *Time and the Conways and Other Plays* (London: Penguin Books, 1969), p.302. See also Beer, *Darwin's Plots*, p.36 & passim: 'Darwin lives in a doubly profuse world – the plenitude of present life, its potential for both development and death, and the recessional and forgotten multitudes which form the ground of the present ...'

4

Reading *Expression*

His hair is grey, & he has a full, grey beard cut square across the upper lip, but the sweetest smile, the sweetest voice, the merriest laugh! And so quick, so keen! He never hears a remark, it seemed to me, but he turns it over, he catches every expression that flits over a face & reads it ... [1]

Anyone may get emotional. If it be a man, 'getting emotional' typically implies tears (Figure 4.1). Tears are public, for all to see. No one who sees a man shed tears can doubt he cried. But why cry? What does his weeping betoken? That is another thing entirely. Even the weeper may wonder. Are his waterworks genuine? Or are they put on—crocodile tears—a bid for sympathy? How might we tell? Was the trigger nontrivial enough to have caused the response? Or was it straw on a camel's back? Are we witnessing an anxiety-attack? A breakdown? A midlife crisis? Or perhaps the cause is banal. A flu bug. Overwork. One too many last night. Yet we discount the profound at our peril. Perhaps his tears tap a buried grief. (Just one kind word may open the sluices.) In which case, might his cloud yet turn silver? Crying as catharsis—an upwelling of relief at honouring some long-forgotten vow, or the glimpse of a prayed-for haven. In more puzzling cases, even months of talking to intimates and therapists do not settle whence a weeper's weeping came. But the tears themselves are undeniable. Sobbing is a body thing.

Darwin's *The Expression of the Emotions in Man and Animals*, treats both of bodily movements, and of their 'cause or meaning'.[2] It stresses what is observable, advancing a study 'of the theory of expression,' *not* a theory of emotion.[3] Separating expression from emotion opens up to question the psychological corollaries of

---

[1] In 1868 Jane Gray (wife of Harvard botanist Asa Gray) wrote this in her journal-letter to her sister Susan Loring, when visiting the 59-year-old Darwin and his family at Down House: Jane Gray, 'Letter to Susan Loring, 28th October to 2nd November,' Darwin Correspondence Project, https://www.darwinproject.ac.uk/people/about-darwin/family-life/visiting-darwins.

[2] E.g. Charles Darwin, 'Letter to Fritz Müller, Feb 22nd,' *Correspondence of Charles Darwin* (Cambridge: Darwin Correspondence Project, 1867), https://www.darwinproject.ac.uk/letter/?docId=letters/DCP-LETT-5410.xml; Charles Darwin, *The Expression of the Emotions in Man and Animals*, 2nd ed. (London: Murray, 1890), p.239; Charles Darwin, 'A Biographical Sketch of an Infant,' *Mind: A Quarterly Review of Psychology and Philosophy* 2, (1877), pp.293–294.

[3] Darwin, *Expression*, p.387. The word 'emotion' is not listed in the book's index.

Darwin's Psychology. Ben Bradley, Oxford University Press (2020). © Oxford University Press.
DOI: 10.1093/oso/9780198708216.001.0001

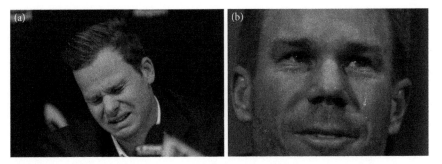

Figure 4.1 Responding to a national furore over accusations of cheating (ball-tampering) during a cricket match in South Africa on 24th March 2018, Australia's Captain Steve Smith (a) and Vice-Captain David Warner (b) make public apologies.
Source: (a) Photo by Brook Mitchell/ Stringer / Getty Images; (b) Photo by Peter Parks / AFP / Getty Images.

expression. Were facial displays best construed in terms of the types of dramas which gave them sense? Or did the physical contours of an expression channel the expresser's internal state, making them mere side effects of emotions?[4]

The most popular contemporary reading of *Expression* makes the book argue that expression *results from* emotion. We hear that Darwin 'believed that a state of mind causes muscular discharge (such as coordinated sets of facial muscle contractions) that expresses said state of mind.'[5] This reading imagines a one-way street: my emotion of, say, anger *causes* my face and posture to display rage. It makes Darwin an *essentialist*, in holding that 'internal essences ("emotions") are externalized via our different facial expressions.'[6] So—is the vision in *Expression*

---

[4] Beatrice De Gelder, *Emotions and the Body* (Oxford: Oxford University Press, 2016), pp.2–3.

[5] Lisa Barrett, 'Was Darwin Wrong about Emotional Expressions?,' *Current Directions in Psychological Science* 20, (2011), p.400. See also: Paulo Jesus, 'The Embodied Nature of Emotions: On Charles Darwin and William James' Legacy,' *Revista da Faculdade de Letras—Série de Filosofia 27–28*, (2010–2011), pp.149–154: 'for Darwin, emotions are states of mind, and the so-called expressions are originally nothing but the serviceable actions provoked by those states of mind ...'; Joel Krueger, 'Dewey's Rejection of the Emotion/Expression Distinction,' in *Neuroscience, Neurophilosophy and Pragmatism: Understanding Brains at Work in the World*, ed. Tibor Solymosi and John Shook (London: Palgrave Macmillan, 2014), p.141: 'According to Darwin, the emotion itself—its affective core, its felt aspect—exists antecedently to and independently of its expression'; Jim Garrison, 'Dewey's Theory of Emotions: The Unity of Thought and Emotion in Naturalistic Functional "Co-Ordination" of Behaviour,' *Transactions of the Charles S. Peirce Society* 39, (2003), pp.406–407; Thomas Dixon, *From Passions to Emotions: The Creation of a Secular Psychological Category* (Cambridge: Cambridge University Press, 2003): p.174, which claims Darwin 'endorse[d] the view that certain movements in animals and humans functioned primarily to express inner feelings'; Maria Gendron and Lisa Barrett, 'Facing the Past: A History of the Face in Psychological Research on Emotion Perception,' in *The Science of Facial Expression*, ed. James Russell and José-Miguel Fernández-Dols (Oxford: Oxford University Press, 2017), p.18: in *Expression*, Darwin subscribed to the view that 'specific facial muscle movements are caused by specific internal states (emotions), and thus they can be used by the perceiver to read the emotions of others.'

[6] Carlos Crivelli and Alan Fridlund, 'Facial Displays Are Tools for Social Influence,' *Trends in Cognitive Sciences* 22, (2018), p.388.

essentialist? Certainly, one can fish out statements from the book which seem to say emotions cause expressions. Yet the wording of Darwin's books often proves ambiguous.

4.1 The Language Problem

Ordinary language brims with psychological assumptions. It was partly for this reason that twentieth-century psychologists developed a technical vocabulary, and a way of reporting findings, which only made sense to a coterie of like-minded fellow-professionals.[7] Here, psychology confronts a difficulty unknown to natural sciences like biochemistry or physics, where there is no common language for the problems that invite investigation: enzymes; sub-atomic particles; chemistry at the synapse. But, in psychology, to sever the argot of research from everyday speech courts incoherence, even fatuity. A finding which casts light on ordinary life must intersect with idiomatic terms (e.g. an *introvert* is 'shy' and 'reticent'). Which makes attempts to insulate psychological terms from ordinary language like King Canute rebuking the waves, or Humpty Dumpty talking down to Alice (Figure 4.2).

Darwin's approach to psychological language was not modern. In Darwin's day, scientific books sold to a broad, non-specialist readership. Arguably these general readers formed Darwin's primary audience, especially for *Expression*: an audience he always treated with respect, seldom subjected to obscure technical terms, and constantly embraced with his predominant pronoun, a participatory 'we'.[8] His language in *Origin* exploits colloquial figures of speech and an informal diction witnessed by the book's refusal to define or delimit the meanings of words so crucial to its argument as *species, instinct*, and *natural selection*.[9] Time and again the book defers to what 'everyone knows.'

[7] Michael Billig, 'Repopulating Social Psychology: A Revised Version of Events,' in *Reconstructing the Psychological Subject: Bodies, Practices and Technologies*, ed. Betty Bayer and John Shotter (London: Sage, 1998), pp.126–151.
[8] Angelique Richardson, '"The Book of the Season": The Conception and Reception of Darwin's Expression,' in *After Darwin: Animals, Emotions, and the Mind*, ed. Angelique Richardson (New York: Rodopi, 2013), pp.51–88.
[9] Cf. Gillian Beer, *Darwin's Plots: Evolutionary Narrative in Darwin, George Eliot and Nineteenth-Century Fiction*, 2nd ed. (Cambridge: Cambridge University Press, 2000); Robert Young, *Darwin's Metaphor: Nature's Place in Victorian Culture* (Cambridge: Cambridge University Press, 1985). E.g. 'I will not attempt any definition of instinct. It would be easy to show that several distinct mental actions are commonly embraced by this term; but every one understands what is meant, when it is said that instinct impels the cuckoo to migrate and to lay her eggs in other birds' nests. An action, which we ourselves should require experience to enable us to perform, when performed by an animal, more especially by a very young one, without any experience, and when performed by many individuals in the same way, without their knowing for what purpose it is performed, is usually said to be instinctive.': p.207 Charles Darwin, *On the Origin of Species by Means of Natural Selection or the Preservation of Favoured Races in the Struggle for Life* (London: Murray, 1859).

Figure 4.2 'When *I* use a word,' Humpty Dumpty said, in rather a scornful tone, 'it means just what I choose it to mean — neither more nor less.'
'The question is,' said Alice, 'whether you *can* make words mean so many different things.'
'The question is,' said Humpty Dumpty, 'which is to be master — that's all.'
No sooner had this conversation finished than egghead Humpty toppled off his high wall and shattered.
Source: Carroll, L. (1897). *Through the Looking-Glass and What Alice Found There*. London: Macmillan and Co. Ltd.
https://birrell.org/andrew/alice/lGlass.pdf, p.81.

Expression defines neither emotion nor expression. The nearest it gets is a footnote referring to Spencer: 'Mr. Herbert Spencer has drawn a clear distinction between emotions and sensations, the latter depending on an individual's agency and so being, "generated in our corporeal framework." '[10] Darwin only notes Spencer's definition of *sensation*, however. He ignores Spencer's definition of emotion (as a 'kind of feeling' which was *not* a direct result of bodily action, being 'independently generated in consciousness').[11] In fact, *Expression* gaily flouts Spencer's careful distinction by referring to 'emotions or sensations' as if these were interchangeable.[12]

---

[10] Darwin, *Expression*, p.28; Herbert Spencer, 'Bain on the Emotions and the Will,' in *Essays: Scientific, Political, and Speculative (Second Series)* (London: Williams and Norgate, 1863), p.138.
[11] Spencer, 'Bain,' p.138.
[12] Darwin, *Expression*, pp.87, 138, 369.

In writing *Expression*, Darwin confronted a deeper problem than that of defining terms, namely, the common sense theories of cognition and emotion embedded in Victorian speech (and in ours). He needed to tell his story 'against the grain of the language available to tell it in.'[13] Aware of this problem, Darwin wrapped up *Expression* by saying:

> Throughout this volume, I have often felt much difficulty about the proper application of the terms, will, consciousness, and intention. Actions, which were at first voluntary, soon become habitual, and at last hereditary, and may then be performed even in opposition to the will. *Although they often reveal the state of the mind, this result was not at first either intended or expected. Even such words as that 'certain movements serve as a means of expression' are apt to mislead, as they imply that this was their primary purpose or object.* This, however, seems rarely or never to have been the case ...[14]

Critics like John Dewey (1859–1952) ignore *Expression*'s warning about the deceptiveness of ordinary language when concluding that, in using the phrase 'expression of emotion,' Darwin espoused essentialism. Dewey carped that Darwin 'begs the question of the relation of emotion to organic peripheral action, in that it assumes the former as prior and the latter as secondary.'[15] Such is the informality of *Expression*'s language, it sometimes seems Darwin ignored his warning too. Like *Origin*, the wording in *Expression* proves neither single nor simple, and was thus capable of being 'extended or reclaimed into a number of conflicting systems.'[16] Unlike *Origin*, the twentieth-century provided no scientific consensus as to a correct reading of *Expression*. Twentieth-century evolutionary biology fixed what *Origin* 'really meant' by assimilating it to the rubric of the Modern Synthesis. Twentieth-century psychology remained as confused about relations between emotions, cognitions, expressions, and circumstances as were the Victorians.

Today we have three competing takes on emotion: *essentialist* (invisible emotions cause visible expressions); *somatic* (emotion is a sensation or reading of one's own bodily changes—I feel frightened because I run away); and *externalist* or *situationist* (emotion is attributed according to how expressive actions interface with their external circumstances).[17] Given the informality of *Expression*'s language, critics adhering to one or other of these views, can easily select quotes from

---

[13] Cf. Beer, *Darwin's Plots*, pp.xviii, 3, 48.

[14] Darwin, *Expression*, p.377. Perhaps, John Dewey, 'The Theory of Emotion (1) Emotional Attitudes,' *Psychological Review* 1, (1894), p.553, did not read this paragraph.

[15] Dewey, 'Emotional Attitudes,' p.553. Dewey's is a facile take. Thus exact phrase does not occur in *Expression*. 'The expression of *the* emotions' occurs twice—*including the title*.

[16] Beer, *Darwin's Plots*, p.3.

[17] José-Miguel Fernandez-Dols, 'Facial Expression and Emotion: A Situationist View,' in *The Social Context of Nonverbal Behaviour*, ed. Pierre Philippot, Robert Feldman, and Erik Coats (Cambridge: Cambridge University Press, 1999), pp.242–261.

Expression's less corseted passages to align it with their ideas. Hence, today we hear that Darwin intended *Expression* to achieve a host of often-incompatible aims.

Commentators may or may not read the book as a 'materialist study,' and disagree over its agenda. Is the book's aim evolutionary or physiological?[18] And whose views did it aim to refute? Grandfather Erasmus Darwin's out-dated theory?[19] Or the natural theology of Charles Bell, in which expressions are 'windows' on the soul?[20] Does the book champion natural selection?[21] Or the inheritance of habits as an explanation for human agency that was (in the 1870s) deemed *superior* to natural selection?[22] As for Darwin's personal feelings, did he write *Expression* to restore his confidence in 'the continuities of emotional gesture across species' which had been shattered by his meeting in 1832 with the 'savages' of Tierra Fuego?[23] Or was it fuelled by his hatred of slavery?[24]

In this chapter, I seek the overall structure of argument and observation composing *Expression*. I find that *Expression* places the same primacy on the theatre of agency as did Darwin's theory of natural selection. Going about their daily lives, humans, like their ancestors and their animal relations, unconsciously move and (more consciously) act in patterned ways, *some* of which have emotional meanings for others. The fact that physical movements have emotional meanings arises more from their interdependency with the interpersonal circumstances in which their expressers act than from an inner state of consciousness.

In evidence for this interpretation, I put most weight on what Darwin *did* when studying expressions: that is, on his methods of research. As for the wording of *Expression*, I have found this becomes less ambiguous the more one reads of the book. For example, when we come across a section headed: 'Contrast between the

---

[18] Gesa Stedman, *Stemming the Torrent: Expression and Control in Victorian Discourses on Emotions, 1830–1872* (Aldershot: Ashgate, 2002), pp.55–58; Paul White, 'Darwin's Emotions: The Scientific Self and the Sentiment of Objectivity,' *Isis* 100, (2009), p.826; Otniel Dror, 'Seeing the Blush: Feeling Emotions,' in *Histories of Scientific Observation*, ed. Lorraine Daston and Elizabeth Lunbeck (Chicago: University of Chicago Press, 2011), p.327.

[19] William Montgomery, 'Charles Darwin's Thought on Expressive Mechanisms in Evolution,' in *The Development of Expressive Behaviour: Biology-Environment Interactions*, ed. Gail Zivin (London: Academic Press, 1985), pp.27–50.

[20] Susan Campbell, 'Emotion as an Explanatory Principle in Early Evolution Theory,' *Studies in History and Philosophy of Science* 28, (1997), pp.453–473; Richard Burkhardt, 'Darwin on Animal Behavior and Evolution,' in *The Darwinian Heritage*, ed. David Kohn (Princeton NJ: Princeton University Press, 1985), p.457; Cf. Charles Bell, *Essays on the Anatomy of Expression in Painting* (London: Longman, Hurst, Rees, and Orme, 1806), p.108.

[21] Roger Smith, *The Fontana History of the Human Sciences* (London: Fontana Books, 1997), p.476.

[22] Gregory Radick, 'Darwin on Language and Selection,' *Selection* 1, (2002), p.14.

[23] Gillian Beer, *Open Fields: Science in Cultural Encounter* (Oxford: Clarendon Press, 1996), pp.23–25.

[24] Gregory Radick, 'Darwin's Puzzling *Expression*,' *Comptes Rendus Biologies* 333, (2010), pp.181–187, passim. For yet more contradictory readings, see: Janet Browne, 'Darwin and the Expression of the Emotions,' in *The Darwinian Heritage*, ed. David Kohn (Princeton NJ: Princeton University Press, 1985), p.309; Paul Ekman, 'An Argument for Basic Emotions,' *Cognition and Emotion* 6, (1992), pp.169–200; Frans De Waal, 'Darwin's Legacy and the Study of Primate Visual Communication,' *Annals of the New York Academy of Siences* 1000, (2003), pp.7–31; Barrett, 'Was Darwin Wrong'; Alan Fridlund, 'Darwin's Anti-Darwinism in the *Expression of the Emotions in Man and Animals*,' *International Review of Studies on Emotion* 2, (1992), pp.117–137.

emotions which cause and do not cause expressive movements,'[25] or a paragraph beginning: 'With all or almost all animals, even with birds, Terror causes the body to tremble'—we should not immediately project onto these isolated snippets the conclusion that, for Darwin, an *internal* emotion (of terror) was *externalized* by an expression (trembling). We should read on. Then we will find that *Expression* traces the cause of things like terrified trembling, not to some *subjective* state, but to a *physiological* disturbance: the excited 'state of the sensorium [the sensory elements of the nervous system], and the consequent undirected overflow ... of nerve-force.'[26]

4.2 A Drama of Surfaces

Darwin conceived the economy of nature as bound together by interdependency: the moth is drawn to the orchid by its blooms' appearance—shape, scent, colour, pattern—to sip its nectar with a specially shaped tongue, just as the flower snugly gloves the outward form of the moth, dusting it with pollen. In like fashion, *Expression* entails a drama of surfaces.[27]

Facial expressions are surface-phenomena, and *Expression* delves into their distinctive anatomical presentations with gusto. Sobbing, eyes shut, brows pulled down and together, tears spilling, posture slumped, lips an unlucky horseshoe (Figure 4.1, a). All such movements get examined in minute physiological detail—Darwin's first focus always being the observable, physical, features of expressions.

Darwin's bent for expression had a long genesis.[28] He had opened the first of his two human-focused 'transmutation' notebooks in July 1838, when he was twenty-nine (see 2.5 Darwin's Approach to the Study of Humans).[29] He labelled it 'M,' noting in his diary that he had 'opened note book, connected with Metaphysical Enquiries.'[30] Three months later it was full. Notebook 'N' was M's sequel, described

---

[25] Darwin, *Expression*, p.69.

[26] Darwin, *Expression*, pp.80–81.

[27] Cf. Tiffany Watt-Smith, 'Darwin's Flinch: Sensation Theatre and Scientific Looking in 1872,' *Journal of Victorian Culture* 15, (2010), pp.101 & passim: *Expression* stages an 'economy of emotional surfaces.'

[28] Examine the working portfolios that Darwin later constructed as resources for his book-writing, and we find the bulk of the pages he cut out from notebooks M and N were used in the portfolio for *Expression* (1872), not for *Descent* (1871)—demonstrating the greater continuity of *Expression* with his earlier interests: 'Table of Location of Excised Pages,' in Paul Barrett et al., eds., *Charles Darwin's Notebooks, 1836–1844* (Ithaca NY: Cornell University Press, 1987), p.632.

[29] Charles Darwin, 'Darwin's Journal,' *Bulletin of the British Museum (Natural History). Historical Series* 2, (1959), p.1; Charles Darwin, 'Notebook M,' (Cambridge: Darwin Online, 1838).

[30] Notebook A dealt with geology. B and C were Darwin's first 'transmutation' notebooks. D and M are both dated 15th July 1838—implying D was a continuation of the old series but Darwin realized he now needed M as well—to start a new, parallel series. Choice of the letter 'M' may, or may not, have alluded to Man and/or Metaphysics (M is also the 13th letter of a 26-letter alphabet). Darwin, 'Notebook B,' p.16r.

as concerning 'Metaphysics & Expression'. When we shut M and N, however, we notice both sport the same word on their front and back covers: 'Expression'.[31]

Given that Darwin *began* by filling M with 'metaphysical inquiries', it appears that the paramount importance of their theme of 'expression' emerged *from* those inquiries. Sure enough, if we read M and N in sequence, we find Darwin's initial enthusiasm for discussing consciousness and metaphysics succumbing to a scepticism about the Mind, as being little more than a 'function of body'.[32] The watershed apparently came in autumn 1838, when Darwin read an essay by James Ferrier on 'The Philosophy of Consciousness'. Ferrier launched a swingeing attack on the very possibility of any psychology, or 'science of the human mind'. Why? Because: 'Man is a "living soul," but science has been trained among the *dead*.'[33]

For Ferrier, *mind* is only an entity worth considering—and even then, only in a colloquial sense—when a person or 'free agent' is *in action*. Compare a dragnet. Trawl, and the net resists, puffed up with water. Out of the water, the net becomes a slack and shapeless tangle of mesh. Likewise in psychology: 'Manifested in ... the fire and rapid combinations of the orator, the memory of the mathematician, the gigantic activities and never-failing resources of the warrior and statesman, or even the manifold powers put forth in every-day life by the most ordinary men', and mind is a productive, living, breathing, reality. But if you imagine the human mind can be dissected as an entity distinct from daily events, you will find that it becomes a 'lifeless abstraction', an 'airy nothing'.[34]

How was the allure of this false parallel sustained, between the living quarry of psychology and the corpse-like objects of natural science? asked Ferrier. By: 'certain curious verbal or grammatical considerations which lie on the very surface of the exposition given of the usual scientific procedure, as applied both to nature and to man.' If you disentangled these grammatical curiosities, you would find: 'that which is called mind is truly an object only in a fictitious sense, and being so, is, therefore, only a fictitious object, and consequently the science of it is also a fiction and an imposture'.[35]

---

[31] The back cover of Notebook N has the word 'Expressions' in the plural. NB A third set of valuable miscellaneous notes examining the human implications of transmutation were packaged originally in a single parcel and labelled by Darwin: 'Old and useless Notes about the moral sense & some metaphysical points written about the year 1837 & earlier.' Actually most of the notes, as indicated by their dates or their watermarks, were written during the years 1837–40.

[32] Charles Darwin, 'Notebook N', ed. Paul Barrett (1838–1839), p.5.

[33] James Ferrier, 'An Introduction to the Philosophy of Consciousness, Part I,' *Blackwood's Edinburgh Magazine* 43, (1838), p.192, his italics. Darwin comments on Ferrier's essay (Parts I and II) are to be found in Darwin, 'Notebook M'. Darwin had at this time just finished re-reading his grandfather's extraordinarily fertile book *Zoonomia* (Erasmus Darwin, *Zoonomia, or, the Laws of Organic Life*, Vol.1 (London: Johnson, 1794) There he would have found a detailed *physiological* associationism that focused on 'fibrous contractions', somewhat after the manner of David Hartley, *Observations on Man: His Frame, His Duty, and His Expectations* (Gainsville FL: Scholars' Facsimiles and Reprints, 1966).

[34] Ferrier, 'Consciousness', pp.192–194.

[35] Ferrier, 'Consciousness', p.197. Darwin seems to have read and possibly re-read Ferrier's essay in August and/or September 1838, just as he was finishing Notebook M. See also, Darwin, 'Notebook N', p.5.

Accordingly, and in contrast to the semantic complexities surrounding topics like 'will, consciousness, and intention,' Darwin noted that expressions were 'the reverse of intellectual, there is no comparison of ideas.'[36] This went for humans, not just animals. Hence, whilst he feared readers would frown at publications arguing worms had intelligence, monkeys imagined, and dogs held beliefs, they might happily accept that humans were like other mammals in their outbreaks of lust, fury, and terror.[37] So, if he could make an evolutionary argument *about expression*, he would ask of his readers a more leapable leap of faith when linking their agency to that of animals.

Unlike the linguistic and introspective contortions required to divine and articulate one's own or others' ideas and intentions, the study of expression played neatly to Darwin's strength: the art and craft of describing. With expression, you could, 'forget the use of language, & judge only by what you see,' he noted.[38] And there were precedents. Witness the writings of Swiss poet and philosopher John Caspar Lavater (1741–1801). Darwin read Lavater closely, both in the 1830s and for citation in *Expression*.[39] Lavater had pioneered *physiognomy*. And physiognomy was *a science of reading*.[40] Physiognomists discerned character by decoding the habitual expressions of faces. From whatever point of view 'Man' (i.e. human beings) was considered, argued Lavater, he could only be known 'by certain external manifestations; by the body, by his surface.'[41]

Darwin talked about what *caused* expressive movements in terms of the observable actions of muscles. A typical example: seeking 'the cause of the obliquity of the eyebrows under suffering,' he summarizes some first-hand experimental observations of his children as showing that an 'involuntary contraction of the pyramidal [i.e. the *pyramidalis nasi*, a muscle running down the bridge of the nose; see D in Figure 4.3] caused the basal part of their noses to be transversely and deeply wrinkled,' and so drew down their eyebrows.[42]

Darwin made no grand claims about the relations between physiology and psychology—though his last words in *Expression* do commend its theory of

---

[36] Darwin, *Expression*, p.377; Darwin, 'Notebook M,' p.108; cf. Charles Darwin, 'Old & Useless Notes,' (Cambridge: Darwin Online, 1838-1840), p.36.

[37] Darwin, 'Notebook M,' p.123: 'Our descent, then, is the origin of our evil passions!! — The Devil under form of Baboon is our grandfather!'

[38] Darwin, 'Notebook M,' p.153e; Darwin, 'Old & Useless Notes About the Moral Sense & Some Metaphysical Points,' p.36.

[39] Lucy Hartley, *Physiognomy and the Meaning of Expression in Nineteenth-Century Culture* (Cambridge: Cambridge University Press, 2001).

[40] John Lavater, *Essays on Physiognomy, Designed to Promote the Knowledge and the Love of Mankind, Vol.I* (London: Murray, Hunter & Holloway, 1789), p.12: 'All people read the countenance pathognomonically; few indeed read it physiognomonically.'

[41] Lavater, *Physiognomy*, p.14. Cf. Scott Juengel, 'Godwin, Lavater, and the Pleasures of Surface,' *Studies in Romanticism* 35, (1996), pp.73–97.

[42] Darwin, *Expression*, pp.196–201. In this he differed from Spencer and Bain: see 4.6 'Resonances.'

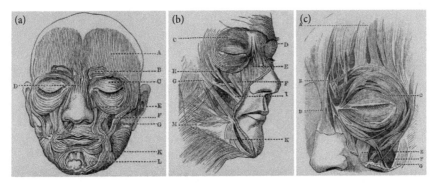

Figure 4.3 A. *Occipito-frontailis*, or frontal muscle; B. *Corrugator supercilii*, or corrugator muscle; C. *Orbicularis palpebrarum*, or orbicular muscles of the eyes; D. *Pyramidalis nasi*, or pyramidal muscle of the nose; E. *Levator labii superioris alæque nasi*; F. *Levator labii proprius*; G. *Zygomatic*; H. *Malaris*; I. *Little zygomatic*; K. *Triangularis oris*, or depressor *anguli oris*; L. *Quadratus menti*; M. *Risorius*, part of the *Platysma myoides*.

Source: Darwin, C. (1890). *The Expression of the Emotions in Man and Animals.*

These are the first three figures (Darwin borrowed the first from Charles Bell and the other two from Jacob Henle), with their legend, from Darwin, *Expression*, pp.24–25.

expression to 'further attention, especially from any able physiologist.'[43] Yet even a flick through the book shows its principal interest lay in morphology and physiology. The first chapter includes three diagrams of a skinned face, with a key to the twelve muscles depicted (Figure 4.3). From there on, we constantly hear about the antics of the *corrugator supercilii*, the *orbicularis palpebrarum*, the *zygomatic*, the depressor *triangular oris*, and so on—discussion being salted with findings from contemporary physiologists.

For Darwin, the causes of muscle action were other muscle actions: the expressions of affection in a dog 'consist in the head and whole body being lowered and thrown into flexuous movements, with the tail extended and wagged from side to side. The ears fall down and are drawn somewhat backwards, which causes the eyelids to be elongated, and alters the whole appearance of the face.'[44] Here, everything lies on one plane: muscle movements cause muscle movements. Moreover, whenever something sets off a sequence of readable expressive movements in *Expression*, this most often occurs, not in the depths of The Mind, but *outside* the body. Darwin would laconically gloss an expression or attitude as being 'of fear.'

---

[43] Darwin, *Expression*, p.387. This contrasts with Bain and Spencer: see Alexander Bain, *The Senses and the Intellect*, 3 ed. (London: Longmans Green, 1868), p.iii; Herbert Spencer, *The Principles of Psychology*, Vol.2, 2nd ed. (London: Williams & Norgate, 1872), p.302.

[44] Darwin, *Expression*, p.124.

But when he described it more fully, he invoked a *mise-en-scène* that his readers would recognize as fear-inducing. Thus: 'Mr. Sutton, the intelligent keeper in the Zoological Gardens, carefully observed for me the Chimpanzee and Orang; and he states that when they are suddenly frightened, *as by a thunderstorm*, or when they are made angry, *as by being teased*, their hair becomes erect.'[45]

Mind is a word *Expression* uses colloquially, not technically. It most often occurs in the phrases, 'states of the mind' or 'frame of mind.' Such states are treated as short-hand for the bodily actions and evoking circumstances typically recognized to constitute a particular expression. So Darwin refers to 'various states of the mind *or body*.'[46] 'Mind' may also connote a person in a recognizable set of circumstances: 'After the mind has suffered from an acute paroxysm of grief'; maternal love is 'the strongest [form of love] of which the mind is capable'; 'when the mind is strongly excited'; etc.[47]

Rage illustrates the way muscles produce meanings *via a context*. Darwin lists several signs of rage: standing tall; face flushed; chest aheave; mouth tight; teeth clenched; glittering eyes; arms tense; fists formed.[48] These may be provoked as easily by—one's own stupidity, a door that won't open, or recalling an insult—as by a real-time confrontation with a flesh-and-blood antagonist. Nevertheless, most symptoms of rage 'represent more or less plainly the act of striking or fighting with an enemy.'[49] So the target-range of one's rage and anger *today* must come down to the effects of association. The *expressive marks* of rage go back to real fighting in a dim evolutionary past—as we shall see (4.4 Evolutionary Purposelessness). But these marks are triggered today by 'analogous circumstances.'[50] And in a way that *bypasses consciousness*:

> it seems probable that some actions, which were at first performed consciously, have become through habit and association converted into reflex actions, and are now so firmly fixed and inherited, that they are performed, even when not of the least use ... In such cases the sensory nerve-cells excite the motor cells, without first communicating with those cells on which our consciousness and volition depend.[51]

So: something happens in the external world which, due to the *unconscious* associations built up by an individual's previous experiences, triggers a reflexive expression of rage. Which gives us a circumstantial or 'situational' account of expression,

---

[45] Darwin, *Expression*, p.95, my italics. This amounts to a 'situational' account of emotional expression: see 4.3.1 'Observations'; 4.6 'Resonances.'

[46] E.g. Darwin, *Expression*, p.44 (my italics).

[47] Darwin, *Expression*, pp.186, 224, 72, 209.

[48] Darwin, *Expression*, pp.78–79.

[49] Darwin, *Expression*, p.78.

[50] Darwin, *Expression*, pp.86, 238.

[51] Darwin, *Expression*, p.39.

not an essentialist one. For Darwin posits no antecedent and independent state of consciousness, *an emotion of rage*, that causes bared teeth, heavy breathing, and so on.

4.3 Methods of Study

If the meaning of an expression were caused by a prior emotion 'inside' the emitter's mind, then the best means of discovery would be introspection. This was the method advocated in Bain's psychology, and in Spencer's.[52] *Expression* makes no mention of introspection, however, instead using *the responses of others* to determine whether and what an expression meant. The studies informing *Expression* took a recognizer's or reader's perspective, not an emoter's perspective. In all, Darwin devised five different research strategies to help identify the meanings of potentially expressive movements, all observational, none introspective. These give a clear vision of how he understood human agency. I will deal with each in turn.

4.3.1 Observations

Darwin distrusted descriptions not built on first-hand experience. Even direct observation was fallible when dealing with expressions: because 'our sympathy is so strongly excited, that close observation is forgotten or rendered almost impossible.' Imagination was a yet greater threat: 'for if from *the nature of the circumstances* we expect to see any expression, we readily imagine its presence.'[53] Hence Darwin adopted several convergent methods to check—or, in today's lingo, *triangulate*— his and others' observations. In particular, he enlisted the help of other experts who could observe phenomena at first hand. Mothers were asked to observe their babies screaming.[54] Doctors were requested to photograph or record the expressions of the insane. Both surgeons and artists were persuaded to observe 'young

---

[52] Alexander Bain, 'The Respective Spheres and Mutual Helps of Introspection and Psychophysical Experiment in Psychology,' *Mind (New Series)* 2, (1893), p.42: 'Introspection is still our main resort—the alpha and the omega of psychological inquiry: it is alone supreme, everything else subsidiary'; Herbert Spencer, *The Principles of Psychology*, 2 vols., vol. 1 (London: Williams & Norgate, 1870), passim; Dixon, *From Passions to Emotions*, pp.154, 192.

[53] Darwin, *Expression*, pp.13–14, my italics: NB Darwin assumes once again it is a reading of the forms of life or social 'circumstances' surrounding an expressive movement that precipitates its interpretation.

[54] E.g. Charles Darwin, 'Letter to Huxley, 30th January,' *Correspondence of Charles Darwin* (Cambridge: Darwin Correspondence Project, 1868), https://www.darwinproject.ac.uk/letter/?docId=letters/DCP-LETT-5817 Cf. Samantha Evans, ed. *Darwin and Women: A Selection of Letters* (Cambridge: Cambridge University Press, 2017), pp.122–135.

and inexperienced girls,' who had been asked to undress, 'and *who at first blush much*, how low down the body the blush extends.'[55]

Such observations targeted 'the causes of the movement of certain muscles' in human and animal expressions.[56] His correspondence from 1867 to 1871 bears this out. Fifty-six letters query the antics of single muscle-groups: the muscle 'which the French call the "Grief muscle"' (that raises the inner corner of the eyebrow), the orbiculars (around the eyes; C in Figure 4.3), and, the *platysma myoides* (a muscle in the neck region; M in Figure 4.3).[57] In 1869, Darwin told the psychiatrist James Crichton-Browne (1840–1938) that the platysma 'has been my *bête noire* for a year or two.'[58] Why? Because a French pioneer of modern neurology, Guillaume-Benjamin-Amand Duchenne (1806–1875), had claimed that the platysma contracted when expressing horror, dubbing it 'the muscle of fright.' Darwin doubted Duchenne. So he consulted fourteen experts to check Duchenne's observation, plus two of his sons. Forty letters were exchanged dealing with this one muscle, generating six pages of text in *Expression*. Darwin's conclusion? That there was no sure connection between expressions of fear and the platysma. The platysma 'often contracts during a shudder,' and a shudder or shiver was *sometimes* associated with fearful situations, real or imagined. But the same muscle also contracted when we shudder from cold on getting out of a warm bed in winter, or at a painful thought.[59]

We now call Darwin's distinctive muscle-based approach to expressions *component analysis*.[60] Darwin extended it from adults to the study of babies, animals, 'the insane,' and 'savages.' All these forms of triangulation helped to show whether an expression was instinctive, or a 'conventional' *gesture* (i.e. fruit of an individual's social experience). Darwin saw an advantage in studying infants here, because their expressions were not only forceful, but 'pure and simple,' that is, unadulterated by convention (see Ch. 9).[61] The mad should also be studied, he thought, 'as they are liable to the strongest passions, and give uncontrolled vent to them.'[62] Which gave Crichton-Browne his value, as he superintended—and therefore could observe and photograph those incarcerated in—the West Riding Lunatic Asylum of Wakefield, Yorkshire.

---

[55] Charles Darwin, 'Letter to Woolner, 7th April,' *Correspondence of Charles Darwin* (Cambridge: Darwin Correspondence Project, 1871), https://www.darwinproject.ac.uk/letter/?docId=letters/DCP-LETT-7665.xml, Darwin's emphasis.

[56] Charles Darwin, 'Letter to Crichton-Browne, 22nd May,' *Correspondence of Charles Darwin* (Cambridge: Darwin Correspondence Project, 1869), https://www.darwinproject.ac.uk/letter/?docId=letters/DCP-LETT-6755.xml.

[57] Darwin, *Expression*, p.16.

[58] Charles Darwin, 'Letter to Crichton-Browne, 8th June,' *Correspondence of Charles Darwin* (Cambridge: Darwin Correspondence Project, 1869), https://www.darwinproject.ac.uk/letter/?docId=letters/DCP-LETT-6779.xml.

[59] Darwin, *Expression*, pp.298–303.

[60] Paul Griffiths, *What Emotions Really Are: The Problem of Psychological Categories* (Chicago: University of Chicago Press, 1997), pp.45–46 & passim.

[61] Darwin, *Expression*, p.13.

[62] Darwin, *Expression*, p.14.

Darwin's approach put him in the mainstream of natural history. Francis Darwin remarked that 'The "Expression of the Emotions" shows how closely he watched his children,' but added, 'it was characteristic of him that (as I have heard him tell), although he was so anxious to observe accurately the expression of a crying child, his sympathy with the grief spoiled his observation.'[63] Darwin's solution to this problem was to describe *photographs* of emotional expressions: 'I have found photographs made by the instantaneous process the best means for observation, as allowing more deliberation' (cf. 2.3 'Instrumental aids to discovery').[64]

Darwin rejected masterpieces of painting and sculpture as clues to expression because, in these, aesthetic concerns meant faces were rarely depicted in the grip of strongly contracted expressions—the 'force and truth' of expressions rather being conveyed by a *mise-en-scène* of 'skilfully given accessories.'[65] Hence the value of photography. In fact, he was one of the earliest authors to include photographs in a scientific publication—and the first author ever to exploit 'heliotyping,' a genuinely photo-mechanical printing process.[66]

Darwin's use of photographs has been called a rhetorical 'narrative strategy.'[67] What Darwin and his son say make it clear that *he* primarily saw photography as an *observational* strategy: a way to distance the observer from the strong and disturbing feelings aroused in the presence of a fellow human—particularly a crying child—when trying dispassionately to note the musculature of the movements producing their expression. Photography was a way of insulating the 'man of science' from feeling the sympathies characteristic of the 'sentimental subject.'[68] Photographs rendered expressions as *specimens* that could receive much the same treatment as had the specimens Darwin collected on *HMS Beagle*.[69] So, what did Darwin's use of photographs imply about expression?

Duchenne's photographs showed people who were having individual facial muscles experimentally ('galvanically') stimulated by electrodes (Figure 4.4a). *Galvanism* enacted a radical severance of expression from 'internal' emotion.[70]

---

[63] Francis Darwin, 'Reminiscences of My Father's Everyday Life,' in *The Life and Letters of Charles Darwin*, ed. Francis Darwin (London: John Murray, 1887), p.132.

[64] Darwin, *Expression*, p.155.

[65] Darwin, *Expression*, pp.15–16.

[66] Phillip Prodger, *Darwin's Camera: Art and Photography in the Theory of Evolution* (Oxford: Oxford University Press, 2009).

[67] Phillip Prodger, 'Illustration as Strategy in Charles Darwin's "the Expression of the Emotions in Man and Animals"', in *Inscribing Science: Scientific Texts and the Materiality of Communication*, ed. Timothy Lenoir (Stanford CA: Stanford University Press, 1998), p.141.

[68] Re photography and scientific subject-positions, see: Paul White, 'The Face of Physiology,' *19: Interdisciplinary Studies in the Long Nineteenth Century* 7, (2008), https://19.bbk.ac.uk/article/id/1567. p.16; Paul White, 'Darwin Wept: Science and the Sentimental Subject,' *Journal of Victorian Culture* 16, (2011), pp.195–213.

[69] But see Paul White, 'The Emotional Specimen,' (2017).

[70] Duchenne treated the people he galvanized as lifeless mannequins, whom he posed and manipulated with his electrical probes. The man pictured in Figure 4.4 was said to suffer from a 'complicated anaesthetic condition' which meant that he could not even feel the electrodes on is face! Guillaume-Benjamin Duchenne, *Mécanisme De La Physionomie Humaine, Ou, Analyse Électro-Physiologique De L'expression Des Passions* (Paris: Jules Renouard, 1862); White, 'The Emotional Specimen,' p.5.

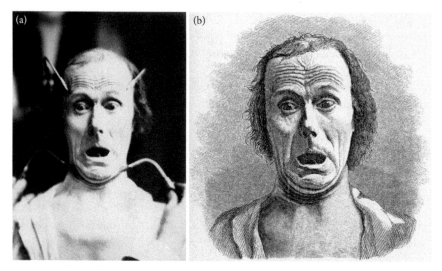

Figure 4.4 'Terror': (a) as portrayed by Duchenne, on left; (b) as re-engraved in *Expression*, at right.

Source: a) Duchenne, G.-B. (1856). *Mécanisme de la physionomie humaine ou analyse électro-physiologique de l'expression des passions*. Photo by adoc-photos/Corbis via Getty Images; b) Darwin, C. (1890). *The Expression of the Emotions in Man and Animals*.

Image credit: Wellcome Collection. Distributed under the terms of the Creative Commons Attribution 4.0 International (CC BY 4.0). https://creativecommons.org/licenses/by/4.0/#_ga=2.241103562.87728 4273.1565857924-514625952.1510241134.

Darwin, *Expression*, p.317.

Even outside Duchenne's laboratory, the limitations of Victorian technology meant that people had theatrically to pose to be photographed. Film was relatively insensitive, and required 'tens of seconds' of exposure to register a scene.[71] This meant that the troublesome 'fleeting' quality of subtle expressions could not be overcome by using cameras.[72] Photographing someone in constant movement would have been hopeless, as the image would blur. So, the photographs in *Expression* were largely simulated or staged—eight by professional actors. In order to illustrate an expression—disdain, say—Darwin's pet photographer Oscar Rejlander (if not posing the expression himself) would instruct his model to imagine she were performing a drama, such as 'tearing up a photograph of a despised lover'—her production of the expression relying on the *mise-en-scène*.[73]

---

[71] Prodger, 'Illustration as Strategy,' p.156.

[72] Darwin, *Expression*, pp.13, 18.

[73] Darwin, *Expression*, p.266. How actors produce emotional expressions—and the place of 'internal states' therein—remains a matter for debate. The possibility that actors do not *experience* the emotions they are acting provided the occasion for the introduction of 'method acting' by Constantin Stanislavski, *An Actor Prepares* (London: Methuen, 1988). Rejlander's imaginative, scenario-based acting-technique differs significantly from Stanislavski's method.

Much selection took place before a photograph was deemed to depict a given expression. First, Darwin chose an expression. Then he would work out how to instruct Rejlander. Next, Rejlander, had to contrive circumstances in which Darwin's instructions were likely to be fulfilled. Then he would reframe these instructions as best to influence a carefully chosen subject—actor or not. Then Rejlander had to pick and catch the decisive moment to open his camera's shutter—presumably requiring several trials. After the shoot, he would choose the best from among his shots to show to Darwin. Some of these might subsequently be manipulated by Rejlander (with Darwin's connivance), being redrawn and then re-photographed, for example, or superimposed on an altered background. Darwin eventually gathered hundreds of such images (including some 150 by Duchenne). From these, Darwin made a *final* selection of the thirty-four photographs incorporated in *Expression*.[74]

As we will now see, Darwin's situational approach to expression intersected with two of his other research methods: *judgement tests*, and his *cross-cultural queries* about expression. For Darwin deemed a *mise-en-scène* or context as central to the *reading* as to the *production* of expressions:

> When I first looked through Dr. Duchenne's photographs, reading at the same time the text, and thus learning what was intended, I was struck with admiration at the truthfulness of all, with only a few exceptions. Nevertheless, if I had examined them without any explanation, no doubt I should have been as much perplexed, in some cases, as other persons have been.[75]

4.3.2 Judgement tests

In 1862 Duchenne had brought out *Mechanism of Human Physiognomy*, accompanied by an atlas of eighty-six photographic plates. His scientific commentary (there was also an aesthetic commentary) argued that electro-physiology could decode the God-given 'language of facial expression,' by experimentally showing which muscles contracted to mark which emotions.[76]

Darwin treasured his copy of Duchenne's atlas, but doubted the Frenchman's equation of single muscle movements with particular meanings: 'a fright muscle'; a 'muscle of lasciviousness' and so on. Darwin dismissed such one-to-one semantic equations as make-believe.[77] Instead, he explored the meanings of the

---

[74] Prodger, *Darwin's Camera*; Prodger, 'Illustration as Strategy.'
[75] Darwin, *Expression*, p.15.
[76] Duchenne, *Mécanisme De La Physionomie*.
[77] Charles Darwin, 'Letter to Crichton-Browne, 8th June,' *Correspondence of Charles Darwin* (Cambridge: Darwin Correspondence Project, 1870), https://www.darwinproject.ac.uk/letter/?docId=letters/DCP-LETT-7224.xml.

photographed expressions in Duchenne's atlas by exhibiting them, without 'any clue to what was intended being given them,' to independent judges. For example, when the image in Figure 4.4a was shown 'to many persons without any explanation and asking what they meant,' it produced near unanimity amongst lay-observers (20 out of 25 people saw it as some version of fear), and so seemed 'well to exhibit extreme fear.'[78] Yet it did not quite settle whether this identification was due to contraction of Duchenne's so-called 'fright muscle' (the platysma).

Darwin tested eleven of Duchenne's photos in this way—those supposed to show: laughter; astonishment; fright, despair/grief; torture/agony; crying; a mix of crying and ('false') laughter; suffering; deep grief; fright with agony; and, hatred (Figure 4.5).[79] Some proved easy to recognize, like laughter. Some proved indecipherable, like hatred. Others were ambiguous, leading Darwin to ponder, for example, what made Duchenne's galvanically induced smile seem 'false' to judges.[80]

The exact procedure of Darwin's experiment remains unclear.[81] He cited eight different sets of results. The numbers of judges tested varied. Eleven, fifteen, twenty-one, twenty-three, or twenty-four, 'educated persons of various ages and both sexes,' are variously reported as having been asked to interpret different photographs. Darwin's conclusion about what a photo 'meant,' and hence, whether or not it was a 'good representation'—of horror or agony, say—drew on how well judges agreed.[82]

Hatred epitomizes Darwin's approach. Six sentences into his chapter on 'Hatred and Anger,' hatred is dismissed. Why? Because, from all his research, Darwin found hatred was 'not clearly expressed by any [specific] movement of the body or features, excepting perhaps by a certain gravity of behaviour, or by some ill-temper.'[83] Neither Duchenne nor Darwin found it easy to produce photographs of hatred—the nearest being 'aggressive hardness,' Duchenne's photograph of which proved unintelligible to Darwin's judges (Figure 4.5). This was because, *in different*

---

[78] Darwin, 'Letter to Crichton-Browne, 22nd May,'; Darwin, 'Letter to Crichton-Browne, 8th June.'

[79] Darwin, *Expression*, p.241.

[80] Darwin, *Expression*, pp.212–213.

[81] Some scholars suggest that Darwin showed the eleven photos to visitors—including family-members—as and when they came to Down House, one judge at a time. Others imagine that he handed them round at a large dinner party: Peter Snyder et al., 'Charles Darwin's Emotional Expression 'Experiment' and His Contribution to Modern Neuropharmacology,' *Journal of the History of the Neurosciences* 15, (2010), pp.158–170; White, 'Darwin's Emotions.' Snyder et al. seem to assume that *all* Darwin's results from his judgement tests are tabulated on the hand-written sheets they reproduce in their article. But *Expression* reports a judgement test on Duchenne's photograph of 'surly reserve' which is not included in the table they discuss. See Darwin, *Expression*, p.242.

[82] Darwin, *Expression*, pp.15, 157, 191, 212–213, 241–242, 294, 316, 324. See also the excellent descriptions and recreation of Darwin's judgement tests at: Anon, 'Emotion Experiment,' (2019), https://www.darwinproject.ac.uk/commentary/human-nature/expression-emotions/emotion-experiment.

[83] Darwin, *Expression*, p.249.

Figure 4.5 Duchenne's picture of 'hatred.' (The hands holding electrodes to the subject's brows are Duchenne's.) Darwin showed this photograph to eleven judges. Only one 'could form an idea of what was intended.'

Source: Artokoloro Quint Lox Limited / Alamy Stock Photo. Guillaume-Benjamin Duchenne, *The Mechanism of Human Facial Expression* (Cambridge: Cambridge University Press, 1990), pp.56–59, Plate 17; Darwin, *Expression*, pp.241–242.

psycho-social contexts, hatred got expressed in different ways: 'if the offending person be quite insignificant, we experience merely disdain or contempt. If, on the other hand, he is all-powerful, then hatred passes into terror, as when a slave thinks about a cruel master, or a savage about a bloodthirsty malignant deity.'[84] Most notably, hatred may not produce *any* action—as in the case of a terrified slave, perhaps. This put the very idea of hatred in question, most of our emotions being 'so closely connected with their expression, that they hardly exist if the body remains passive.'[85] Even if one's 'bodily frame' *were* moved, the manner of its movement would depend on the target, and the context of the hater's hatred. Whilst suppressed or unsuppressed rage was hatred's most usual 'outward sign'—such signs could be so various, that there existed no definitive bodily indication of hatred.[86]

---

[84] Darwin, *Expression*, p.249.
[85] Darwin, *Expression*, pp.249–250.
[86] Darwin, *Expression*, pp.83, 254.

4.3.3 Cross-cultural queries

Shortly after *Origin*'s publication, Darwin mailed off a list of nine questions about expression, seeking information from 'the Fuegians or Patagonians.'[87] The list reached a young Anglican missionary in Patagonia, Thomas Bridges (1842–1898). Bridges' answers did not get back to Darwin until January 1867. This was good timing, as Darwin had just sent *Variation* to press, and was beginning to plan a book on human evolution. By February 1867, Darwin had expanded these 'Queries on Expression,' and was circulating them widely. He amended the Queries often, if slightly, producing seven versions in all—the last set being printed and published in a journal, in the hope of attracting widespread responses. The Queries reached at least 48 people. Darwin hoped to enrol (educated Caucasian) observers in this re-search, placed in as varied a mix of locations as possible, with special weight given to 'observations on natives who have had little communication with Europeans'—as this would be 'of course the most valuable.' *Expression* reports the survey elicited 36 replies. These came from Europe (England), Australasia (Australia, New Zealand,) Asia (Borneo, Malaysia, China, Calcutta, Sri Lanka), southern and western Africa, and North and South America. Some focused on children. And two focused on primates.[88]

In most versions, sixteen Queries are numbered, though a single number might entail several questions. They cover: astonishment; shame and blushing; indig-nation or defiance; 'considering deeply'; 'low spirits'; 'good spirits'[89]; sneering or snarling; 'a dogged or obstinate expression'; contempt; disgust; extreme fear; laughter; contexts for shrugging; children when sulky; guilty or sly or jealous ex-pressions; expressions of affirmation and negation. The occasional use of emotion-words in the queries is apparently to be seen as synoptic of a situation. Eleven of the queries use no emotion-words. Six begin with 'when': 'When considering deeply on any subject ...'; 'When a man wishes to show that he cannot prevent something being done ...'; 'When a man is indignant or defiant does he frown, hold his body and head erect, square his shoulders and clench his fists?'[90] Darwin wanted obser-vers to report on more than expressions themselves, only having confidence in an-swers where, 'the circumstances have been recorded under which each expression

---

[87] For more on the back-story on Darwin's cross-cultural survey of expression, see Gregory Radick, 'How and Why Darwin Got Emotional About Race: An Essay in Historical and Deep-Historical Reconstruction,' in *Historicizing Humans: Deep Time, Evolution, and Race in Nineteenth-Century British Sciences*, ed. Efram Sera-Shriar (Pittsburgh PA: University of Pittsburgh Press, 2018), pp.138–171.

[88] Darwin, *Expression*, pp.16–18.

[89] The only time Darwin volunteers a *definition* of an emotion in *Expression* is situational: 'I heard a child, a little under four years old, when asked what was meant by being in good spirits, answer, "It is laughing, talking, and kissing." It would be difficult to give a truer and more practical defin-ition': Darwin, *Expression*, p.222.

[90] Darwin, *Expression*, pp.16–17.

was observed.' (Where the answers were simply yes or no, 'I have always received them with caution.')[91]

Despite Darwin's stress on observing the circumstances of expression, *Expression* largely amputated contexts of observation when *presenting* his results. Had these been included, they would have larded Darwin's results in contingency. For Darwin's observers were almost all in a position of colonial authority over the 'savages' and 'natives' whose expressions they saw—which demonstrably queered the pitch for the display of 'pure' expressions. The natives in northern Sri Lanka, wrote the colonial chaplain at Trincomalee, 'all endeavour to drill their countenances so as to express as little emotion as possible before Europeans.' A hundred miles south of Trincomalee, the botanist George Thwaites, perceived something similar: 'really there is usually very little expression at all even when they are talking together, but there is sometimes slyness & sometimes vindictiveness very evidently indicated.'[92] In India, the botanist John Scott wrote: 'Shame is … expressed by an averted and declined head with a wavering unsteady askant, eye … though indeed I am rather disposed to regard these as indications of fear than shame—the brute-like dread of corporal punishment and not the susceptibilities of a moral nature.'[93]

The results of Darwin's Queries, coming from the six inhabited continents of the world, were cross-checked against observations from infants, animals, the insane, physiological experiments, and judgement tests. His heaviest use of the Queries was to show the pan-cultural universality of blushing (see Ch. 5). Other expressive displays occurred in most or all of the six continents: tears in laughter and distress; expressions of grief, surprise, fear, and high spirits; contraction of the *depressor anguli oris* (which pulls down the corners of the mouth; see K in Figure 4.3a) in low spirits; frowning when perplexed; lip-protrusion when angry and in sulky children; shrugging when helpless; open mouth and covering of the mouth in astonishment; and, in wonder, the expression of lifting the arms with open, out-stretched hands. Conversely, certain human gestures were not pan-continental: kneeling in devotion; squaring the elbows and clenching the fists when aggressive or indignant; and kissing. As for shaking the head to deny and nodding to affirm, Darwin shrugged: his results proved equivocal.[94]

The Queries were not well designed by today's standards. The questions were 'leading' in that they included the answers that Darwin was looking for, rather than

[91] Darwin, *Expression*, p.18.

[92] George Thwaites, 'Letter to Darwin, 22nd July,' *Correspondence of Charles Darwin* (Cambridge: Darwin Correspondence Project, 1868), https://www.darwinproject.ac.uk/letter/?docId=letters/DCP-LETT-6285.xml; Revd. S.O. Glenie (enclosure in Thwaites letter).

[93] John Scott, 'Letter to Darwin, 4th May,' *Correspondence of Charles Darwin* (Cambridge: Cambridge Correspondence Project, 1868), https://www.darwinproject.ac.uk/letter/?docId=letters/DCP-LETT-6160.

[94] Darwin, *Expression*, pp.172, 195, 201–202, 223, 228, 234, 243, 245, 265, 275, 277, 288, 291, 304, 311, 333, 372, 231, 286, 373.

asking observers what they saw without any clues as to what Darwin thought.[95]
And they were also sometimes clumsy or unclear—as Darwin acknowledged.

4.3.4 Experiments, natural, contrived, and physiological

Darwin was an inventive and experienced experimenter. Curious about the agency
of infants, he directed a 'wondrous number of strange noises, & stranger grimaces'
at his firstborn son during the early months of life, to see how he would respond
(see Ch. 9).[96] Curious about the muscular protection of the eyes when sneezing: 'I
gave a small pinch of snuff to a monkey . . . and it closed its eyelids whilst sneezing.'[97]
Interested in the limits of will-power over the dictates of the nervous system:

> I put my face close to the thick glass-plate in front of a puff-adder in the
> Zoological Gardens, with the firm determination of not starting back if the snake
> struck at me; but, as soon as the blow was struck, my resolution went for nothing,
> and I jumped a yard or two backwards with astonishing rapidity. My will and
> reason were powerless against the imagination of a danger which had never been
> experienced.[98]

Darwin also discussed the expressions of Laura Bridgman (1829–1889), an
American woman who had suffered an attack of scarlet fever when aged two, and
been left blind, deaf, and with no sense of taste or smell. In spring 1871, he got
his friend Asa Gray to send the Queries to Bridgman's carer. Whenever Bridgman
was noted to have made an expression, Darwin concluded that it was 'innate,' 'instinctive,' or 'natural.'

Expression reports findings from many leading physiologists from Darwin's day.
Thus, one hypothesis central to its arguments about weeping was that the orbicular
muscles contracted 'for the protection of the eyes during violent expiration' (Figure
4.3C).[99] Darwin hypothesized such contraction to be fundamental to several of
our most important expressions. But, did the pressure in the eyes really increase
when violently exhaling? And did squinting or closing the eyes really relieve that

[95] We should note that Darwin had long been aware of the danger of asking 'leading questions' in sur-
veys, as shown by his remarks when working on a questionnaire put out by the Committee for the British
Association for the Advancement of Science in 1839: see p.8e, Charles Darwin, "D", (1838); Richard
Freeman, 'Queries About Expression,' (1977), http://darwin-online.org.uk/EditorialIntroductions/
Freeman_QueriesaboutExpression.html.
[96] Charles Darwin and Emma Darwin, *Notebook of Observations on the Darwin Children*
(Cambridge: Darwin Online, 1839–1856), http://darwin-online.org.uk/content/frameset?
itemID=CUL-DAR210.11.37. p.17r.
[97] Darwin, *Expression*, p.171.
[98] Darwin, *Expression*, p.40. Cf. Watt-Smith, 'Darwin's Flinch.'
[99] E.g. Darwin, *Expression*, pp.167ff.

pressure? Bell said so. Darwin sought proof. So he enlisted the help of the eminent Dutch physiologist, Franciscus Donders (1818–1889). Donders used the most sophisticated ophthalmological instruments of his day (some of his own invention) to conduct a series of meticulous experiments for Darwin, on rabbits, dogs, and humans. Donders' results, published in 1870, not only confirmed, but added considerable detail to Bell's proposal, buttressing Darwin's case.[100]

Darwin's attention to physiological research in *Expression* helped shift the mind-set and language of the science of emotions. By the 1870s, instrumentation allowed fine discriminations of physiological change, adding an unprecedented precision and depth to Darwin's acute, observational, approach to expression. Instead of qualitative generalities framed in terms of feeling, *Expression* put a focus on measurable quantities, observed at the level of the internal milieu and the viscera: heart rate; blood pressure; movements of the facial muscles; angles of the brow; developmental stages in the individual or the species.[101]

4.3.5 Comparative evidence

As mentioned earlier, one reason Darwin took to studying expressions was that 'no one can doubt' a connection between the expressions of 'man & Brutes.' Hence his frequent trips to London Zoo from the 1830s on. Taken from Francis Bacon (1516–1626), a key constraint on Darwin's case for evolution was that it must have been gradual and continuous, because nature makes no jumps ('natura non facit saltum'). Hence there should be no unbridgeable gap between human abilities and those of animals (see 3.4.1, Living as agency; Ch. 6).[102] As we saw earlier, Darwin reasoned that homologies between human and animal expressions would be far easier to demonstrate by observation than animal–human links between other forms of mentality. Hence *Expression* often describes the expressions of humans alongside 'the several passions in some of the commoner animals.' The importance of this, Darwin hastened to add, was 'not of course for deciding how far in man certain expressions are characteristic of *certain states of mind*'—which would have been an essentialist project. Rather, comparative evidence provided 'the safest basis for generalisation' about the evolutionary origin of the various muscular movements making an expression.[103]

---

[100] Darwin, *Expression*, pp.167–168; Franciscus Donders, 'On the Action of the Eye-Lids in Determination of Blood from Expiratory Effort,' *Beale's Archives of Medicine* 5, (1870), p.20.

[101] White, 'Darwin's Emotions,' p.826; Dror, 'Seeing the Blush: Feeling Emotions,' p.327.

[102] Charles Darwin, *The Origin of Species by Means of Natural Selection, or the Preservation of Favoured Races in the Struggle for Life*, 6th ed. (London: Murray, 1876), pp.156, 166, 234, 414; Francis Bacon, *Novum Organum Scientiarum*, (1620), http://www.gutenberg.org/files/45988/45988-h/45988-h.htm#Footnote-142: where Bacon argues for 'a general law of continuity that seems to pervade all nature, and which has been aptly embodied in the sentence, "natura non agit per saltum." '

[103] Darwin, *Expression*, p.18, my italics.

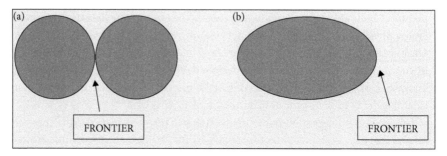

Figure 4.6 Two ideas of a 'frontier instance': (a) Bacon's—via Herschel (at left); (b) Darwin's (at right).

In his twenties, Darwin had worked out a distinctive rationale for the use of comparative evidence. This adapted the idea of *frontier instances* put forward by Bacon—via Darwin's reading of John Herschel's book on scientific method.[104] Bacon had argued that the continuity of nature allowed a scientist to perceive 'an intimate connection' between facts which at first 'seemed hostile to each other.' What he called frontier instances were cases in point. Herschel illustrated them with examples of physical phenomena: water boiling to become steam; opaque gold leaf being pared away until it transmitted light. Such observations lead us to a knowledge of 'intimate connection' where we should never have expected to find it: bridging differences between liquid and gas; or linking the opacity of metals with the transparency of 'pellucid' bodies.[105]

What Darwin drew from this passage in Herschel was not the lesson intended.[106] According to Herschel's examples, the frontier fell *between* two (apparently) distinct domains, as with liquid and gas (Figure 4.6a). If Darwin had slavishly followed Herschel, saying 'man' was a *frontier instance* would have implied *Homo sapiens* inhabited a borderland between two otherwise-divided kingdoms—beast and superman, perhaps.[107] For Darwin, however, 'man' was *a* frontier of a *single* domain of manimality—a more democratic and inclusive vision (Figure 4.6b). For him humans were, *in certain limited regards*, the *ne plus ultra*, the leading edge and paragon—positioned at the furthest reach of one great concourse of organic variability: most notably, *with regard to intellect*. In other respects, non-human

[104] John Herschel, *A Preliminary Discourse on the Study of Natural Philosophy* (London: Longman & Co, 1830).

[105] Herschel, *Preliminary Discourse*, p.188.

[106] Darwin, 'Notebook N,' p.49: 'Arguing from man to animals is philosophical, viz. (man is not a cause like a deity, as M. Cousin says.) because if so ourang outang. — oyster & zoophyte: it is (I presume — see p. 188 of Herschel's Treatise) a "travelling instance" a—"frontier instance".—for it can be shown that the life & will of a conferva [an algae] is not an antagonist quality to life & mind of man. — & we do not suppose an hydatid [a tapeworm] to be a cause of itself.'

[107] As in Friedrich Nietzsche, *Thus Spoke Zarathustra* (Harmondsworth: Penguin Books, 1975).

organisms would be 'higher' and hence *they* would be the frontier instance of natural diversity: for ocular acuity, 'an eagle's eye' would be the nonpareil.[108]

Investigating the human intellect, therefore, gave the clearest possible insight into the ways reasoning and thought might, more dimly, play a part in the life of *any* class of organism.[109] This strategy featured strongly in the just-published *Descent*, where Darwin sought to show that aspects of agency previously assumed unique to humans—such as belief in God, a moral sense, language—had less-developed parallels in other species (see Ch. 7). Contrariwise, to investigate the dramatically vivid, passionate expressions of baboons and chimpanzees—or infants (see Ch. 9)—would be to illuminate less-obvious features of the more refined emotional displays in civilized creatures like ourselves.

4.4 Evolutionary Purposelessness

Thus far, this chapter has concerned the immediate dramas which comprise the short-term axis of Darwin's stereoscopic view of life. Comparisons between the expressive movements of humans and animals point to the second, long-term axis of his vision: common descent. In examining the first axis, *Expression* implies facial and other movements gain meaning by playing their parts in different settings of circumstance (*mise-en-scènes*) which make them recognizable to witnesses as hatred, anger, joy, and so on. *Expression*'s theses on common descent strongly reinforce this reading, underlining as they do the fundamental lack of biological function, or *purposelessness*, of the movements we find expressive of emotion. Evolution has not *designed* tears and smiles to communicate sadness and good cheer. These associations are precisely that: *associations*—links between movement and circumstance formed by accident. If agency is *purposive* movement (see 3.1 Contours of Agency), then *Expression* shows how psychological significance can accrue as a side-effect to purpose*less* movement.

As already noted, Darwin principally aimed his texts to convince his contemporaries that humans were descended from animal forebears, not just in body but in mind (see 3.3 Deriving Evolution). In this regard, *Expression* was his most radical publication. Given the success of *Origin*[110] at convincing men of science

---

[108] Darwin, *Origin 1876*, p.145. 'We consider those [organisms], when the intellectual faculties [/] cerebral structure most developed, as highest. — A bee doubtless would when the instincts were': Charles Darwin, 'Notebook B,' (1837–1838), p.74.

[109] This argument could be run in reverse. If there are no absolute discontinuities in the natural world, the common conclusion that 'lower' animals do not have free will implies humans do not have it either: '... we do not suppose an hydatid [a tapeworm] to be a cause of itself. — (by my theory no animal, as now existing can be cause of itself) & hence there is great probability against free action. — on my view of free will, no one could discover he had not it': Darwin, 'Notebook N,' p.49. This did not mean that humans had *no* power of will or decision-making, however—as we will see in the next section.

[110] Which was in its sixth and final edition by 1872, the year *Expression* came out.

to take evolution seriously, if not always to accept it, and the relief of having dispatched *Descent* to the printers, *Expression* engaged with popular sensibilities more directly, inventively, and with more panache, than its predecessors.[111] In contrast to *Origin* and *Descent*, which tirelessly highlight natural and sexual selection, *Expression* hardly mentions either. This confirms that Darwin's main push as a man of science was not to champion a particular means of common descent, but the doctrine of common descent itself. This is *Expression*'s central message. If, when attacked by a fearsome dog, we find our hair bristling, says Darwin, or if we notice we have bared our teeth when infuriated by someone else's bad behaviour: these purposeless movements 'can hardly be understood, except on the belief that man once existed in an much lower and animal-like condition.'[112]

So, how to explain such common descent?

First, if expressive movements had evolved, they must be heritable or innate. So they must appear early in life. Hence Darwin's study of infant expressions. Yet, *Expression* also argues that the facial and bodily movements associated with emotion in humans have *not* originated by affording their possessors communicative advantages—as such an endowment might imply divine origins.[113] Rejecting the idea that expressions have a communicative function makes it hard to explain their evolution by Darwin's law of natural selection, however. For if expressions were *adaptations for communication*, they would be 'purposeful,' and so—as natural theologians like Charles Bell held—invite explanation as God-given 'apparatus for the purpose of enabling ['man'] to communicate with his fellow-creatures.'[114] Conversely, if it could be shown that emotional expressions were largely uniform across all cultures, yet *not* communicative, their uniformity would imply common descent.[115] An anti-selective argument for the common descent of universal human expressions held out a further prize to Darwin: of proving the different human races all belonged to one species, thereby kiboshing the polygenic rationale for slavery (see 2.6.1 Natural histories of man).[116]

Steering clear of natural selection, *Expression* devises three explanatory principles for the reconstruction of the evolutionary history of the movements we find most expressive of emotion. All three make the case for their lack of

[111] White, 'Darwin Wept'; Richardson, 'The Book of the Season.'

[112] Darwin, *Expression*, p.13.

[113] Darwin, *Expression*, p.11; See Charles Darwin, 'Darwin to Wallace, 12th–17th March,' *Correspondence of Charles Darwin* (Cambridge: Darwin Correspondence Project, 1867), https://www.darwinproject.ac.uk/letter/?docId=letters/DCP-LETT-5440.xml, when describing his aims for his work on expression: 'I want, anyhow, to upset Sir C. Bell's view, given in his most interesting work "the anatomy of Expression" that certain muscles have been given to man solely that he may reveal to other men his feelings.'

[114] Charles Bell, *The Anatomy and Philosophy of Expression as Connected with the Fine Arts*, 3 ed. (London: John Murray, 1844), p.121.

[115] i.e. if they were functionless, they could not have become similar in different races due to convergent evolution.

[116] Radick, 'Darwin's Puzzling.'

function: although expressive movements 'often reveal the state of the mind,' concedes *Expression*, 'this result was not at first either intended or expected.'[117] It is because expressive movements were, in Darwin's word, *purposeless* (not 'serviceable,' functionless) that they could only have gained meaning through *how they were read*. Hence, for an expression to have evolved, or 'become instinctive,' its 'recognition' would also have to have evolved and 'likewise have become instinctive.'[118]

In evidence that the capacity *to read* expressions was innate, *Expression* refers to Darwin's social judgement tests, where, in most cases, nearly all the judges 'instantly recognised ... so many shades of expression ... without any conscious process of analysis.'[119] It also describes four observations from Darwin's study of his infant son, 'Doddy.' For example, when Doddy was only four months old, Darwin 'made in his presence many odd noises and strange grimaces,' but these did not scare Doddy, being 'taken as good jokes,' because they were, 'preceded or accompanied by smiles'—Darwin's smile being innately legible to Doddy as making humorous the circumstances in which it occurred (see Ch. 9).[120]

The first of the three principles *Expression* elaborates to account for the evolutionary origin of expressions guarantees that movements, like raising one's eyebrows when trying to remember something, are of no service to the expresser, however informative they may be to the conspecifics who see them.[121] This principle makes expressions result from *serviceable associated habits*: if some movements were originally tied as corollaries to useful circumstantial reactions in an animal's long-dead ancestors—they may continue to be triggered today by *analogous* circumstances, even though useless. To interested observers, these movements may thus serve to indicate the 'state of mind' (i.e. link them to the characteristic circumstances) which initially evoked the reactions of which they were corollaries. There is a logic of physiological association at work here—as we saw earlier for displays of anger (cf. 4.2 A drama of surfaces).

Darwin traced several human (and animal) expressions to the operation of this principle: most notably, weeping. Tears are vestiges from the clenching of the (orbicular) muscles around the eyeball, which *usefully* protects it when the eye is likely to be damaged, by bright light, or violent exhalation. Weeping is an accidental and

---

[117] Darwin, *Expression*, p.377.
[118] Darwin, *Expression*, p.378.
[119] Darwin, *Expression*, p.380.
[120] Darwin, *Expression*, p.379.
[121] Darwin, *Expression*, pp.10, 34, 74–75. NB 1. *More than one* of the principles I now go on to discuss might be used to explain the same expression: 'It is, however, often impossible to decide how much weight ought to be attributed, in each particular case, to one of our principles,' as Darwin put it, ibid, p.87; 2. Darwin formulated the first two of these principles aged 29—the first expanding on an early claim of Bell's, and the second being of Darwin's own invention [see pp.107, 146e–147, Darwin, 'Notebook M,']. The third principle, which Bain and Spencer probably got from the same source, was taken from Johannes Muller, *Elements of Physiology* (London: Taylor and Walton, 1842), p.3. When explaining expressions, Darwin sometimes also confusingly wrote of three *other* principles: of 'habit, association, and inheritance': see e.g. Darwin, *Expression*, p.325.

useless side-effect: orbicular tightening inevitably squeezes water out of nearby, tear-producing, glands. While in our ancestors, tears were produced only when the danger of eye-damage was real—when screaming in fear, for example—they are now produced under *analogous* circumstances. Thus, when abandoned, even adults may weep.

Gregory Radick sums up Darwin's second principle as: 'reverse habits come for free.'[122] Darwin dubbed it a principle of *antithesis*. When we find ourselves in 'directly opposite' circumstances to those which produced serviceable associated habits—as per principle one—there is 'a strong and involuntary tendency to the performance of movements of a directly opposite nature, though these have never been of any service.' For good reasons, we square our shoulders when defiant (in preparation for fisticuffs). For no good reason, we shrug or slouch when impotent. We self-protectively lower our brows when suspicious. So, gratuitously, we raise them in welcome. A cat purposefully crouches and growls when menacing, sweeping its tail sinuously from side to side. Conversely, it arches its back and purrs when acting affectionately, its tail held stiffly upright.

This principle of explanation for expressive movements is a special case of a more general physiological principle, said Darwin. Under 'opposite impulses of the will,' we naturally perform *opposite* muscular movements: when lifting a weight versus lowering it; when pulling a door versus pushing; when turning right versus left. This produces a habit of muscular reversal when in reverse circumstances. Hence humble expressions become the muscular antithesis of expressions of pride, and open-faced, expansive astonishment has as antithesis a limp-armed, closed-down look of indifference.

Principle three is that certain movements, 'which we recognise as expressive,' are 'the direct result of the construction of the nervous system.' These movements are, from the first, 'independent of the will, and, to a large extent, of habit,' being side-effects of exorbitant arousal.[123] Purposelessness comes again to the fore. The nervous system has evolved for good, adaptive reasons. But expression is not one of those reasons. Think of the nervous system as a dam that can spill: the 'undirected overflow of nerveforce' that results from extreme circumstances will produce gross movements, with no intrinsic meaning—except by association with readable circumstances.[124] Darwin's examples include: shouting, clapping, or dancing in joy; blanching of hair in terror or grief; trembling in fear, joy, or rage; writhing and groaning in agony; pacing or drumming the fingers in anxiety and impatience; sweating or grinding the teeth in pain; and fainting under various kinds of duress—a Victorian speciality.

[122] Radick, 'Darwin's Puzzling,' p.183; Darwin, *Expression*, pp.52, 67.
[123] Darwin, *Expression*, p.69.
[124] Darwin, *Expression*, pp.10, 33, 75, 84.

4.4.1 The will

In *Expression*, an expressive movement may be called involuntary, reflex, innate, inherited, or instinctive, if it is not under control of *the will*. However, the will does play a part in *Expression*. To start with, when promulgating its first principle of serviceable associated habits, the book holds that, back in the mists of our ancestry, before informative actions became automatic, they had been *consciously* performed, only later being 'rendered easy through long-continued habit so as at last to be performed unconsciously, or independently of the cerebral hemispheres.' For example, 'starting' in fear would have been originally acquired 'by the habit of [deliberately] jumping away as quickly as possible from danger.' In rare cases, where Darwin thought that a movement could never have been willed—as with contractions of the iris—the inheritance of acquired characters did not work as an evolutionary explanation, so he invoked natural selection.[125]

Animals and humans may also modify or *augment* what are at first involuntary reactions. In panic, I run madly. But I run *away* from the danger.[126] The hair on the back of an angry dog bristles automatically. But he may also bite my ankle. Beyond this is the capacity of humans, in particular, voluntarily to use expressive movements that are otherwise involuntary. One may *deliberately* raise one's brows to signal disbelief; smile to beguile; grin to hide grief.

In humans, informative movements, though automatically associated with certain circumstances, can be 'partially repressed by the will.'[127] Even then, 'the strictly involuntary muscles, as well as those which are least under the separate control of the will, are liable still to act; and their action is often highly expressive.' *Expression*'s main exhibit in this regard is infants' 'habit of screaming' and weeping. With age, children will learn to control their screams, from self-consciousness and the assimilation of etiquette. But they cannot so easily repress the respiratory aspects of screaming. So they sob. Self-conscious adults may even manage to repress sobbing, but they will still frown, being unable wholly to control the muscles in their brows. *Slyness* cannot be detected from the face, said Darwin, but 'by movements about the eyes; for these are less under the control of the will.'[128] A novice lecturer may manage to speak brightly, but will not always quell quivering legs and thumping heart.

Note that the will was a contentious topic in psycho-physiological discussions during the 1860s and 1870s, because it was a hold-out against the rise of materialism. In 'the expanding empire of deterministic scientific explanation,' the category of mind 'conceived as self-determining ego, posed the most formidable

---

[125] Darwin, *Expression*, pp.42, 43.
[126] Darwin, *Expression*, pp.69, 33, 45, 74, 30.
[127] Darwin, *Expression*, pp.29, 51, 371.
[128] Darwin, *Expression*, p.276.

obstacle to that advance.'[129] Darwin never explicitly took up a position in this debate. But his early notebooks argue that one can consistently talk about the operation of will without deeming it *free*, unmotivated, or lacking in neural basis. Thus what most lacks in the asylum, says *Expression*, is that the insane respond without willed restraint to the situations that evoke their passions, creating 'unbridled' displays.[130] Their lack of will was susceptible to neurological explanation, however.[131] Which implies will has a neural basis.

4.5 Kinks and Muddles

There is an amiable lack of pedantry about Darwin's writing. Yet, as discussed further in Chapter 6, such informality invites ambiguity and prejudice. The coherence of *Expression* suffers from a further tension, caused by its large-scale renunciation of natural and sexual selection as explanations for humans' non-verbal behaviour. Various kinks of logic and muddled arguments result—some of which require notice.

As we have seen, Darwin argued that, while expressions have no communicative purpose, they can still reveal states of mind to onlookers. Which yields the argument that *recognition* of expressions is innate. Such was Darwin's principal conclusion from studying babies: 'an infant understands to a certain extent, and as I believe at a very early period, the meaning or feelings of those who tend him, by the expression of their features' (see Ch. 9).[132] Note, however, that Darwin's word 'recognition' does not imply a veridical *decoding* of the expresser's intention: a grimace which means, 'reveals,' or *is read as* fearful, need not have had the intention or purpose of expressing fear. Similarly, to an intelligent onlooker flashes of lightning 'mean thunder'—without imputing intentions to the sky.

While *Expression* largely circumvents arguments that would make natural selection a viable way of explaining the evolution of expressions, it occasionally slips. 'The power of intercommunication is certainly of high service to many animals,' it admits at one point.[133] Moreover, 'reflex actions are in all probability liable to slight variations, as are all corporeal structures and instincts.'[134] Which opens the door to arguing that any intelligible expression, and its recognition, could be explained by

---

[129] Lorraine Daston, 'The Theory of Will Versus the Science of Mind,' in *The Problematic Science: Psychology in Nineteenth-Century Thought*, ed. William Woodward and Mitchell Ash, (New York: Praeger, 1982), p.96. Cf. Roger Smith, *Free Will and the Human Sciences in Britain 1870–1910* (Pittsburgh PA: University of Pittsburgh Press, 2016).

[130] Darwin, *Expression*, pp.254, 14. Cf. Darwin, 'Notebook M,' pp.26–31, 72–74, 118–125; Darwin, 'Notebook N,' p.49; Darwin, 'Old & Useless Notes About the Moral Sense & Some Metaphysical Points,' pp.25–28.

[131] Darwin, *Expression*, p.163.

[132] Darwin, 'Biographical Sketch,' pp.293–294.

[133] Darwin, *Expression*, pp.62–63.

[134] Darwin, *Expression*, p.43.

natural selection. At times, this becomes explicit: male animals who make themselves appear larger and more 'terrible to their rivals' by inhaling, fluffing up feathers, or erecting hair, will have avoided injurious fights, and hence 'on an average
have left more offspring to inherit their characteristic qualities,' than other males.[135]

Likewise, *Expression*'s principle of 'serviceable associated habits,' sometimes
implies an expression is functional today, and was functional in antiquity. For example, screaming and crying in the young of members of any social animal, will
long have been advantageous in attracting 'mutual aid.'[136] Yet, the biological advantages which attend screaming do not lead Darwin to invoke natural selection as
explaining the origin of human frowning and weeping—as they might.

Darwin's struggles to deny communicative intent as the origin of facial and
other gestures sometimes results in absurdity, especially with regard to antithetical
expressions.[137] For example, about apologetic shrugging—supposedly derived as
the antithesis to the erect posture associated with human pride—we read:

> The gesture is sometimes used consciously and voluntarily, but it is extremely im-
> probable that it was at first deliberately invented, and afterwards fixed by habit;
> for not only do young children sometimes shrug their shoulders under the above
> states of mind, but the movement is accompanied ... by various subordinate
> movements, which not one man in a thousand is aware of, unless he has specially
> attended to the subject.[138]

Here two bizarre ad hoc arguments are used to refute a basis in intentionality.

One is that children shrug. But this makes no sense as a refutation of a long-
ago intentionality, because plenty of other expressions (e.g. of fear), which are now
innate, began as intentional habits, according to *Expression*. The book states this
clearly: 'the fact of the gestures being now innate, would be no valid objection to
the belief that they were at first intentional; for if practised during many generations, they would probably at last be inherited.'[139]

The second thrust of Darwin's rebuttal is that, when we shrug, we use muscles about which most of us are ignorant. This implies that we cannot ever have
shrugged voluntarily, because the anatomy involved is too complicated for us deliberately to have willed a shrug, muscle by muscle. Yet to say I cannot voluntarily
make a soufflé because I do not have a sufficient anatomical understanding of all
the 'various subordinate movements' involved in soufflé-making (which I don't)
is untrue. Moreover, this directly contradicts Darwin's principle of serviceable

---

[135] Darwin, *Expression*, pp.109–110.
[136] Darwin, *Expression*, p.76.
[137] Darwin, *Expression*, pp.62–63.
[138] Darwin, *Expression*, p.65.
[139] Darwin, *Expression*, p.63.

associated habits: (a) inherited habits comprise 'complex actions or movements'; (b) these were first performed 'voluntarily'.[140]

Later on, *Expression* tries to maintain that nodding the head in affirmation and shaking it in negation are instinctive, even though Darwin's cross-cultural research cast doubt on the universality of this gesture. So he cites additional evidence from the deaf and blind Laura Bridgman, 'constantly accompanying her yes with the common affirmative nod, and her no with our negative shake of the head.' This made it 'highly probable' that such gestures were innate. Which hardly convinces, because, contrary to Darwin's description of the acutely intelligent Bridgman as 'born in this condition,' Bridgman did not lose her sight and hearing until she caught scarlet fever, aged two, by which time she had presumably had plentiful opportunities to learn the conventional meanings of gestures like nods and shakes of the head.[141]

Bridgman's case underlines that Darwin had to handle more than one time-scale within the framework of *Expression*. One is evolutionary or phylogenetic, involves biological inheritance, and straddles countless generations. The other is ontogenetic, involves learning, and covers one life-span. Sometimes *Expression* invokes both these scales. For example, we are told adults weep because, in infancy, humans scream, thereby giving an 'incidental' squeeze to the lachrymal glands as a result of tightening the protective, orbicular muscles around the eyes (meanwhile also contracting the depressor *anguli oris* around the mouth: Figure 4.3K).[142] Tears are thus a useless side-effect of screaming. Later, 'with the advancing age of the individual,' tears will fall whenever something occurs which the individual associates with the circumstances that once produced screaming.

This appears to be an ontogenetic explanation, which should work today just as well as it did for our ancestors 'numberless generations' ago.[143] Which exposes a crack in Darwin's reasoning: *Expression* sometimes invokes an evolutionary explanation for an expression when it may have no need to, because ontogeny may be explanation enough. Darwin cements over this crack by arguing that ontogeny becomes phylogeny over large stretches of time:[144] 'As the habit of contracting the brows has been followed by infants during innumerable generations, at the

---

[140] Darwin, *Expression*, pp.29, 182–183.

[141] Darwin, *Expression*, pp.287, 329. Darwin's source states: 'Laura Bridgeman, [was] a female endowed with a peculiarly active mind, but deprived *from earliest infancy* of sight and hearing' (my italics): Francis Lieber, 'A Paper on the Vocal Sounds of Laura Bridgeman the Blind Deaf-Mute at Boston; Compared with the Elements of Phonetic Language,' *Contributions to Knowledge* 2, (1851), https://archive.org/details/101299133.nlm.nih.gov/mode/2up.

[142] Darwin, *Expression*, p.185.

[143] Darwin, *Expression*, pp.182, 184.

[144] Note that Darwin's argument here has a logic opposite to that claimed by scholars who find that a 'recapitulatory' thesis (i.e. of 'phylogeny' *causing* 'ontogeny'; see 9.2.4 'Founding father?') underlies Darwin's psychology, e.g. Evelleen Richards, *Darwin and the Making of Sexual Selection* (Chicago: Chicago University Press, 2017), pp.286, 448.

commencement of every crying or screaming fit, it has become firmly associated with the incipient sense of something distressing or disagreeable.'[145]

The crack re-emerges when *Expression* invokes ontogeny to explain the *repression* of expressions with age, such as the reduced screaming and shedding of tears in older children and adults, and in men compared to women. This brings in a third time-scale, that of 'the advancing culture of the race.'[146] Darwin holds that the expressions of 'savages' remain in a 'primordial condition,' whereas 'civilized Europeans' have more restrained expressions.[147] Some of this restraint is attributed to habits *that have become inherited* by Europeans, particularly male Europeans. But mostly, it is attributed to the ontogenetic learning of cultural conventions and gestures as 'civilized' individuals grow up (for more on Darwin's take on gender and culture, see Chs 6–8).

A final muddle attends *Expression*'s denial of *sexual* adaptiveness to expressions. Thus, blushing might easily be explained as a sexual come-on, because, as Darwin admits, 'a slight blush adds to the beauty of a maiden's face.' Yet explanations for blushing by sexual selection is ruled out because 'dark-coloured races blush ... *in an invisible manner.*'[148] Fair enough, did not *Expression* claim elsewhere that the observational reports gathered in Darwin's cross-cultural research show that blushing was 'common to most, probably to all, of the races of man'—including, Polynesians, Chinese, American Indians, and 'negroes.'[149] How do we know this? Because:

> Several trustworthy observers have assured me that they *have seen on the faces of negroes* an appearance resembling a blush, under circumstances which would have excited one in us, though their skins were of an ebony-black tint. Some describe it as blushing brown, but most say that the blackness becomes more intense.[150]

Not so invisible after all, then.[151]

---

[145] Darwin, *Expression*, p.236. After its chapter on the topic, *Expression* only twice alludes to an evolutionary or phylogenetic basis for weeping.

[146] Darwin, *Expression*, p.183.

[147] Darwin, *Expression*, p.245.

[148] Darwin, *Expression*, p.338, my italics.

[149] Darwin, *Expression*, pp.340, 357. Darwin became so confident that all humans blush that he was prepared to discount some observational reports to the contrary: ibid, p.335, 'Lady Gordon is mistaken when she says Malays and Mulattoes never blush.' Captain Osborn and Mr Geach had seen several Malays blush.

[150] Darwin, *Expression*, p.338.

[151] Paul White, 'Reading the Blush,' *Configurations* 24, (2016), pp.289–290 reads *Expression*'s treatment of blushing—in the light of an ungrammatical and self-correcting note Darwin made in 1839—as being a sign of 'healthy [sexual] desire.' But White's interpretation of this note fails to observe a distinction Darwin was groping for in 1839, and made explicitly in *Expression* in 1872: between *blushing* (because 'each sex thinks more of what another thinks of him'); and *flushing* ('like erection'): cf. Darwin, 'Notebook N,' pp.51–52; see 5.1 Phenomena of Blushing.

4.6 Resonances

Acutely tuned to his audiences, Darwin had written *Expression* as an invitation to readers to *participate in* the study of expression. Not only did he write engagingly and include unprecedented riches of illustration. He was constantly suggesting ways that readers could test for themselves what he was saying. And *Expression* sold rapidly. All sorts bought it, from eminent scientists and famous novelists, like George Eliot and Lewis Carroll, to the general reader. Newspapers welcomed it as 'one of the most eagerly anticipated books of the season'; being 'the most popular, and yet the most profound' of all Darwin's works—not just informative, but entertaining.[152]

As for its scientific reception, *Expression* is often lumped in with works by contemporaries, Alexander Bain (1811–1877), and Herbert Spencer (1820–1903; Figure 4.7). By all three, we hear, 'facts about bodies (especially about nerves, lower animals, infants, 'savages' and the insane) were given a privileged place,' and all three implied 'by their methodology and language' that the real business of emotions 'went on at the physiological and neurological levels.'[153]

This is only superficially true. Bain's *The Emotions and the Will* (1865), Spencer's *The Principles of Psychology* (1872), and Darwin's *Expression* (1872) do all mention similar topics, and do all defer to physiology. But the ways they walk their talk differ. Bain and Spencer advance theoretical and empirical claims about emotion, but make no move to pit their claims against first-hand observation, or even to state their claims in observable form.[154] For example, Bain stated: 'The life of infant humanity is overshadowed with terrors; the wild gleams of rejoicing shoot out of a diffused blackness'—hardly an empirical proposition.[155] Spencer's remarks on babies also smack more of clairvoyance than observational rigour: 'To the incipient intelligence of an infant, noise does not involve any conception of body. In an oft-recurring echo, the sound has come to have an existence separate from the original concussion.'[156]

In contrast—whether talking about infants, nerves, muscles, 'savages,' the insane, or lower animals—Darwin had either himself painstakingly observed the phenomena he discussed, or he had gotten other trustworthy informants to make

---

[152] *Expression* was published on the 26th November 1872. See *Argus*, 23rd January, 1873; *New York Tribune*, 14th January 1873; both quoted in Richardson, 'The Book of the Season,' p.69.

[153] Dixon, *From Passions to Emotions*, pp.143–144.

[154] Robert Young, *Mind, Brain, and Adaptation in the Nineteenth Century: Cerebral Localization and Its Biological Context from Gall to Ferrier*, History of Neuroscience (New York: Oxford University Press, 1990), p.191: 'He [Spencer] advocated comparative and developmental studies but conducted none.'

[155] Alexander Bain, *The Emotions and the Will*, 2nd ed. (London: Longmans Green, 1865), pp.57, 62.

[156] Spencer, *The Principles of Psychology, Vol.2*, p.141.

Figure 4.7 (a) Alexander Bain (left); (b) Herbert Spencer (right).

Source: (a) Bain, A. (1898). Photogravure by Synnberg Photo-gravure Co. Credit: Wellcome Collection. Distributed under the terms of the Creative Commons Attribution 4.0 International (CC BY 4.0). https://creativecommons.org/licenses/by/4.0; (b) Spencer, H. (1901). Photogravure, after Sir H. von Herkomer. Credit: Wellcome Collection. Distributed under the terms of the Creative Commons Attribution 4.0 International (CC BY 4.0). https://creativecommons.org/licenses/by/4.0/#_ga=2.20701 36.877284273.1565857924-514625952.1510241134.

first-hand observations for him, or both. (Witness, for example, Darwin's trail-blazing observations of babies' expressions discussed in Chapter 9.)[157]

Darwin's approach was that of an active naturalist, not of a philosopher, armchair psychologist, or metaphysician. The need to experiment and to observe, not just to generalize and theorize, was imperative. Hence, his irritation with Spencer's grandiloquent conclusions: 'Here would be a fine subject for half-a-dozen years' work,' he exclaimed. (see 2.4 Empirical Orientation; 2.5 Darwin's Approach to the Study of Humans). And his retort to Bain—whose appendix to the 1873 edition of *The Senses and the Intellect* accused *Expression* of stealing Bain's ideas. Darwin responded by politely if wryly apologizing to Bain that, unfortunately, he was unable 'to grasp' Bain's ideas about the 'principle of spontaneity, as well as some other of your points, *so as to apply them* [i.e. empirically] *to special cases*'—adding: 'But as we look at everything from different points of view, it is not likely that we should agree closely.'[158]

---

[157] Darwin, 'Biographical Sketch'; Darwin and Darwin, *Notebook of Observations on the Darwin Children*.

[158] Charles Darwin, 'Letter to Bain, 9th October,' *Correspondence of Charles Darwin* (Cambridge: Darwin Correspondence Project, 1873), https://www.darwinproject.ac.uk/letter/?docId=letters/DCP-LETT-9092.xml, my italics.

Both Bain and Spencer were essentialists where emotion was concerned. Their explanations of expressive movements invoked an abstruse ping-pong of conscious inner states, which Bain called, an '*internal mechanism* that we all possess alike—of the sensations and emotions, intellectual faculties and volitions, of which we are every one of us conscious.'[159] Both men clung to an associationist account of emotions as one class of 'ideas' among others. Trains of mental states, studied via introspection, would *cause* the visible expressions that were their 'external manifestations.' Thus, a feeling of pleasure might, through association, evoke a memory or the representation of another sensation—a 'recurrence in the trains of thought'—such that, 'the discovery of good feelings in self causes a glow of mingled admiration and fond love.'[160] *Expression* nowhere invokes an internal mechanism of mental states, and makes no mention of introspection.

In the twentieth century, *Expression* shared in the eclipse of Darwin's psychology that accompanied the hegemony of the Modern Synthesis in biology (see 1.1 Darwin and Modernism). Several writers in later decades of last century, who expatiated on Darwin and psychology, skipped over *Expression*—mostly, I believe, because the book spelt out 'no biological function for the expression of the emotions and had no role for natural selection.'[161] For like reasons, the fifty-five chapters of the latest *Handbook of Evolutionary Psychology* refer copiously to Darwin's work on natural selection (more than a hundred times), but make one lone reference to *Expression*.[162] Nevertheless, the last century did not entirely bypass *Expression*. Indeed *Expression* influenced three men hailed as founders of modern psychology: Wilhelm Wundt (1832–1920), James Mark Baldwin (1861–1934), and George Herbert Mead (1863–1931).[163]

4.6.1   Wundt, Baldwin, Mead

Wundt's fame has traditionally ridden on his foundation of 'the first ever' psychological laboratory (see 2.3 'Instrumental Aids to Discovery'). The fact that Wundt elaborated a sophisticated *social* psychology has gained less attention.[164]

---

[159] Alexander Bain, 'Phrenology and Psychology,' *Fraser's Magazine* 61, (1860), p.699, my italics; cf. Spencer, 'Bain.'

[160] Bain, *Emotions 1865*, pp.viii, 23, 102.

[161] Robert Richards (personal communication, 11th Jan 2017).

[162] David Buss, *Handbook of Evolutionary Psychology*, 2 vols. (Hoboken NJ: Wiley, 2016).

[163] Robert Farr, 'On Reading Darwin and Discovering Social Psychology,' in *The Development of Social Psychology*, ed. Robert Gilmour and Steven Duck (London: Academic, 1980).

[164] Arthur Blumenthal, 'Wilhelm Wundt and Early American Psychology: A Clash of Cultures,' in *Wilhelm Wundt and the Making of Scientific Psychology*, ed. Robert Rieber (New York: Plenum, 1980), p.vii. As a younger man, Wundt had written reputedly the first textbook on experimental psychology: *Principles of Physiological Psychology* (1874). This proclaimed itself 'deeply imbued with those far-reaching conceptions which, by [Darwin's] labours, have become an inalienable possession of natural science'—referring to *Origin*. Wilhelm Wundt, *Principles of Physiological Psychology*, trans. Edward Titchener (New York: Macmillan, 1910). Titchener's translation left out the latter two-thirds

Open his monumental, ten-volume *Social Psychology: An Investigation of the Laws of Evolution of Language, Myth, and Custom* (i.e. *Völkerpsychologie. Eine Untersuchung der Entwicklungsgesetze von Sprache, Mythus und Sitte*, 1910–1920), and you find that Wundt's distinctive take on human sociability launches itself with an appropriation of the concept of gesture (conventional expression) from *Expression*.[165] This 'folk psychology' influenced key figures in the social sciences, including Bronislaw Malinowski, Franz Boas, William Thomas, Émile Durkheim, and Sigmund Freud.

Mead's social behaviourism also set out from an appropriation of *gesture* and *attitude* from *Expression* —as in the bristling attitude of a dog who 'intends to fight'.[166] In particular, Mead took up the notion that a gesture was a part-action which others completed (an attitude was the beginning of an act). Following Darwin, the *meaning* of gestures, or any social action, was to be found 'in the response which it elicits from others,' said Mead: 'for an act to be a gesture it must be perceived by, and evoke a response from, another organism.'[167]

Baldwin often acknowledged Darwin as a source for his ideas—including of the so-called 'Baldwin effect,' which he repackaged from *Origin*'s discussion of 'transitional habits' (see 3.3.4 Agency directs selection). Baldwin's notions of adaptive *assimilation* (modes of action in which old habits are imposed *associatively* on new content) and, to a lesser extent, *accommodation* (as when an old habit or pattern of action is altered to adapt to new content or attain a new purpose) also derive from *Expression*. For example, referring to Darwin's principle of serviceable associated habits, Baldwin wrote:

> The principle of Assimilation ... is a direct reflection in consciousness of this aspect of the law of habit. And this is only to say, as Darwin said, that we ought to find, in certain states of mind, attitudes struck which have arisen, not for use in this condition of mind, but in conditions of mind which *feel like it* in any respect.[168]

The principles of assimilation and accommodation later became central features of Jean Piaget's (1896–1980) developmental psychology.[169]

---

of the German original, which was where Wundt discussed Darwin's work: see Richard Anderson, 'The Untranslated Content of Wundt's *Grundzüge Der Physiologischen Psychologie*,' *Journal of the History of the Behavioural Sciences* 11, (1975), pp.381–386.

[165] E.g. Wilhelm Wundt, *Völkerpsychologie: Eine Untersuchung Der Entwicklungsgesetze Von Sprache, Mythus Und Sitte, Teil 1: Sprache* (Leipzig: Wilhelm Engelmann, 1904).

[166] Darwin, *Expression*, p.53.

[167] Farr, 'Darwin and Social Psychology,' p.130.

[168] James Baldwin, *Mental Development in the Child and Race: Methods and Processes* (New York: Macmillan, 1895), p.257.

[169] John Broughton and John Freeman-Moir, eds., *The Cognitive-Developmental Psychology of James Mark Baldwin: Current Theory and Research in Genetic Epistemology* (Norwood NJ: Ablex, 1982).

4.6.2 Against essentialism

Nowadays, we increasingly hear resonances with *Expression*'s treatment of movements as gaining salience through the interpretative capacities of their beholders.[170] This treatment extends Darwin's 'horizontal', relational understanding of coadaptation, where conspecific and cross-species interdependencies help constitute the economy of nature. The *relational* construction of expressions in *Expression* does not end with the inter-personal. Reading expressions is fully contextual: I understand your behavioural attitude as angry because I read it as responsive to your circumstances.[171] This too harks back to Darwin's immersion in the long tradition of natural history, where the interdependencies that make up the economy of nature in a given habitat relate an organism to both the animate *and the inanimate* features of the locality in which all live: soil, climate, terrain.

This *externalism* or *situationism* resonates powerfully with current debates about expression and emotion, both in psychology and in philosophy. First come chimes with the increasing stress on embodiment in cultural theory. Cognition and abstract thought are not obviously 'of the body.' Emotional expression is. Hence, what has been called the *turn to affect* in psychology and the neurosciences.[172] Darwin saw emotionality as interdependent with our corporeal framework:

> Most of our emotions are so closely connected with their expression, that they hardly exist if the body remains passive—the nature of the expression depending in chief part on the nature of the actions which have been habitually performed under this particular state of the mind. A man, for instance, may know that his life is in the extremest peril, and may strongly desire to save it, yet may exclaim as did Louis XVI, when surrounded by a fierce mob, "Am I afraid? feel my pulse." So a man may intensely hate another, but until his bodily frame is affected he cannot be said to be enraged.[173]

Expression is primarily about embodied movements, their physiology, and musculature—or what Joel Krueger calls their 'situated agency.'[174] The book's

---

[170] Paul Griffiths and Andrea Scarantino, 'Emotions in the Wild: The Situated Perspective on Emotion', in *The Cambridge Handbook of Situated Cognition*, ed. Murat and Robbins Aydede, Philip (Cambridge: Cambridge University Press, 2009); Joel Krueger, 'Varieties of Extended Emotions', *Phenomenology and the Cognitive Sciences* 13, (2014), pp.533–555.

[171] José-Miguel Fernández-Dols, 'Natural Facial Expression: A View from Psychological Constructionism and Pragmatics', in *The Science of Facial Expression*, ed. James Russell and José-Miguel Fernández-Dols (Oxford: Oxford University Press, 2017), pp.457–477.

[172] Ruth Leys, 'The Turn to Affect: A Critique', *Critical Inquiry* 37, (2011), pp.434–472; John Cromby, *Feeling Bodies: Embodying Psychology* (London: Palgrave-Macmillan, 2015); De Gelder, *Emotions*; Hank Stam, ed. *The Body and Psychology* (London: Sage, 1998). Margaret Wetherell, *Affect and Emotion: A New Social Science Understanding* (London: Sage, 2012).

[173] Darwin, *Expression*, pp.249–250.

[174] Albeit, attributing this idea to Dewey, not Darwin: Krueger, 'Dewey's Rejection', p.141.

last paragraph enlists the scene in *Hamlet* where an actor sheds tears whilst impersonating Hecuba mourning her murdered husband.[175] Such is 'the intimate relation which exists between almost all the emotions and their outward manifestations,' Darwin remarks, that 'even the simulation of emotion tends to arouse it in our minds.'[176]

Emotions thus become secondary: being *attributed to* or *resulting from* bodily expressions. This makes *Expression* the trail-blazer for William James' and Carl Lange's famous 'discharge' theory of emotion (and its successors, e.g. Dewey's theory, and Stanley Schachter's 'two-factor' theory of emotion): 'that we feel sorry because we cry, angry because we strike, afraid because we tremble, and not that we cry, strike, or tremble, because we are sorry, angry, or fearful ...'[177] James held that the bodily changes that we call an emotional expression '*follow directly the PERCEPTION of the exciting fact, and that our feeling of the same* [bodily] *changes as they occur IS the emotion.*'[178] Darwin's approach chimes too with research decoding the meanings of *animal* gestures via the responses of the recipients of those gestures.[179]

Note that the nexus between expression and recognition made Darwin's analysis of *all* expressions unavoidably social. Against this, contemporary writers sometimes distinguish between *basic* human emotions and *social* emotions. Basic emotions are supposed to be biologically based and hence, pan-cultural and shared with animals. (For example: joy, fear, anger, surprise, sadness, disgust— though lists of core emotions vary.) Social emotions, on the other hand, 'directly refer to and reflect a property of our interaction with others.' They are *relational* and *external*: 'part of interactions that take place not only in the human mind but

---

[175] William Shakespeare, *Shakespeare's Hamlet* (New York: Henry Holt, 1914), Scene 2 Act 2. The final chapter of *Expression* quotes Hamlet's expostulation on seeing an actor, who—in an impromptu speech describing the queen's grief at her husband's murder—'Could force his soul so to his own conceit,/ That from her working all his visage wann'd,/ Tears in his eyes, distraction in's aspect,/ A broken voice, and his whole function suiting,/ With forms to his conceit ...' Physical signs of grief convincingly distort the player's face, yet he has no real emotion to express. He grieves 'for nothing,' as Hamlet exclaims—'in a fiction, in a *dream* of passion.'

[176] Darwin, *Expression*, pp.386–387.

[177] Campbell, 'Emotion as an Explanatory Principle'; William James, 'What Is an Emotion?', *Mind (New Series)* 9, (1884), p.190; Dewey, 'Emotional Attitudes'; Stanley Schachter and Jerome Singer, 'Cognitive, Social, and Physiological Determinants of Emotional State,' *Psychological Review* 69, (1962), pp.379–399.

[178] James, 'What Is an Emotion?', pp.189–190, capitals and italics his. NB James soon elaborated a far more sophisticated understanding of emotion than this. However, Campbell, 'Emotion as an Explanatory Principle' has shown strong parallels between James and Darwin—especially in the 'failure' of both to elaborate a theory of emotion. Cf. Phoebe Ellsworth, 'William James and Emotion: Is a Century of Fame Worth a Century of Misunderstanding?', *Psychological Review* 101, (1994), pp.222–229.

[179] John Smith, 'Message, Meaning, and Context in Ethology,' *American Naturalist* 99, (1965), pp.405–409; Kirsty Graham, Takeshi Furuichi, and Richard Byrne, 'The Gestural Repertoire of the Wild Bonobo (*Pan Paniscus*): A Mutually Understood Communication System,' *Animal Cognition* 20, (2017), pp.171–177.

also in the external world and that have their roots in the exchanges between the interacting agents.' Pride and shame are examples.[180] But for Darwin the biological *includes* the social (see 3.2 Agency Implies Interdependence). And *all* emotional movements are social.[181] The fact that they function as expressions at all depends on the presence of someone else to recognize the expression.

Does *Expression* chime with the hypothesis that emotions may be *extended*? Does Darwin find 'cases where emotions literally extend beyond the agent's body in that they are partially constituted by factors and feedback external to the agent'? Yes, because in *Expression*'s presentation of emotional expression, the interpretation of movements as showing a particular emotion depends on there being a recognizer to interpret *both*, the bodily changes comprising it, *and* the situation in which they have occurred.[182]

Beyond psychology, philosophical assaults on dualism have led a number of thinkers to take up expression as a topic.[183] Some draw on the later philosophy of Ludwig Wittgenstein (1889–1951). Wittgenstein maintained that our feelings are not contained *within* us somewhere. Rather, 'the characteristic mark of all "feelings" *is that there is expression of them*, i.e., facial expression, gestures, of feeling.'[184] Hence, when we look at a face, 'we *see* emotion.' We do not 'see facial contortions *and make inferences from them* (like a doctor framing a diagnosis) to joy, grief, boredom. We describe a face *immediately* as sad, radiant, bored, even when we are unable to give any other description of the features.'[185]

[180] Paul Ekman and Daniel Cordaro, 'What Is Meant by Calling an Emotion Basic,' *Emotion Review* 3, (2011), pp.364–370; Jaak Panksepp, 'Affective Consciousness: Core Emotional Feelings in Animals and Humans,' *Consciousness and Cognition* 14, (2005), pp.81–88; De Gelder, *Emotions*, p.222.

[181] For a new move (back) to the idea that *emotions* are 'social affairs,' see Clara Fischer, 'Feminist Philosophy, Pragmatism, and the "Turn to Affect": A Genealogical Critique,' *Hypatia* 31, (2016), pp.810–826.

[182] Krueger, 'Extended Emotions,' p.536; Griffiths and Scarantino, 'Emotions in the Wild.'

[183] These followed earlier and far more concerted attacks on expression in the arts. Oscar Wilde, *The Decay of Lying*, (1889), http://cogweb.ucla.edu/Abstracts/Wilde_1889.html. p.95, was among the leaders here, with his comment in *The Decay of Lying* (1889): 'Art never expresses anything but itself. This is the principle of the new aesthetics.' Cf. Cocteau: 'Dance must *express nothing*'; Stravinsky: 'I consider music powerless to *express* anything: a feeling, an attitude, a psychological state, a natural phenomenon'; and Samuel Beckett: 'expression is an impossible act,' etc. Francis Steegmuller, *Cocteau: A Life* (London: Macmillan, 1970), p.95; Alex Ross, *The Rest Is Noise: Listening to the Twentieth Century* (London: Harper, 2009), p.117; Samuel Beckett, 'Three Dialogues with Georges Duthuit,' in *Proust* (London: Calder & Boyars, 1970), p.121.

[184] Ludwig Wittgenstein, *Zettel* (Berkeley CA: University of California Press, 1967), p.90e, my italics; John Shotter, 'Agential Realism, Social Constructionism, and Our Living Relations to Our Surroundings: Sensing Similarities Rather Than Seeing Patterns,' *Theory & Psychology* 24, (2014), pp.308, 318, 322. Cf. Kenneth Gergen, *Relational Being: Beyond Self and Community* (Oxford: Oxford University Press, 2009), p.102: 'It is not that we "feel emotions" so much as we *do* them. And this doing is only intelligible within a particular tradition of relationship.'

[185] Wittgenstein, *Zettel*, p.41e, barring '*see*,' italics mine.

Others philosophers draw in Alfred North Whitehead's (1861–1947) philosophy of organism. Whitehead rejected the bifurcation of nature into two separate realms: of *primary* (physical) qualities, and *secondary* (subjective) qualities. For him, emotion could never be something subjective, 'inside' the individual, which caused an 'outward' physical expression. Perhaps this is not surprising, as Whitehead's philosophy is seen to descend directly from Darwin's biology[186]—in particular, invoking the horizontal dimension of synchronous interdependency that constituted for Darwin the economy of nature.

Various philosophers have followed Whitehead's lead in their discussions of expression, including Maurice Merleau-Ponty, Gilles Deleuze, and Hannah Arendt. Thus, Arendt argued that 'nothing and nobody exists in this world whose very being does not suppose a *spectator*.' In other words, 'nothing that is, insofar as it appears, exists in the singular; everything that is is meant to be perceived by somebody. Not Man but men inhabit this planet. Plurality is the law of the earth.'[187] Any creature depends upon a world which 'solidly appears as the location for its own appearance, on fellow-creatures to play with, and on spectators to acknowledge and recognize its existence,' Arendt went on. Hence, the *expressiveness* of an appearance, did not press out from 'something inside—an idea, a thought, an emotion.' It was of a different order: 'it "expresses" nothing but itself, that is, it exhibits or displays ... it becomes manifest.' To imagine the outside *expresses* what is inside was mistaken. Speaking literally, *inside* we are all (relatively) simple and alike: 'the science of physiology and medicine relies on the sameness of our inner organs.'[188] But, on the surface, we are all fascinatingly different, and unique. Any attempt in psychology to reduce our surface-complexity to a plonking sameness is therefore bankrupt.

Arendt's examples include the contrast between the sexual urge—something psychologists may reduce to the operation of genes, instincts, or glands—and the marvellous, coruscating complexities of everyday amorous interaction around the globe: 'Without the sexual urge, arising out of our reproductive organs, love would not be possible; but while the urge is always the same, how great is the variety in the actual appearances of love!'[189]

---

[186] Bruno Latour, 'What Is Given in Experience? A Review of Isabelle Stengers, *Penser Avec Whitehead*,' *Boundary* 2, (2005), pp.222–237.
[187] Hannah Arendt, *The Life of the Mind: Vol.1 Thinking* (London: Secker & Warburg, 1978), p.19; Maurice Merleau-Ponty, *La Nature. Notes—Cours Au Collège De France* (Paris: Seuil, 1995) Gilles Deleuze, *Expressionism in Philosophy: Spinoza* (New York: Zone Books, 1990); Len Lawlor, 'The End of Phenomenology: Expressionism in Deleuze and Merleau-Ponty,' *Continental Philosophy Review* 31, (1998), pp.15–34.
[188] Arendt, *Life of the Mind*, pp.30, 31.
[189] Arendt, *Life of the Mind*, pp.30–35.

4.6.3 Universal emotions?

Nowadays, *Expression*'s most discussed claim concerns the *universality* of some emotional expressions. The fact that what *Expression* calls *true* emotional expressions were to be found in all kinds of human—and often in animals too (weeping and blushing were the exceptions)—was central to Darwin's argument that both mind and body in *Homo sapiens* are products of common descent.[190] Today, this remains a hot potato.

Few contemporary psychologists doubt humans have evolved. In question is, first, the extent to which human behaviour can be attributed to culture, and, secondly, the extent to which culture escapes evolutionary explanation (see Chs 3, 7, and 8). Whether or not emotions (not, usually, expressions) are universal to all human beings—or even to all apes—has become a battleground in this debate. The Santa Barbara school of Evolutionary Psychology, for example, holds that our emotions date back to the Stone Age. Indeed, the very idea that there are *basic* emotions implies that these form part of a phylogenetically 'given' human nature, biologically inscribed in every healthy member of our species.[191]

Today's debate over basic emotions entails confusion between three claims:

1. The same *expressive movements* are seen in all human beings;
2. Some such movements are recognized or *read the same way* across all cultures;
3. All human beings have the same set of *emotions*.

Such confusion occurs in the evidence of Paul Ekman and his colleagues for so-called basic emotions, research which never empirically investigates emotions as such—or any other subjective state—even though Ekman & co. constantly define emotions as subjective phenomena (e.g. 'contempt' is 'feeling morally superior to another person').[192] Ekman's data show there are discrete patterns of observable activity (i.e. expressions)—'facial, vocal, autonomic physiology'—each with distinct initiating conditions (*mise-en-scènes*).[193] Also, adds Ekman, 'the

---

[190] Darwin, *Expression,* only twice uses the phrase 'true expression': once glossed as 'universally recognized' (p.48); and once as 'innate' (p.52). As usual, he did not fetishize definition.

[191] Tim Lewens, 'Human Nature: The Very Idea,' *Philosophy & Technology* 25, (2012), pp.459–474; Tim Lewens, *Cultural Evolution: Conceptual Challenges* (Oxford: Oxford University Press, 2015); John Tooby and Leda Cosmides, 'The Evolutionary Psychology of the Emotions and Their Relationship to Internal Regulatory Variables,' in *Handbook of Emotions,* ed. Michael Lewis (New York: Guilford Press, 2008); Alan Fridlund, 'Evolution and Facial Action in Reflex, Social Motive, and Paralanguage,' *Biological Psychology* 32, (1991), pp.3–100; Hillary Elfenbein and Nalini Ambady, 'On the Universality and Cultural Specificity of Emotion Recognition: A Meta-Analysis,' *Psychological Bulletin* 128, (2002), pp.203–235.

[192] Ekman and Cordaro, 'Basic Emotion,' p.365.

[193] 'Emotions' are also *basic* because they 'have evolved through adaptation to our surroundings' (ibid, p.364)—although this proposition seldom gets properly tested (3.4.6 'Ancestry').

archetypal expressions' for the basic emotions are all 'universally recognized.'[194] Unfortunately, however, as Darwin knew, the kinds of data to which Ekman & co. appeal could only ever sustain *a theory of expression*—not of emotion—however perfect the studies which produced them. Because, without some method of examining the subjective feel of an emotion—which is independent of the ways one studies both that emotion's purported expressions, and its recognition—one has no independent basis upon which to test emotion's role in one's theory. Hence, Ekman's claim to have shown there are basic *emotions* remains unsupported.

Even claims that some expressions are *read* or *recognized* the same way by all human beings fail when tested. For example, in 2012, Rachael Jack and her colleagues published an ingenious study: 'Facial Expressions of Emotion are not Culturally Universal.'[195] Based on the known muscular anatomy of the human face, they created all the 4,800 possible human facial expressions using a life-like computer graphics-generating package.[196] Then, in a version of Darwin's 'judgment test,' they got fifteen Western Caucasians and fifteen Eastern Asians independently to name each of the 4,800 expressions. The thirty participants were asked to use only Ekman's six 'basic emotion' terms (anger, sad, happy, disgust, surprise, fear), plus 'don't know' (see Figure 4.8).

The naming-maps that emerged from this exercise differed markedly between the two cultural groups. While Ekman's six emotion-names distinguished six fairly discrete clusters of expression *for the Caucasians*, the same six terms could not capture the representational space used by East Asian judges. Which suggests, contrary to Ekman, that facial expressions are *not* universally recognized.

Did Darwin posit *basic* emotions? *Expression* does once distinguish, by implication, between emotional attitudes that are identifiable with a distinct expression (roughly comprising Ekman's list of 'basic' expressions; see Figure 4.8), and those which are 'complex' and not 'revealed by any fixed expression, sufficiently distinct to be described or delineated,' namely: 'Jealousy, Envy, Avarice, Revenge, Suspicion, Deceit, Slyness, Guilt, Vanity, Conceit, Ambition, Pride, Humility, &c.'[197] When attributing these more complex states of mind, Darwin says we draw on what we know of context and history to make sense of what we see and

---

[194] Ekman and Cordaro, 'Basic Emotion,' p.369.
[195] Rachael Jack et al., 'Facial Expressions of Emotion Are Not Culturally Universal,' *Proceedings of the National Academy of Sciences of the United States of America* 109, (2012), pp.7241–7244.
[196] This is an awful lot of judgements per participant, though not half as many as the 12,000 made by each participant in Rachael Jack, Roberto Caldara, and Philippe Schyns, 'Internal Representations Reveal Cultural Diversity in Expectations of Facial Expressions of Emotion,' *Journal of Experimental Psychology (General)* 141, (2012), pp.19–25. Dr Jack has informed me that, depending on the task, each participant completed the experiment over anything up to 12 hours. This was split into several separate 45-minute sessions with at least 1-hour break in between each session to avoid fatigue. Each session also had several mini breaks—e.g. after each set of 50 trials.
[197] Darwin, *Expression*, pp.275–276.

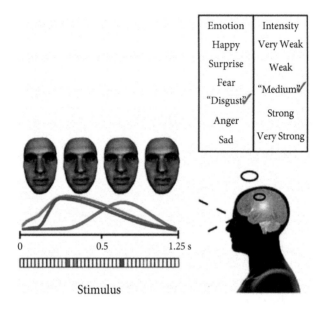

Emotion	Intensity
Happy	Very Weak
Surprise	Weak
Fear	"Medium?"✓
"Disgust?"✓	
Anger	Strong
Sad	Very Strong

Stimulus

Figure 4.8 Detail from Rachael Jack et al.'s illustration of how artificial facial expressions were randomly generated by a computer graphics programme in a virtual face as if by the combined movement of a subset of three muscles (here, Action Units 17, 10L & 9), at varying intensities (coloured lines), and then individually judged as representations of six 'basic' emotions, with varying intensities (here, judged as 'disgust' at 'medium intensity').

Source: Reproduced from Jack, R.E., Garrod, O.G. B., Yu, H., Caldara, R., and Schyns, P.G. Facial expressions of emotion are not culturally universal. *PNAS*, 109 (19): 7241–7244. Copyright © 2012 by National Academy of Sciences. doi.org/10.1073/pnas.1200155109.

hear: 'our previous knowledge of the persons or circumstances.' Most often, however, Darwin used the word *fixed* as a synonym for the durability lent to a character by habit or inheritance 'during many generations,' rather as one might have spoken last century of a photographic image being 'fixed' by chemicals (i.e. stabilized). Darwin did not claim that any expressions were 'fixed' in the sense of being rigidly stereotyped.[198] Expressions of happiness or of horror, even when innate, showed many 'gradations': from a gentle smile to 'excessive laughter,' or from terror to 'mere attention.' In this sense, expressions were neither fixed nor genuinely reflex.[199]

---

[198] Darwin, *Expression*, p.220.

[199] For Darwin no character is ever absolutely fixed, but characters or behaviours can be *relatively* fixed and so have the *quality* of reflexes. Foregoing quotes from Darwin, *Expression*, pp.40, 217–222, 274–275, 325.

4.6.4 The three principles

Expression's first principle, the idea that expressions like those of fear originated as the side-effects of functional habits, is the closest Darwin's explanation for human expressions gets to contemporary claims that emotional expressions evolved via natural selection, through being adaptive.[200] But Darwin added something to the idea that expressions originated as 'serviceable habits' which, over generations, became hereditary: the word *associated*. This makes facial expressions far more complicated than does today's idea that they express biologically-ingrained and therefore *basic* emotions. An unusual history of association can make a face express fear under almost any circumstance—as many experienced psychotherapists could attest: upon the reception of a gift; when complimented; when attracted; faced by an unusually delicious meal; and so on.[201]

Darwin's second principle, of antithesis, proposes that there is an inherited tendency to reverse muscular movements when making opposed choices. What is 'opposite' about an action is referred by Darwin either to the will (we will oppose things), or to our being in 'directly opposite circumstances': eyes open wide when we are excited; eyelids droop when we are tired or depressed. To date, I have not found any contemporary resonance with Darwin's second principle. The nearest is a study, by Peter Hurd and colleagues, which used computers to simulate the co-evolution of two different signals, along with a receiving neural network that was programmed to give different responses to the two signals. As the simulation evolved, the forms of the signals became polarized, exaggerated, and antithetical. Hence, *the recognition of expression* could push expressions to become antithetical—a rather different explanation from the one Darwin proposed, and one assuming value for the expresser in being better-understood by recognizers.[202]

Regarding *Expression*'s third principle, what psychologists call the hydraulic model of nervous action has had a long innings.[203] Darwin was neither the first nor the last to subscribe to it. Its longevity may partly have to do with the conceptual consequences of common speech, as illustrated by figures that equate 'energy' with

---

[200] Ekman and Cordaro, 'Basic Emotion'; Tooby and Cosmides, 'Emotions,' Karen Schmidt and Jeffrey Cohn, 'Human Facial Expressions as Adaptations: Evolutionary Questions in Facial Expression Research,' *Yearbook of Physical Anthropology* 44, (2001), pp.S3–S24.; Alan Fridlund, *Human Facial Expression: An Evolutionary View* (San Diego: Academic Press, 1994).

[201] Regarding the contingency of the sense of facial expressions, see Darwin, *Expression*, pp.183–184.

[202] Peter Hurd, Carl-Adam Wachtmeister, and Magnus Enquist, 'Darwin's Principle of Antithesis Revisited: A Role for Perceptual Biases in the Evolution of Intraspecific Signals,' *Proceedings of the Royal Society B* 259, (1995), pp.201–205.

[203] Historians have argued that hydraulic models in physiology (usually traced to Descartes and/or Boerhaave) were gradually replaced with an energetic model by the 19th century, involving electrical circuits, nervous energy, current, or force conducted through channels. Metaphors invoking fluids (current, channel) feature in both, however: see Christopher Lawrence, 'The Nervous System and Society in the Scottish Enlightenment,' in *Natural Order: Historical Studies of Scientific Culture*, ed. Barry Barnes and Steven Shapin (London: Sage, 1979), pp.19–40.

a liquid that can flow easily, be dammed up, released, discharged, or run out. So I can say I am 'overflowing with energy,' or 'drained of energy.' Ethologists Konrad Lorenz and Niko Tinbergen subscribed to a version of this model when talking about *displacement* activities: I drum my fingers while awaiting an interview; a dog scratches itself when confused; a bird preens when caught between approach and avoidance. Displacement activities supposedly resulted from two contrasting drives (e.g. approach–withdrawal) blocking each other so that surplus energy spills over into, or is 'displaced' onto, some seemingly incongruous lower-order activity.

Ethologists abandoned the hydraulic model of 'drive' in the 1950s. But it still crops up in psychology. The model is particularly persistent in the language of research on sexual behaviour, where terms like 'arousal,' 'appetitive,' and 'consummatory' behaviour, suggest a hydraulic model. Thus it still makes sense to say that, when someone is 'full of sexual energy' (aroused), some of that behaviour will 'overflow' (in 'appetitive behaviours,' such as intimate touching, flirtatious looking, blushing, genital engorgement), until the energy is 'spent' or 'relieved' (as in 'catharsis' or 'consummatory behaviour,' such as sublimation or intercourse). Whether or not such a vocabulary remains scientifically useful remains contentious.[204]

4.7 Conclusion

Expression is a complex book. In some regards, it is a text of its time. Its engagement with popular tropes and sentiments, its invitation to a do-it-yourself study of expression, its delight in sometimes-sentimental narrative, all draw on Victorian discourses and structures of feeling.[205] But the methods Darwin devised and adapted to the study of expression, his side-lining of emotion as an explanatory concept, the research-base of his book in cutting-edge physiology, the observational emphasis, all mark *Expression* down as a book of the future. Given this complexity, and the untechnical, sometimes slapdash, language in which Darwin's arguments and observations are clothed, it is also a book that requires close and forgiving attention if it is to be understood—as today's copious, contradictory glosses of *Expression* prove.

The first three chapters in this book show how *Origin* posed long-term genealogical change—evolution—as being generated through synchronous interdependencies which make up the 'infinitely complex relations' of individuals 'to

---

[204] Sigmund Freud, 'Instincts and Their Vicissitudes,' ed. James Strachey, *The Standard Edition of the Complete Psychological Works of Sigmund Freud, Volume XIV* (London: Hogarth, 1957); Roger Tourangeau and Phoebe Ellsworth, 'The Role of Facial Response in the Experience of Emotion,' *Journal of Social and Personality Psychology* 37, (1979), pp.1519–1531; Gregory Ball and Jacques Balthazart, 'How Useful Is the Appetitive and Consummatory Distinction for Our Understanding of the Neuroendocrine Control of Sexual Behaviour?,' *Hormones and Behaviour* 53, (2008), pp.307–318; Robert Hinde, 'Ethological Models and the Concept of 'Drive',' *British Journal for the Philosophy of Science* 6, (1956), pp.321–331.

[205] White, 'Reading the Blush'; White, 'Darwin Wept'; Richardson, 'The Book of the Season.'

other organic beings and to their physical conditions of life.'[206] Darwin's stereo-scopic vision of nature had two axes: a horizontal or synchronous theatre of agency, and a vertical or diachronic record of slow-moving, one-way, organic transform-ation. In this vision, dramas in the theatre of agency *drive* phylogenetic changes, which comprise their unintended effects.

This chapter shows that Darwin's understanding of expression, and his psych-ology, place the same primacy on a theatre of agency as did his theory of natural selection. Going about their daily lives, humans, like their ancestors and animal relations, unconsciously move and (more consciously) act in patterned ways, *some* of which have emotional meanings for the others with whom they stand in relation—so that, in Mead's reading, the meanings of each individual's actions ini-tially *emerge from* his or her interdependency with others. They do not arrive at the surface of the body already dyed with meanings brewed, like an inner essence, deep in the hidden, immaterial recesses of the expresser's mind.

The stress *Expression* places on others' interpretations of movements-*in-situ* establishes the direction in which Darwin's psychology must grow as it addresses more complex forms of agency. If our understanding of movement as emo-tional primarily derives from the ways we read others' actions, then a second order of complexity will be generated when this form of agency rebounds on the agent: when I read you *reading me*. Darwin's theory of this more complex kind of recognition is most clearly worked out in his analysis of blushing, as we shall see in Chapter 5.

---

[206] E.g. Darwin, *Origin 1876*, p.49.

5

Being Read

Why did I blush tonight? Sitting among colleagues, our Head of Department had asked the meeting to weigh two options: give a small grant of money to needy students for their research; or, we each get a similar sum to fund our own studies. A fair choice? Once publish a promise to students, and consumer law applies. There can be no going back. Contrariwise, the university's senior managers are a skittish bunch. They constantly cry poor, and long since abandoned any obligation to provide *all* staff with research funds. (Staff must compete!) So, up goes my hand. Option two is only worth considering, I say, if *here and now* staff get a guarantee that the funds will not tomorrow be sucked back into university coffers—to cover 'more pressing' needs. The Head, suddenly animated, runs off at a tangent, finds a new topic, and we get no guarantee. I look down. I am blushing! Do I worry I look a fool? Or do I feel I have outed my boss as less than frank—such that, in his place, *I* would feel shamed?

The remarkable fact that *my* burning cheeks can, against my wishes, express *another's* predicament, helped shape Darwin's understanding of our 'higher psychical faculties,' an understanding best approached via the longest chapter in *Expression*: on 'Blushing.'[1] This chapter will deal in turn with his theory's three pillars: duality of the self; sympathy; and their product, self-attention. I then examine how Darwin related his theory to the physiology and evolution of blushing. But first we must review the observations which Darwin's chapter set out to explain.

5.1 Phenomena of Blushing

Expression brims with observations about blushing, only the most distinctive of which do I summarize now. Most obviously, blushing crimsons the skin, though, in most Europeans, solely the face, ears and neck redden. Nevertheless, while blushing intensely, many people 'feel that their whole bodies grow hot and tingle; and this shows that the entire surface must be in some manner affected.'[2] Most blushes start on the cheeks, afterwards spreading to the ears and neck. Sometimes, blushers report that they feel they have reddened, when in fact they have turned pale.[3]

---

[1] Charles Darwin, *The Expression of the Emotions in Man and Animals*, 2nd ed. (London: Murray, 1890) The chapter's full title reads: 'Self-Attention—Shame—Shyness—Modesty: Blushing.' pp.328–367.
[2] Darwin, *Expression*, p.330.
[3] Darwin, *Expression*, pp.330–331.

Darwin's Psychology. Ben Bradley, Oxford University Press (2020). © Oxford University Press.
DOI: 10.1093/oso/9780198708216.001.0001

Blushing is associated with other movements, most often a turning away of the body from observers, from whom the blushers 'almost invariably' cast down their eyes or 'look askance,' displaying a restless gaze. Sometimes blushes produce tears, if not 'the oddest trick' of blinking with 'extraordinary rapidity.'[4]

Secondly, blushing is the 'most human' of all expressions, states Darwin. Monkeys redden from passion, but 'it would require an overwhelming amount of evidence to make us believe that any animal could blush.'[5] Hence *Expression*'s crucial distinction between *blushing* and *flushing*. Any strong emotion that makes the heart beat faster will redden the face: whether in circumstances provoking rage or fear, high spirits or great joy. The face also reddens from desire: 'when lovers meet, we know that their hearts beat quickly, their breathing is hurried, and their faces flush.'[6] Blushing, on the other hand, has no obvious connection to stimuli that quicken the heart (see 5.2.4 Physiology)—or indeed to any physical stimulus:

> We can cause laughing by tickling the skin, weeping or frowning by a blow, trembling from the fear of pain, and so forth; but we cannot cause a blush ... by any physical means,—that is by any action on the body. *It is the mind which must be affected.* Blushing is not only involuntary; but the wish to restrain it ... actually increases the tendency.[7]

Blushing proves the most psychical of expressions. Hence the challenge it presents to explanation.[8]

If blushing were quintessentially human, all humans should blush. But do they? The second of the sixteen 'queries about expression' that, over the 1860s, Darwin mailed out to observers around the world, read: 'Does shame excite a blush when the colour of the skin allows it to be visible? and especially how low down the body does the blush extend?'[9] From the answers he received, Darwin concluded that blushing was common to 'most, probably to all, of the races of man.'[10] Not that observers of the same groups all agreed, especially about people with dark skin. For example, Darwin had eight informants who reported on blushing in Australian Aborigines. Four said Aborigines 'never blush.' One prevaricated, noting that 'only a very strong blush could be seen, on account of the dirty state of their skins.' And three stated that Aborigines do blush. Darwin went with the three—correctly, according to today's pundits (cf. 4.5 'Kinks and muddles').[11]

---

[4] Darwin, *Expression*, p.340.
[5] Darwin, *Expression*, p.328.
[6] Darwin, *Expression*, p.83.
[7] Darwin, *Expression*, pp.328–329, my italics.
[8] A challenge taken up by neither Spencer, nor Bain. See 5.3 Resonances, for the continuing psychological controversy over blushing.
[9] Darwin, *Expression*, p.16.
[10] Darwin, *Expression*, p.340.
[11] Darwin, *Expression*, p.339. All contemporary commentators aver that, once infancy has passed, all humans blush: e.g. Robert Edelmann, 'Embarrassment and Blushing,' in *Encyclopaedia of Human*

Further evidence about blushing came from the Reverend R.H. Blair, founder of Worcester College for the Blind. Blair reported that three children born blind, out of the seven or eight then in his asylum, were 'great blushers.' Such children would blush when corrected.[12] Blair reported they were 'not at first conscious that they were observed,' and it was a 'most important part of their education,' to impress them with this knowledge: 'the impression thus gained would greatly strengthen their tendency to blush.'[13]

When we come to *Expression*'s treatment of the causes of blushing, we find what we should expect to find, given Darwin's general approach to expression as described in Chapter Four. The examples *Expression* elaborates in its section called 'The Nature of the Mental States which induce Blushing,' describe circumstances and inter-personal dynamics, rather than any hidden ping-pong of mental states. Examples partly come from received opinion and informal observation, partly from other observers, and partly from a kind of interview-study in which Darwin asked women whom he identified as blushers about the conditions under which they blushed. He collated what he learnt from these sources under the headings of: shyness; shame or guilt; and modesty.

Shyness, wrote Darwin, seemed to depend on sensitiveness to the opinion of others—whether good or bad—most often with respect to *external appearance*. While strangers do not care about our conduct or character, *Expression* says, they may, and often do, criticize our appearance. Hence shy people were particularly apt to blush in the presence of strangers. Apprehension of anything peculiar, or even new, in apparel, or 'any slight blemish on the person, and more especially on the face—points which are likely to attract the attention of strangers—makes the shy intolerably shy.'[14]

Unlike matters of appearance, on the other hand, a sense of *improper conduct* most often caused blushing with known people whom the shy value, not strangers: 'A physician told me that a young man, a wealthy duke, with whom he had travelled as medical attendant, blushed like a girl, when he paid him his fee; yet this young man probably would not have blushed and been shy, had he been paying a bill to a tradesman.'[15]

Behaviour, ed. Vilayanur Ramachandran (London: Academic Press, 2012). Yet I have found only one empirical study that *tests* whether blushing is universal—and that is not observational, but linguistic: Michael Casimir and Michael Schnegg, 'Shame across Cultures: The Evolution, Ontogeny and Function of a 'Moral Emotion', in *Between Culture and Biology: Perspectives on Ontogenetic Development*, ed. Heidi Keller, Ype Poortinga, and Axel Schölmerich (Cambridge: Cambridge University Press, 2002), pp.270–300. Casimir and Schnegg conclude that blushing is 'panhuman.'

[12] William Darwin, 'Letter to Darwin, 25th March,' *Correspondence of Charles Darwin* (Cambridge: Darwin Correspondence Project, 1868), https://www.darwinproject.ac.uk/letter/?docId=letters/DCP-LETT-6069.xml.

[13] Darwin, *Expression*, pp.329–330.

[14] Darwin, *Expression*, p.349.

[15] Darwin, *Expression*, p.349.

While shy people mainly blushed from the apprehension of censure, praise did sometimes make blushers blush.[16] A humble person would blush at praise as much as at their own or others' indecorum. Such blushes might occur in solitude, or even in the dark. If so, 'the cause almost always relates to the thoughts of others about us—to acts done in their presence, or suspected by them; or again when we reflect what others would have thought of us had they known of the act.'[17]

Shame, says *Expression*, typically arises from 'moral causes.' But *blushing from shame* has rather different dynamics. One might be conscious of a fault before God, but not blush—one of Darwin's female informants told him.[18] Likewise, a guilty man would not blush when alone. So it was not the sense of guilt itself which crimsoned the face, 'but the thought that others think or know us to be guilty.'[19] In fact, even someone innocent may blush if she or he feels others suspect impropriety: Even the thought, observes *Expression*, that others perceive me to have made an unkind or stupid remark, was 'amply sufficient to cause a blush, although we know all the time that we have been completely misunderstood.'[20] Beyond this stands impropriety—and breaches of etiquette—of the kind which opened this chapter: 'a sensitive person, as a lady has assured me, will sometimes blush at a flagrant breach of etiquette by a perfect stranger, though the act may in no way concern her.'[21]

One of Darwin's literary sources, *The Fable of the Bees* by Bernard Mandeville, describes a particularly baroque source of blushing. A 'virtuous young woman' will blush, if, in her presence, obscene words are spoken, notes Mandeville. Yet the same young woman who hears the same bawdy talk from another room, 'where she is sure that she is undiscovered ... [will] hearken to it, without blushing at all.'[22] Here, the young woman's blushes result from recognition that others recognize that she has recognized something that it was improper for her to have recognized.

Expression calls blushes at indelicacy *modest*: 'modesty frequently relates to acts of indelicacy; and indelicacy is an affair of etiquette, as we clearly see with the nations that go altogether or nearly naked.' Such blushes often relate to the opposite sex, and so are 'apt to be intense.' This is not so much because young lovers feel strong desires—desire being a source of *flushing*, not blushing.[23] Love intensifies blushing, wrote Darwin, because lovers value each other's admiration and love 'more than anything else in the world.'[24]

---

[16] Darwin, *Expression*, p.350.
[17] Darwin, *Expression*, p.355.
[18] Darwin, *Expression*, p.352.
[19] Darwin, *Expression*, p.366.
[20] Darwin, *Expression*, pp.352–353.
[21] Darwin, *Expression*, p.353.
[22] Bernard Mandeville, *The Fable of the Bees; or, Private Vices, Public Benefits*, vol. 1 (Indianapolis: Online Liberty Fund, 2011), p.87. Darwin reported 'extracting' from Mandeville's book (Vol.2) on the 24th April, 1840: p.7r, Charles Darwin, "Books to Be Read' and 'Books Read' Notebook,' in *Darwin Manuscripts* (Cambridge: Cambridge University Library, 1838-1851).
[23] Darwin, *Expression*, p.83.
[24] Darwin, *Expression*, p.347.

Whatever the circumstances, *Expression* notes, some people blush more than others. This is partly a matter of character: blushing is tied to shyness, and 'the conceited are rarely shy, for they value themselves much too highly to expect depreciation.'[25] And it is partly a matter of inheritance, at least as regards its patterning. One of Darwin's medical informants, Sir James Paget, was struck by a girl's 'singular manner of blushing': a big splash of red appeared first on one cheek, then other splashes scattered over the face and neck. Sir James asked the girl's mother whether her daughter always blushed in this way. The mother answered, 'Yes, she takes after me,' and blushed—showing 'the same peculiarity as her daughter.'[26]

Beyond character and inheritance come gender and age. *Expression* states that women blush more than men, and the young blush more than the elderly. Once again, this had to do with degrees of sensitivity to (imagined) censure. Women were 'much more sensitive about their personal appearance than men,' wrote Darwin, 'especially elderly women in comparison with elderly men, and they blush much more freely.'[27] Likewise, compared to older adults, the young are 'highly sensitive to the opinion of each other with reference to their personal appearance.'[28]

Expression finds two exceptions to the law that all humans blush: infants and idiots. Here Darwin drew on reports of observers whom he had specifically asked to examine blushing. He called the absence of blushing in infants remarkable, as we know that even tiny infants 'redden [flush] from passion.' The youngest blushers credibly reported to Darwin were two-year-olds. He commented: 'It appears that the mental powers of infants are not as yet sufficiently developed to allow of their blushing. Hence, also, it is that [e.g. microcephalic] idiots rarely blush.' Observations of microcephaly came from asylum-superintendent James Crichton Browne who 'observed for me those ['idiots'] under his care, but never saw a genuine blush, though he has seen their faces flush, apparently from joy, when food was placed before them, and from anger.'[29]

---

[25] Darwin, *Expression*, p.350.

[26] Darwin, *Expression*, p.330.

[27] Darwin, *Expression*, pp.329, 346.

[28] Darwin, *Expression*, pp.346–347, quotes Thomas Henry Burgess, *The Physiology or Mechanism of Blushing: Illustrative of the Influence of Mental Emotion on the Capillary Circulation; with a General View of the Sympathies* [i.e. the sympathetic nervous system], *and the Organic Relations of Those Structures with Which They Seem to Be Connected* (London: John Churchill, 1839), p.33—as supporting a view of women as 'blushing more freely than men.' I can find no such statement in Burgess. To the contrary, Burgess wrote (pp.54, 57) that men blush as much as women. Not that Burgess related a systematic study to support his views. Today's experts disagree with Darwin about women blushing more than men, but agree that there is 'a peak period of embarrassment ... during adolescence. There is then a decline in reported embarrassment in later adult life': Edelmann, 'Embarrassment and Blushing,' pp.27–28. I have found no recent studies bearing on Darwin's statements about young lovers.

[29] All quotes in this paragraph from Darwin, *Expression*, p.329.

5.2 Explanations for Blushing

Darwin's explanations for blushing are both psychological and physiological. I will first describe the three psychological planks of his analysis. I then examine how *Expression* treats the physiology of blushing, before discussing Darwin's take on how blushing evolved.

5.2.1 The dual self

Darwin's early notebooks contain copious naturalistic documentation of human characteristics and idiosyncrasies. As if still at large in the Brazilian rainforest with its bewildering array of gaudy insects, strange trees, and stranger flowers—Darwin's jottings luxuriate in fascinating details about the insane and the senile, infants and drunks, eruptions of passion and the content of dreams, sexual desire and castles in the air, autobiographical memory and the dynamics of forgetfulness. Comparisons with animals abound, but so do facts gleaned from his own self-observation and from his questioning of others—plus those plucked from his voracious reading.

The main organizing principle for all this material was Darwin's novel take on the concept of *double consciousness*. Double consciousness was widely discussed in Britain in the 1820s and 1830s.[30] One early case that Darwin cited had been discussed by the Scottish physician John Abercrombie (1780-1844). An 'ignorant' servant girl, when in the throes of somnambulism, 'showed an astonishing knowledge of geography and astronomy'—knowledge she had supposedly picked up from overhearing a tutor giving instructions to the young people of the family she served.[31] Today, we might be prone to seek a diagnostic category of mental disorder for such a case.[32] But Darwin was interested in what such cases implied about human action in general. Double consciousness suggested a way to systematize all his observations about agency. Commenting on another case of somnambulism, Darwin noted: 'the young lady almost equally in her senses in either state.' He went on: 'does this throw light on instinct, showing what trains of action may be done unconsciously as far as the ordinary state is concerned?'[33] This line of questioning harks back to Darwin's medical training at Edinburgh, where it had long been

---

[30] Ian Hacking, 'Double Consciousness in Britain, 1815–1875,' *Dissociation* 4, (1991), pp.134–146.

[31] John Abercrombie, *Inquiries Concerning the Intellectual Powers and the Investigation of Truth*, 8th ed. (London: John Murray, 1838), pp.301–302. The case was first reported by a Dr Dewar in 1822; see Hacking, 'Double Consciousness in Britain, 1815–1875.'

[32] Controversially, Hacking, 'Double Consciousness in Britain, 1815–1875,' pp.134ff, equates cases like these with 'multiple personality disorder.'

[33] Charles Darwin, 'Notebook M,' (Cambridge: Darwin Online, 1838), p.110.

taught that there were two strands of human feeling, *sensibility* and *sentience*: the former lacking, the latter having, consciousness.[34]

The twenty-nine-year-old Darwin's notebooks show he assessed with care a diversity of evidence bearing on the hypothesis that the waking mind comprised two different trains of thought or action. Train One was more or less unconscious, being equated with spontaneous, unthinking habit, or even with instinct. Thus when a senile woman, Miss Cogan, repeatedly sang songs that she had only heard in her early childhood, Darwin compared this to 'birds singing, or some instinctive sounds.' He noted that 'Miss C. memory cannot be called memory, because she did not remember, it was an habitual action of thought-secreting organs.'

Train One unreflectively threw up thoughts or committed acts that were easily forgotten, resembling 'the instincts of animals.' Examples in humans included: momentary pleasures; insane acts; things done when drunk; and unreflective imaginative absorption as when reading Dickens, or daydreaming, or night-dreaming. 'There seems no distinction between enthusiasm passion & madness,' one note observes. The fact that Train One phenomena were so forgettable suggested that the natural cure for both passion and madness was forgetfulness: Train One phenomena were like an 'insubstantial pageant ... rounded with a sleep.' Building on these links between dreams, castles in the air, senility, passion, and madness, Darwin quoted his doctor-father as saying that: 'there is perfect gradation between sound people and insane.— that everybody is insane, at some time.'[35]

All this went to underline the plausibility

> ... of the brain having whole train of thoughts, feeling & perception separate from the ordinary state of mind, [as] probably analogous to the double individuality implied by habit, when one acts unconsciously with respect to more energetic self, & likewise one forgets, what one performs habitually.[36]

Such a 'double self' was also manifest in the case of the insanity of a medical friend of Darwin's paternal grandfather, who 'struggled as it were with a second & unreasonable man.'[37]

Train Two was what Darwin sometimes called *reflective* consciousness or *self*-consciousness.[38] Darwin associated Train Two with: long-lasting,

---

[34] Christopher Lawrence, 'The Nervous System and Society in the Scottish Enlightenment,' in *Natural Order: Historical Studies of Scientific Culture*, ed. Barry Barnes and Steven Shapin (London: Sage, 1979), pp.26–27.

[35] Previous quotes from Darwin, 'Notebook M,' pp.8, 13, 18, 81, 113 & passim; cf. William Shakespeare, *The Tempest* (1611), Scene I.

[36] Darwin, 'Notebook M,' p.81.

[37] Darwin, 'Notebook M,' pp.19, 81, 116–117.

[38] Darwin, 'Notebook M,' p.115; Darwin, *Expression*, pp.346, 366. This echoed James Ferrier, 'An Introduction to the Philosophy of Consciousness, Part I,' *Blackwood's Edinburgh Magazine* 43, (1838), for whom all consciousness—which he differentiated from *states of mind*—was self-consciousness (see Ch. 4).

'philosophical' happiness;[39] efforts of will; conscious recollection; reasoning; comparing; judging; 'the intellectual faculty'; choosing; making links between present, past, and future; acquiring new ideas; and 'original, inventive thought.'[40] Comparison was the key: 'it is solely the comparison, with past ideas, which makes consciousness.'[41]

Relations between Trains One and Two were not always friendly. Darwin described a variety of situations in which he struggled to maintain Train Two thought: 'one cause of the intense labour of original inventive thought is that none of the ideas are habitual, nor recalled by obvious associations.' He observed a natural tendency of Train One thought to overpower Train Two: 'Fear must be simple instinctive feeling: I have awakened in the night, being slightly unwell & felt so much afraid though my reason was laughing & told me there was nothing.' On the other hand, once its impetus was spent, Train One thought evanesced. Thus, young Darwin deplored 'the facility with which a castle in the air is interrupted & utterly forgotten —, so as to feel a severe disappointment': the *content* of the reverie being lost, even while the fact that one has had a reverie is retained.[42] So, in maturity—to Freud's appreciation—Darwin crafted a golden rule: 'whenever a published fact, a new observation or thought came across me, which was opposed to my general results, [I undertook] to make a memorandum of it without fail and at once; for I had found by experience that such facts and thoughts were far more apt to escape from the memory than favourable ones.'[43]

An echo of Darwin's 'two trains' hypothesis appears in what *Expression*'s calls the *confusion of mind* associated with blushing, as when blushers 'lose their presence of mind, and utter singularly inappropriate remarks.' One of Darwin's interviewees 'a young lady, who blushes excessively,' told him that, when blushing, 'she does not even know what she is saying.'[44] Blushing confuses because, in it, a Train One process derails a Train Two process. Blushing differs from all other expressions because its *production* involves Train Two self-consciousness—which explains its absence in infants, idiots, and animals.

---

[39] Darwin's discussion of the intellectual and moral basis for lasting happiness is largely to be found in Charles Darwin, 'Old & Useless Notes,' (Cambridge1837–1840): e.g. p.27, which talks of a 'man who has thought very much. & he will know his happiness lays in doing good & being perfect.' See also Darwin's marginalia to Abercrombie, *Inquiries Concerning the Intellectual Powers and the Investigation of Truth*.

[40] Darwin, 'Notebook M,' pp.87, 91.

[41] Darwin, 'Notebook M,' pp.34, 103.

[42] Darwin, 'Notebook M,' pp.34, 35, 54, 63e, 87, 103, 115.

[43] Charles Darwin, *The Autobiography of Charles Darwin, 1809–1882: With Original Omissions Restored*, ed. Nora Barlow (London: Collins, 1958), p.123; Sigmund Freud, *The Psychopathology of Everyday Life* (New York: Macmillan, 1914), pp.154–155.

[44] Darwin, *Expression*, pp.341–342.

5.2.2   Sympathy

Sympathy plays a central part in Darwin's theory of blushing, as it does across his whole psychology. Today *empathy* has largely displaced sympathy in the interests of psychological researchers.[45] But the term 'empathy' was not coined until 1909—and its meanings remain confused.[46] So we must pause to examine what Darwin may have meant when referring to sympathy.

The main dimensions now said to distinguish empathy stress: conscious effort; theory of mind; role-taking; and the need to achieve cognitive congruence between the mental states of target and empathizer. Empathy represents the deliberate attempt to reach out so as mentally to 'walk in the shoes' of another person, and thus gain access to their experience, whatever that may be—one may empathize as much with the thought-process of a sadistic killer as with that of his victim. By contrast, sympathy connotes an immediate, unstoppable *bodily* response to another's distress.[47] 'In sympathy the sympathiser is "moved by" the other person'—often, to alleviate their suffering. Empathy is more cerebral, less visceral: 'a way of "knowing." ' Sympathy is agentive, 'a way of "relating." '[48] A common synonym for sympathy today is compassion, conceived as 'a feeling of concern for another person's suffering which is accompanied by the motivation to help.'[49] However, this equation clashes with views that attaining compassion requires rigorous intellectual and spiritual discipline. The Dalai Lama describes compassion (concern for others) as *the goal* of Buddhist practice, not a natural given. In which case, compassion and sympathy are best *not* viewed as synonyms. For Darwin called sympathy a 'distinct emotion,' 'instinct,' or 'innate feeling,' that did not require reasoning or practice.[50]

Darwin's usage of the word *sympathetic* in *Expression* melds the physiological with the psychological. This was nothing new. For centuries, sympathy had been used in science as a term to refer to 'a relation between two bodily organs or parts (or between two persons) such that disorder, or any condition, of the one induces a corresponding condition in the other.'[51] This concept of sympathy helped found the distinctive approach to physiology taught to medical students at Edinburgh University, where sympathy was held to be a special case of neural *sensibility* (Train

---

[45] According to the journal database, PsycINFO, by 2018, more than twice as many psychological publications have targeted empathy (15720) than sympathy (7595), in the previous decade.

[46] Benjamin Cuff et al., 'Empathy: A Review of the Concept,' *Emotion Review* 8, (2016); Lauren Wispé, 'The Distinction between Sympathy and Empathy: To Call Forth a Concept, a Word Is Needed,' *Journal of Personality and Social Psychology* 50, (1986), pp.144–153.

[47] Cuff et al., 'Empathy'p.145.

[48] Wispé, 'Sympathy and Empathy'p.318. For an opposite view, that empathy is visceral and sympathy cerebral, see https://chopra.com/articles/whats-the-difference-between-empathy-sympathy-and-compassion.

[49] Tania Singer and Olga Klimecki, 'Empathy and Compassion,' *Current Biology* 24, (2014), p.R875.

[50] Darwin, *Expression*, pp.227, 379.

[51] '1b. Sympathy' in Anon, *Oxford English Dictionary Online*, (Oxford: Oxford University Press, 2018).

Figure 5.1 Detail of 'Kit's Writing Lesson,' by Robert Martineau, 1852.
Source: Robert Braithwaite Martineau, *Kit's Writing Lesson*, 1852. Tate Britain, London, UK. Painters / Alamy Stock Photo.

One) in which feeling was communicated between different bodily organs, 'manifested by one bodily organ when another was stimulated.'[52] In this vein, Darwin described as an instance of sympathetic movements the coordination of action in an attentive dog: 'a young dog with straight ears, whose master, from a distance, shows him an appetising piece of meat, avidly fixes his eyes on that object and follows its every movement, and while his eyes look, his two ears stick forward as though the object could be heard.'[53]

Similar cases of sympathetic movement occur in humans. For example, when someone cuts cloth with scissors, they 'may be seen to move their jaws simultaneously with the blades.' Children learning to write 'often twist about their tongues as their fingers move, in a ridiculous fashion' (Figure 5.1). When watching sporting events, such as 'leaping matches'—or, in our day, boxing bouts, American football, or rugby league: 'as the performer makes his spring [or hits an opponent], many of the spectators,' cannot help 'move their feet' and bodies in sync.[54] Darwin referred to these kinds of sympathy as an 'habitual co-action of . . . muscles.'[55] He also used

[52] Lawrence, 'Nervous System and Society,' pp.27–30.
[53] Darwin, *Expression*, p.7.
[54] Darwin, *Expression*, pp.35–36.
[55] Darwin, *Expression*, p.250.

the term in a more purely physiological sense, as when writing about blushing, of 'the intimate sympathy which exists between the capillary circulation of the surface of the head and face, and that of the brain' (see 5.2.4 Physiology).[56]

According to eighteenth century Edinburgh physiology, sympathy, while generally an unconscious physiological coordinator, would sometimes breach consciousness, as in onlookers affected by shocking sights: 'people subject to hysteric fits will often by sympathy fall into a fit by seeing another person fall into the same.'[57] In these cases, sympathy transcended the individual. 'Between two persons,' sympathy had an immediacy which undercut reason, says *Expression*, as when Darwin's six-month-old son 'instantly assumed a melancholy expression,' on seeing his nurse pretend to cry.[58] Equally, despite naturalists' aspirations to detached observation, 'when we witness any deep emotion, our sympathy is so strongly excited, that close observation is forgotten or rendered almost impossible.' (Hence Darwin's use of photographs in his research on expression: see 4.3.1 Observation).[59] Such witnessing did not solely mark distress. In contrast to modern usage, which generally points to feelings aroused by others' grief—tears could also be evoked by 'sympathy with the *happiness* of others, as with that of a lover.'[60] *Expression* repeatedly states that we sympathize with joy as much as distress.

Darwin had argued in *Descent* that sympathy evolved from 'parental or filial affections.'[61] This explains why we 'sympathise far more deeply with a beloved than with an indifferent person; and the sympathy of the one gives us far more relief than that of the other.'[62] That sympathy 'gives relief' to others underlines another feature distinguishing *Expression*'s treatment of sympathy from modern approaches, namely, its effects on the people who are its objects.[63] When someone sympathizes with us: 'our sufferings are thus mitigated and our pleasures increased; and mutual good feeling is thus strengthened.'[64]

One thing missing from *Expression*, when compared with Darwin's early psychological notebooks, is its neglect of sympathy as a prompt to mutual aid. Sympathy had been synonymous with *congruent* fellow-feeling for several of Darwin's Scottish predecessors, including Adam Smith (1723–1790; see 5.3 Resonances).

---

[56] Darwin, *Expression*, p.343.

[57] John Gregory (1769) quoted in Lawrence, 'Nervous System and Society,' pp.27–28.

[58] Darwin, *Expression*, p.379.

[59] Darwin, *Expression*, pp.13, 18; cf. Paul White, 'Sympathy under the Knife: Experimentation and Emotion in Late Victorian Medicine,' in *Medicine, Emotion and Disease, 1700–1950*, ed. Fay Alberti (Basingstoke: Palgrave-Macmillan, 2006), pp.100–124.

[60] Darwin, *Expression*, p.227, my italics.

[61] Charles Darwin, *The Descent of Man, and Selection in Relation to Sex* (London: Murray, 1874), p.105.

[62] Darwin, *Expression*, p.228.

[63] This has been shown even in babies. See Mitzi-Jane Liddle, Ben Bradley, and Andrew McGrath, 'Baby Empathy and Peer Prosocial Responses,' *Infant Mental Health Journal* 36, (2015), pp.446–458; NB this study would have been better called a study of 'baby sympathy' than of empathy, I now realize.

[64] Darwin, *Expression*, p.379.

Against this, Darwin had argued in the 1830s that sympathy with the pain of others involves a 'sorrowful *delight*': people gain pleasure from the pain of others. 'The extreme pleasure children show in the naughtiness of brother children shows that sympathy is based,' wrote the young Darwin, 'on pleasure in beholding the misfortunes of others.' Without this pleasure, people would not go to the aid of, but shun victims of adversity.[65] This argument is not rehearsed in *Expression*, (though the link between sympathy and mutual aid does feature in *Descent*; see Ch. 7). Young Darwin's denial of symmetry between the feelings of sympathizers and those of their objects—the 'eternal reciprocity of tears'— *does* figure in *Expression*, however.[66] 'So strong… is the power of sympathy' that, sometimes, a person will 'blush at a flagrant breach of etiquette by a perfect stranger,' *who does not himself feel he has erred*—as we saw at the start of this chapter. Such asymmetry is also seen in sympathy with imaginary characters—the imaginary being an important strand in Darwin's theory of self-attention.[67]

5.2.3 Self-attention: reading you reading me

Darwin's explanation for blushing combined his thinking about the dual self with his concept of sympathy. Chapter 4 showed that, for Darwin, the meaning of expressions arose through the way one person's movements were *read* by others— the capacity for such recognition having become instinctive over generations. Our ability to read meaning into others' movements gains a second layer, however, through reflective (Train Two) consciousness, and so rebounds: we read others *as reading us*. From this, blushing results—albeit, under specific circumstances. In Darwin's words, *self-attention* proves the 'essential element' of blushing: 'It is not the simple act of reflecting on our own appearance, but the thinking what others think of us, which excites a blush.'[68] And this goes for the moral appearance of our conduct, not just for our physical looks—as shown by many of the quotes from *Expression* in preceding sections of this chapter.

The power of others' eyes to cause blushing is as much imaginary as real. We blush 'whenever we know, *or imagine*, that any one is blaming, though in silence,

---

[65] Darwin, 'Notebook M,' pp.51, 88–89 (my italics), 108. The twenty-nine-year-old Darwin had rejected Smith's theory that 'we can only know what others think by putting ourselves in their situation, & then we feel like them,' because it did not 'like Burke explain [the] pleasure' that attracts us to calamities. See Adam Smith, *The Theory of Moral Sentiments*, 6th ed. (Sao Paulo: Soares, 2006), pp.4–8; Edmund Burke, *A Philosophical Inquiry into the Origin of Our Ideas of the Sublime and the Beautiful: With an Introductory Discourse Concerning Taste* (London: Nimmo, 1887), pp.118–119. Delight in others' misfortune was later dubbed *schadenfreude* by Thomas Carlyle, 'Shooting Niagara--and After?,' *Macmillan's Magazine* 16, (1867), p.677.

[66] Wilfred Owen, 'Insensibility,' in *The Collected Poems of Wilfred Owen*, ed. Cecil Day Lewis (London: Chatto & Windus, 1963), p.38.

[67] Darwin, *Expression*, p.227.

[68] Darwin, *Expression*, p.345.

our actions, thoughts, or character.'[69] It is thus with blushing in solitude. As already quoted, the source of such hidden reddening 'almost always relates to the thoughts of others about us—to acts done in their presence, or suspected by them; or again when we reflect what others would have thought of us had they known of the act.' I blush because I sympathize with those who I imagine are reading me.[70]

5.2.4 Physiology

The forerunner, and principal antagonist, for Darwin's theory of blushing was the surgeon Thomas Burgess. In 1839, Burgess had published *The Physiology or Mechanism of Blushing*. A pioneering monograph in its detailed focus on the physiology of expression, Burgess nevertheless saw the blush as 'palpable evidence of Design ... another convincing argument proclaiming to the hearts of men that "*the hand which made them is Divine.*"' Why? Because blushing was primarily caused by shame and so provided an ever-open window on 'internal emotions' exhibiting themselves, whether the blusher liked it or not. Blushing thus served as a check on the conscience, which prevented the moral faculties 'from being infringed upon, or deviating from their allotted path.'[71]

As with the other expressions he discussed, Darwin aimed to show that blushing had no divine purpose. One proof invoked mental confusion. Those who believed in design, would find it hard to account for shyness being 'the most frequent and efficient of all the causes of blushing,' as it made the blusher 'to suffer and the beholder uncomfortable, without being of the least service to either of them.'[72] Another was the observation: 'Many a person has blushed intensely when accused of some crime, though completely innocent of it.'[73]

As usual, however, *Expression*'s main game was to prove blushing a physiological accident, with no possible ulterior purpose. In outline, its theory assumed that, as blushing results from self-attention, then, 'whenever we believe that others are depreciating or even considering our personal appearance,' our attention must be:

> ... vividly directed to the outer and visible parts of our bodies; and of all such parts we are most sensitive about our faces, as no doubt has been the case during many past generations. Therefore, *assuming for the moment that the capillary vessels can be acted on by close attention*, those of the face will have become eminently susceptible. Through the force of association, the same effects will tend to

---

[69] Darwin, *Expression*, p.365, my italics.
[70] Darwin, *Expression*, p.227.
[71] Burgess, *Physiology of Blushing*, pp.11, 25.
[72] Darwin, *Expression*, p.358.
[73] Darwin, *Expression*, p.352.

follow whenever we think that others are considering or censuring our actions or character.[74]

Blushing would thus become a useless side-effect of, first, the structure of the sympathetic nervous system, and second, the effects of attention on the diameter of blood-vessels. These were physiological matters. So Darwin turned for help to trail-blazers in the burgeoning new science of experimental physiology.

The first step in understanding the physiology of the vascular events that lead to blushing was its distinction from flushing. Any strong emotion *that made the heart beat faster* could flush the face with blood: whether lust, anger, or joy. However, Darwin reported, a form of reddening which 'resembles blushing in almost every detail,' could be experimentally produced by various chemicals (all psychoactive). Inhaling amyl nitrite—now used as a euphoria-inducing party popper—proved particularly effective. But so too were laughing gas and the sedative, chloral hydrate. All three drugs dilated the blood vessels, including the capillaries in the cheeks, without necessarily raising heart-rate.[75]

These chemical experiments—by Crichton-Browne on the asylum inmates under his care[76]—gave strength to Darwin's view that the physiology of blushing depended on sympathetic action between the capillaries supplying the skin of the face with the blood-supply to the brain more than the heart.[77] A link to the brain tallied with his theory that blushing arose from a form of self-consciousness: one's awareness—real or imaginary—that someone else was inspecting one's own appearance or actions. But, if blushing were to result from how one imagined others were viewing oneself, how could such self-attention be proved to dilate the capillaries in one's cheeks?

When the first edition of *Expression* came out in 1872, Darwin's physiological evidence was non-specific and circumstantial. Darwin lined up five eminent physiologists who had either stated or shown that 'attention closely directed to *any* part of the body tends to interfere with the ordinary and tonic [gradual, continuous] contraction of the small arteries of that part.'[78] For example, Thomas Laycock had written: 'when the attention is directed to any portion of the body, innervation and circulation are excited locally, and the functional activity of that portion developed.'[79]

---

[74] Darwin, *Expression*, p.358, my italics—to stress the assumption for which Darwin most needed physiological evidence.

[75] Darwin, *Expression*, pp.343–345; James Crichton-Browne, 'Letter to Darwin, 16th April,' *Correspondence of Charles Darwin* (Cambridge: Darwin Correspondence Project, 1871), https://www.darwinproject.ac.uk/letter/?docId=letters/DCP-LETT-7689.

[76] There were no 'ethics committees' in those days.

[77] Except in extreme cases: 'When a person is much ashamed or very shy, and blushes intensely, his heart beats rapidly and his breathing is disturbed': Darwin, *Expression*, p.342.

[78] Darwin, *Expression*, p.357.

[79] Darwin, *Expression*, pp.359–360.

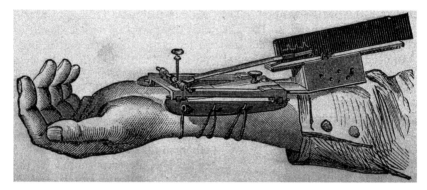

Figure 5.2 An early sphygmograph.

Credit: Wellcome Collection.

Distributed under the terms of the Creative Commons Attribution 4.0 International (CC BY 4.0) license. https://creativecommons.org/licenses/by/4.0.

From here, habit and association would take over. Humans' age-old concern for how others judged their personal appearance, particularly their faces, and particularly when young and courting, meant that they had, 'during endless generations', become concerned about provoking critical—and to a lesser extent, *any* (even positive)—attention.[80] This, through the effects of attention on circulation, would have produced blushing, and, 'through the force of association and habit', blushing would have been triggered 'in relation to the opinion of others on our conduct.'[81] So now, even if I am on my own and am only imagining or remembering some faux pas, I will blush.

Darwin wanted more direct confirmation of the effect of attention on circulation, however. Anticipating a second edition of *Expression*, he wrote to the physiologist John Burden Sanderson in April 1876, asking him to conduct a bespoke laboratory-experiment. Sanderson should first attach one of physiology's newfangled mechanical gadgets to a subject's arm. This was a 'sphygmograph', and would accurately record small changes in blood-pressure (Figure 5.2). Next, the subject should be asked to look at his feet. After a period of time had elapsed, the subject would then be made to focus his attention intently on his arm, 'by some humbug viz. by asking him to attend to any creeping sensation, or of heat or cold, over the whole surface of the arm.'[82] Unfortunately, the first edition of *Expression* outlasted Darwin, so the sphygmographic experiment seems never to have been performed.

---

[80] Darwin, *Expression*, p.364.
[81] Darwin, *Expression*, p.356.
[82] Paul White, 'Reading the Blush', *Configurations* 24, (2016), pp.293–294.

Both informal and planned observations strengthened the conclusion that blushing responded to the (imagined or real) exposure of the blusher's person to another's gaze and judgement. First, the main site of blushing is the face, which is 'the chief seat of beauty and of ugliness, and throughout the world is the most ornamented.' The face, therefore, would have been subjected 'during many generations to much closer and more earnest self-attention than any other part of the body.' Secondly, hands are less scrutinized than faces by would-be critics, and therefore by self-attention, and 'hands rarely blush,' although being 'well supplied' with nerves and small blood-vessels, and equally exposed to the elements as the face. Thirdly, responses to Darwin's cross-cultural survey of expression had shown that 'men of certain races, who habitually go nearly naked, often blush over their arms and chests and even down to their waists.' Was this true in cultures which went fully clothed?[83]

Three eminent surgeons were recruited to look into the matter. The observations garnered all focused on completely or partially undressed women. Thus: '[Dr Charles] Langstaff examined a woman lately, who blushed most intensely on the face, but not the least on the thighs.'[84] The fact that doctors could legitimately demand women to strip was the point: 'If ever you come across an extra blushing damsel do not forget the downward extent of the blush. I am assured that some half naked natives of Eastern races blush over arms & legs!'[85] Darwin's second medical informant (Sir James Paget) took up the gauntlet with glee: 'I have begun to make as many people as possible blush in conditions favourable for inspection.'[86]

Darwin's third source was James Crichton-Browne, who reported the case of a married woman, aged twenty-seven, who suffered from epilepsy:

On the morning after her arrival in the Asylum, Dr Browne, together with his assistants, visited her whilst she was in bed. The moment that he approached, she blushed deeply over her cheeks and temples; and the blush spread quickly to her ears. She was much agitated and tremulous. He unfastened the collar of her chemise in order to examine the state of her lungs; and then a brilliant blush rushed over her chest, in an arched line over the upper third of each breast, and extended downwards between the breasts nearly to the ensiform cartilage of the sternum. This case is interesting, as the blush did not thus extend downwards until it

[83] Darwin, *Expression*, p.333.

[84] Darwin, *Expression*, p.329; William Darwin, 'Letter to Charles Darwin, 22nd April,' *Correspondence of Charles Darwin* (Cambridge: Darwin Correspondence Project, 1868), https://www.darwinproject.ac.uk/letter/?docId=letters/DCP-LETT-6137.

[85] Charles Darwin, 'Letter to Sir James Paget, 29th April,' *Correspondence of Charles Darwin* (Cambridge: Darwin Correspondence Project, 1869), https://www.darwinproject.ac.uk/letter/?docId=letters/DCP-LETT-6716.

[86] James Paget, 'Letter to Darwin, 9th July,' *Correspondence of Charles Darwin* (Cambridge: Darwin Correspondence Project, 1867), https://www.darwinproject.ac.uk/letter/?docId=letters/DCP-LETT-5582.

became intense by her attention being drawn to this part of her person. As the examination proceeded she became composed, and the blush disappeared; but on several subsequent occasions the same phenomena were observed.[87]

5.2.5 Evolution

Expression holds the instinctive, involuntary quality of blushing as evidence for its evolution. Given the uniqueness of blushes to humanity, this claim could not be supported by evidence from animals. And given their lack in young infants, it could not easily be called innate. So the book proposes that the evolution of blushing has come about through a mix of physiological happenstance and the inheritance of habit. Many people—shy, modest, or ashamed—become concerned about how others judge their looks, especially their faces. Such people cannot help sympathizing with these others' views, becoming worried about (and so directing attention to) themselves, especially their own faces. Once grant this, and the 'incipient tendency in the facial capillaries' to be affected by such self-attention, 'will have become in the course of time greatly strengthened through ... nerve-force passing readily along accustomed channels, and inherited habit.'[88] Then, by 'frequent reiteration during numberless generations, the process will have become so habitual, in association with the belief that others are thinking of us, that even a suspicion of their depreciation suffices to relax the capillaries, without any conscious thought about our faces.'[89] Thus the faces of the self-conscious purposelessly redden.

This explanation oddly avoids questions about the evolutionary origin of self-attention. Yet sexual or natural selection might easily explain how the young of both sexes, in particular, came to be so concerned to gain the approval of others regarding their personal appearance—because, the more one pleases potential mates, the more likely one is to attract mates and leave offspring. *Expression* also summarily dismisses the possibility that blushing evolved as a sexual ornament, notwithstanding its concession that 'a slight blush adds to the beauty of a maiden's face.'[90] Nevertheless, take a step beyond *Expression* and grant selective advantages

---

[87] Another potential source was the Pre-Raphaelite sculptor and poet, Thomas Woolner (1825–1892), to whom Darwin wrote: 'Could you persuade some *trustworthy* men to observe young and inexperienced girls who serve as models, and *who at first blush much*, how low down the body the blush extends,' (Darwin's emphasis). *Expression* contains no data from this source. Instead, Darwin quoted Lavater's editor, Moreau de la Sarthe (1771–1826), who had confided, 'on the authority of a celebrated painter, that the chest, shoulders, arms, and whole body of a girl, who unwillingly consented to serve as a model, reddened when she was first divested of her clothes': Darwin, *Expression*, pp.332–333; Charles Darwin, 'Letter to Woolner, 7th April,' *Correspondence of Charles Darwin* (Cambridge: Darwin Correspondence Project, 1871), https://www.darwinproject.ac.uk/letter/?docId=letters/DCP-LETT-7665.xml.
[88] Darwin, *Expression*, p.364.
[89] Darwin, *Expression*, p.365.
[90] Darwin, *Expression*, p.357.

to self-attention, plus an instinctive capacity to read faces (see Ch. 4), and we could understand all the better why our ancestors will 'during endless generations have had their attention often and earnestly directed' to their own personal appearance, and especially to their faces.

5.3 Resonances

As with expression in general, Darwin's treatment of the evolution of blushing highlights inheritance of acquired habits. This emphasis has led *Expression* to be passed over by many recent commentators (see 4.6 Resonances). However, with the rise of a twenty-first century biology of developmental plasticity, championed by the likes of West-Eberhard, *Expression*'s arguments about the evolution of blushing may be due another look. Either way, the complex theatre of agency that Darwin's treatment of blushing sets out to describe remains the 'what'—the starting-point— for any argument about its evolutionary 'how come.'

Blushing has not received much attention from researchers since *Expression* came out. This may be due to a lack of appropriate physiological gadgets and techniques for measuring blushes, even today. Hence, the physiology of blushing is still a puzzle. In particular, physiologists still debate what distinguishes blushes from the flushing that accompanies other emotional events like lust and anger.[91]

Amongst psychologists, three main theories of blushing are current, emphasizing: social repair; social anxiety; and self-attention.[92] Regarding *social repair*, it has been hypothesized that blushing serves to remedy a social situation in which the blusher has broken a social taboo, or feels they have behaved inappropriately. Here, the blush is supposed to signal the blusher's awareness of their transgression, thus placating onlookers who may be evaluating them. While there is no observational research supporting this view, research using fictional vignettes of blushing scenarios does suggest that onlookers are imagined to be more forgiving of a transgression if the person who committed it blushes. In this case, blushing would be a communication that eased social intercourse, and so could be imagined to have evolved by natural selection. This theory has difficulty in explaining why people sometimes blush when complimented.[93] Also, the judgement of vignettes

---

[91] Ray Crozier and Peter De Jong, 'The Study of the Blush: Darwin and After,' in *The Psychological Significance of the Blush*, ed. Peter de Jong and W. Ray Crozier (Cambridge: Cambridge University Press, 2012), pp.1–12.

[92] Mark Leary and Kaitlin Toner, 'Psychological Theories of the Blush,' in *The Psychological Significance of the Blush*, ed. ed. Peter de Jong and W. Ray Crozier (Cambridge: Cambridge University Press, 2012), pp.26–152.

[93] Unless to be seen to do well is to transgress—a problem which can haunt successful women: Jamila Rizvi, *Why Women Do the Work and Don't Take the Credit*, podcast audio, Big Ideas 2018, http://www.abc.net.au/radionational/programs/bigideas/why-women-do-the-work-and-dont-take-the-credit/10051596.

describing blushing in *ambiguous* circumstances suggests onlookers may judge blushers *more* culpable than non-blushers: reading the blush as a sign of guilt. Here, blushing is no social remedy. Finally, it is hard to understand why, as Darwin observed, blushers often try to hide their blushes, by lowering their heads, covering their faces, looking away, or running out of a room—if blushing had evolved to be a social remedy. Blushes need to be seen if they are to repair a social breach.[94]

The *social attention* or *social anxiety* hypothesis holds that blushing occurs as a result of unwanted social attention. This makes better sense of the fact that blushers act to hide their blushes than the remedial theory. And it can also explain blushing as a response to (unwanted) positive attention.[95] It is supported by an experiment on women who had been asked to sing six verses of 'Old MacDonald had a Farm' (more embarrassing) or read a text on 'the brain' out loud (less embarrassing). Findings showed that, when singing, skin-temperature and blood-flow were significantly heightened on one side of the face, when a female experimenter stared at that side of their face, as compared to when the experimenter did not look at them. While it is hard to see how this theory could explain the evolution of blushing by natural selection, its proponents cite anecdotal evidence that blushing discourages others from staring at the blusher. What survival value this might have, though, escapes me.[96]

The strength of Darwin's *theory of self-attention* is in explaining blushing in non-social situations, when the blusher is alone—which neither of the other theories can do. Self-attention actually underpins both the social repair and the social anxiety theories: blushers must have read their own actions as transgressive through the eyes of others before sensing the need for repair; and others' attention can only arouse anxiety if blushers perceive it as directed at themselves. A blind singer who could not see an experimenter staring at one side of her face would presumably not blush asymmetrically. Yet, there is very little empirical evidence bearing on Darwin's theory—and none that successfully tests it against the other two theories. For example, one study has shown that women who had high levels of social awareness, as measured by high scores on a Social Phobia scale, blushed more in a difficult social situation than women with low phobia scores. This result could equally well be explained by heightened self-attention, by unwanted social attention, and by the need to effect social repair.[97]

---

[94] Corine Dijk and Peter De Jong, 'Fear of Blushing: No Overestimation of Negative Anticipated Effects but a High Subjective Probability of Blushing,' *Cognitive Therapy and Research* 33, (2009)

[95] Leary and Toner, 'Theories of the Blush.'

[96] Peter Drummond and Nadia Mirco, 'Staring at One Side of the Face Increases Blood Flow on That Side of the Face,' *Psychophysiology* 41, (2004), pp.281–287.

[97] Susan Bögels, Rijsemus, Wendy and de Jong, Peter, 'Self-Focused Attention and Social Anxiety: The Effects of Experimentally Heightened Self-Awareness on Fear, Blushing, Cognitions, and Social Skills,' *Cognitive Therapy and Research* 26, (2002), pp.461–472.

5.3.1 Two types of thought

Darwin's idea that humans manifest two kinds of thought-process is ancient, but also modern. From antiquity, a fast, unconscious, almost automatic, process of *intuition* or instinct had been contrasted with a far slower, more laborious process of *reasoning*. Current interest in dual-process theories of cognition revolves around the work of Nobel-prize-winners Daniel Kahneman and Amos Tversky.[98] Most cognitive psychologists now subscribe to some kind of dual process theory of mind, although there are more than a dozen such theories. Many of these share with Darwin the belief that fast, 'system 1' thinking is un- or pre-conscious, non-verbal, evolutionarily old, and shared with animals whereas slow, 'system 2' thinking is more recently evolved, conscious, and uniquely human, having links to language.

Perhaps the best-known dual-process theories of mind are psychoanalytic. Darwin frequently used the qualifiers 'unconsciously' and 'unconscious'—though he never made *unconscious* a noun. We saw earlier how he theorized a 'double individuality' that was implied by habit, where one 'acts unconsciously with respect to more energetic self, & likewise one forgets, what one performs habitually.'[99] This *double self* comprised a reflective self which generally forgets the machinations of the other, instinctive, 'unreasonable' self. While the irrational self does have memory, it could not 'be called memory,' because it was a habitual action of 'thought-secreting organs,' brought to light by 'morbid action.'[100] Hence Darwin arrived at a view that there was a species of memory which 'cannot be called memory,' but was connected by association—as in dreams for example—with waking thought.[101] This unrecognized kind of memory was physically based in the brain, and operated unconsciously, often asserting itself in direct opposition to the will[102]—not just via habits, but in dreams, fantasy, madness, intoxication, and our dotage.

Darwin never published these early views on the double self, though they help structure his analyses of blushing, and of conscience (see Ch. 7). They do strongly resonate with the work of two Viennese doctors—Sigmund Freud and Josef Breuer—who were attempting to treat neuroses in the 1890s, and believed that their patients' problems arose from 'the unconscious,' something they defined as 'a category of memory that is absolutely not recognised as memory.'[103]

---

[98] Daniel Kahneman and Shane Frederick, 'Representativeness Revisited: Attribute Substitution in Intuitive Judgement,' in *Heuristics and Biases: The Psychology of Intuitive Judgement*, ed. Thomas Gilovich, Griffin, Dale and Kahneman, Daniel (New York: Cambridge University Press, 2002), pp.49–81; Jonathan Evans, 'Dual-Processing Accounts of Reasoning, Judgement, and Social Cognition,' *Annual Review of Psychology* 59, (2008), pp.255–278.

[99] Darwin, 'Notebook M,' p.80.

[100] Darwin, 'Notebook M,' p.8 (my italics).

[101] Darwin, 'Notebook M,' pp.102e, p.46: 'recollection of anything is solely by association, & association is probably a physical effect of brain.'

[102] Darwin, *Expression*, p.377.

[103] Sigmund Freud and Josef Breuer, *Studies on Hysteria* (Harmondsworth: Penguin Books, 1977), p.300.

5.3.2 Symbolic interaction

Darwin did not use the word *behaviour* in the first edition of *Origin*, but he used variants of the word *habit* hundreds of times. In his sixties (the 1870s), he began to write of people, animals, and plants 'behaving', if sparingly. Habit remained his word of choice. With the rise and hegemony of behaviourism in the last century, behaviour became the chief focus for psychologists, or what are still sometimes called *behavioural* scientists. Insofar as behaviour equated to an organism's object-ively measurable adjustments to its external environment, (self-) consciousness then ceased to be a primary psychological concern.[104]

Nevertheless, viewed according to the history of its ordinary usage, the word *behave* implies reflexivity. Be-have originated as meaning *to have or to hold how one is*: 'to have or bear oneself (in a specified way)', as in German: *sich behaben*. It referred to *proper* conduct.[105] ('Behave!' my parents would hiss, if I fooled around in public.) Defining behaviour *behaviouristically*, as mere activity, misses the cru-cial place of propriety in how humans act. Such reflexivity was the dimension upon which Darwin insisted in his understanding of (self-) conscious agency— the dimension that most marks humans off from animals. His analysis of blushing opened up the possibility that everything we do is shaped by how we think we ap-pear to others. In which case all human action is intrinsically moral or social (see Chs 7–8).

Not sure about this? Then go to a common-room where people live with senility. Inmates will be losing just that sense of propriety which *behaving* once implied—of knowing and so controlling how one appears to others—a concern that has be-come so ingrained in you and me, so much second nature, that we forget (like be-haviourists) it has anything to do with conducting everyday life. Study those who are slumped, gaping in their chairs, asleep or vacant. (Is that a whiff of piss?) Hear the solitary crooning. See the gaffer who sits in a corner, chewing a gardening glove: 'The people in the home were once like you and me – adhering to and validating a very limited set of behaviours.' Now they have strayed, making us feel uncomfortable—thus giving the lie to the idea that there is anything wholly un-reflexive about the ways we ordinarily adjust to our environments.[106]

It was partly on these grounds that George Herbert Mead sought an anti-dote to the psychological behaviourism championed by Watson and Skinner. He called it *social* behaviourism.[107] This, as we saw in Chapter 4 (4.6 'Resonances'),

---

[104] Lisa Blackman, 'Habit and Affect: Revitalizing a Forgotten History,' *Body & Society* 19, (2013), pp.186–216.
[105] Edwin Ardener, "Behaviour'—a Social Anthropological Criticism,' in *The Voice of Prophecy and Other Essays*, ed. Michael Chapman (Oxford: Blackwell, 1989), pp.105–108.
[106] Greta Stoddart, 'Who's There?,' *Poetry News*, (2017).
[107] George Mead, *Mind, Self, and Society: From the Standpoint of a Social Behaviourist* (Chicago: University of Chicago Press, 1934).

was partly inspired by *Expression*. Hence *Expression*, and particularly its treatment of blushing, has been hailed as the text that inaugurated social psychology.[108] Darwin's psychology is undoubtedly social. Not only did what Darwin call an *attitude* operate as an expression by being meaningful to others. But Darwin's natural history brought him to question the very idea of an autonomous individual (see 3.2 Agency Implies Interdependence). Hence, we find James Mark Baldwin saying the 'psychological unit' in Darwin's psychology was a group-based *socius*— meaning by socius a *being-with*, 'fitted to enter into fruitful social relations' with other group-members.[109]

The implied dynamics of blushing, as being the consequence of me recognizing that I am being read by others, seems to be original to Darwin in its specifics. But it resonates vibrantly with texts that both precede and succeed *Expression*. Both Adam Smith and Charles Horton Cooley (1864–1929) suggested that it is how we are read by others that produces our sense of self. 'We suppose ourselves the spectators of our own behaviour,' Smith wrote, 'and endeavour to imagine what effect it would, in this light, produce upon us. This is the only looking-glass by which we can, in some measure, with the eyes of other people, scrutinize the propriety of our own conduct.'[110] This same view is to be found in Baldwin: we judge ourselves 'by the meed of reproach or commendation which we receive from others.'[111] Nowadays, Smith's, Darwin's, and Cooley's contention that individuals' self-judgements result from assimilating the judgments of significant others sounds like a truism. Both sociological literature and symbolic interactionists hold that 'our identity is on the surface, visible to others, hidden from ourselves,' and 'that identity is as much our reaction to others' reactions to our self-presentations as anything else.'[112]

Expression gives movements the status of gestures—if they connote something emotional to someone else. Darwin's book explicitly denies that a gesture, like the snarl of a dog, is *intended* to mean what it is read as meaning. Wundt appropriated this conception of *gesture* to track how 'vocal gestures' might turn into language (see 4.6 Resonances). More influentially, Mead saw that, to explain how language

---

[108] Robert Farr, 'On Reading Darwin and Discovering Social Psychology,' in *The Development of Social Psychology*, ed. Robert Gilmour and Steven Duck (London: Academic, 1980), pp.111–136.

[109] Farr, 'Darwin and Social Psychology'; James Baldwin, *Darwin and the Humanities* (Baltimore: Review Publishing, 1909), p.44. On 'being-with' (*mitsein*), see Hubert Dreyfus, *Being-in-the-World: A Commentary on Heidegger's Being and Time Division I* (Cambridge MA: MIT Press, 1991), pp.147ff; cf. Kenneth Gergen, *Relational Being: Beyond Self and Community* (Oxford: Oxford University Press, 2009).

[110] Smith, *The Theory of Moral Sentiments*, p.101; Charles Cooley, *Human Nature and the Social Order* (New York: Scribner's, 1902).

[111] James Baldwin, *Social and Ethical Interpretations in Mental Development: A Study in Social Psychology* (London: Macmillan, 1897), p.312.

[112] David Lundgren, 'Social Feedback and Self-Appraisals: Current Status of the Mead-Cooley Hypothesis,' *Symbolic Interaction* 27, (2004), pp.267–286. Timothy Adams, *Telling Lies in Modern American Autobiography* (Chapel Hill NC: University of North Carolina Press, 1990), p.108. Note too Gertrude Stein's aphorism: 'I am I because my little dog knows me' (ibid, p.33).

came about, he needed to explain the origin of *symbols*, that is, gestures which *mean the same thing to the expresser as to their interlocutors*. Only then could gestural displays communicate:

> The vocal gesture becomes a significant symbol ... when it has the same effect on the individual making it that it has on the individual to whom it is addressed or who explicitly responds to it, and thus involves a reference to the self of the individual making it.[113]

Here, *Expression*'s treatment of blushing helped Mead. For, it was only in blushing that the significance of an expression for its expresser potentially reflected its reading by observers.[114] In Darwin's analysis of blushing, Mead found not just a 'conversation of gestures', as in a dog-fight, but the door to genuinely symbolic interaction—providing him with an insight which opened up the fertile branch of social science called symbolic interactionism.[115]

Mead's analysis of symbolic communication has three elements: the two people interacting and the tradition of symbolic meanings they both inhabit, but did not originate. This third element was something that, Mead proposed, the individual eventually synthesized into the standpoint of a 'generalised other'—the presumed attitude of the whole community towards oneself.[116] There is now a growing variety of theories like Mead's, of what is called *thirdness*—a term first introduced by the semiotician and pragmatist Charles Sanders Peirce (1839–1914), and often equated with culture or social context.[117] Thus, one way of understanding culture is that it provides what Homi Bhabha has called a *third space* which constitutes the indispensable background to the use of otherwise-ambiguous signs ('the discursive conditions of enunciation') so that they mean one thing rather than another to the members of that culture.[118]

This third element most obviously figures in Darwin's treatment of blushing and etiquette. Etiquette consisted for Darwin in rules that could be quite arbitrary or meaningless, so far as their moral sense was concerned, being fixed solely by

---

[113] Mead, *Mind, Self, and Society*, p.46.

[114] The importance of the imaginary dimension in symbolic interaction chimes with psychoanalytic writings on inter-subjectivity by Jacques Lacan: see Ben Bradley, 'Experiencing Symbols', in *Symbolic Transformations: Toward an Interdisciplinary Science of Symbols*, ed. Brady Wagoner (London: Routledge, 2010), pp.93–119.

[115] Alex Dennis and Greg Smith, 'Interactionism, Symbolic', ed. James Wright, 2nd ed., vol. 12, *International Encyclopedia of the Social & Behavioral Sciences* (Amsterdam: Elsevier, 2015), https://doi. org/10.1016/B978-0-08-097086-8.32079-7.

[116] Mead, *Mind, Self, and Society*, passim.

[117] Ben Bradley, 'Jealousy in Infant-Peer Trios: From Narcissism to Culture', in *Handbook of Jealousy: Theories, Principles and Multidisciplinary Approaches*, ed. Sybil Hart & Marie Legerstee (Hoboken NJ: Wiley-Blackwell, 2010), pp.192–234; John Shotter, *Images of Man in Psychological Research* (London: Methuen, 1975), p.136.

[118] Homi K Bhabha, *The Location of Culture* (New York: Routledge, 1994), p.37; Bradley, 'Jealousy', passim.

custom.[119] Yet a breach of etiquette was just as effective a trigger for blushing as shame for a genuine crime. In short, the meaning of the rules of etiquette was relative to the culture into which the blusher had been inducted, something for which the origin was independent of both evolution and individual characteristics (see Ch. 8). In Darwin's analysis of blushing and etiquette, expression became indivisible from social organization—a point famously developed in Erving Goffman's dramaturgical conception of social life.[120]

Note here that even explanations of the physiological aspects of blushing, such as innervation and blood-flow, cannot be (entirely) physiological. *Which* portions of the body redden *when* cannot be known without reference to the prism of the 'generalised others' that different blushers have adopted over their lifetime, and through which they now imagine their own conduct and appearance is viewed. From the perspective of an alien culture, such blushing may appear, as Darwin calls it, 'meaningless.' Nevertheless, he goes on, *whatever* the culture to which we belong, 'the fixed custom[s] of our equals and superiors, whose opinion we highly regard, ... are considered almost as binding as are the laws of honour to a gentleman.'[121] Hence, if a Victorian gentleman were to lose his clothing when taking tea with others in a polite drawing-room, his sense of embarrassment, and the likelihood of him blushing, would be far more acute than would be the case in 'the nations that go altogether or nearly naked.'[122]

Conversely, as we saw ('Physiology'), 'men of certain races, who habitually go nearly naked, often blush over their arms and chests and even down to their waists.' This was not true of the British, (who mainly blushed in the face and neck, these being the areas of the body most likely to attract the praise or disapprobation of others in their milieux)—or, at least, not *unless and until* usually hidden portions of their body were exposed to others' gaze, as planned observations by Darwin's medical informants proved.

I say this to underline the psycho-socio-cultural lability of the physiology of blushing in Darwin's account: the dependence of blushing as a bodily response on the social theatre of meanings which determines its occasions. This shows that, in *Expression*, blushing is in no simple way 'hard-wired,' instinctive, or simply physiological, even though it is pan-human and often involuntary. Nor can *Expression* be seen as proposing a simple empiricism in its approach to blushing, because, insofar as they leave the specificity of the interpersonal and socio-cultural meanings out of account, the refinements of micro-measurement lead the observer 'below the level of significant phenomena.'[123]

---

[119] Darwin, *Expression*, p.353.

[120] Erving Goffman, 'Embarrassment and Social Organization,' *American Journal of Sociology*, (1956), pp.264–271.

[121] Darwin, *Expression*, p.353.

[122] Darwin, *Expression*, p.354.

[123] Edwin Ardener, 'The New Anthropology and Its Critics,' in *The Voice of Prophecy and Other Essays*, ed. Michael Chapman (New York: Berghahn, 2018), pp.47–48.

5.4 Conclusion

Darwin's analysis of blushing presents the clearest working-out of the principles inspiring his understanding of human action. It endows with a whole new layer of complexity the theatre of agency which humans sustain. His analysis of the efficacy of all facial expressions as being due to the recognition of bodily attitudes and gestures by others is shown to rebound upon the expresser—thus a woman may find herself involuntarily responding to *her own* reading of the meanings that *others,* imagined or real, read into how she looks and acts. Blushing is one result of being so read. Later chapters show that the same looking-glass dynamic underpins Darwin's understandings of erotic attraction (see Ch. 6), and of the functioning of some of humans' highest psychical faculties, most notably conscience (see Ch. 7).[124] While the topic of blushing might strike some as mere hobby-horse, it has much significance when viewing Darwin's psychology as a whole.

In addition, Darwin's account of blushing illustrates his general understanding of the task of (Train Two) consciousness, or more precisely, *self*-consciousness. *Expression* does not posit consciousness as an abstract, empty, Descartes-inspired, 'I think therefore I am.' My consciousness is constituted *reflectively*, through consideration of my acts and their effects, observed and imagined. No one knows themselves or their consciousness immediately—via psychological evidence, intellectual intuition, or mystical vision—contrary to what the method of introspection implies.[125] People know themselves through reflection on what is mediated by the utterances, actions, works, institutions, and monuments of others that give concrete testimony to their agency and its effects.

---

[124] Darwin, *Descent*, p.98.
[125] Paul Ricoeur, *Freud and Philosophy: An Essay on Interpretation* (New Haven CT: Yale University Press, 1970), pp.42ff.

6

Sex, Difference, and Desire

Bell tolling, Roman arch lit up inside, the Midland city's oldest church draws in its flock. A new academic year beckons and the just-arrived students, filled with the spirit of God or just curious, pack into the University church for its annual university service. They can expect nothing much beyond regular evensong, a favourite hymn-tune perhaps, a mention in the prayers. Save the sermon, the sermon is always a wildcard.

Sure enough, at his appointed moment, a tall stooped layman steps out from a forward pew, nods at the altar, and laboriously climbs three steps to the pulpit. Benign, humorous eyes look out for a moment across the congregation from a handsome craggy face. He has chosen the topic of sex, he says—and a text from the Old Testament: 'I sleep, but my heart waketh, it is the voice of my beloved that knocketh, saying, Open to me, my sister, my love, my dove, my undefiled ...' He gives some context, a brief exegesis, then come the lessons from his experience, the unsung ecstasies of the long-married. Like climbing a mountain, he rhapsodizes—what we took for the summit at the start, turns out just a ridge. For the path leads ever higher, climbing up, up, to peaks far beyond our first imagining.

In the pew below, a woman has dropped a book. The preacher's wife. Beside her sit two children, a teenage girl, and a boy. The girl looks up at her father, face flushed, adoring and defiant. The boy looks down at his hassock, frowning, as his mother reins in her agitation. What had she told him last month? That sex lost its attraction for her once she had had her children. And his Dad too. Had he not said: 'sex has played a smaller part in my life than I once hoped'? Yet *now* ... what was he talking about?

Are we not all confused about sex? And if we aren't, weren't our parents? Or *their* parents? And if you really don't know of anyone perplexed by sex, and you haven't seen *Cat on a Hot Tin Roof*, then take Darwin.[1] *The Descent of Man and Selection in Relation to Sex* has undoubtedly proven his most baffling book, or, at least, the parts of it dealing with sex contain, and have spawned, as much contradiction as anything he published.

An example. Nowadays, commentators typically find that *Descent* attributes an *essential passivity* to women where sex is concerned.[2] We hear that Darwin's book

---

[1] Tennesee Williams, *Cat on a Hot Tin Roof* (New York: New Directions Books, 2004).

[2] E.g. Evelleen Richards, *Darwin and the Making of Sexual Selection* (Chicago: Chicago University Press, 2017), pp.216, 519.

Darwin's Psychology. Ben Bradley, Oxford University Press (2020). © Oxford University Press.
DOI: 10.1093/oso/9780198708216.001.0001

portrays the fair sex as coy and careful, 'passive, passionless, and maternal.'[3] This goes for animals too. 'Whether writing of birds or humans,' asserts Jim Endersby, 'Darwin always described females as 'coy' and modest.'[4] Yet, if we look for the word *modest* in *Descent*, it occurs just once, applied to a dog (a 'he'), when 'begging too often for food.'[5] *Descent* uses the word *coy* three times, but never to connote passivity, and never to refer to women. Darwin rarely chose the word *passive* when discussing sex (just five times), and always qualified it, as in the phrase 'comparatively passive'—applied once to males and once to females.[6] *Descent*'s most sweeping statement about females and passivity runs:

> The sexual struggle is of two kinds; in the one it is between the individuals of the same sex, generally the males, in order to drive away or kill their rivals, the females remaining passive; whilst in the other, the struggle is likewise between the individuals of the same sex, in order to excite or charm those of the opposite sex, generally the females, *which no longer remain passive*, but select the more agreeable partners.[7]

Travel back to 1871, and we find an opposite response to *Descent*. Among its first reviews were those equating Darwin's work with Algernon Swinburne's licentious poetry and Pre-Raphaelite decadence (Figure 6.1). William Boyd Dawkins rejected outright any scientific grounds for *Descent*'s insistence on sexual selection in humans, then went further. 'We do him no injustice,' wrote an outraged Dawkins, 'in ascribing to him the theory ...

> ... that Venus [goddess of venereal love] is the creative power of the world, and that the mysterious law of reproduction, with the passions which belong to it, is the dominant force of life. He appears to see nothing beyond or above it. In a heathen poet such doctrines appear gross and degrading, if not vicious. We know not

---

[3] Richards, *Darwin*, p.482; Sarah Hrdy, 'Raising Darwin's Consciousness: Female Sexuality and the Prehominid Origins of Patriarchy,' *Human Nature* 8, (1997), pp.8–9; Georgina Montgomery, 'Darwin and Gender,' in *The Cambridge Encyclopaedia of Darwin and Evolutionary Thought*, ed. Michael Ruse (Cambridge: Cambridge University Press, 2013), p.445.

[4] Jim Endersby, 'Darwin on Generation, Pangenesis and Sexual Selection,' in *The Cambridge Companion to Darwin*, ed. Jonathan Hodge and Gregory Radick (Cambridge: Cambridge University Press, 2009), pp.73–74; cf. Gowan Dawson, *Darwin, Literature and Victorian Respectability* (Cambridge: Cambridge University Press, 2007), p.76.

[5] Charles Darwin, *The Descent of Man, and Selection in Relation to Sex* (London: Murray, 1874), p.71.

[6] Darwin, *Descent*, pp.222, 225, and see also p.300: 'the female does not remain passive.' Darwin used *passive* seven times in *Descent*, but two mentions relate to non-sexual topics.

[7] Darwin, *Descent*, p.614, my italics. NB the remark 'the females remaining passive' merely underlines the fact that females *did not participate* in same-sex fights between males—hardly a necessary statement, did it not rhetorically prepare readers for the second kind of struggle, in which females emphatically *do* participate.

Figure 6.1 'Laus Veneris' ('In Praise of Venus') by Edward Burne-Jones (1878), said to depict 'stifling, enslaving, erotic love.' The painting was partly inspired by Swinburne's poem of the same name (1866).

Source: Universal Images Group North America LLC / Alamy Stock Photo.

Though this poem may also have been inspired by an 1861 drawing by Burne-Jones: George Landow, Laus Veneris by Sir Edward Coley Burne-Jones, 2006, http://www.victorianweb.org/painting/bj/paintings/21.html.

how to characterise them in an English naturalist, well known for the purity and elevation of his own life and character.[8]

Here Dawkins accused *Descent* of purveying a vision of sex so vicious, gross, and degrading that he could not even discuss it in print—for at *Descent*'s heart was the view that both male *and female* sensual desires ('Venus') and reproductive passions, were the spring of life. By aligning Darwin's vision with Swinburne's ('a heathen poet'), Dawkins cast *Descent* as imputing a predatory lewdness to females, where—to take the vision from Swinburne's verses 'Laus Veneris' as example—the insatiable woman lays hold upon her lover with 'lips luxurious' and cleaves to him 'clinging as a fire that clings.'[9]

---

[8] Dawkins quoted by Dawson, *Victorian Respectability*, p.49. For Venus as seducer, see e.g. William Shakespeare, *Venus and Adonis* (Hammersmith: Doves Press, 1912).

[9] Dawson, *Victorian Respectability*, p.45; Algernon Swinburne, *Laus Veneris* (Portland ME: Thomas B Mosher, 1909).

So: does *Descent* draw a picture of the female as a lusty, voracious siren; or as passive, maternal, and modest? Examine the text carefully, and the answer is arguably: 'Both.'[10] Because the book has at least as many voices as *Origin*, albeit less harmoniously orchestrated—and deals with several demanding topics. Leaving Part I aside for the time being (until Ch. 7), the book focuses on three complicated and controversial debates: about the evolution of sex; about sexual differences; and about their relations to desire—and discusses each topic, both with regard to animals (in Part II of *Descent*), and to humans (in Part III).

This chapter deals first with Darwin's lifelong deliberations about the evolutionary advantages of sex, and how he understood sexual selection to operate. I then go on to consider his discussion of difference and desire in animals. After that, I critically consider his treatment of difference and desire in humans. Finally, I point out some kinks and muddles in his account.

6.1 Desire's Effects

Darwin's interests in common descent, natural selection, the processes of heredity, mating and maturation, the benefits of inter-crossing, the dimorphism of plants, and a host of other topics, can all be subsumed within a single, larger enterprise: his theorizing about *generation*.[11] His concept of generation could cover everything from the repair of injuries in individual organisms to the propagation of new species, and, indeed, to the genesis of the whole tree of life.[12] Only some of the topics Darwin dealt with under the umbrella of generation are germane to his thinking about agency, however—most of which track back to his arguments for the evolutionary advantages of sexual over asexual reproduction.

Like many of his predecessors, including Erasmus Darwin, Charles Darwin sought to explain, not just why, but *how* sex had come about. He argued that species with two sexes had evolved from ancestors exhibiting hermaphroditism or androgyny—something he had closely studied in lower animals like barnacles, and in plants like orchids. For instance, puzzling over why men have nipples, he hypothesizes that: 'long after the progenitors of the whole mammalian class had ceased to be androgynous, both sexes yielded milk, and thus nourished their young.'[13] Hermaphrodites had, in turn, evolved from species that had only one sex—female, according to Darwin. These single-sex creatures perforce bred asexually, by splitting off identical copies of themselves.

---

[10] Or, if one is pedantic about sticking to *Darwin*'s words: 'Neither.'

[11] Jonathon Hodge, 'Darwin as a Lifelong Generation Theorist,' in *The Darwinian Heritage*, ed. David Kohn (Princeton, NJ: Princeton University Press, 1985), pp.208–209.

[12] Charles Darwin, *The Variation of Animals and Plants under Domestication*, 2 vols., vol. 2 (London: John Murray, 1875), pp.284, 352, 376, 379.

[13] Darwin, *Descent*, p.163.

For psychology, the important strand in the evolution from sexually undifferentiated creatures to androgyny, and from androgyny to species with males and females, was the growing role of *difference*. The evolutionary advantages of sex were all to do with difference: 'Generation being means to propagate & perpetuate differences, (of body, mind & constitution),' noted the twenty-nine-year-old Darwin.[14] Individual differences, or variations in character, were to prove critical to adaptation in *Origin*, and hence to natural selection. Because, when reproducing sexually, not only do mates differ from each other, but they also produce offspring *unlike either parent*. Sex mixed up the characters of different individuals, producing new combinations that were potentially '*infinite* in *number*.'[15]

The more combinations of character, even if aimlessly made, the greater the chance that some of the young in a populace would be born better adapted than their parents to a world in flux. Moreover, as Darwin observed, because the products of sexual mating were unlike their parents, each had to blaze his or her unique path to maturity. This emphasized the crucial role of *development* in the production of heritable variations (cf. 3.3.5 Transmission and development). As Darwin had noted right at the start of his thinking about evolution, in July 1837, aged twenty-eight: if organisms lived for ever, they would be frozen in form, and the effect of 'accidental injuries' would be everlasting. The old were static in habit and physique. The young were fresh and malleable. Maturation affords plasticity. Hence the biological value of death. Hence the need for generation to beget generation. Hence the need for the life-cycle of reproduction: 'to adapt & alter the race to changing world.'[16]

At the same time, cross-fertilization or the 'beautiful law of intermarriages' ensured that members of a species all continued to draw from the same pool of characters, for otherwise 'there would be as many species, as individuals.'[17] Sex also damped random change, rendering transmutation of habit, physiology, and form stable and adaptive: 'Physical changes should act not on individuals, but on masses of individuals. — so that the changes should be slow & bear relation to the whole changes of country, & not to the local changes — this could only be effected by sexes.'[18] Sexual mixing obliterated the 'minute changes' constantly flooding into a population accidentally. 'The great changes of nature are slow,' he wrote, and many of the myriad individual variations haphazardly affecting individuals 'would not be fitted to the slow great changes really in progress.'[19]

---

[14] Charles Darwin, "D," (1838), p.177.
[15] Charles Darwin, 'Notebook B,' (1837–1838), p.5, emphasis Darwin's.
[16] Darwin, 'Notebook B,' pp.3–5.
[17] Charles Darwin, "E," (1838–1839), pp.48ff.
[18] Darwin, 'Notebook E,' pp.50–51. The pros and cons of in-breeding was a topic that later became of personal interest to Darwin, as he married his first cousin: Emma Wedgwood. He later published a book demonstrating the benefits of out-breeding: Charles Darwin, *The Effects of Cross and Self Fertilisation in the Vegetable Kingdom*, 2nd ed. (London: John Murray, 1878).
[19] Darwin, 'Notebook D,' p.167.

The virtues of sex were, thus, primarily the virtues of moderated difference. Darwin had long believed that cross-fertilization produced healthier and more vigorous offspring than self-fertilization: out-breeding strengthens, whereas 'marrying in [in-breeding] *deteriorates* a race.'[20] His later experiments in plant-breeding considerably amplified this virtue: 'the offspring from the union of two distinct individuals … have an immense advantage in height, weight, constitutional vigour and fertility over the self-fertilized offspring from one of the same parents.' And this fact was 'amply sufficient to account for the development of the sexual elements, that is, for the genesis of the two sexes.'[21] One issue remained, however. How was sexual reproduction to be effected in higher animals? Or more precisely: How was *mating* to be effected?

This question underpins the last two-thirds of *Descent*. Darwin's first answer was given in 1837: 'there is instinct for opposites to like each other.'[22] *Difference creates desire.* Hence: 'Desire lost when male & female too closely related.'[23] Too much difference—as between members of two different species—also caused a failure in 'the life of passion.'[24] So: (moderate) differences fuel desire. And desire effects mating. The dynamics of desire thus precipitate the advantages of sexual reproduction. For, animals—like many humans—do not, so far as immediate motivation is concerned, mate *in order to* breed. They mate from a hot desire to couple with a particular other. So *Descent*'s treatment of sex takes for granted that, when two creatures come together and copulate, they will hardly have the need to produce offspring at 'front of mind.'

In this regard, reproduction is an unintended long-term consequence of the short-lived dramas of desire. Though having an entirely different time-scale—ranging from seconds to hours[25]—lust serves as sole guarantor for the cycle of generations, a process which takes years in the human case, as generation includes the development of the young through maturity. Beyond this, lust has an even longer-term consequence in sexual selection, which takes effect over 'a long course of generations.'[26] Hence an extraordinarily diverse theatre of animal and human sexual agency provides the focus for Part II (on animals), and then Part III (on humans), of *Descent*.

Just as the law of natural selection had, according to Darwin, *resulted from* processes producing variability in organisms and their struggles for life—so sexual

---

[20] Darwin, 'Notebook B,' pp.4–6, Darwin's emphasis.
[21] Darwin, *Cross Fertilisation*, p.467.
[22] Darwin, 'Notebook B,' p.6.
[23] Darwin, 'Notebook D,' p.176.
[24] Darwin, 'Notebook D,' p.177. Though Darwin, *Descent*, pp.414ff, also remarks the not-infrequent production of inter-species hybrids in the wild, and under domestication (cf. James Mallett, 'Hybrid Speciation,' *Nature* 446, (2007), pp.279–283), bespeaking attractions that jump the species-line.
[25] Or days in the Tasmanian Devil.
[26] Darwin, *Descent*, pp.214, 318 & passim.

selection had been effected by the intersections of desire and difference with breeding. While *some* differences between male and female creatures had resulted from natural selection—most notably, the physiology and anatomy of male and female reproductive organs (called 'primary' sexual characters)—others had not. Wherever the males and females of any creature had the same general habits of life, but differed in structure, colour, or ornament, such differences, *Descent* argues, had been 'developed through sexual selection.'[27] This was important because, at times, the advantages that led to the evolution of 'secondary' sexual characters by sexual selection apparently ran counter to the kinds of adaptiveness identified with natural selection.

The peacock's tail, and that of the Argus pheasant, provided two of Darwin's favourite examples (Figure 6.2). The peacock's tail was (and still is) held up by Bible-believers as a natural fact so glorious that it could only be product of God's craft.[28] Unwieldy and conspicuous, the peacock's huge decorative train must surely require an unusual investment of developmental and muscular energy, plus added risks of predation, when compared to the relatively petite and inconspicuous peahen. As such the male's enormous fan should have proven a hindrance in the struggle for life that effects natural selection, insofar as its bearer's survival required easy locomotion and escape from predation. So the cock's huge tail-piece must have promoted benefits according to some other law of effects than natural selection, namely, sexual benefits. Peahens must have preferred to mate with cocks with more elaborate trains.[29] And hence, generation after generation, males with more gorgeous trains must have left more, or more viable, offspring than drabber rivals. So the peacock's tail had originated through sexual selection.

As with characters produced by natural selection, the evolution of secondary sexual characters depended on their inheritance. According to *Descent*, characters which had inflamed the sexual proclivities of the opposite sex needed, both, to become biologically ingrained, that is, immured in 'gemmules,' so as to be *transmitted* to subsequent generations, and they needed *to develop*. Development was the key, however, because *all* heritable sexual characters were transmitted to *both* sexes, even though they were usually only developed in one.

The presence in both sexes of gemmules that seeded all ('male' and 'female') secondary characters could be shown from hybrids. When two species, having strongly-marked sexual characters, were crossed, they produced hybrid males having 'male' sexual characters from *both* species; and likewise for female hybrids.

---

[27] Darwin, *Descent*, p.210.

[28] Stuart Burgess, 'The Beauty of the Peacock Tail and the Problems with the Theory of Sexual Selection,' *Journal of Creation* 15, (2001), pp.94–102.

[29] Note that some, but not all, contemporary evidence shows peahens prefer peacocks with more elaborate tails: Adeline Loyau et al., 'Do Peahens Not Prefer Peacocks with More Elaborate Trains?,' *Animal Behaviour* 76, (2008), pp.e5–e9.

Figure 6.2 'Side view of male Argus pheasant, whilst displaying before the female,' showing 'the extraordinary attitudes assumed ... during the act of courtship, by which the wonderful beauty of his plumage is fully displayed.'

Source: Darwin, C. (1871). *The Descent of Man and Selection in Relation to Sex*. (London: John Murray, Albemarle Street).

Darwin, *Descent*, pp.399–400.

This meant each parent transmitted the characters 'proper to its own male and female sex to the hybrid offspring of either sex.' The presence in all individuals of the gemmules for *both* male- *and* female-linked characters could also be shown by what we now call *gender plasticity*, as illustrated in: castrated males, who develop female characteristics; or in old age or illness, when women grow beards, or the common hen assumes 'the flowing tail-feathers, hackles, comb, spurs, voice, and even pugnacity of the cock.'[30]

The possibility of gender plasticity[31] necessitated further conceptual regularization, however, before *Descent* could make sense of sexual differences. Ground zero for his thinking was the marked sexual dimorphism of many birds, to which *Descent* Part II gives almost three times as much space as to any other group (mammals included). The contrast between the highly ornamented cocks, as compared to the far drabber hens, in many bird species was captured by *Descent*'s most fundamental law: that any secondary sexual character like the peacock's tail, which first appear relatively late, namely, in maturity, *typically appear in only one sex*: 'variations which first appear in either sex at a late period of life, tend to be developed in the same sex alone.' This was because the variation itself had presumably first occurred relatively late in the cock's development—not least because sexual selection can only affect adults old enough to court. And if, as a result, novel characters were first acquired at a given stage of life, when inherited, they emerged in inheritors at the corresponding developmental phase.[32] If acquired early in life, a character would reappear at the same age and last for the same time. If acquired when old, the character would not appear until its inheritors were old.

This principle had a rider. Characters which develop in *both* sexes: 'first appear early in life.'[33] For example, in most deer, bucks have much larger horns than does do—and bucks' horns only develop in maturity. Reindeer are the one deer species in which does and stags have similar antlers. And reindeer are the only deer in which horns appear in the very young, 'and at the same time in both sexes.'[34]

Taken together, these two general observations—Darwin called them *principles of inheritance*—have the corollary that the sex which *does not* develop a secondary character (usually the female, said Darwin) will resemble the young more than the sex which *does*. Unfortunately, though, Darwin knew so many exceptions to this law, he had to invent another law to account for them. This was the law of 'equal transmission,' though *equal development* or *equal expression* would have been more apt.[35] In many species, secondary sexual characters apparently acquired by one sex, came to be expressed in all members of *both* sexes. By way of illustration, *Descent* instances cases where females exhibited, 'more or less perfectly, characters proper to the male, in whom they must have been first developed, and then transferred

---

[30] Darwin, *Descent*, p.227.

[31] Often under the control of hormones (e.g. the 'castrated males' Darwin mentioned), gender plasticity is found in many species: Mary Jane West-Eberhard, *Developmental Plasticity and Evolution* (New York: Oxford University Press, 2003), pp.260–295. Darwin also noted 'cross-sexual transfer' in plants: Darwin, *Variation, Vol.2*, 2, pp.386ff; Darwin, *Descent*, pp.227–228.

[32] Darwin, *Descent*, pp.228–229. As they resulted from changes to *time-sensitive* developmental processes.

[33] Darwin, *Descent*, pp.232–239. As the 'tissues' of the sexes are most alike when young.

[34] Darwin, *Descent*, p.234.

[35] Because, as we have seen, all sexual characters are equally transmitted to all offspring, no matter when or how asymmetrically they are expressed. Cf. Darwin, *Descent*, pp.230ff, 342, 542ff, passim.

to [i.e. *developed in*] the female.'[36] This was often true in mammals, as in the coloration of some antelope species, tigers, and zebras. Most notably, it occurred in primates, *including humans*—where most ornamental sexual characters, presumed (by Darwin) to have first evolved in the male, develop equally in both sexes.[37] Non-mammalian examples of equal development included the peacock wrasse (a fish), the scarlet ibis, the European jay, kingfishers, crows, swans, storks, parakeets, and many herons and seabirds. *Descent* also illustrates the converse case: of 'the first development of characters in the female and of transference to [transmission to then development in] the male.' An example was the bumblebee, where 'the pollen-collecting apparatus is used by the female alone for gathering pollen for the larvae, yet in most of the species it is partially developed in the males to whom it is quite useless.'[38] Nipples in men might seem to be another example, although Darwin argued differently: that during 'a former prolonged period' male mammals had nursed offspring, but that, due to some change in circumstances (e.g. fewer offspring), had stopped helping—so the male mammary glands, including nipples, now remained in their pre-pubescent state, and did not develop into functioning breasts.[39]

6.2 Animal Plotlines

Across the living world, the long-term results of sexual selection—sex differences—have been effected by the dramas of desire and difference that have made up the coruscating and inconceivably complex theatre of sexual agency over millenia. The weave of agency with structure is nowhere more intricate than in the dynamics of mating. Secondary sexual characters are the structural results of sexual selection. But they would never have evolved, had they not somehow embellished the sexual displays of their possessors. Evelleen Richards makes much of the way Darwin transposed the fashionable coquetry of Victorian women—gorgeously attired in gaudy bodices and bonnets, corsets and crinolines—onto animals, thus endowing *male* birds, in particular, with, what in humans, would be a 'sense of *female* agency in dress and decoration, an agency that is aesthetic in impulse and is designed to attract . . . sexual attention.'[40] Her gloss beautifully evokes the way dramas of desire

[36] Darwin, *Descent*, pp.227–228, glosses 'transferred to' by saying: 'but in truth they are simply developed in the female for in every breed each detail . . . is transmitted through the female to her male offspring.'

[37] Sometimes, both animals will have the same character, but it will be more pronounced in the male than the female, as with the antlers of the reindeer: Darwin, *Descent*, pp.230–231, passim.

[38] Darwin, *Descent*, p.228.

[39] Darwin, *Descent*, pp.163–164.

[40] Richards, *Darwin*, p.244, my italics; See e.g. Darwin, *Descent*, p.385: 'As any fleeting fashion in dress comes to be admired by man, so with birds a change of almost any kind in the structure or colouring of the feathers in the male appears to have been admired by the female.'

produce sexual selection in *Descent*. 'Courage, pugnacity, perseverance, strength and size of body, weapons of all kinds, musical organs, both vocal and instrumental, bright colours and ornamental appendages,' wrote Darwin, had all been indirectly gained by the one sex or the other, through 'the exertion of choice, the influence of love and jealousy, and the appreciation of the beautiful in sound, colour or form.'[41]

How could the voice box of Australian magpies have developed, did they not sing? What point has a peacock's flamboyant train unless it be suddenly lifted and fanned, quivering and iridescent, then stilled to be seen in all its glory? 'Ornaments of all kinds, whether permanently or temporarily gained, *are sedulously displayed by the males*'—to a target, often but not necessarily, female (Figure 6.2; see 6.2.2 Male-on-male plot).[42] Nor is that the end of the matter. For the consummatory value of sexual ornaments depends twice-over on agency. Not only must the males 'display their charms with elaborate care and to the best effect' *to* females.[43] Females must themselves actively get involved. They must not only be present and attentive to the displays. They must appreciate the beauty of the males, and respond to them.[44] To achieve consummation, the courtship behaviour or 'love antics and dances' in animals must, to a lesser or greater extent, involve both males and females as participants.[45]

Conversely, unless it be known how animals display themselves, interpretation of their ornamentation remains speculative. For example, *Descent* describes a moth that has a pale greyish-ochreous top side to its fore-wing, while the wing's 'magnificently ornamented' lower surface sports an eye-spot or *ocellus* of 'cobalt-blue, placed in the midst of a black mark, surrounded by orange-yellow, and this by bluish-white' (Figure 6.3). However, *Descent* laments, the habits of these moths 'are unknown; so that no explanation can be given of their unusual style of colouring.'[46] This became an acute problem in Darwin's discussion of mammalian sex, because,

> with mammals we do not at present possess any evidence that the males take pains to display their charms before the female; and the elaborate manner in which this is performed by male birds and other animals, is the strongest argument in favour of the belief that the females admire, or are excited by, the ornaments and colours displayed before them.[47]

[41] Darwin, *Descent*, p.617.
[42] Darwin, *Descent*, p.394, my italics: 'when not in the presence of the females … the peacock … evidently wishes for a spectator of some kind, and, as I have often seen, will shew off his finery before poultry, or even pigs.'
[43] Darwin, *Descent*, p.394.
[44] Darwin, *Descent*, p.496.
[45] E.g. Darwin, *Descent*, pp.380, 486.
[46] Darwin, *Descent*, p.315.
[47] Darwin, *Descent*, p.541.

Figure 6.3 The Fallen Bark Looper (*Gastrophora henricari*) (a); showing (b) the eye-spots on the underwing which Darwin could not interpret.

Source: a) Donald Hobern / Flickr. Distributed under the terms of the Creative Commons Attribution 2.0 Generic (CC BY 2.0). https://creativecommons.org/licenses/by/2.0/; b) Boisduval & Guenée: Uranides et Phalénites. (1857). *Histoire naturelle des Insectes. Spécies général des Lépidoptères*.

The dynamics of agency governing sexual display give the key to sexual ornamentation because it is partly the active plasticity of display that hooks the desires of the opposite sex in the first place: 'slight changes for the sake of change, have sometimes acted on female birds as a charm . . .'[48] Novel action would over time have produced novel anatomy, and hence, even more than with natural selection, sexual selection 'primarily depends on variability'—sexual desire being itself aroused by caprice.[49]

As was Darwin's wont, Part II of *Descent* elaborates the dynamics of sexual agency by describing displays in hundreds of species of animal. Four types of drama emerge from these analyses, each having its own plotline and cast of characters. Each character-type or attribute exists on a sliding scale—from promiscuous to faithful, for example. I call the four plots: Female-controlled; Male-on-male; Inverted; and Symmetrical. I discuss each in turn.

6.2.1   Female-controlled plot

6.2.1.1   Female characters

According to *Descent*, females play the decisive role in the most common plotline for animal sex. Their agency is complex. First and foremost comes their sexual

---

[48] Darwin, *Descent*, p.494.
[49] Darwin, *Descent*, p.319.

flammability. Mating would never occur did not females have the capacity *to get excited* ('allured,' 'seduced') by male displays, and so, yearn to involve themselves in the antics of their inamorato, thus coming to a mutual understanding with him 'by means of certain gestures,' as prelude to their copulation.[50] *Descent* describes a multitude of love-antics in which both male and female animals participate, albeit acting different roles.[51]

6.2.1.1.1 Saucy

As I have said, the word *coy* occurs just three times in *Descent*. One reference is to the candy stripe spider, and one to a small insect-like creature called a springtail. The only time 'coy' refers to female animals in general, it glosses a (truncated) quote from the naturalist John Hunter (1728–1793). Hunter's original text runs:

> It would appear that the female is not so desirous for copulation as the male. We find in most animals, if not in all, that the male always courts the female; that she requires being courted to give her desires, otherwise she would not have them [i.e. copulate with the males] so often.[52]

Hunter's quote states clearly that females have active sexual desires—something especially obvious when they are denied consummation: 'My brown cow generally wanted [to hump with] the bull every three weeks,' when Hunter separated cow and bull.[53]

Darwin glossed the 'illustrious Hunter' as saying that the female, with the rarest exceptions, 'is less eager than the male,' she, generally: ' "requires to be courted"; she is coy, and may often be seen endeavouring for a long time to escape from the male.'[54] *Less* eager does not mean *passionless*.[55] Likewise, Darwin's word *coy* does not connote reticence so much as *arch* reticence. Thus, when describing the candy stripe spider, Darwin wrote: 'The male is generally much smaller than the female, sometimes to an extraordinary degree, and he is forced to be extremely cautious in making his advances, as the female often carries her coyness to a dangerous pitch.'[56]

Regarding the springtail (Figure 6.4), Darwin quoted John Lubbock as saying:

---

[50] Darwin, *Descent*, pp.262, 380–381, 415, 417, 420, 497, passim. Summarizing his discussion of musical courtship in insects, spiders, fishes, amphibians, and birds, Darwin wrote: 'unless the females were able to appreciate such sounds and were excited or charmed by them, the persevering efforts of the males, and the complex structures often possessed by them alone, would be useless': p.569.

[51] See for example the 'great magpie marriage' celebration. Darwin, *Descent*. p.406. Cf. Tim Birkhead, 'Studies of West Palearctic Birds. 189. Magpie,' *British Birds* 82, (1989), pp.583–600.

[52] John Hunter, *Essays and Observations on Natural History, Anatomy, Physiology, Psychology, and Geology, Vol.I* (London: Van Voorst, 1861), p.194.

[53] Hunter, *Observations on Natural History*, p.194.

[54] Darwin, *Descent*, p.222.

[55] Contra Richards, *Darwin*, p.217; see also her pp.364, 482.

[56] Darwin, *Descent*, p.273. Dangerous, because she may devour the male rather than copulate with him.

Figure 6.4 Springtails 'coquetting,' whips at the ready.
Source: thatmacroguy / Shutterstock.com.

it is very amusing to see these little creatures (*Smynthurus luteus*) coquetting to-gether ... the female pretends to run away and the male runs after her with a queer appearance of anger, gets in front and stands facing her again; then she turns coyly round, but he, quicker and more active, scuttles round too, and seems to whip her with his antennae; then for a bit they stand face to face, play with their antennae, and seem to be all in all to one another.[57]

More extreme cases could be found where females 'not only exert a choice, but ... court the male, or even fight together for his possession' (see 6.2.3 Inverted plot). This was seen in hen bullfinches, and in peafowl, amongst whom the first advances were 'always made by the female.' Something similar takes place, said Darwin, with 'the older females of the wild turkey.'[58]

6.2.1.1.2 Fanciful
Coquetry and aggressive lust aside, the passional engagement of females in animal courtship was most universally shown by the fact that 'strong individual antip-athies and preferences' were frequent in females. Darwin's best evidence for the de-cisive sexual proclivities of females came from domestic breeds, where the nuances of sexual antics were most easily observed. He also noted that a non-instinctive

---

[57] Darwin, *Descent*, pp.279–280. See also Sarah Hrdy, 'Empathy, Polyandry, and the Myth of the Coy Female,' in *Feminist Approaches to Science*, ed. Ruth Bleier (New York: Pergamon Press, 1986), pp.119–146.
[58] Darwin, *Descent*, pp.419–420, on the authority of John Audubon.

partiality must drive the cross-species mating productive of hybrids in nature, now known to be far more common than formerly thought.[59] More generally, *Descent* stressed that courtship in the wild 'almost invariably' involved females being pursued by many males. Did females mate at random under such conditions? No, said Darwin, citing copious observations by the American ornithologist and illustrator John James Audubon (1785–1851), who 'does not doubt that the female deliberately chooses her mate.' Thus, in woodpeckers, the hen was 'followed by half-a-dozen gay suitors, who continue performing strange antics,' until a marked preference was shewn for one. Likewise with nightjars, red-winged starlings, vultures, and Canada geese. More generally, insofar as sexual passion equated to the strength of an individual animal's antipathy or lust for a particular mate, this was particularly exhibited 'more commonly by the female than by the male' among mammals.[60]

6.2.1.1.3 Ms Epicure

Obviously, for his argument to explain the iridescent glamour of hummingbird plumage or a nightingale's melodious song, Darwin needed female desire to have been swayed by the *aesthetic* qualities of male displays and accessories. Only if females were 'unconsciously excited by *the more beautiful* males,' would males 'slowly but surely be rendered more and more attractive through sexual selection.'[61] The females needed to be so pernickety about colour and form that their preferences could hardly be automatic or instinctive: implying 'powers of discrimination and taste on the part of the female which will at first appear extremely improbable.'[62] So Darwin went out of his way to show that, particularly in birds, females could draw on 'acute powers of observation,' and had 'some taste for the beautiful both in colour and sound.'

Female arbitration might be aesthetically based. But it was not effete. Her choices were hot-blooded. Females did not 'consciously deliberate.' The lucky male was the one 'she is most excited or attracted by.' His beauty *aroused* her. So was she 'struck only by the general effect' of male display and gallantry? Perhaps. Yet, perhaps not:

> after hearing how carefully the male Argus pheasant displays his elegant primary wing-feathers, and erects his ocellated plumes in the right position for their full effect; or again, how the male goldfinch alternately displays his gold-bespangled wings, we ought not to feel too sure that the female does not attend to each detail of beauty.[63]

[59] Darwin, *Descent*, pp.414ff, Mallett, 'Hybrid Speciation.'
[60] Darwin, *Descent*, p.525.
[61] Darwin, *Descent*, pp.496–497, my italics.
[62] Darwin, *Descent*, p.211.
[63] Darwin, *Descent*, p.421.

6.2.1.2 Male characters

Part II of *Descent* is justly celebrated for its extraordinary parade of coloured patterns, ornate songs, musky odours, and a host of otherwise useless sexual 'ornaments', mostly found in male animals—including extravagantly elaborated and gaudily decorated tails, crests, combs, knobs, frills, fins, faces, feathers, skin, hairstyles, horns, air-distended sacs, topknots, naked shafts, throat-pouches, tentacles, wattles, and rumps. The largest division of these had resulted from males' phenotypic responses to the sexual whims of their females:

6.2.1.2.1 Dandies

If females were to be the final arbiters of male display, then males would have to shape their actions to win female approval—which put males in a position requiring a form of *meta*-recognition: he must present his behaviour in a way to be read by her so as to titillate (cf. 5.2.3 Self-attention: reading you reading me). Insofar as a male's success in winning a female was not accidental, it resulted from *him* acting in a way that fulfilled or excited *her* desire. The peacock with his long train, appeared 'like a dandy', wrote Darwin.[64] Accordingly, *Descent* debates whether males *knowingly* display their charms before the female: 'When we behold ... several male birds displaying their gorgeous plumage, and performing strange antics before an assembled body of females, we cannot doubt that, though led by instinct, they know what they are about, and consciously exert their mental and bodily powers.'[65]

Often male deliberation, not to say vanity, seemed patent:

> Sufficient facts have now been given to shew with what care male birds display their various charms, and this they do with the utmost skill. Whilst preening their feathers, they have frequent opportunities for admiring themselves, and of studying how best to exhibit their beauty.[66]

Yet this confident statement immediately gets modulated, if not withdrawn. *Descent*'s next words read: 'But as all the males of the same species display themselves in exactly the same manner, it appears that actions, at first perhaps intentional, have become instinctive.'[67] Following this logic: whenever males *do not* court females in 'exactly the same manner', the males must be conscious of what they are doing. Thus, the 'play-houses' built in Australia by bower-birds vary from

---

[64] Darwin, *Descent*, p.364.
[65] Darwin, *Descent*, p.211.
[66] Darwin, *Descent*, p.402.
[67] Darwin, *Descent*, p.402.

Figure 6.5 Spotted Bower-birds preparing to perform in their purpose-built 'playhouse' or theatre.

Source: Darwin, C. (1871). *The Descent of Man and Selection in Relation to Sex*. (London: John Murray, Albemarle Street).

Darwin, *Descent*, pp.381–382.

bird to bird (see this book's cover; Figure 6.5). Depending on the kinds of material available in their immediate vicinity, a great deal of attention, skill, and discrimination must be exercised.[68] As Darwin observed:

> The bower of the Spotted bower-bird 'is beautifully lined with tall grasses, so disposed that the heads nearly meet, and the decorations are very profuse.' Round stones are used to keep the grass-stems in their proper places, and to make divergent paths leading to the bower. The stones and shells are often brought from a great distance. The Regent bird, as described by Mr. Ramsay, ornaments its short

[68] Darwin, *Descent*, p.381. NB in this book's cover-photograph, the colour-coordinated clothes peg chosen to decorate the stage on which the hen Satin bower-bird is bowing to the male. One species of Australian bower-bird (*Scenopoeetes dentirostris*) is known as the Stage-maker. For courtship, it clears a display-court: a flat area, decorated with green leaves turned bottom-up, and containing at least one tree trunk used by the cock for perching. Upon the approach of a hen the cock drops to the floor and dances: Peter Slater, *A Field Guide to Australian Birds, Vol.2. Passerines* (Adelaide: Rigby, 1974), p.284.

bower with bleached land-shells belonging to five or six species, and with 'berries of various colours, blue, red, and black, which give it when fresh, a very pretty appearance … the whole shewing a decided taste for the beautiful.'[69]

The point that bower-birds sometimes bring decorations *from a great distance* implies conscious discrimination, as do the various colours birds choose. Note that aesthetic taste here is demonstrated by the bower-bird cocks, not just the hens.[70] In Darwin's treatment, the bower-bird cocks consciously exercise their architectural taste *in order to* satisfy or attract what they sense will be the aesthetic-libidinal proclivities of a hen. *Both* sexes accordingly make aesthetic choices, but to succeed, the male's desire must be subordinated to, and second-guess, the female's desire.[71] He must do *what she would like* him to do to succeed.[72]

6.2.1.2.2 Mr Ever Eager

Given female control over mating, males cannot usually afford to be choosy. If the female proves willing, the male is 'so eager that he will accept any female, and does not, as far as we can judge, prefer one to the other.'[73] The opening chapter of *Descent* Part II, in particular, stresses greater male than female activity in courtship in most kinds of animal, giving rise to the 'notorious' eagerness of males in pursuing females.[74]

Darwin explained the greater eagerness of male than female animals to mate by reference to the greater 'matter and force' required by females in even relatively lowly creatures in the formation of her ova. Female gametes were larger than male; produced in far smaller numbers; needed subsequent nourishment or protection; and hence, were less easily transported. Darwin held that the two sexes expended nearly equal energies in reproduction. Balancing the larger force required for

[69] Darwin, *Descent*, pp.413–414, quoting John Gould, *Handbook to the Birds of Australia*, 2 vols., vol. 1st (London: Taylor & Francis, 1865), pp.444–461; Edward Ramsay, 'Letter to the Editor,' *Ibis* 9, (1867).

[70] Recent experimental research has shown that the males of this species do *deliberately* vary the design of their bowers, which, hence, differ signally from individual to individual: Laura Kelley and John Endler, 'Male Great Bowerbirds Create Forced Perspective Illusions with Consistently Different Individual Quality,' *Proceedings of the National Academy of Sciences of the United States of America* 109, (2012), pp.20980–20985.

[71] This is an almost comically apt instantiation of what Jacques Lacan called 'mirror stage' subjectivity, something he illustrated from the behaviour of cockerels: Jacques Lacan, 'The Mirror-Stage as Formative of the Function of the I as Revealed in Psychoanalytic Experience,' in *Ecrits: A Selection*, ed. Alan Sheridan (London: Tavistock, 1977), pp.1–8; Jacques Lacan, *Speech and Language in Psychoanalysis*, ed. Anthony Wilden (Baltimore: Johns Hopkins University Press, 1981)

[72] And this must eternally remain an open question, because females, according to Darwin, are fanciful. Which would explain why Freud's 'great question … "What does a woman [or female animal in this case] want?" … has never been [and can never be] answered': p.421, Ernest Jones, *Sigmund Freud: Life and Work, Vol.2* (London: Hogarth, 1953).

[73] Characteristically, Darwin immediately went on from here to describe cases where males *do* prefer one female over others.

[74] E.g. Darwin, *Descent*, p.221.

females to produce and nurture their eggs *in situ*, the male put his reproductive efforts into 'fierce contests with his rivals, in wandering about in search of the female, in exerting his voice, pouring out odoriferous secretions, &c.'—that is, into male eagerness (see 6.5 Resonances).[75]

6.2.2 Male-on-male plot

Some of the most glaring male–female differences in animals turn out, behaviourally, not to serve in male displays to females, but in male–male competition 'for the possession of females.'[76] Cue, a whole subdivision of the flamboyant Mardi Gras of masculine peculiarity that makes up *Descent* Part II—here featuring the tools of macho violence: big bodies; augmented physical strength; over-size, 'cutting or tearing' teeth; tusks; snapping beaks; twisted horns; branched antlers; crenelated skulls; vicious spurs; spikey spines; defensive shields and manes; 'enormously developed mandibles'; ferocious claws; and 'sword-like front-limbs, like hussars with their sabres.' What does all this imply about the roles males play in the lead-up to intercourse?

6.2.2.1 Male characters
6.2.2.1.1 Warriors
'The season of love is that of battle,' wrote Darwin.[77] Darwin's *law of battle* fills *Descent* with copious examples of males fighting 'desperately' with other males, to get access to females—as evidenced by 'mortal wounds,' pierced eyes, hides 'covered with scars,' 'the snow all bloody,' and feathers that 'fly in every direction.' *Descent* describes the dynamics of male–male combat as accentuating boldness, courage, and pugnacity, which therefore comprise peculiarly masculine characters.[78] He also observed that certain animals would chirp or otherwise vocalize 'in triumph' over a defeated rival. It was particularly with regard to battle that Darwin could assert: 'the males of almost all animals [have] stronger passions than the females' (but see 6.2.1.1.1 Saucy, p. 188–190; 6.4 Kinks and Muddles).

The advantages of winning prove non-obvious. Because, when the sexes exist in equal numbers, even the 'worst-endowed' males will ultimately find females and 'leave as many offspring, as well fitted for their general habits of life, as the

---

[75] Darwin, *Descent*, pp.209, 221–225. This foreshadows what is now called 'parental investment theory' (see 'Resonances: Evolutionary Psychology'): David Geary, 'Evolution of Parental Investment,' in *Handbook of Evolutionary Psychology*, ed. David Buss (Hoboken NJ: Wiley, 2016), pp.524–541.

[76] E.g. Darwin, *Descent*, p.268.

[77] Darwin, *Descent*, p.366.

[78] Depending on whether or not he chose to invoke the 'law of equal transmission,' which generally, he did not. For quotes see Darwin, *Descent*, pp.221, 363, 368, 500–501.

best-endowed males' (except where polygamy prevails).[79] Yet, for sexual selection to work, there had to be some advantage gained by fighting. So Darwin proposed that victorious males—'the strongest' and 'best armed'—would have the pick of the females and would mate earlier than defeated males. At the same time, 'the more vigorous and better-nourished' females would be ready to breed in the spring before the others. Such 'vigorous pairs' would surely rear a larger number of offspring 'than the retarded females,' who could only mate with losers.[80]

6.2.2.1.2 Poseurs

The battles among the cocks of the Ruffed Grouse 'are all a sham,' being performed to show the cocks off to the greatest advantage, confides *Descent*, quoting a 'good observer'—who testified in evidence: 'I have never been able to find a maimed hero, and seldom more than a broken feather.' But, show themselves off *to whom*? The good observer implied that the cocks performed their sham to impress 'the admiring females who assemble around.'[81] Darwin proposed a more intricate drama.

First Darwin stated more than once that the aim of males in battle was to '*drive away* or kill their rivals.'[82] There are obvious advantages in driving away rivals, rather than fighting them to the death, namely, reduction in the risk of injury. Hence Darwin noted that males sometimes threaten but do not necessarily fight with rivals. For example, the roaring of the lion 'may be of some service to him by striking terror into his adversary; for when enraged he likewise erects his mane and thus instinctively tries to make himself appear as terrible as possible.'[83] (This theme was elaborated in *Expression*.)[84]

What we now call *threat displays* could be seen in the erected neck-hackles of a rooster, and in the raised collars of cock Ruffs.[85] Sound-making was the most common way of competing in this more peaceful kind of contest, as evident in the cries of several different groups of animals, including the agile gibbon who 'delight in their own music, and try to excel each other.'[86] Most notably, *Descent* cites the view of 'many naturalists' that the singing of birds is almost exclusively 'the effect of

---

[79] Darwin, *Descent*, p.214.

[80] Darwin, *Descent*, pp.214–215. NB this quote directly contradicts the previous one, which comes from the preceding paragraph: do the losers 'leave as many offspring, as well fitted for their general habits of life, as the best-endowed males,' or not?

[81] Darwin, *Descent*, p.367.

[82] E.g. Darwin, *Descent*, p.614, my italics.

[83] Darwin, *Descent*, p.526.

[84] E.g. Charles Darwin, *The Expression of the Emotions in Man and Animals*, 2nd ed. (London: Murray, 1890), pp.100, 110ff.

[85] Darwin, *Expression*, p.100.

[86] Darwin, *Descent*, pp.527, 553.

rivalry and emulation,' and not for the sake of charming potential mates.[87] Darwin agreed, writing: 'It is certain that there is an intense degree of rivalry between the males in their singing.' And he added: 'That male birds should sing from emulation as well as for charming the female, is not at all incompatible.'[88]

Which suggests that male desire has a *triangular* form: the male, *A*, sings or displays to achieve his own desired ends by outdoing, *B*, (one or more) other males, *but also*, *C*, to win the female. This triangle founds jealousy. Thus, a migratory locust stridulates 'to call or excite the mute females.' But he will make just the same sound 'from anger or jealousy, if approached by other males ... whilst coupled with the female.'[89] Likewise Darwin described songbirds as being jealous of their rivals' singing. Some cock-birds also 'pay particular attention to the colours of other birds' out of jealousy.[90]

All of which creates a nexus between Darwin's two main scenarios for animal sexual attraction—mediated by a now-ambiguous concept of *display*. We saw in the standard female-controlled plotline that, ultimately, *female* desire determines who mates with whom—which implies that a male's sexual agency must be styled to satisfy females. But male display does not *solely* have females as its designated audience, after all. Thus we find Darwin writing, unexpectedly (according to the standard female-controlled plot), that 'the strongly-marked colours and other ornamental characters of male quadrupeds are beneficial to them *in their rivalry with other males*.'[91]

Nevertheless, despite the added dimension same-sex emulation gave to male display, Darwin stressed that female choice remains its primary driver. Even a fight to the death does not mean the victors obtained 'possession of the females, independently of the choice of the latter.'[92] While victorious males may try to drive away or kill their rivals before they pair, it did not appear that the females 'invariably prefer the victorious males.' Males must first fight, and then, *also* 'endeavour to charm or excite their mates by love-notes, songs, and antics,' not always with success.[93] For this reason, courtship was 'often a prolonged, delicate, and troublesome affair.'[94] While the strongest males were totally absorbed in same-sex warfare, a female capercaillie, for example, would sometimes steal away with a young male 'who has not dared to enter the arena with the older cocks,' just as some does of the red deer in Scotland did.[95]

---

[87] 'Difference is initially exciting, but ultimately more exciting is finding any thing in which you can see your self': Kathryn Maris, 'The Death of Empiricism,' *Poetry Review* 108, (2018), p.34.
[88] Darwin, *Descent*, p.369.
[89] Darwin, *Descent*, p.283.
[90] Darwin, *Descent*, p.412.
[91] Darwin, *Descent*, p.540, my italics.
[92] Darwin, *Descent*, p.214.
[93] Darwin, *Descent*, p.367.
[94] Darwin, *Descent*, p.407.
[95] Darwin, *Descent*, p.367.

The male-on-male plot adds a refinement to the mirror-dynamic of the standard female-controlled plot which subordinates it. Now, in the first instance, cocks sing to outrival a like cock, not to please the imagined desires of a bird from the opposite sex: 'That the habit of singing is sometimes quite independent of love is clear, for a sterile, hybrid canary-bird has been described as singing whilst viewing itself in a mirror, and then dashing at its own image.'[96]

The step from displaying *to outdo* a male conspecific to displaying *to allure* a female conspecific is the step that opens the door for the whimsicality of taste (and cultural mediation in humans) to enter into the formation of sexual display and ornamentation, thus creating the 'extreme diversity' that so delighted and exercised Darwin (see Ch. 8).[97] *Descent* frames this as a law: that secondary sexual characters are 'often widely different in closely-allied forms' or species, when subject to the caprices of female fancy—and, thereby, to sexual selection.[98] This was where Darwin's parallel with human fashions takes effect. For example:

> ... mere novelty, or slight changes for the sake of change, have sometimes acted on female birds as a charm, like changes of fashion with us. Thus the males of some parrots can hardly be said to be more beautiful than the females, at least according to our taste, but they differ in such points, as in having a rose-coloured collar instead of 'a bright emeraldine narrow green collar;' or in the male having a black collar instead of 'a yellow demi-collar in front,' with a pale roseate instead of a plum-blue head.[99]

6.2.3 Inverted plot

Darwin documented several animal species in which the female-controlled plot of sexual desire was partly or completely inverted. For example, in wild turkeys and certain species of grouse, it is the hens who court the cocks.[100] Female Rhesus monkeys are more brightly coloured than males, as are the females of some bees and butterflies.[101] And in horses, stallions sometimes show the same caprice in mate-choice that the standard model attributes to females. Caged female moths, Emperors and Oak Eggars, can attract 'vast numbers' of males—some of whom

---

[96] Darwin, *Descent*, p.369 & p.413: 'When birds gaze at themselves in a looking-glass (of which many instances have been recorded) we cannot feel sure that it is not from jealousy of a supposed rival.' Cf. Jacques Lacan, 'Aggressivity in Psychoanalysis,' in *Écrits: A Selection* (London: Tavistock Publications, 1977), pp.9–32.

[97] Darwin, *Descent*, p.297.

[98] Darwin, *Descent*, p.476; Charles Darwin, *The Descent of Man and Selection in Relation to Sex, Vol.2* (London: Murray, 1871), p.230.

[99] Darwin, *Descent*, p.494.

[100] Darwin, *Descent*, pp.475–480.

[101] Darwin, *Descent*, pp.293, 319, 539.

'will even come down the chimney to her.'[102] Complete inversion of the female-controlled plot is found most often in birds: *Descent* has six pages describing cases where the hen was 'more conspicuous than the adult male,' and, the young of both sexes 'in their first plumage resemble the adult male.' Examples include a species of quail, the emu, the cassowary, a tree-creeper, a nightjar—all Australians—plus the painted snipe, phalaropes, and dotterels. But other taxonomic groups show the same 'complete inversion,' including a pipefish, the tawny mining bee, and the horntail wasp. In such cases, we may find that 'an almost complete transposition of the instincts, habits, disposition, colour size, and of some points of structure, has been effected between the two sexes.'[103]

Inversion may affect parenting too. In emus and cassowaries, mouth-breeding fish, gobies, and sticklebacks, it is the males who tend the young.[104] It may also affect fighting, where females are larger and more pugnacious than the males—as in barred buttonquail, emus, and cassowaries. In some such cases the females 'apparently have acquired their greater size and strength for the sake of conquering other females and obtaining possession of the males.'[105]

Descent turns inversion to good effect. It gives proof to Darwin that, whichever sex sports the more 'masculine' characters, the characters in the drama remain complementary, such that males correspondingly show the expected *feminine* characters, proving both sets result from 'the action of one common cause, namely sexual selection.'[106]

6.2.4 Symmetrical plot

Lodged quietly in the interstices of *Descent*'s examples, we find a degree of symmetry between the roles played out by some animals to fulfil their desires. Many animals are not sexually dimorphic. *Descent* sometimes invokes its law of equal transmission to explain such cases, but not always. *Both* male and female bowerbirds help build their bower, and both use it for their different displays—displays which must therefore have developed independently.[107]

The indiscriminate randiness of males does not always prevail: both sexes sometimes proving equally picky, such that males and females of the same species, who inhabit the same district 'do not always please each other, and consequently do

[102] Darwin, *Descent*, p.252. The males are attracted by the female's smell (pheromones). Darwin, *Descent*, p.265, hypothesized a similar form of chemical lure in small female sea-bed-dwelling crustaceans.
[103] Darwin, *Descent*, pp.265, 346, 475–480, 614.
[104] Darwin, *Descent*, pp.345ff.
[105] Darwin, *Descent*, p.362.
[106] Darwin, *Descent*, pp.370, 614.
[107] Darwin, *Descent*, p.406.

not pair.'[108] Many species are monogamous, which implies equal fidelity in both sexes,[109] and often a similarity in appearance—as in most corvids, many parrots, and some seals.[110] Certain strictly monogamous species show similar courtship behaviour also, for example, holding 'nuptial assemblages ... where it is believed both sexes assemble,' as in magpies, Scandinavian ptarmigans, and the Australian lyre-bird.[111] Both sexes of the saddle-backed bush cricket have musical apparatus, though the organs 'differ in the male and female to a certain extent.' Hence, they must have been 'independently developed in the two sexes, which no doubt mutually call to each other during the season of love.'[112] Once bonds form, both male and female partners may reinforce them by mutually calling and displaying—as in gibbons, bower-birds, and beavers.[113]

Beyond this, Darwin briefly proffers the hypothesis that, in some cases of similarity between the sexes, 'a double process of selection has been carried on,' the males having selected the more attractive females, and the females the more attractive males. But he soon withdraws this view as 'hardly probable, for the male is generally eager to pair with any female' (cf. 6.4.5 Two-handers).[114]

6.3 The Human

If Darwin's account of animal desires and differences in *Descent* Part II comprises many plots and characters, and overflows with exceptions, it pales by comparison to the complications—and muddles—comprising Part III, 'Sexual Selection in Relation to Man.' These complications revolve around several nodal concerns, concerns I now review. I shall then move on to examine how *Descent* portrays the human theatre of desire, and its results.

6.3.1 Nodal concerns

Gillian Beer comments on Part III of *Descent*, that, once Darwin 'substitutes the term "Man" for "male," his descriptor for all other species, a rush of social

---

[108] Darwin, *Descent*, p.407.

[109] Darwin's evidence for an equality of restraint was that, when a male, or a female, member of a magpie pair was shot, the widowed survivor would almost immediately pair with another bird. He reported similar findings with jays, owls, sparrows, peregrine falcons, ptarmigans, chaffinches, redstarts, nightingales, bullfinches, and starlings. Darwin, *Descent*, pp.407–410.

[110] Darwin, *Descent*, pp.218, 457, 516.

[111] Darwin, *Descent*, p.406.

[112] Darwin, *Descent*, p.288.

[113] E.g. Darwin, *Descent*, pp.406, 412, 523, 527.

[114] Darwin, *Descent*, pp.225–226. Darwin is too pat here, given that, according to his own evidence, the males in many species *do not* evince a promiscuous eagerness to mate (e.g. see previous paragraph).

assumptions gathers behind his statements.'[115] And not just social assumptions, but deep personal investments. For decades Darwin had held to his theory of sexual selection, in the teeth of sustained criticism: 'I will not give up,' he exclaimed to one of its chief doubters, Alfred Wallace.[116] Wallace later called sexual selection, '*altogether your own subject.*'[117] It had indeed been altogether Darwin's pet concern, from its first public outing in their joint publication of 1858, when Darwin was forty-nine, through to its definitive exposition in the second edition of *Descent* sixteen years later. The design of *Descent* aimed to lay down in Part II, on animals, the plotlines of desire and sexual difference for extrapolation to human sexuality in Part III. The elegance of this design was beset, however, by a swarm of other concerns.

When Darwin first mooted writing 'a little essay on the Origin of Mankind' to Wallace in February 1867, he explained: 'I still strongly think ... that sexual selection has been the main agent in forming the races of Man.'[118] Darwin believed his theory of sexual selection would help *Descent* argue that superficial differences in 'external appearance' such as those between 'the races of man,' could vary independently of the qualities essential to being human, such as morality, high intellectual powers, and bone-structure ('monogenism').[119] So the book set out to torpedo arguments for slavery, resurgent during the American Civil War (1861–1865), which held that slaves and their masters came from different species (a thesis called 'polygenism': see 2.6.1 Natural histories of man). However, Darwin hardly held a consistently benevolent attitude towards non-Caucasians.

Whilst sometimes lauded for having challenged the idea that there was a hierarchy or 'scale' built into nature—making some species lower, and others higher[120]—the author of *Descent* was not chary of comparing 'the lower races of man' with the 'higher' or 'civilised races.' The 'lowest' race, closest to the gorilla, was either 'the negro or Australian.'[121] The superiority of the 'higher' over these

---

[115] Gillian Beer, 'Late Darwin and the Problem of the Human,' (2010), https://nationalhumanities center.org/on-the-human/2010/06/late-darwin-and-the-problem-of-the-human/.

[116] Charles Darwin, 'Letter to Wallace, 15th June,' *Correspondence of Charles Darwin* (Cambridge: Darwin Correspondence Project, 1864), https://www.darwinproject.ac.uk/letter/ ?docId=letters/DCP-LETT-4535.

[117] Alfred Wallace, 'Letter to Darwin, May 1st,' *Correspondence of Charles Darwin* (Cambridge: Darwin Correspondence Project, 1867), https://www.darwinproject.ac.uk/letter/?docId=letters/DCP-LETT-5522, Wallace's emphasis.

[118] Charles Darwin, 'Letter to Wallace, 26th February,' *Correspondence of Charles Darwin* (Cambridge: Darwin Correspondence Project, 1867), https://www.darwinproject.ac.uk/letter/ ?docId=letters/DCP-LETT-5420. Three weeks later, he again told Wallace: 'my sole reason for taking [his 'Man Essay'] up is that I am pretty well convinced that sexual selection has played an important part in the formation of races': Charles Darwin, 'Darwin to Wallace, 12th–17th March,' *Correspondence of Charles Darwin* (Cambridge: Darwin Correspondence Project, 1867), https://www.darwinproject. ac.uk/letter/?docId=letters/DCP-LETT-5440.xml.

[119] Darwin, *Descent*, pp.178, 606. Cf. Adrian Desmond and James Moore, *Darwin's Sacred Cause: Race, Slavery and the Quest for Human Origins* (London: Penguin, 2009).

[120] E.g. Gillian Beer, *Darwin's Plots: Evolutionary Narrative in Darwin, George Eliot and Nineteenth-Century Fiction*, 2nd ed. (Cambridge: Cambridge University Press, 2000), p.54.

[121] Darwin, *Descent* p.156.

'lowest barbarians' proved so evident that: 'At some future period, not very distant as measured by centuries, the civilised races of man will almost certainly exterminate and replace throughout the world the savage races.'[122] Conversely, of civilized peoples, the English were the paragon.[123]

Darwin's attitude to what Victorians called *savages* emerged in his response to the inhabitants of Tierra del Fuego, whom he had seen when twenty-three—one hesitates to say 'met'—from the *Beagle* in 1832. The finale of his published *Journal* of the voyage recalls:

> Of individual objects, perhaps nothing is more certain to create astonishment than the first sight in his native haunt of a barbarian,—of man in his lowest and most savage state. One's mind hurries back over past centuries, and then asks, could our progenitors have been men like these?—men, whose very signs and expressions are less intelligible to us than those of the domesticated animals; men, who do not possess the instinct of those animals, nor yet appear to boast of human reason, or at least of arts consequent on that reason.[124]

Darwin's private, on-board diary reveals that the Fuegians disturbed him far more profoundly and viscerally than this passage implies. They 'were the most abject and miserable creatures I anywhere beheld ... stunted in growth, their hideous faces bedaubed with paint, their skin filthy and greasy, their hair entangled, their voices discordant, their gestures violent.'[125] Drawing more from tall tales and hearsay—all now refuted—than direct observation, he found the people he met on Wollaston Island (Yamans-Yahgans) particularly revolting.[126] These 'savages' had no sense of home and lacked 'domestic affection,'[127] 'for the husband is to the wife a brutal master to a laborious slave.'[128] They had no religion, but were ludicrously superstitious. In war, they practised cannibalism. Likewise in famine: 'they kill and devour their old women before they kill their dogs.'[129] Their language 'scarcely deserves to be called articulate.'[130] They slept on the wet ground 'coiled up like animals.'[131] They looked like 'so many demoniacs.'[132] He doubted they were capable

---

[122] Darwin, *Descent*, pp.65, 156.

[123] Darwin, *Descent*, p.142.

[124] Charles Darwin, Journal of Researches into the Natural History and Geology of the Countries Visited During the Voyage of H.M.S. Beagle Round the World, under the Command of Capt. Fitz Roy R.N., *Journal of Researches* (London: John Murray, 2006) Darwin, *Journal*, p.504.

[125] Charles Darwin, *Diary of the Yoyage of H.M.S. Beagle*, ed. Paul Barrett and Richard Freeman, vol. I, The Works of Charles Darwin (London: William Pickering, 1986) Darwin, *Voyage*, pp.209–210.

[126] For a thoroughgoing refutation of pretty much every negative element of Darwin's views about the Fuegians, see Anne Chapman, *Darwin in Tierra Del Fuego* (Buenos Aires: Imago Mundi, 2006).

[127] Darwin, *Journal of Researches*, p.223.

[128] Darwin, *Journal of Researches*, p.216.

[129] Darwin, *Journal of Researches*, p.214.

[130] Darwin, *Journal of Researches*, pp.205–206.

[131] Darwin, *Journal of Researches*, p.213.

[132] Darwin, *Journal of Researches*, p.221.

of self-improvement: 'Their skill in some respects may be compared to the instinct of animals; for it is not improved by experience: the canoe, their most ingenious work, poor as it is, has remained the same, as we know from Drake, for the last two hundred and fifty years.'[133] Even the physical conditions in which they lived were repugnant. Darwin execrated their 'useless forests' and 'endless storms.'[134]

Aligned with the clash between Darwin's laudable opposition to slavery, and his disgust at so-called savages, was his need for people like the Fuegians and Australian aborigines to play two conflicting roles in *Descent*.[135] One was to demonstrate animal–human continuity—*natura non facit saltum*—linking 'the highest men of the highest races' to apes: 'we may trace a perfect gradation from the mind of an utter idiot, lower than that of an animal low in the scale, to the mind of a Newton,' he wrote.[136] Here, the range of mental powers in humans overlaps hugely with animals, going right down to those 'low in the scale' (even snakes had 'some reasoning power,' Darwin had observed).[137]

However, as we have just seen, aiming to remove any scientific grounds for slavery, Darwin felt a tug to prove that all humans, however different, still had so many similarities that they could all be considered as belonging to one species which was *distinct* from animals. Here, rather than the dullest human being excelled by snakes, Darwin claimed a vast gulf between even the 'lowest [stupidest] savage' and the brightest primate:

> No doubt the difference [in mental powers between animals and humans] is enormous, even if we compare the mind of one of the lowest savages, who has no words to express any number higher than four, and who uses hardly any abstract terms for common objects or for the affections, with that of the most highly organised ape.[138]

Exacerbating the conflicting parts played in *Descent*'s argument for humans' gradual evolution from animals, and his argument against slavery, was Darwin's representation of hominid history. *Descent* distinguishes *three* periods in the evolution of human sexuality. Furthest back in time was Era 1: a 'most ancient,' 'primeval,' 'very remote epoch'—when our ancestors were 'semi-human.' Era 2 was a fully human period, comprising 'long ages' of savagery. Era 3 was that of 'civilised nations,' culminating in Victorian society. Era 3 seems to have begun well after the

---

[133] Darwin, *Journal of Researches*, p.216.
[134] Darwin, *Journal of Researches*, p.215.
[135] Cannon Schmitt, *Darwin and the Memory of the Human: Evolution, Savages, and South America* (Cambridge: Cambridge University Press, 2009).
[136] Darwin, *Descent*, pp.66, 127.
[137] Darwin, *Descent*, p.353.
[138] Darwin, *Descent*, p.65.

flowering of ancient Greece (800 BC), and possibly, some time after the demise of the Spanish Inquisition (1478–1834), that is, moments ago on an evolutionary time-scale.[139]

Of these three periods, *Descent* Part III tells us that it is 'chiefly concerned with primeval times': Era 1. Its focus wavers, however. One reason for this vacillation is that Darwin used knowledge about then-*contemporary* 'savages' to construct *both* Era 1 *and* Era 2![140] The two Eras differ markedly in Darwin's account. Yet the very same data—about the current inhabitants of Tierra del Fuego, for example—were used to prove, for example: both that our earliest, 'half-human,' ancestors *did not* treat women as slaves (in Era 1); and that, during the *second* Era of prolonged 'barbarism,' our fully human but uncivilised ancestors *did* treat women as slaves! Furthermore, because some humans still lived 'in the savage state,' Darwin's Era 2 overlapped chronologically with his Era 3.[141]

Confusion also results from *Descent* making the 'civilisation' of Era 3 both absolute, *and* a matter of degree. 'Civilised nations'—where the word 'civilised' has no qualifier or other demur—crowns the cultural achievements of educated white Victorians and their peers.[142] This is the more absolute sense. On the other hand—and leaving aside the few doomed breeds of 'savage' extant in Darwin's day—the rest of then-contemporary nations were 'more,' 'less,' or 'half,' civilized. In *less* civilized nations (or 'races'), for example, 'reason often errs, and many bad customs and base superstitions … are then esteemed as high virtues, and their breach as heavy crimes.'[143] In the *more* civilized, on the other hand (which often merged with the *absolutely* civilized): 'the conviction of the existence of an all-seeing Deity has had a potent influence on the advance of morality.'[144] The 'old Greeks,' who stood 'some grades higher in intellect than any race that has ever existed,' temporarily represented a summit of civilization, but later 'retrograded.'[145] Their summit had since been rescaled by 'the Western nations of Europe,' and, most particularly, the English.[146] Not that this Victorian peak might not eventually be bettered by 'a more civilised state, as we may hope, even than the Caucasian' (see Ch. 7).[147]

---

[139] Darwin, *Descent*, pp.605, 40, 47, 571, 596, 563, 584 & passim.

[140] Darwin, *Descent*, p.573: '… we are chiefly concerned with primeval times, and our only means of forming a judgment on this subject is to study the habits of existing semi-civilised and savage nations.' (Darwin's other source of evidence for Era 1 behaviour was higher primates, or the 'Quadrumana.')

[141] Darwin, *Descent*, pp.587, 593, 573. On the 'double and contradictory place' of the savage in *Descent*, see Schmitt, *Darwin and the Human*, p.51; Richards, *Darwin*, p.426 & passim.

[142] Some would add 'male' to these qualifiers. However, if morality were the acme of civilization, then women were more civilized than men according to one strand of Darwin's argument—as I show later.

[143] Darwin, *Descent*, p.611. Part I gives human sacrifice, and mediaeval trials for witchcraft as examples: p.96.

[144] Darwin, *Descent*, p.612.

[145] Darwin, *Descent*, pp.140–141.

[146] Darwin, *Descent*, pp.141–142, e.g. 'The remarkable success of the English as colonists, compared to other European nations, has been ascribed to their "daring and persistent energy;" a result which is well illustrated by comparing the progress of the Canadians of English and French extraction …'

[147] Darwin, *Descent*, p.156.

Darwin usually held that countless generations were required before modifications could evolve in a species. Yet in *Descent* Part III, some human changes are said to have been wrought (or maintained) by selection pressures solely operating during Era 3. This had only lasted between 400 and 3000 years (i.e. when dated from the Spanish Inquisition, versus, from the fall of ancient Greece). For instance, Darwin held that 'our aristocracy,' from having chosen, 'during many generations from all classes the more beautiful women as their wives,' had become 'handsomer, according to the European standard, than the middle classes.'[148] Likewise among 'civilized people,' men's greater strength than women had supposedly been kept up by natural selection, because men laboured disproportionately more than women 'for their joint subsistence.'[149]

Nested in Era 3 is an even shorter time-scale: of fads. Here, 'mere novelty, or slight changes for the sake of change,' had sometimes acted 'as a charm.'[150] This briefest of brief time-scales comprehends 'capricious changes' in customs and fashions.[151] For even the most fleeting 'fashion in dress' might come to be admired by men.[152] The transience and exaggerations of fashion were most excessive in civilized countries. In 'fashions of our own dress we see ... [a] desire to carry every point to an extreme.' In contrast, the fashions of 'savages' were 'far more permanent than ours.'[153]

Confusions about race and history were tangled further by *Descent*'s contrasting takes on human sexual difference. Notoriously, and true to our image of the Victorian patriarch, Darwin adopted what he called 'the common rule' of the male excelling the female.[154] The standard female-controlled plotline *Descent*, outlined for animals in Part II, neatly enrolled this patriarchal truism among the laws of nature.[155] Countless descriptions of males with larger bodies, more fearsome teeth or claws or horns, more developed whiskers, louder bellows, stronger smells, more prominent super-ciliary ridges and, of course, brighter ornamentation than females, ensure that, by the time *Descent*'s readers totter to the end of Part II, they will have become accustomed to Darwin's 'common rule.' Nor do such 'external attractions' exhaust the scope of male supremacy, because the standard plot means that males' 'vigour, courage, and other *mental* qualities' also get heightened by sexual selection.[156] Hence, when Part III embarks on its discussion of 'Difference in the

---

[148] Darwin, *Descent*, p.586. This elitist view of English beauty was strongly contested by Alfred Wallace, 'Letter to Darwin, May 29th,' *Correspondence of Charles Darwin* (Cambridge: Darwin Correspondence Project, 1864), https://www.darwinproject.ac.uk/letter/?docId=letters/DCP-LETT-4514. See Richards, *Darwin*, passim.

[149] Darwin, *Descent*, p.563.

[150] Darwin, *Descent*, p.494.

[151] Darwin, *Descent*, p.93.

[152] Darwin, *Descent*, p.385.

[153] Darwin, *Descent*, p.583.

[154] Darwin, *Descent*, p.539.

[155] David Ingleby, 'The Job Psychologists Do,' in *Reconstructing Social Psychology*, ed. Nigel Armistead (Harmondsworth: Penguin Books, 1974), p.318.

[156] Darwin, *Descent*, p.405.

Mental Powers of the two Sexes,' it may not surprise us to find the common rule up in lights. A brief aside to remind us that the bull 'differs in disposition from the cow, the wild-boar from the sow, the stallion from the mare'; and then we get this:

> The chief distinction in the intellectual powers of the two sexes is shewn by man's attaining to a higher eminence, in whatever he takes up, than can woman— whether requiring deep thought, reason, or imagination, or merely the use of the senses and hands. If two lists were made of the most eminent men and women in poetry, painting, sculpture, music (inclusive both of composition and perform- ance), history, science, and philosophy, with half-a-dozen names under each subject, the two lists would not bear comparison. We may also infer, from the law of the deviation from averages, so well illustrated by Mr. Galton, in his work on 'Hereditary Genius,' that if men are capable of a decided pre-eminence over women in many subjects, the average of mental power in man must be above that of woman.[157]

Should we have doubts—remembering perhaps that Victorian girls grew up having vastly different experiences from their brothers, with: fewer rights (in marriage for example); higher mortality rates; and much less access to wealth, education, influ- ence, and employment—Darwin quells them. The species-wide intellectual super- iority of men over women arises from exactly the same laws of nature that ensure cock umbrella birds sport their immense top-knots: 'With respect to differences of this nature between man and woman, it is probable that sexual selection has played a highly important part.'[158]

Enough said? No. Because, cutting across *Descent*'s 'common rule' of male su- periority, the popular Victorian image of the middle-class wife as 'the angel in the house'—sympathetic, unselfish, and self-sacrificing—contributes a counter-flow of *feminine* superiority.[159] The text makes clear that Darwin rated sympathy above selfishness, and, 'woman seems to differ from man ... in her greater tenderness and less selfishness'—qualities *Descent* attributes to 'her maternal instincts.' 'Man,' on the other hand, was 'the rival of other men.' He delighted in competition, and this led to ambition which 'passes too easily into selfishness.' These latter qualities seemed to be his 'unfortunate birthright.'[160] One of Darwin's last letters stated ex- plicitly: women were 'generally superior to men' in moral qualities.[161] Witness

---

[157] Darwin, *Descent*, pp.563–564.

[158] Darwin, *Descent*, pp.373, 563.

[159] Coventry Patmore, *The Angel in the House* (Boston: Ticknor and Fields, 1856); Virginia Woolf, 'Professions for Women,' in *The Death of the Moth and Other Essays* (New York: Harcourt, Brace & Company, 1942).

[160] Darwin, *Descent*, p.563.

[161] Charles Darwin, 'Letter to Caroline Kennard, 9th January,' *Correspondence of Charles Darwin* (Cambridge: Darwin Correspondence Project, 1882), https://www.darwinproject.ac.uk/letter/ ?docId=letters/DCP-LETT-13607.

how he described his own domestic angel in his late *Autobiography* (which was addressed to his children):

> You all know well your Mother, and what a good Mother she has ever been to all of you. She has been my greatest blessing, and I can declare that in my whole life I have never heard her utter one word which I had rather have been unsaid. She has never failed in the kindest sympathy towards me, and has borne with the utmost patience my frequent complaints from ill-health and discomfort. I do not believe she has ever missed an opportunity of doing a kind action to anyone near her. I marvel at my good fortune that she, so infinitely my superior in every single moral quality, consented to be my wife. She has been my wise adviser and cheerful comforter throughout life, which without her would have been during a very long period a miserable one from ill-health. She has earned the love and admiration of every soul near her.[162]

This all sounds utterly conventional. Yet, if we take *Descent* as a whole, *explaining* the moral supremacy of women plunges it into contradiction. Typically, today's commentators represent Darwin's views on human sexuality by extracting quotations solely from *Descent* Part III.[163] This over-simplifies. Darwin only began the psychological chapters (3 and 5) of Part I *after* he had completed Parts II and III of the book, in June 1869. These late-written chapters cast the claim of women's moral superiority in different light, by arguing that moral qualities—*not* the intellect *per se*—formed 'one of the highest psychical faculties of man'.[164] By the time he returned to Part III, to write the concluding chapter of the book (completed August 1870), he unequivocally called moral qualities 'the highest part of man's nature,'[165] explaining them as being 'advanced, either directly or indirectly, much more through the effects of habit, the reasoning powers, instruction, religion, &c., than through natural selection.'[166]

Given Victorian women's lesser education—which Darwin *sometimes* acknowledged (see 6.4 Kinks and Muddles)—it is hard to see how they could exceed men in the 'reasoning powers' and educability required for moral excellence if they were men's intellectual inferiors. In fact, Darwin stated in a late letter that he thought women *would* rival the intellectual achievements of men if they became 'as regular "bread-winners" as are men.' He worried about granting occupational

[162] Charles Darwin, *The Autobiography of Charles Darwin, 1809–1882: With Original Omissions Restored*, ed. Nora Barlow (London: Collins, 1958), pp.96–97.

[163] E.g. Beer, 'Late Darwin and the Problem of the Human.'; Richards, *Darwin*; Montgomery, 'Gender.'

[164] Charles Darwin, *The Descent of Man and Selection in Relation to Sex Vol.1* (London: Murray, 1871), p.71.

[165] Darwin, *Descent of Man Vol.2 1871*, pp.403–404. This was augmented in the second and final (1874) edition of Darwin, *Descent*, p.611, by the statement: 'The moral faculties are generally and justly esteemed as of higher value than the intellectual powers.'

[166] Darwin, *Descent*, p.618; Darwin, *Descent of Man Vol.2 1871*, pp.403–404.

equality to women, however, as this would jeopardize their angelic influence over the home: 'the early education of our children, not to mention the happiness of our homes, would in this case greatly suffer.'[167]

A final concern aroused when dealing with human sexuality was that Darwin, like his friend Thomas Huxley, aligned himself with 'men of science.'[168] Victorian men of science (unlike the new-fangled, Americanized 'scientists') had garnered a goodly quotient of public respect through their 'broad learning, and moral gravity,' thereby earning the right to pronounce on matters of general interest.[169] They had therefore vigilantly to safeguard their respectability, so as to preserve their moral authority. On these grounds, Darwin's publisher John Murray had concerns about *Descent*. He vetoed one 'objectionable' adjective in its proposed title—presumably the word 'sexual.'[170] And requested another passage be toned down, as it contained sentences 'liable to the imputation of indelicacy.'[171] This was the section in *Descent* describing female mammals' 'Preference or Choice in Pairing.'[172] Even after toning-down, Darwin concluded that she quadrupeds were 'much more commonly' 'desirous' of males than vice versa.[173] Murray complained: 'you scarcely do justice to the females' in mammal species. Particularly perhaps, when bringing out feminine promiscuity, for example: 'Mr. Blenkiron has never known a mare to reject a horse.'[174]

Gowan Dawson has investigated Darwin's pursuit of respectability in detail. He points out that the most salacious passages in *Descent* were very properly veiled in Latin, the better to baffle the prurience of less educated readers.[175] Quotations were also secretly amended to remove scandalous details, as in: the 'admiration' of Hottentot men for the big projecting bottoms of their women; and the deliberate elongation by Hottentot women of the lips of their vulvas so that they were able,

---

[167] Darwin, 'Letter to Caroline Kennard, 9th January.'

[168] E.g. Charles Darwin, 'Letter to Falconer, April 22nd,' *Correspondence of Charles Darwin* (Cambridge: Darwin Correspondence Project, 1863), https://www.darwinproject.ac.uk/letter/?docId=letters/DCP-LETT-4121.xml.

[169] Paul White, *Thomas Huxley: Making the "Man of Science"* (Cambridge: Cambridge University Press, 2003), p.1.

[170] John Murray, 'Letter to Darwin, July 1st,' *Correspondence of Charles Darwin* (Cambridge: Darwin Correspondence Project, 1870), https://www.darwinproject.ac.uk/letter/?docId=letters/DCP-LETT-7339.

[171] John Murray, 'Letter to Darwin, 28th September,' *Correspondence of Charles Darwin* (Cambridge: Darwin Correspondence Project, 1870), https://www.darwinproject.ac.uk/letter/?docId=letters/DCP-LETT-7329.

[172] Darwin, *Descent of Man Vol.2 1871*, pp.268–273.

[173] Darwin, *Descent*, pp.523-525. Darwin appears to have cut out some attribution of impropriety to 'the she cat'—censured in Murray's letter, and now absent in *Descent*.

[174] Darwin, *Descent*, p.272; Murray, 'Letter to Darwin, 28th September.'

[175] John Murray, 'Letter to Darwin, October 10th,' *Correspondence of Charles Darwin* (Cambridge: Darwin Correspondence Project, 1870), https://www.darwinproject.ac.uk/letter/?docId=letters/DCP-LETT-7339.

during intercourse, 'to encircle the man's loins and fix him by them until the appetite of both was thoroughly satisfied.'[176]

Given the diversity of cultural and personal concerns aroused by Darwin's public discussion of sex in humans, how closely do the dramas of human desire *Descent* Part III portrays tally with the plotlines describing sex in animals from Part II? As we will now see, some human plotlines do reflect those found in animals. Others do not. The differences most immediately arise from his three Eras of human history. While Darwin could assume that very similar dynamics of desire and difference had acted on animals in their distant evolutionary past as still acted in the present, he could *not* assume the same of humans. Culture had intervened.

6.3.2 Woman-led plot

Descent Part II holds females mostly to have played the decisive role in the theatre of animal desire. However much males might compete between themselves for access to females, nothing guaranteed that females would choose to copulate with the victor. Frequently therefore, males had to embark on post-battle displays to charm and excite females, by singing, dancing, and other antics. At this point, the female's often-capricious libidinal antipathies and preferences came into play—urges that proved particularly strong in female mammals, as compared to their males.

To the extent that this animal plot pertains to humans, we would expect women to play the deciding role in courtship. A man will need to work hard to win a woman's favours, ornamenting himself, displaying, and otherwise impressing her with his charms. His success will be to inflame her desires—female desires ultimately determining who mates with whom. As a result of this dynamic, men will gain secondary sexual characters while women will not. And, to the extent that it has governed the evolution of human sex differences, adult women will resemble children more than adult men do.

6.3.2.1 Desirous women

Descent contains quantities of evidence that the sexual desires of women *have* shaped men's behaviour according to the standard animal female-controlled plot—and (hence, or by implication) their inbuilt mental and physical qualities. Leaving aside the larger physique of men—which *Descent* attributes to sexual selection via the law of battle (see 6.3.3 Macho match-ups)[177]—certain physical and

---

[176] Dawson, *Victorian Respectability*, p.36; Andrew Smith, 'Letter to Darwin, 26th March,' *Correspondence of Charles Darwin* (Cambridge: Darwin Correspondence Project, 1867), https://www.darwinproject.ac.uk/letter/?docId=letters/DCP-LETT-5465; Darwin, *Descent*, pp.8, 578.

[177] *Darwin, Descent*, pp.596–597. NB in Part I (p.107), men's larger, stronger physique is attributed to *natural* selection; see 'Kinks and muddles.'

mental 'ornaments' that differentiate men from women bespeak feminine libidinal caprice. As just noted, such caprice is most clearly demonstrated where the character in question differentiates men from both women *and children*.[178] Thus when *Descent* makes the following blanket statement, it is acknowledging the ubiquity of the standard female-controlled plot in humans: 'Male and female children resemble each other closely, like the young of so many other animals in which the adult sexes differ widely; they likewise resemble the mature female much more closely than the mature male.'[179]

For his most extended illustration of this plotline, the grandly bearded Darwin pointed to facial hair—something both children and nubile women lack.[180] In further evidence that the woman-led plot applied to humans, Darwin *could* have pointed to men having relatively hairy *bodies*—something that also differentiated them from the relative nudity of *both* women *and* children.[181] But he spun this paradoxically, making women's nudity a sexual character, not male body-hair (see 6.4 Kinks and Muddles). Darwin noted too that men, like male primates, had longer vocal chords, and hence lower voices, than both women and children. But he spun this differently too (see 6.4.5 Two-handers).[182]

As for cultural evidence of the importance of the women-led plot, Darwin lavished fascinated attention on the 'Influence of Beauty in determining the Marriages of Mankind.'[183] 'Savages,' he said, paid the greatest attention to their personal appearance. Their passion for ornament was 'notorious.' Indeed, he said, it seemed plausible that clothes were first made 'for ornament and not for warmth.' Everywhere, wrote Darwin, members of tribal cultures 'deck themselves with plumes, necklaces, armlets, ear-rings,' and 'paint themselves in the most diversified manner':

> In one part of Africa the eyelids are coloured black; in another the nails are coloured yellow or purple. In many places the hair is dyed of various tints. In different countries the teeth are stained black, red, blue, &c., and in the Malay Archipelago it is thought shameful to have white teeth 'like those of a dog.' Not one great country can be named, from the Polar regions in the north to New Zealand in the south, in which the aborigines do not tattoo themselves. This practice was followed by the Jews of old, and by the ancient Britons. In Africa some of the natives tattoo themselves, but it is a much more common practice to raise protuberances by rubbing salt into incisions made in various parts of the body; and these are considered by the inhabitants of Kordofan and Darfur 'to be great personal

178 Darwin, *Descent*, p.232.
179 Darwin, *Descent*, p.557.
180 Darwin, *Descent*, pp.602–604.
181 Darwin, *Descent*, p.556.
182 Darwin, *Descent*, pp.600–601, 566ff.
183 Darwin, *Descent*, pp.573ff.

attractions.' In the Arab countries no beauty can be perfect until the cheeks 'or temples have been gashed' ...[184]

And so on, for page after page. Furthermore: 'In most, but not all parts of the world, *the men are more ornamented than the women*, and ... use the finest ornaments.'[185]

This again tallies with the female-controlled plotline which is standard among animals, especially birds, where hen-birds 'admire, or are excited by, the ornaments and colours displayed before them.'[186] Here, male display would be governed by 'preference on the part of the women.' When steadily acting in any one direction, female taste and desires would ultimately affect the character of the tribe, for the women would generally choose 'the handsomest men, according to their standard of taste.'[187]

6.3.2.2 Randy men

Back in the day, if existing tribal cultures set the pattern, then men—like most male animals, according to Darwin—would have had many more sexual partners than did women. Polygamy, said Darwin, summarizing ethnographic evidence, 'is almost universally followed by the leading men in every tribe.'[188] Darwin showed uncertainty whether such polygamy had obtained in the semi-human 'primeval times' of Era 1. He was surer that polygamy reigned in the fully-human but savage Era 2, though even then, the picture clouded, because Darwin had little evidence to gainsay Victorian anthropologists' conclusion that both our male *and female* ancestors had once engaged in 'licentious' communal marriages.[189]

In Era 2 savagery, male promiscuity appeared to have extended beyond marriage, however, polygamous or not, because, 'with all savages, female slaves serve as concubines.'[190] The same appeared true even in Era 3. Thus *Descent* explicitly deplores the slowness of the spread of sexual 'virtue' (faithful monogamy) to the male sex, even 'at the present day.'[191] On the face of it, this implied that 'civilized'

---

[184] Darwin, *Descent*, pp.573–574.

[185] Darwin, *Descent*, p.577, my italics.

[186] Darwin, *Descent*, p.541, NB Darwin amassed far more ethnographic evidence for the existence of women's choice in tribal cultures (8 tribes from 4 continents), than he did for men's choice in the same cultures (2 tribes from 2 continents; see 'Role-reversal').

[187] Darwin, *Descent*, p.599.

[188] Darwin, *Descent*, pp.591, 595: 'At present the chiefs of nearly every tribe throughout the world succeed in obtaining more than one wife.' Darwin did not publish about promiscuity or 'profligacy' *in his own class* of 'civilized' societies. Part I of *Descent* (p.139) equates promiscuity more with 'lower classes' as when it refers to 'the intemperate, profligate, and criminal classes.' He seems to have thought marriage a check on promiscuity: Charles Darwin, 'Letter to Gaskell, November 15th,' *Correspondence of Charles Darwin* (Cambridge: Darwin Correspondence Project, 1878), https://www.darwinproject.ac.uk/letter/?docId=letters/DCP-LETT-11745.

[189] Darwin, *Descent*, p.591.

[190] Darwin, *Descent*, pp.587ff.

[191] Darwin, *Descent*, p.119.

Figure 6.6 Emma Darwin in 1840 (a); and Charles Darwin in 1840 (b). Who chose whom?

Source: (a) Watercolour of Emma Darwin by George Richmond (1840). Science History Images / Alamy Stock Phot.; (b) Watercolour of Charles Darwin by George Richmond (1840). FineArt / Alamy Stock Photo.

Both portraits were painted by George Richmond https://www.researchgate.net/publication/24028827_Bravo_Emma_Music_in_the_life_and_work_of_Charles_Darwin/figures?lo=1; https://en.wikipedia.org/wiki/Charles_Darwin#/media/File:Charles_Darwin_by_G._Richmond.png.

Victorian men remained promiscuous, ever eager to hump with whomsoever—so that, even in Darwin's own drawing-room, who ended up mating with whom was ultimately subject to women's whims (Figure 6.6).

6.3.3 Macho match-ups

Descent reads directly into the evolution of men the dynamic that Darwin held to have produced the greater size, strength, courage, and pugnacity of non-human males, especially in primates: the male-on-male plot of combat and emulous display. Indeed, Part III launches by asserting 'man on an average is considerably taller, heavier, and stronger than woman, with squarer shoulders and more plainly-pronounced muscles ... Man is more courageous, pugnacious and energetic than woman, and has a more inventive genius ... As with animals of all classes, so with man.'[192] *Descent* Part III

---

[192] Darwin, *Descent*, pp.556ff.

treats all these differences, physical and mental, as secondary sexual characters, crediting them to sexual selection.

6.3.3.1 Sticks and stones

In particular, Part III extrapolates to humans the *law of battle* earlier found in beetles, birds, and baboons. This law modulates according to the era of human evolution considered. Drawing analogies from gorillas and one or two tribal cultures—for instance in Australia, where the women were 'the constant cause of war both between members of the same tribe and between distinct tribes'—Darwin held that the law of battle had still prevailed with man 'during the early stages of his development,' that is, in Era 1.[193] As our ancestors became more human, and began to walk on two legs, freeing their arms, men would have fought each other with 'sticks and stones' instead of jaws and teeth—requiring less strength and less full-blooded ferocity. It was thus to their earliest 'half-human' forbears that Victorian men most owed their superiority over women, as measured by breadth of shoulder, muscle definition, 'rugged outline of body,' courage, and pugnacity.[194]

During the second era of human evolution, when our ancestors were fully human but still 'savage,' 'contests for wives' continued—preserving or even augmenting men's advantages. Even in civilized times (Era 3), when 'the arbitrament of battle for the possession of the women has long ceased,' men would still have kept up their advantage in strength over women, wrote Darwin, because, 'the men, as a general rule, have to work harder than the women for their joint subsistence'— a claim which invokes natural, not sexual, selection (cf. 6.4 Kinks and Muddles).[195]

6.3.3.2 Winsome ways

To the extent that women had the power to choose among their male suitors, the law of battle would not suffice to determine who mated with whom—just as it had not in animals. Men could not simply vanquish their male rivals to fulfil their desires; they had also to win over a member of the *opposite* sex. Violence alone did not suffice: hence, savage men's ardour for adornment. Motives for such adornment varied, wrote Darwin, but 'self-adornment, vanity, and the admiration of others, seem to be the commonest motives.'[196] Akin to the rivalry between displaying cock capercaillie, 'man is the rival of other men; he delights in competition.'[197] But women must be the ultimate targets of masculine display. Thus Darwin argued that music and impassioned speech had two functions. They aided our half-human forebears in same-sex contests 'of jealousy, rivalry, and triumph.' But they

---

[193] Darwin, *Descent*, pp.561–562.
[194] Darwin, *Descent*, pp.561–563.
[195] Darwin, *Descent*, p.563.
[196] Darwin, *Descent*, p.576.
[197] Darwin, *Descent*, p.563.

also functioned as 'courtship, when animals of all kinds are excited ... by love'—so being pulled into the triangle that yields jealousy (see 6.2.2.1.2 Poseurs).[198]

6.3.4 Role-reversal

Several commentators summarize *Descent* Part III's take on human sexuality as an 'inversion', 'simply reversing the roles of the sexes' played out by animals.[199] I hope this chapter shows that such 'simple' summaries travesty the variety of contrasting plotlines described in Parts II and III of *Descent*. Nevertheless, as we have just seen, Darwin did sometimes propose that Era 2 men had gained the power of selecting sexual partners, thereby reversing the stress on female choice in the female-controlled animal plot. So, how many other inverted features of this plot are invoked in Part III? Can role-reversal in humans attain to the 'almost complete transposition of the instincts, habits, disposition, colour, size, and of some points of structure' that Darwin found in pipefish or cassowaries (see 6.2.3 Inverted plot)?

6.3.4.1 She's so vain
To usher in its claim that men (in Era 2) 'have gained the power of selection' in humans, *Descent* does briefly gesture towards animal cases in which males 'are the selecters, instead of having been the selected.'[200] Role-reversal quickly extends to female vanity. Just as peacocks and their feathery brethren took 'frequent opportunities for admiring themselves,' so women were 'everywhere conscious of the value of their own beauty; and when they have the means, they take more delight in decorating themselves with all sorts of ornaments than do men.' In the tribal cultures of Era 2, feminine vanity often became actively murderous. Women recognized that child-rearing meant 'loss of beauty.' Hence savage mothers 'very commonly' killed their babies, due to 'the desire felt by the women to retain their good looks.'[201]

6.3.4.2 Sirens
Taking further their appropriation of the characters of males from the female-controlled plot in animals, women's vain self-ornamentation likewise blurs same-sex rivalry with the seduction of men. Recounting how, both in 'barbarous' and in European nations, women 'added to their naturally bright colours'—in the latter, by 'rouge and white cosmetics'—and 'exaggerate whatever characters nature may

---

[198] Darwin, *Descent*, p.572.

[199] John Durant, 'The Ascent of Nature in Darwin's the Descent of Man,' in *The Darwinian Heritage*, ed. David Kohn (Princeton, NJ: Princeton University Press, 1985), p.300; Dawson, *Victorian Respectability*, p.33; Richards, *Darwin*, pp.xxi, 252.

[200] Darwin, *Descent*, p.597.

[201] Darwin, *Descent*, pp.591–593, 577.

have given,' even to the extent of self-mutilation in some tribal cultures, *Descent* underlines that, 'in the fashions of our own dress we see exactly the same principle and the same desire to carry every point to an extreme.' Women show 'the same spirit of emulation' as when male animals compete with other males, Darwin added. For example, Arab women 'vie *with each other* in the superlativeness of their own style': a woman–woman dynamic.[202]

The fast-changing fads of feminine fashion were not solely a matter of same-sex emulation however. Voicing the presumed desires of his Era 3 male peers, Darwin saw the push to extremes in European dress as having the same aim as the exotic ornaments, charms, and love-philtres worn by Era 2 women—'to gain the affections of the men':[203]

> If all our women were to become as beautiful as the Venus de' Medici, we should for a time be charmed; but we should soon wish for variety; and as soon as we had obtained variety, we should wish to see certain characters a little exaggerated beyond the then existing common standard.[204]

6.3.4.3 Male coercion

Harking back to the rumours, if not first-hand observations, he had absorbed as a young man in Tierra del Fuego—of husbands abusing wives as 'a brutal master' treats 'a laborious slave,' with no hint of 'domestic affection'—Darwin *sometimes* identified his second era, of human 'savagery,' as one during which men kept women 'in a far more abject state of bondage, than does the male of any other animal.'[205] This went beyond a role-reversal that produced male *choice*. Intercourse was coerced by men, that is, men *raped* women. In this vein, Darwin caustically extolled Era 1 over Era 2, as occurring long ago, 'before our ancestors had become sufficiently human to treat and value their women merely as useful slaves.'[206] Era 2 women would get no choice. Men had 'gained the power of selection,' so that marriage equated to a man 'keeping possession' of a woman 'by the law of might.'[207]

6.3.4.4 Missing

Beyond female vanity and seductiveness, and male 'choice' (i.e. coercion)—some male characters from the female-controlled plot in animals are not

[202] Darwin, *Descent*, pp.583–584, my italics.
[203] Darwin, *Descent*, p.577.
[204] Darwin, *Descent*, p.585.
[205] Darwin, *Descent*, p.597. Elsewhere, Darwin *rebuts* such allegations, e.g. p.577: 'I have heard it maintained that savages are quite indifferent about the beauty of their women, valuing them solely as slaves; it may therefore be well to observe that this conclusion does not at all agree with the care which the women take in ornamenting themselves, or with their vanity.'
[206] Darwin, *Descent*, p.573.
[207] Darwin, *Descent*, pp.597, 588.

explicitly reversed to apply to women, according to *Descent*. For example, reversal of standard male eagerness to mate was liable to render every Victorian damsel a jezebel—including Darwin's wife and daughters—a portrait apt to scandalize Victorian mores.[208] Darwin did sometimes countenance promiscuity in women, however, at least in Era 2. Reading between the lines of *Descent*, Darwin also seems to have feared that, even in Era 3 of civilized society, 'unmarried females' lagged behind their married sisters in 'female virtue'—especially as, even in marriage, wives' virtue was only guaranteed by their husbands' jealousy, not by a want of polyandrous desires.[209]

This fear became explicit in Darwin's private thinking. Witness a letter he wrote in 1878, concerning the dangers of contraception: 'If it were universally known that the birth of children could be prevented, and this were not thought immoral by married persons, would there not be great danger of extreme profligacy amongst unmarried women, and might we not become like the "arreoi" societies in the Pacific?' The 'arreoi' were the Tahitians, who were (until missionaries arrived) a byword for 'the profligacy of the women.'[210]

Completely missing from the plot of role-reversal in *Descent* Part III, was any admission that men were more like children than women—which *should* have been the result of male choice. Such female superiority should have extended as much to mental as to physical qualities, because, unlike male animals, men 'highly value [the] mental charms and virtues' of the opposite sex.[211]

6.4.5 Two-handers

While the first chapter in *Descent* Part III opens by stressing men's choice, its second chapter opens by introducing a concept of two-way choice, given that 'in civilised nations women have free or almost free choice.'[212] Even in Eras 1 and 2, Darwin suggested 'well-endowed' pairs selected *each other*, asserting that this double form of selection 'seems actually to have occurred, especially during the earlier periods of our long history.'[213] An example was the evolution of human music (which grounded the evolution of language, Darwin argued; see Ch. 7): 'it appears probable that the progenitors of man, either the males or females or both sexes, before

---

[208] Dawson, *Victorian Respectability*, passim.

[209] Darwin, *Descent*, p.119.

[210] Darwin, *Descent*, p.188; Darwin, 'Letter to Gaskell, November 15th,' comma added. See too Charles Darwin, 'Letter to Bradlaugh, June 6th,' *Correspondence of Charles Darwin* (Cambridge: Darwin Correspondence Project, 1877), https://www.darwinproject.ac.uk/letter/?docId=letters/DCP-LETT-10988 in which Darwin refused to support contraception on the grounds that it 'would in time spread to unmarried women & w^d destroy chastity, on which the family bond depends; & the weakening of this bond would be the greatest of all possible evils to mankind.'

[211] Darwin, *Descent*, p.617.

[212] Darwin, *Descent*, p.586.

[213] Darwin, *Descent*, p.599.

acquiring the power of expressing their mutual love in articulate language, endeav-oured to charm each other with musical notes and rhythm.'[214]

A similar symmetry applied elsewhere. According to Victorian ethnographic data, many savages appeared 'utterly licentious,' as we have seen, enjoying 'communal marriages,' where, as Darwin quaintly put it, 'all the men and women in the tribe are husbands and wives to one another.'[215] This implied that the basic pattern of human sexuality in both men *and* women was promiscuous.

Darwin veered to and fro on human promiscuity. He approvingly quoted Henry Sidgwick that most living 'savages' were promiscuous, deferring to the three ethno-logical authors who had studied the marriage-tie 'most closely'—John McLennan, Edward Tylor, and John Lubbock. All three had concluded that: 'very loose inter-course was once extremely common throughout the world.' Darwin then de-murred, arguing that, 'from what we know of the jealousy of all male quadrupeds, armed, as many of them are, with special weapons for battling with their rivals, that promiscuous intercourse in a state of nature is extremely improbable.'[216] Finally, having canvassed inconclusive findings about polyandry and polygamy in apes and 'savages,' he summed-up agnostically: 'Whether savages who now enter into some form of marriage, either polygamous or monogamous, have retained this habit from primeval times, or whether they have returned to some form of mar-riage, after passing through a stage of promiscuous intercourse, I will not pretend to conjecture.'[217]

6.4.6 Cultural plotlines

Much in the theatre of human desire was framed by culture, according to *Descent*.[218] This was particularly important in Darwin's argument that the differ-ences of skin-colour and physiognomy which Victorians[219] used as marks of race were superficial and hence inessential to humanity. Cultural forces were also pre-sumably responsible for freeing up women's choice of sexual partners in the civil-ized societies of Era 3.

Descent mostly emphasizes culture as cause of the stark contrasts in how dif-ferent peoples around the world set their standards of beauty. How else to explain that, 'the Chinese of the interior' thought Europeans 'hideous, with their white

[214] Darwin, *Descent*, pp.599–600.
[215] Darwin, *Descent*, pp.587–589.
[216] Darwin, *Descent*, p.590. NB this was a weak argument, however, as female monogamy *enforced* by male jealousy implies no lack of polyandrous desires in female apes.
[217] Darwin, *Descent*, p.591.
[218] Note, however, that Darwin only began using the word 'culture' in *Descent* after he had finished writing about sexual selection (see Chs 7–8). Before that, his nearest synonym where humans were con-cerned was 'habits of life.'
[219] And post-Victorians.

skins and prominent noses,' while, when the 'Batokas knock out ... the two upper incisors' to enhance their beauty, Europeans think this 'gives the face a hideous appearance'?[220] *Descent* answers with a thought-experiment. It asks us to imagine a tribe colonizing an unoccupied continent. The tribe would 'soon split up into distinct hordes,' separated from each other by both physical barriers, and still more effectually, by 'the incessant wars between all barbarous nations.' Each horde would then experience slightly different conditions and habits of life, and would sooner or later 'come to differ in some small degree.' As a result, each isolated tribe would form for itself a slightly different standard of beauty; and then, 'unconscious selection would come into action through the more powerful and leading men preferring certain women to others.' Thus the skin-deep but inessential differences between the tribes, 'at first very slight, would gradually and inevitably be more or less increased'—producing distinct races, but not different species.[221]

Culture also explains, or might ameliorate, some secondary sexual differences. For example, to afford women intellectual equality with men, they ought, *when adult*, 'to be trained to energy and perseverance,' and have their 'reason and imagination exercised to the highest point.'[222] They would then transmit these qualities to their adult daughters. And, provided such well-schooled daughters were more fecund than less educated women—a somewhat bizarre proposition—women would eventually rival men in their achievements.

6.4 Kinks and Muddles

If the previous sections show anything, they show that *Descent* does not tell a simple story about desire and sexual difference. I have drawn out four plotlines that structure the theatre of sexual agency in animals, as described in *Descent* Part II, and five, sometimes-related plots relating to human sexuality (from *Descent* Part III). No sooner do we drill down to Darwin's actual examples, however, and there is yet greater profusion of stories and characters.

The link between *Descent*'s Parts II and III remains largely implicit. The most charitable way to read the book is to assume that Darwin put faith in his readers to *themselves* extrapolate, or as he says 'extend,' the standard animal plot from Part II and so make sense of, say, the elaborate self-ornamentation of savage men—by invoking female choice, in that case.[223] Generally, academic commentators have

---

[220] Darwin, *Descent*, pp.575, 578.

[221] Darwin, *Descent*, p.596.

[222] Darwin, *Descent*, p.565. As we saw earlier, Darwin's later, and simpler—though equally cultural—solution to women's under-achievement was: let them become 'bread-winners.'

[223] The 'Introduction' to *Descent* says nothing about the relation between Parts II and III. The 'General Summary and Conclusion' says only this: 'The reader who has taken the trouble to go through the several chapters devoted to sexual selection, will be able to judge how far the conclusions at which I have arrived are supported by sufficient evidence. If he accepts these conclusions he may, I think, safely extend them to mankind': p.617.

lacked such charity. Hence, given the narrative complexity of the book, any inter-preter who tries to impose a spurious consistency on it rides for a fall. Either s/he must leave out the large swathes of *Descent* which conflict with his or her pet gloss.[224] Or s/he must invest heavily in the idea that *any* reading of any book inevit-ably has numerous exceptions—exceptions which, should they prove too onerous, can always be blamed on Darwin.

Beyond doubt, the search for a single consistent message in *Descent* can make it contradict itself in the starkest manner. Did Victorian men have 'power of selec-tion' in sexual matters? Cherry-pick your quotations, and the answer can be made a bald Yes, or a resounding No. Was Era 2 one of endemic rape? Again, both Yes and No answers make themselves available. Were men intellectually superior to women in Darwin's view? Or vice versa? Either answer can be rendered plausible. Thus are 'the contradictions piled up.'[225] When the blame for them gets dumped at Darwin's door, his book may be belittled or dismissed. On the other hand, even when we recognize the multivalence of *Descent*'s account of sex, and distinguish the many contrasting plotlines structuring the theatre of sexual agency it portrays, we cannot escape the suspicion that the muddles and contradictions in *Descent* self-serve.

Run a ruler over the logic in Darwin's discussions of male and female, women and men, and *a pattern* of gaps and fallacies emerges. Darwin truncates, or simply does not pursue, searches for evidence and argument that look like leading him into *obvious* self-contradiction—or into unpatriarchal impropriety—especially concerning what he wanted to say about women. Alternatively, where evidence lacked, he sometimes assumed that what he could not illustrate conformed to the conclusion he wanted to draw. This is true even in Part II. For example: 'It is not known that male crustaceans fight together for the possession of the females, but it is probably the case; for with most animals when the male is larger than the female, he seems to owe his greater size to his ancestors having fought with other males during many generations.'[226] Part III far outdoes Part II in this type of complication.

The fact that different plotlines may be applied to the same sex difference cre-ates muddles for readers who seek a single message in *Descent*. These are exacer-bated in the human case by its 'general rule' for mammals that 'characters of all kinds are inherited equally by the males and females.' 'We might therefore ex-pect,' it goes on, 'that with mankind any characters gained by the females or by the males through sexual selection, would commonly be transferred to the offspring of both sexes.'[227]

[224] For example, Desmond and Moore, *Darwin's Sacred Cause: Race, Slavery and the Quest for Human Origins*(p.xix) hold that Darwin's 'core project' in *Descent* was to combat slavery—a premise that hardly explains the contents of Part I (see Ch. 7), which they therefore dismiss as 'a side engagement' (p.370).

[225] Richards, *Darwin*, p.452.

[226] Darwin, *Descent*, p.268.

[227] Darwin, *Descent*, p.585.

This rule creates a logical problem. Even when an ornament does appear 'useless', or harmful to success in the general struggle for existence, and therefore likely to be a product of sexual selection—to which sex is the reproductive benefit of the ornament to be attributed? It may first have been acquired by men, for some reason, and then transferred to women. Or it may have been acquired by women, presumably for a different reason, and then been transferred to men. For example, as already seen, Darwin suggested that the relative hairlessness of women, as compared to men, was acquired by women to charm men—referencing the hirsute character of other terrestrial mammals and therefore arguing that women's greater nakedness was the more evolved character. But the relative hairiness of men might equally have evolved to excite women, *especially as human children are also relatively hairless*. This fact Darwin neglects to mention, implying as it does—according to his bird-based rules of inheritance—that men's body-hair evolved to titillate women.

When an inter-sexual difference is only one of degree—as all human differences are—*Descent* frequently forgets that selection affecting either sex, or both, is equally plausible. Thus, following the standard female-controlled plot, Darwin keenly sought evidence that men were physically more 'variable' in their secondary sexual characters than women. Extrapolating from animals, this would allow him to conclude that men were more 'modified' or *evolved* than women, sometimes perhaps resulting from macho match-ups, but largely from women's capricious standards of taste in different tribes.[228] So we find *Descent* arguing:

> In the excellent observations made on board the [Austro-Hungarian frigate] *Novara*, the male Australians were found to exceed the females by only 65 millim. in height, whilst with the Javans the average excess was 218 millim.; so that in this latter race the difference in height between the sexes is more than thrice as great as with the Australians ... these measurements shew that the males differ much more from one another than do the females. This fact indicates that, as far as these characters are concerned, it is the male which has been chiefly modified [evolved], since the several races diverged from their common stock.[229]

Yet this difference in variability is not between the *absolute* range of the heights of women, compared with the *absolute* range in men. The two numbers are measures of the *relative, subtracted* differences, *between* the sexes, to which variations in the absolute heights *of both men and women* may have contributed. From the evidence collected on *Novara*, it would be perfectly possible for the two measures of inter-sexual difference to have been entirely due to the *women* varying massively in stature, while all the men were of identical height!

---

[228] E.g. Darwin, *Descent*, p.224: men's ears vary more than women's. This implies women have 'the power of selection', not men.
[229] Darwin, *Descent*, pp.559–560.

This kind of *confirmation bias*—plumping for the interpretation you want over other equally plausible ones—riddles *Descent*. One more example. Part II sports a whole section on sex-smells in mammals, with one overriding aim: to survey species in which 'the male emits a strong odour during the breeding-season ... to excite or allure the female.'[230] Elephants, goats, musk ox, antelopes, and musk-deer all get mentions, all being species in which scent-glands are either smaller or absent in females. The fact that male scent-glands are 'large and complex' shows they must be 'of considerable importance to the male,' wrote Darwin. And this is made intelligible by sexual selection: 'the most odoriferous males are the most successful in winning the females, and in leaving offspring to inherit their gradually perfected glands and odours.' End of section.

This discussion brims with ellipses and silences. It admits there are many mammalian species in which 'the glands are of the same size in both sexes' (e.g. the beaver), but immediately breaks off this line of evidence, saying: 'but their uses are not known.'[231] Even in the species it does highlight, it notes that, in most cases, the females also have scent glands.

So, given *Descent* acknowledges the existence of homologous scent-glands in the males and females of 'many' species, why is the topic of female scent-glands ignored? Include it, and Darwin could scarcely have avoided finding that females 'allure and excite' males by their smell, just as males do females. Which harks back to a comment in Darwin's early notebooks: 'We need not feel so much surprise at male animals smelling vaginae of females when it is recollected that smell of ones own pud [pudendum, genitals, is] not disagree[able].'[232]

Why does this line of observation not appear in *Descent*? Might it be because Darwin thereby avoided highlighting active desire in females?

Confirmation bias underlines the destructive, befuddling impact of the social concerns that besieged Darwin's thinking about sex (6.3.1 Nodal concerns). Presumably, it was these concerns that produced the extraordinary contortions of logic and writing that erupt in Part III of *Descent*. We may suppose that, given the calibre of Darwin's other work, the Victorian mores surrounding sex were so threatening as to make certain conclusions about male–female differences and relations unthinkable to him (and/or his peers)—especially when we lay some of the statements in *Descent* alongside Darwin's less-public attitudes and actions. Darwin's letters show him to have been almost always respectful and laudatory to women who were fellow-naturalists. He asked them to conduct experiments for him, reported their results in his books, and assisted in the publication of their findings.[233]

---

[230] Darwin, *Descent*, pp.528–530.
[231] Darwin, *Descent*, p.529.
[232] Charles Darwin, 'Notebook M,' (Cambridge: Darwin Online, 1838), p.85.
[233] See Samantha Evans, ed. *Darwin and Women: A Selection of Letters* (Cambridge: Cambridge University Press, 2017), e.g. p.116

Samantha Evans argues that, writing in 1871, Darwin 'couldn't have been unaware of the problems of bias and social disadvantage' experienced by Victorian women, which were hotly contested in the 1860s and 1870s, even among Darwin's close family. His brother Erasmus was a successful advocate of women's education, and his wife and daughters were well versed in feminism.[234]

Darwin's befuddlement can be observed at various levels in *Descent*'s arguments, from the broad-brush historical to the fine-grained grammatical. For example, the fact that he used contemporary ethnographic evidence about 'savages' to reconstruct *both* Era 1 *and* Era 2 of human evolutionary history invited a host of slippages into his argument. Note, for instance, Darwin's beguiling ab/use of tense in stating: 'Man is more powerful in body and mind than woman, and in the savage state he keeps her in a far more abject state of bondage, than does the male of any other animal; therefore it is not surprising that he should have gained the power of selection.'[235] Because he held, *both* that humans still existed 'in the savage state' in 1874 (e.g. in Australia and Tierra del Fuego), *and* that Era 2 characteristics laid the groundwork for Era 3 sex differences, he could use the present tense ('Man is') followed by the present perfect tense ('should have gained'), thereby encouraging the misleading implication—contradicting several other human plotlines, and his distinction between Era 2 savagery and Era 3 civilization—that *all* men *continue* to retain 'the power of selection,' Era 3 Britons included.[236] Yet this contradicts statements *Descent* makes elsewhere, for example, that 'in civilised nations women have free or almost free choice.'[237]

Perhaps the most tortuous grammar in *Descent* treats of the question whether a better education for Victorian girls could help raise them to men's 'superior' level of intellect. The relevant paragraph begins by referring to two predominantly bird-based rules: that characters appearing in only one sex—like the peacock's tail—tend to be first expressed during maturity; whereas characters found in both sexes, tend to appear early in development. However, the passage goes on, these avian rules do not always obtain, or 'hold good,' particularly in mammals:

> *If they always held good*, we might conclude ... that the inherited effects of the early education of boys and girls *would be* transmitted [i.e. developed] equally [in] both sexes; so that the present inequality in mental power between the sexes *would not be* effaced by a similar course of early training; *nor can it have been caused by their dissimilar early training.*[238]

---

[234] Samantha Evans, 'Preface,' in *Darwin and Women: A Selection of Letters*, ed. Samantha Evans (Cambridge: Cambridge University Press, 2017), p.xxv. The National Society for Women's Suffrage was founded in 1869: Endersby, 'Darwin on Sex,' p.85.

[235] Darwin, *Descent*, p.597.

[236] NB The next sentence in this paragraph also continues in the present tense: 'Women *are* everywhere conscious of the value of their own beauty ...'

[237] Darwin, *Descent*, p.586.

[238] Darwin, *Descent*, p.565.

Descent begins the above quote using a conditional tense: *IF* the avian rules of inheritance *did* always hold good with humans, *THEN* the 'superior' intelligence of men *would be* developed in both men and women, because the great educational advantages enjoyed by Victorian boys occur *early*, before puberty. Use of the conditional tense here makes it sound like Darwin was open to the avian rules being jettisoned in the human case. In this vein, he should have gone on to consider alternative *non-avian* explanations for Victorian men outperforming their sisters, namely: that Victorian boys got better schooling than did girls. *But this avenue of alternatives is summarily barred from discussion.*

The barring results from Darwin's unwarranted grammatical shift at the end of the quoted sentence from the conditional to an indicative, factual, assertion: 'nor CAN [*can* should read *could*] it have been caused by their dissimilar early training.' If Darwin *had* examined both the options his grammar at first invoked—that the avian rules might or *might not* hold in humans—he would have had to entertain the possibility that women's purported intellectual under-achievement when compared to men *was due* to Victorian girls' 'dissimilar early training,' compared to boys. For Darwin, this was evidently a bridge too far.

As for the law of sexual selection itself, *Descent* tells us that the characters it describes have arisen independently of those subsumed under natural selection, because secondary sexual characters have added no advantage in the struggle for survival, as in: the evolution of men's beards; or the relative 'roundness' of women's faces.[239] At other times, both types of selection may apply to a single character, as for example when hen-birds do not share the gaudy colours of their cocks, and thus escape notice when sitting on a nest. Other secondary sexual characters had evolved *despite* the processes giving rise to natural selection, as with the enormous, unwieldly antlers of stags, or the relatively naked, unprotected, skin of humans.

The fact that the laws of natural and sexual selection may sometimes apply to the same sexual difference creates confusion, because the dynamics producing sexual and natural selection have different implications for sex differences. With animals, Darwin would admit that the two kinds of selection could not always be disentangled, as with 'the disappearance of the spots and stripes' on the bodies of piglets and fawns, when they mature.[240] (This might be natural selection, if loss of the markings made the adult animals less conspicuous; or it might be sexual selection, if the adult coats were deemed sexual ornaments.) But when discussing human sex differences, Darwin often failed to mention that there were two competing explanations for how a character had evolved. For example, the strength, courage, and pugnacity that *Descent* Part III attributes more to men than women, is explained as a result of the 'law of battle' between men to gain possession of women: that is, sexual selection. Yet *Descent*

---

[239] Darwin, *Descent*, p.578.
[240] Darwin, *Descent*, p.547.

also claimed, especially in Part I—but also occasionally in Part III—that men's greater strength, pugnacity, and courage than women fell under the law of natural selection, because, over long epochs, primeval and savage men would have benefited from strength and pugnacity when defending and protecting the women and young of their communities against marauders.[241] Battle is also said in Part III to have been the origin of men's intellectual superiority over women. So this is a significant ambiguity.

The picture is further muddied by a case Darwin made when discussing 'civilized people': that natural selection covered men's 'superiority' over women, because, over countless generations, men had, as *Descent* repeatedly remarks, 'worked harder than woman for his own subsistence and that of his family.'[242] Yet elsewhere we hear that this was only a recent phenomenon: 'women are made by [contemporary] savages to perform the greatest share of the work'—a state of affairs that must therefore have obtained over a considerable portion, if not the vast majority, of human history, if Eras 1 and 2 were modelled on 'savages.'[243] In which case, as hard work breeds intelligence and strength—which Darwin claimed for men—women should exceed men intellectually and physically.

6.5 Resonances

Descent's treatment of sexual selection has resounded widely across contemporary scholarship, and hence intellectual and popular culture—not always positively. I will now draw out some key resonances.

6.5.1 Feminism

Resonances of *Descent* within feminism are legion. Early on, Antoinette Blackwell (1825–1921) and Eliza Gamble (1841–1920) both deplored the inferiority *Descent* projected onto women, arguing the book had employed a method tainted by male bias. This showed, they said, that the study of women should in future be conducted *by women*. Blackwell herself reworked Darwin's understanding of evolution to conclude that women and men were different but equal. Gamble argued that Darwin had failed to recognize that the power of female choice in animals extended to humans. A third Victorian feminist, Charlotte Perkins (1860–1935), took a different tack. She used Darwin's work to stress the moral superiority of women over men

---

[241] Darwin, *Descent*, pp.107, 564.
[242] Darwin, *Descent*, p.563.
[243] Darwin, *Descent*, p.577.

with regard to cooperation and altruism. She concluded that human progress depended on the removal of barriers to the education of women.[244]

Today we see equally various views. From the 1970s onwards, feminists have cast Darwin's take on human sexuality as aimed to immure men in a position of unshakeable biological superiority over women. Some like Ruth Hubbard label Darwin's views 'blatant sexism.' Others deem such charges simplistic—though even they find in *Descent* an 'embedded conventional objectification of passive female sexuality.'[245]

At odds with such views, feminist philosopher Elizabeth Grosz applauds *Descent* as birthing a new kind of science, where sexual difference becomes relative, not absolute, being locally produced—and productive—in multifarious ways, such that male and female have no in-built essences.[246] Most importantly, writes Grosz, *Descent* focuses primarily on erotic attraction and excess: the forces of desire bend less on reproduction than on gaining access to those objects of desire that lusting animals deem most attractive.[247] *Descent* thus makes the 'incalculable force' of desire and sexual appeal responsible for all the rich variety of life on earth, and particularly its 'perceptible beauty and charm.'[248] So Darwin is enrolled, 'unbeknownst to himself,' as the first feminist of difference![249]

6.5.2 Anthropomorphism

Critics who find *Descent*'s treatment of gender to have resulted from projection onto nature of Victorian conventions of sexuality and courtship, propose anthropomorphism as the scientific disqualification of such projection.[250] This is a long-lived criticism that has aged badly.[251] True, for all but the final decade of the twentieth century, anthropomorphism was widely branded as a cardinal sin in science—particularly in psychology.[252] Today however, the tables have turned,

[244] Montgomery, 'Gender,' pp.443–450; Rosaleen Love, 'Darwinism and Feminism: The Woman Question in the Life and Work of Olive Schreiner and Charlotte Perkins Gilman,' in *The Wider Domain of Evolutionary Thought*, ed. David Oldroyd and Ian Langham (Dordrecht: Reidl, 1983), pp.113–132.
[245] Montgomery, 'Gender'; Richards, *Darwin*, pp.216–217, 526.
[246] Elizabeth Grosz, *Becoming Undone: Darwinian Reflections on Life, Politics and Art* (Durham N.C.: Duke University Press, 2011), p.117.
[247] Grosz, *Becoming Undone*, p.124.
[248] Grosz, *Becoming Undone*, pp.141–142.
[249] Grosz, *Becoming Undone*, p.142.
[250] E.g. Dawson, *Victorian Respectability*, pp.35, 76; Richards, *Darwin*, pp.440, 452.
[251] However popular it remains: see also Roger Smith, *Free Will and the Human Sciences in Britain 1870–1910* (Pittsburgh PA: University of Pittsburgh Press, 2016), p.8.
[252] The growth of *scientific* psychology was premised on a rejection of the kind of natural histories constructed by Darwin and his peers, partly *because* these seemed so anthropomorphic: James Angell, 'The Influence of Darwin on Psychology,' *Psychological Review* 16, (1909), pp.152–169; Donald Hebb, 'Emotion in Man and Animal: An Analysis of the Intuitive Processes of Recognition,' *Psychological Review* 53, (1946), pp.88–106; Alan Costall, 'How Lloyd Morgan's Canon Backfired,' *Journal of the History of the Behavioural Sciences* 29, (1993), pp.113–122.

making it anthropo*centric* to assume that humans alone have sophisticated mental capacities.[253] This brings us back to Darwin's argument that both humans and animals comprise a single domain of explanation. Humans may serve as heuristically valuable *frontier instances* of this domain in some regards: for example, as illuminating most clearly the relationship between intellectual operations and agency in the organic world. Other species serve as frontier instances in other regards (see 4.3.5 Comparative evidence). But, overall, all creatures are netted together in the same fine mess.[254]

This argument converges with another. We can only think anthropomorphism a viable concept or fault if we already know 'what the mental differences are between limpet and Locke.'[255] Which we do not. In the meantime, given that scientists primarily use language to communicate with other scientists, and that human language principally serves to make meaning *for* humans *about* humans, any worded description of animal action cannot help but be anthropomorphic. *Descent* makes exactly this point when defending its extension of terms like *choice* to animals—arguing that, in the absence of a common language with animals, we are like 'an inhabitant of another planet' looking at their courtship: 'we can judge of choice being exerted, only by analogy.'[256] Hence, using 'hugs infant' and 'infant trusts'—rather than some cryptic code like '¥174'—to tag my observation of a baboon cuddling its baby, may well constitute the most practical way of communicating to other scientists what I observe to be going on. Furthermore, given that the complexities of trust are something humans know about, such a description is more likely to throw up questions, doubts, tests, and hypotheses than labelling the baboon's action '¥174'[257]—especially as '¥174' would eventually have to be defined in ordinary words—at which point science falls back into the same old anthropomorphic mire.

Witness, for example, Bert Hölldobler and Edward Wilson's devotion of a whole section of their beautiful book on ants to refuting Pierre Huber's (1777–1840)

---

[253] Frank Beach, 'The Snark Was a Boojum,' *American Psychologist* 5, (1950), pp.115–124; Frans De Waal, 'Are We in Anthropodenial?,' *Discover Magazine*, no. July 1st (1997), http://discovermagazine. com/1997/jul/areweinanthropod1180. See also Graham Richards, *On Psychological Language and the Physiomorphic Basis of Human Nature* (London: Routledge, 1989); Lorraine Daston and Gregg Mitman, eds., *Thinking with Animals: New Perspectives on Anthropomorphism* (New York: Columbia University Press, 2005). For a rear-guard defence of the traditional behaviourist position, see Clive Wynne, 'What Are Animals? Why Anthropomorphism Is Still Not a Scientific Approach to Behaviour,' *Comparative Cognition & Behavior Reviews* 2, (2007), pp.125–135 and attached peer commentaries. For an earlier argument *for* anthropomorphism, see Maurice Mandelbaum, 'A Note on 'Anthropomorphism' in Psychology,' *Journal of Philosophy* 40, (1943), p.47: 'As long as we believe that there is a significant continuity between animal and human behaviour, we have as much right to approach animal behaviour through what we can discover about human behaviour as to proceed in the opposite direction.'

[254] Darwin, 'Notebook B,' p.232.

[255] Richards, *Psychological Language*, p.3.

[256] Darwin, *Descent*, pp.420–421; see also Richards, *Psychological Language*, passim.

[257] For an example of the way male baboons who are 'potential losers' foster 'trust'-based 'friendships' with infants that help them ward off aggression from other males, see: Shirley Strum, 'Darwin's Monkey: Why Baboons Can't Become Human,' *Yearbook of Physical Anthropology* 55, (2012), pp.S3–23.

labelling as *play* his descriptions of ants: 'embracing one another'; 'falling and scrambling up again'; 'seizing a mandible, a leg, or antenna, and letting it go immediately ... without appearing to inflict an injury.'[258] Darwin delightedly glossed Huber's description as 'ants chasing and pretending to bite each other, like so many puppies.'[259] The section in Hölldobler and Wilson's book called 'Ants Do Not Play' does not dispute Huber's *data*: it recasts them as 'territorial war' fought 'in deadly earnest'—which is an equally anthropomorphic description![260]

Finally, critics who slight the scientific quality of *Descent* on grounds of its anthropomorphism set themselves the problem of explaining how—if anthropomorphism is such a grievous fault in science—many of the findings and principles *Descent* espouses still retain credibility in biology today (see 6.5.4 Evolutionary biology; 6.5.5 Evolutionary Psychology; Ch. 7).

6.5.3 Psychoanalysis

Frank Sulloway has catalogued the resonances of early psychoanalysis with *Descent*. His list includes: the periodicity of physiological functions (e.g. incubation in birds and menstruation in humans) as being traced back to the intertidal habitat of the larval sea squirts from which vertebrates were descended, according to *Descent*; Darwin's close interest in child development and its uptake by his protégé George Romanes and others to suggest various animal-like stages of development, including early sexual stages, and the early expression of animal-like instincts; Freud's and Darwin's common emphasis on the primitive hermaphroditism or bisexuality of ancestral (and embryonic) mammals and humans; Darwin's stress on two often-conflicting evolutionary imperatives—self-preservation (ego-preservative instincts) and mating (sexual instincts); the associated idea of instinctual conflict (and struggle) as basic to psychic life; the past as interwoven in the present[261]; the notion of instinctual 'fixation' as a result of physiological anomaly or early experience/trauma; and Darwin's stress on the irrational side of human behaviour (see Ch. 7).[262]

---

[258] Huber's words taken from Bert Hölldobler and Edward Wilson, *The Ants* (Berlin: Springer-Verlag, 1990), pp.398–9, 406–413 & passim.

[259] Darwin, *Descent*, p.69.

[260] The way Hölldobler and Wilson, *Ants*, pp.406–413 re-describe Huber's data is further undermined by their admission that the 'territorial wars' said to be conducted by many ant species are more often 'display tournaments'—like sporting competitions—in which 'almost no physical fights occur.' Such tournaments typically involve hundreds of ants who perform mass fly-bys, or ritualised 'jerking displays,' where ants act 'like dancers' moving on 'stilt-like legs.' All of which takes us back to Huber, albeit with the word *display* replacing *play*.

[261] See 'counter-transference' and 'deferred action' in Jean Laplanche and Jean-Bertrand Pontalis, *The Language of Psycho-Analysis* (London: Hogarth Press, 1983).

[262] Frank Sulloway, *Freud, Biologist of the Mind: Beyond the Psychoanalytic Legend* (Cambridge MA: Harvard University Press, 1992), pp.238–276 & passim.

6.5.4 Evolutionary biology

While not all evolutionary biologists see eye-to-eye on sexual selection, both of Darwin's proposed dynamics—intra-sexual rivalry, and inter-sexual courtship-and-choice—have gained support over the last three decades of research on animals.[263] But controversy boils on.[264] Wallace never subscribed to Darwin's theory of sexual selection. Likewise, Julian Huxley argued in 1938 that sexual selection was 'merely an aspect of natural selection,' the aspect concerned with characters which 'subserve mating, and are usually sex-limited.'[265] More recent biologists argue that effective secondary sexual characters are *honest signals*, because they show the attractive animal has 'good genes.'[266] This argument also subsumes sexual under natural selection—though being light on supporting evidence.[267]

Critics of the hypothesis that sexual ornaments act as honest signals insist that natural and sexual selection involve distinct processes, sexual selection resulting from processes having intrinsically social dynamics, unlike natural selection.[268] Yet *Descent* does occasionally hint at an overlap of sexual with natural selection, as when it notes that females often 'prefer the more vigorous and lively males,' adding that, 'the great vigour of the male during the season of love seems often to intensify his colours.'[269] In game-cocks, for instance, 'the female almost invariably prefers the most vigorous, defiant, and mettlesome male.'[270]

For Darwin, sexual selection was unlike natural selection in that it showed *no limit* to the amount of advantageous modification it could produce, 'so that

---

[263] Richard Prum, *The Evolution of Beauty: How Darwin's Forgotten Theory of Mate Choice Shapes the Animal World—and Us* (New York: Doubleday, 2017); Mary Jane West-Eberhard, 'Darwin's Forgotten Idea: The Social Essence of Sexual Selection,' *Neuroscience and Biobehavioral Reviews* 46, (2014), pp.501–508; Malte Andersson and Leigh Simmons, 'Sexual Selection and Mate Choice,' *Trends in Ecology and Evolution* 21, (2006), pp.296–302; Geoff Parker and Tommaso Pizzari, 'Sexual Selection: The Logical Imperative,' in *Current Perspectives on Sexual Selection: What's Left after Darwin?*, ed. Thierry Hoquet (Dordrecht: Springer, 2015), pp.119–164.

[264] E.g. Gail Patricelli, Eileen Hebets, and Tamara Mendelson, 'Book Review of R.O. Prum, 'The Evolution of Beauty,' *Evolution* 73, (2018), pp.115–124; Ferris Jabr, 'How Beauty Is Making Scientists Rethink Evolution,' *New York Times Magazine*, January 9th 2019.

[265] Julian Huxley, 'The Present Standing of the Theory of Sexual Selection,' in *Evolution: Essays on Aspects of Evolutionary Biology*, ed. Gavin De Beer (Oxford: Clarendon Press, 1938), p.34.

[266] John Maynard-Smith, 'Theories of Sexual Selection,' *Trends in Ecology and Evolution*, (1991), pp.146–151; Christopher Chandler, Charles Ofria, and Ian Dworkin, 'Runaway Sexual Selection Leads to Good Genes,' *Evolution* 67, (2012), pp.110–119

[267] E.g. Diego Gil and Manfred Gahr, 'The Honesty of Bird Song: Multiple Constraints for Multiple Traits,' *Trends in Ecology and Evolution* 17, (2002), pp.133–141; Anne Peters et al., 'No Evidence for General Condition-Dependence of Structural Plumage Colour in Blue Tits: An Experiment,' *Journal of Evolutionary Biology* 24, (2011), pp.976–987; Richard Prum, 'The Lande–Kirkpatrick Mechanism Is the Null Model of Evolution by Intersexual Selection: Implications for Meaning, Honesty, and Design in Intersexual Signals,' *Evolution* 64, (2010), pp.3085–3100.

[268] West-Eberhard, 'The Social Essence.'

[269] Darwin, *Descent*, p.224.

[270] Darwin, *Descent*, p.417.

as long as the proper variations arise, the work of sexual selection will go on.' This circumstance helped to explain the 'extraordinary' amount of variability seen in secondary sexual characters, and the excessiveness of such ornaments as the peacock's hyperbolic queue.[271] *Descent* also instances the 'wonderful extreme' to which the development of antlers in the stag has gone, making them 'singularly ill-fitted for fighting'—their ostensible purpose. Stags' antlers had gradually been rendered 'longer and longer, through sexual selection, until they acquired their present extraordinary length and position.'[272] They are now so big that stags fight kneeling down![273] In 1930, Ronald Fisher suggested a mathematical model for this continuous 'exaggeration' of ornaments through sexual selection. The result has been called *runaway selection*. Fisher's formulation has since been mathematically refined, but is hard to confirm with data from real life.[274]

Elsewhere in *Descent*, Darwin examined the evolutionary importance of the ratios of males to females in a species. He had initially assumed that, because males typically compete with each other for access to females, most animal populations where secondary masculine sexual characters were well developed would show a ratio in which the males would 'considerably exceed' the females. His extensive research disproved this assumption.[275] He largely found a 1:1 male–female ratio. Some departures from this ratio could be accounted for as natural selection on grounds of a range of miscellaneous causes, but, overall, the picture he drew proved so intricate as to elude general explanation.

In recent decades, debates in evolutionary biology about sex ratios have grown hot. Many species have been found to have a preponderance of females (e.g. water-dwelling crustaceans called Copepods). Indeed, the 'unbeatable strategy' for the highest average fitness in a population is sometimes said to be a moderately *female*-biased sex ratio. This involves a compromise between a 1:1 ratio, as favoured by individual-based selection, and a *heavily* female-biased ratio, as favoured by group selection. A higher proportion of females than males is favourable because population growth is limited by the number of eggs, not the number of sperm. Yet, while, a female-weighted sex ratio benefits the group or population as a whole, it does not benefit the *relative* fitness of individuals within the group. Hence the need for a compromise.[276]

[271] Darwin, *Descent*, p.226.
[272] Darwin, *Descent*, pp.227, 510.
[273] Darwin, *Descent*, p.509.
[274] Chandler, Ofria, and Dworkin, 'Runaway Sex.'
[275] Darwin, *Descent*, pp.213, 242–260.
[276] Elliott and Wilson Sober, David, *Unto Others: The Evolution and Psychology of Unselfish Behaviour* (Cambridge MA: Harvard University Press, 1998), pp.38–43.

6.5.5 Evolutionary psychology

Traditionally, psychology texts make little space for sex. Sex *differences* have always attracted comment. But the dynamics of lust have long been treated as almost taboo, except in psychoanalysis (which is itself viewed askance by many experimental psychologists). The recent rise of Evolutionary Psychology (particularly, the Santa Barbara version discussed in Ch. 3) has changed all this. It has been estimated that three-quarters of research by evolutionary psychologists deals with sex.[277] Chapters 3 and 8 underline some key differences between the theses in *Descent* and the claims of the Santa Barbara school—particularly surrounding their different takes on genes and culture. However, there are also commonalities. For example:

Darwin's female-controlled plot for sexual selection in animals derived the greater activity or eagerness of males from an argument about relative *physiological* investment in relatively simple animals and plants ('Mr Ever Eager'). In a similar vein, the fact that women get pregnant and men do not has been made the crux of what is today called *parental investment theory*.[278] This proposes that animals in which the relative investment per offspring varies between males and females will show a concomitant dimorphism in mating strategy. In humans, Stone Age men are held to have invested so little in bearing and raising each of their children that they mated almost indiscriminately (a short-term strategy: the more seeds one sows, the more children will result). Stone Age women, on the other hand, will not only have had to sustain nine months' pregnancy per child, but had to breast-feed their babies, and (it is assumed) also took most of the responsibility for raising each child until he or she became independent. A mother thus would have invested much more time and physiological energy when raising her child than did its father. Women will therefore have evolved to be far more circumspect about whom they mate with than have men: they will be keen to find a mate who will stick around and support them as they embark on child-rearing. Hence, unlike men, they will not be innately promiscuous but adopt a long-term mating strategy, focused on finding and keeping a well-resourced and reliable man.

Notable in the language of today's parental investment theory is a disconnect between courtship and parenting. First put forward by Robert Trivers in 1972, the theory states: '*what governs the operation of sexual selection* is the relative parental investment of the sexes *in their offspring*.' Thus Trivers goes on to observe that 'males usually invest almost nothing in their offspring.'[279] This implies that the dynamics of courtship are determined by *post-conception* parental investment.

---

[277] David Buss, ed. *Handbook of Evolutionary Psychology* (Hoboken NJ: Wiley, 2005).

[278] Geary, 'Parental Investment.'

[279] Robert Trivers, 'Parental Investment and Sexual Selection,' in *Sexual Selection and the Descent of Man, 1871–1971*, ed. Bernard Campbell (Chicago: Aldine Press, 1972), p.141, italics removed.

In contrast, *Descent* argues that courting males have to put far more effort than females into rushing around to fight, or posturing to scare off, their same-sex rivals—after which they then have to perform all sorts of energetic antic to arouse a female—*before they ever get to mate*. Meanwhile, their brides-to-be typically expend far less energy, neither fighting nor flamboyantly parading their charms. *After* conception, investment reverses, with females spending more energy than males in the nourishment and protection of their young. Thus Darwin focuses on the energetics of the reproductive cycle *as a whole*, not simply on investment in parenting. Recent research goes against Trivers—and for Darwin—showing that courtship and child-care *do* form a single package: the more energetic the competition among one sex (often males) to copulate, the less parental investment by that sex subsequent to mating, and *vice versa*.[280]

6.6 Conclusion

A cat may look at a king yet never relinquish its own catty world of hunger and grace, animus and the pursuit of bodily comforts—remaining oblivious to the concerns of state, marital tensions, parental responsibilities, and whatever other preoccupations trouble the monarch. Today we must read charitably and with an enlarged curiosity, if we are to decentre sufficiently to make sense of how *Descent* portrays sexuality. Because here, as in Dickens, George Eliot, and the *Origin*, only more so: 'the unruly superfluity of Darwin's material at first gives an impression of superfecundity without design. Only gradually and retrospectively does the force of the argument emerge from the profusion of example.'[281]

Descent's accounting for sexual selection is not simple but manifold. Four contrasting plotlines structure its portrayals of animal desire—female-controlled, male-against-male, inverted, and symmetrical—each with its own character-types. While these plots implicitly extend from Part II into Part III, they get changed and augmented in Darwin's accounting for the complexities of human sexual agency and difference. Darwin's stress on the interdependencies of sexed anatomy with the antics of display, and the interweaving of same-sex emulation with heterosexual desires, makes *Descent* pre-eminently the exposition of a psychology which has irrationality and the sociality of libido at its fore. Accordingly, sexual selection has recently been re-framed in evolutionary biology as a form of *social* selection. This chimes with the central theme of this chapter: 'Signals under sexual and other kinds of social selection ... are favoured *because of their effects on others*, not their

---

[280] Tim Janicke et al., 'Darwinian Sex Roles Confirmed across the Animal Kingdom', *Science Advances* 2, (2016), e1500983; Hanna Kokko and Michael Jennions, 'Parental Investment, Sexual Selection and Sex Ratios', *Journal of Evolutionary Biology* 21, (2008), pp.919–948.

[281] Beer, *Darwin's Plots*, p.42.

value as indicators of quality in other contexts.' To talk of signals having efficacy through their effects on others implies a dependence of sexual selection both on a process of *reading signals* by others, and of the need to act in a manner so as to *be read* in a wished-for way by others.[282] Which links Darwin's account of desire to both his theory of blushing (see Ch. 5), and, as we shall see in Chapter 7, to his theory of conscience.

Humans were for Darwin quintessentially social animals. And sex, via sociality, was responsible for all that made humans 'higher.' If organisms reproduced solely by selfing, wrote the twenty-nine-year-old Darwin, 'there would not be social animals, hence not social instincts, which as I hope to show is probably the foundation of all that is most beautiful in the moral sentiments of the animated beings.'[283] Accordingly, Part I of *Descent* argues at length that social instincts—particularly, sympathy—ground all that is most distinctively human, and humane, about our actions and mental lives. No eros: no mind: no humanity. The implications of sociality for moral action will take centre stage in Chapter 7.

[282] West-Eberhard, 'The Social Essence,' p.502, italics in original, some italics and bolding removed.
[283] Darwin, 'Notebook E,' p.49.

7

Life in Groups

'You brute.' 'You beast.' 'You animal!' The kinds of agency humans share with non-humans are easily disparaged, especially when embodying lust—whether or not we link libido to delight in fine feathers and soaring song. How then to understand human actions of a more refined ilk? Part I of *The Descent of Man* addresses this question by elaborating a natural history embracing all those aspects of agency supposedly unique to *Homo sapiens,* notably: highly advanced intellectual feats; cultural mores; the worship of God; refined aesthetic taste; and use of language. Most crucial for Darwin was to devise a naturalistic account for the highly complex sentiment of conscience, and of the humane morality that, ideally, would extend sympathetic action to 'men of all races, to the imbecile, maimed, and other useless members of society, and finally to the lower animals.'[1]

Sociality forged the master-key to Darwin's take on what he had, as a twenty-nine-year-old, called 'all that is most beautiful in the moral sentiments of the animated beings' (see Ch. 6).[2] Over the years, Darwin identified numerous species as what he called *social animals.* Sociality could lift both the production of natural selection, and agency, to new levels of sophistication—as humans demonstrated. In particular, the fact that we and our ancestors get born into, live, and die *in groups,* gave the open sesame to what most distinguished our psychology. Paralleling *Descent* Part I, this chapter pays most attention to the complex characters of conscience—once I have introduced Darwin's conceptions of social animals, social instincts, and social forms of natural selection.

7.1 Social Animals

Staying with his in-laws, aged thirty, the newly married Darwin—pondering over the origins of human self-knowledge—unexpectedly refers to man as 'a socialist.' This was not a political remark. Darwin was emphasizing the essential sociality of human beings—exploiting a now-archaic sense of the word *socialist.* This remained a truism for him throughout his career. Thirty-five years later, *The Descent*

---

[1] Charles Darwin, *The Descent of Man, and Selection in Relation to Sex* (London: Murray, 1874), pp.124–126, 133.
[2] Charles Darwin, "E," (1838-1839), pp.48–49.

Darwin's Psychology. Ben Bradley, Oxford University Press (2020). © Oxford University Press.
DOI: 10.1093/oso/9780198708216.001.0001

Figure 7.1 Social plants: Grasses (a); Thistles (b).
Source: (a) Aleksander Bolbot / Shutterstock.com; (b) A_Lesik / Shutterstock.com.

of Man cited Marcus Aurelius to the effect that, 'the prime principle of man's moral constitution' is 'social.'[3]

Humans were not the only socialists. In 1835, Charles Lyell's *Principles of Geology* had proposed that cows, dogs, horses, and sheep were social, as only 'social animals' could be domesticated. Solitary species could be *tamed*, but never 'afforded true domestic races.' This was because social animals formed herds, which meant they naturally recognized and obeyed the member who 'by its superiority' had become chief of the herd. Domestication consisted in inducing herd animals to recognize 'man ... as the chief of its herd.'[4]

From his earliest forays in the study of animals, Darwin had shown a liking for social or colonial species: sea pens and coral amongst them. As we saw in Chapter 3, he even applied the adjective *social* to certain species of plant: especially those found in large numbers wherever they grow, even at the limits of their range. Examples were most grasses, and 'cardoons & thistles on the plains of La Plata'

---

[3] Charles Darwin, 'Notebook N,' ed. Paul Barrett (1838–1839), p.109; Darwin, *Descent* p.126; Marcus Aurelius, *The Thoughts of the Emperor Marcus Aurelius Antonius* (London: Bell & Sons, 1880), pp.53, 61, e.g. 'a social state is manifestly the Natural State of man, the state for which Nature fits him ...'

[4] Charles Lyell, *Principles of Geology, Vol.2* (London: Murray, 1835), pp.458–459. James Mackintosh (1765–1832), a family friend of the Darwins who instructed the young Charles in ethical theories, also stated that 'man is ... a social animal,' this being the foundation of his exposition of Grotius: James Mackintosh, *Dissertation on the Progress of Ethical Philosophy, Chiefly During the Seventeenth and Eighteenth Centuries* (Edinburgh: Adam and Charles Black, 1862), p.66. Half a century earlier, Gilbert White, who inspired the boy Darwin to study birds, had penned a whole essay on the topic of 'the wonderful spirit of sociality in the brute creation.' Examples included large winter flocks of gregarious birds, like wild geese, wild ducks, and cranes. But also domesticated species: horses, cows, and sheep—all of whom 'will not bear to be left alone.' In captivity, isolated individuals of a social species would form mutual attachments with members of incongruous orders, as when a horse and a hen made friends: Gilbert White, *The Natural History of Selborne* (Oxford: Oxford University Press, 2013), p.267, and Letter 24 to Daines Barrington (August 15th, 1775), pp.227–229. Cf. Darwin, *Descent*, p.100: 'Everyone must have noticed how miserable horses, dogs, sheep, &c., are when separated from their companions, and what strong mutual affection the two former kinds, at least, shew on their reunion.'

(Figure 7.1).[5] Darwin called plants social, because, in a 'somewhat strained sense', they 'help each other' by increasing their mutual chances of cross-fertilization (and hence vigour), and by reducing the depredations of their 'devourers' (e.g. buffalo; birds eating their seeds). This meant that, 'if they did not live in numbers, they could not live at all.'[6]

Sociality in plants was a limit-case, showing the central importance of 'mutual aid' to Darwin's concept of the social.[7] Several advantages accrued to social organisms: ease of access to many different sexual partners;[8] the enhanced vigour due to the resultant outbreeding;[9] surviving predation;[10] and the efficiencies due to the specialization afforded by a division of labour.[11] The first three of these affected plants. All were to be found in animals. Mutual aid formed a superordinate category which subsumed the division of labour and coordinated defence against enemies.[12]

As with plants, some of the mutual services social animals rendered each other were passive: the more animals in a group, the more likely one among them would notice an approaching predator and by its agitation (unintentionally) alert its fellows. A lone wildebeest spotted by lions will almost certainly die. Most

[5] Charles Darwin, *Charles Darwin's Natural Selection: Being the Second Part of His Big Species Book Written from 1856 to 1858* (Cambridge: Cambridge University Press, 1975), p.203. Cardoons are the edible 'artichoke thistle'.

[6] Charles Darwin, *The Origin of Species by Means of Natural Selection, or the Preservation of Favoured Races in the Struggle for Life*, 6th ed. (London: Murray, 1876), pp.53–55; Darwin, *Natural Selection*, pp.129, 203–205. We now know that trees in forests meet Darwin's criterion for sociality—mutual aid—far more fulsomely than he could have known (see 7.5 Resonances).

[7] Darwin, *Descent*, p.129. Cf. Martin Nowak, 'Why We Help,' *Scientific American* 307, (2012), pp.34–39.

[8] Dioecious plants (i.e. those species having separate males and females) which rely on the wind or passing mammals, not insects, for fertilization, 'could seed well only when growing in masses': Darwin, *Natural Selection*, p.205.

[9] Darwin's experimental research on plants proved that cross-fertilization enhanced the constitutional vigour of offspring, while self-fertilization was injurious to the young. Noting that 'trees are very apt to grow together or to be social,' he stated that: 'single trees would interbreed [self-fertilise] & would produce seedlings not so well able to struggle with surrounding vegetation, as the crossed offspring of the same species, & therefore the species might be able to take root & grow only where several individuals existed': Darwin, *Natural Selection*, p.61; Charles Darwin, *The Effects of Cross and Self Fertilisation in the Vegetable Kingdom*, 2nd ed. (London: John Murray, 1878), p.436.

[10] Presuming that 'enemies' of plants and animals often did not travel in groups, Darwin argued that we could 'see why a plant or animal may exist in large numbers in one spot & not spread; for when once established in numbers it might escape destruction by its enemies, but when thinly scattered in colonies, (owing to the severe struggle going on) all might easily perish.' Crowding could be a mixed blessing, however, being sometimes 'carried to an injurious excess.' For example, 'the antelopes in S. Africa & the Passenger Pigeons in N. America are followed by hosts of carnivorous beasts & birds, which could hardly be supported in such numbers if their prey was scattered': Darwin, *Natural Selection*, pp.205, 523–524.

[11] Well-illustrated by social insects.

[12] Darwin, *Descent*, pp.50, 63, 100, 108, 127, 129. Cf. Charles Darwin, *The Expression of the Emotions in Man and Animals*, 2nd ed. (London: Murray, 1890), p.76. NB 'mutual aid' is the name of a famous book by Peter Kropotkin that came out in 1902. Kropotkin got his title (and the germ of his ideas) from *Descent*.

Figure 7.2 A herd of wildebeest.
Source: Joubert Tulleken / Shutterstock.com.

wildebeest in a large herd attacked by lions survive (Figure 7.2). Or again, provided they attended to their fellows, scavengers like condors, and the other carrion vultures that Darwin had observed on his *Beagle* voyage, would swiftly 'gain the intelligence' that an animal 'is killed in the country ... & congregate in an inexplicable manner' around the corpse.[13]

Beyond this, Darwin knew that many social species of bird and mammal *actively* signalled danger, some even posting sentinels to warn the group of approaching enemies. Thus rabbits stamp their hind-feet, and female seals act as look-outs.[14] Social creatures may also actively groom each other, removing parasites, or licking each other's wounds. Animals like wolves, killer whales, and pelicans hunt in concert, sometimes with a combined strategy. Social animals mutually defend each other too, and thereby show their 'heroism.'[15] For example, the naturalist Alfred Brehm had seen an eagle seize a young Guenon monkey, which, by clinging to a branch, was not at once carried off. It cried loudly for assistance, upon which the other members of the troop, 'with much uproar, rushed to the rescue, surrounded the eagle, and pulled out so many feathers, that he no longer thought of his prey, but only how to escape.'[16]

As for the reach of social sympathies, a truly social animal sought society *beyond* its own family.[17] Gorillas, lions, and tigers were not social in Darwin's sense, because, while they 'no doubt' felt sympathy for the suffering of their young, they

---

[13] This social intelligence was probably the result of passive observation by spectator birds, but might possibly have resulted from active signalling by the corpse-spotter. Charles Darwin, 'Darwin's Ornithological Notes [1837]. Edited with an Introduction, Notes, and Appendix by Nora Barlow,' *Bulletin of the British Museum (Natural History) Historical Series* 2, (1963), p.243.

[14] Darwin, *Descent*, pp.100–101.

[15] Darwin, *Descent*, pp.101–103.

[16] Darwin, *Descent*, pp.102–103.

[17] Darwin, *Descent*, p.108.

did not sympathize with 'any other animal.'[18] On the other hand, though animal sociability went beyond the family, it did not extend to *all* conspecifics. A social animal only sympathized with familiars, that is, with members of its own group. Darwin's generic terms for these supra-familial social groups were *tribe, horde, clan*, or *community*. Members of alien tribes would typically be treated as enemies.

Descent's account of mutual aid among animals invokes 'social instincts.' Amongst these, sympathy held most importance, as distinct from a second social instinct, love. Many social animals manifest love or attachment to one another, Darwin tells us, but not necessarily sympathy. Thus a herd of animals bound together by mutual love may nonetheless expel a wounded animal from their group or 'gore or worry it to death,' which was 'almost the blackest fact' in natural history.[19] Likewise, a bitch may love her master, 'howl dismally' when separated from him, yet never *sympathize* with him—because she can never genuinely enter into his concerns, as when gripped by a chess problem, say.[20] Not that animal love is a one-way street (although this would alone keep a herd together): *it looks for recompense*. Animals have a desire *to be loved*. This is one way jealousy arises. Even dogs and pet monkeys may protest or try to intervene when their master's affection is 'lavished on any other creature.'[21]

In contrast with love, sympathy shows no drive for congruence or requital between sympathizer and sympathized-with other (see 5.2.2 Sympathy). It typically prompts action to assuage the other's sufferings—the best evidence for sympathy in animals relating to pain and anxiety, rather than pleasure. Defence of troop-members against predators by their comrades in baboons and monkeys; dogs fighting off attacks on their masters; a blind pelican being fed by its cronies; and the homely scene of a dog bestowing friendly licks on an ailing cat figure among *Descent's* illustrations.[22]

Sympathy and love have shadows: hatred and envy. *Descent* says little about these darker instincts, which it calls 'the complement and converse' of true social instincts. Yet it notes that just a small step separates loving from hating a companion. Envy formed a sub-division of hatred, being defined as hatred of another 'for some excellence or success.' Evidence for hate came from dogs, who were very apt to hate both strange men and strange dogs, 'especially if they live near at hand,' but did not belong to the same 'family, tribe, or clan.' Such hate would 'seem to be innate.'[23]

---

[18] Darwin, *Descent*, p.106.
[19] Darwin, *Descent*, p.102.
[20] Darwin, *Descent*, p.101.
[21] Darwin, *Descent*, p.71.
[22] Darwin, *Descent*, pp.102–103.
[23] All quotes in this paragraph from a footnote added to the 2nd edition of Darwin, *Descent*, p.112.

7.2 Social Forms of Natural Selection

How could evolution explain the features of social, as opposed to non-social, animals—particularly, mutual aid, but also the specialized roles furnishing a division of labour in social insects (e.g. queen, drones, workers, soldiers)?

Origin's chapter on instinct gives a sketch of how natural selection might cover the evolution of specialized roles in communities of social insects, focusing on sterile ants.[24] It proposed that sterile members of insect communities—who are all related to each other—could have evolved by what we nowadays call *kin selection*.[25] As the first edition of *Origin* put it: 'selection may be applied to the family, as well as to the individual, and may thus gain the desired end. Thus, a well-flavoured vegetable is cooked, and the individual is destroyed; but the horticulturist sows seeds of the same stock, and confidently expects to get nearly the same variety.'[26]

Descent briefly notes that family-based selection might have been an important factor in the group-based evolution of human qualities like intellect. Even if some particularly ingenious or 'sagacious' tribe-member left *no* children to inherit their mental superiority, the tribe would still include their blood-relations, 'and it has been ascertained by agriculturists that by preserving and breeding from the family of an animal, which when slaughtered was found to be valuable, the desired character has been obtained.'[27]

Proto-human tribes comprised *several* families, however—according to *Descent*. This meant family-based selection would not explain the evolution of mutual aid. So Darwin proposed a second process, which today's biologists would call *reciprocal* altruism: I'll scratch your back, if you'll scratch mine.[28] 'As the reasoning powers and foresight of the members became improved,' wrote Darwin, each member of a tribe would soon realize that, 'if he aided his fellow-men, he would commonly receive aid in return.' This implies a norm of reciprocity in proto-human clans, but could be reduced to a form of selfishness: I help you because I will benefit in the long run. So Darwin called reliance on reciprocity a 'low motive' for helping others, though one that might prompt an individual to 'acquire the habit of aiding his fellows.' He added that 'the habit of performing benevolent actions certainly strengthens the feeling of sympathy which gives the first impulse to benevolent actions.' Thus a low motive might produce moral improvement, which, if continued over long periods, could be inherited.[29]

---

[24] Charles Darwin, *On the Origin of Species by Means of Natural Selection or the Preservation of Favoured Races in the Struggle for Life* (London: Murray, 1859), pp.235–242.

[25] John Maynard-Smith, 'Group Selection and Kin Selection,' *Nature* 201, (1964), pp.1145–1147.

[26] Darwin, *Origin 1859*, pp.237–238.

[27] Darwin, *Descent*, p.129.

[28] Robert Trivers, 'The Evolution of Reciprocal Altruism,' *Quarterly Review of Biology* 46, (1971), pp.35–37.

[29] Darwin, *Descent*, pp.130–131.

Darwin's favourite dynamic for producing natural selection of a kind that would lead to the advance of sociality in human groups seems to have emerged from an intense debate with Alfred Wallace over the evolution of sterility in cross-species hybrids in plants.[30] While Wallace denied the existence of social animals, he proposed to Darwin in 1868 that the sterility of hybrids in the offspring of adjoining populations of primulas would be naturally selected, not through advantages to individual plants, but through advantages to the two different 'groups'—because it would keep both species 'pure,' so maintaining the integrity of their existing adaptations.[31] Darwin consistently rebutted Wallace's argument during their correspondence on this topic. But the next year, when finalizing his corrections to the penultimate edition of *Origin*, Darwin introduced a new claim. From 1859 to 1866, editions one to four of the book had stated that, in communities of social animals, natural selection 'will adapt the structure of each individual for the benefit of the community; if each [individual] in consequence profits by the selected change.'[32] The fifth (1869) and sixth (1872) editions of *Origin* show a crucial change, stating that: 'In social animals [natural selection] will adapt the structure of each individual for the benefit of the whole community; if *the community* profits by the selected change.'[33] This kind of community selection provided *Descent* with a strong means for explaining the most human of human faculties, language and conscience.

7.3 Human Sociality

Whilst *Descent* Part I deals with the most complex forms of human agency, a strategy pivotal to its evolutionary message is the use of observations of animal

---

[30] Malcolm Kottler, 'Charles Darwin and Alfred Russell Wallace: Two Decades of Debate over Natural Selection,' in *The Darwinian Heritage*, ed. David Kohn (Princeton NJ: Princeton University Press, 1985), pp.407–410. This was a problem that Darwin never fully solved, one which led even his friend and ally Thomas Huxley to question Darwin's theory: Paul White, *Thomas Huxley: Making the "Man of Science"* (Cambridge: Cambridge University Press, 2003).

[31] The sterility of hybrids would also nullify their 'hybrid vigour' argued Wallace: Alfred Wallace, 'Letter to Darwin, March 15th,' *Correspondence of Charles Darwin* (Cambridge: Darwin Correspondence Project, 1868), https://www.darwinproject.ac.uk/letter/?docId=letters/DCP-LETT-6012 Alfred Wallace, 'Letter to Darwin, February 24th,' *Correspondence of Charles Darwin* (Cambridge: Darwin Correspondence Project, 1868), https://www.darwinproject.ac.uk/letter/?docId=letters/DCP-LETT-5922; Alfred Wallace, 'Letter to Darwin, March 1st,' *Correspondence of Charles Darwin* (Cambridge: Darwin Correspondence Project, 1868), https://www.darwinproject.ac.uk/letter/?docId=letters/DCP-LETT-5966. Wallace soon withdrew the primula example. See also Alfred Wallace, 'The Origin of Human Races and the Antiquity of Man Deduced from the Theory of "Natural Selection",' *Journal of the Anthropological Society of London* 2, (1864), p.clxii, where Wallace recognized that animals could be gregarious, but denied that they were 'social and sympathetic.' Only humans were genuinely social. In animals, there was, 'as a general rule, no mutual assistance between adults ... Neither is there any division of labour.' It is puzzling that Wallace acknowledged no division of labour among social insects.

[32] Charles Darwin, *On the Origin of Species*, 4th ed. (London: Murray, 1866), pp.98–99.

[33] Charles Darwin, *On the Origin of Species*, 5th ed. (London: Murray, 1869), p.99; Darwin, *Origin 1876*, p.67; italics added.

behaviour to prove that, even at their most distinctive, human actions have animal roots. Having listed all the shibboleths, supposedly unique to *Homo sapiens*, Part I walks the reader through rebuttals of nearly all these claims (see Table 7.1).[34] The rebuttals have two forms. One is that animals can be shown to possess the 'uniquely human' capacity outright, or, at a minimum, some rudimentary form of that capacity. For example, the supposedly human 'prerogative' of imagination is linked to the observation that dogs, cats, horses, and even birds, 'have vivid dreams.' This is 'shewn by their movements and the sounds uttered' when asleep. They therefore have 'some power of imagination.'[35] Alternatively, Darwin might query whether a supposedly unique human capacity existed in *all* humans. If it were absent in some, it could not be called a distinctive feature of humanity as such.

Table 7.1. *Descent*'s refutation of the supposed uniqueness of higher human attributes[36]

Attribute	Evidence animals have this?	All humans show this?	Where discussed
Attention	Cats stalking prey; courting capercaillie oblivious to hunter's approach; trainable monkeys have good attention.		pp.73–74, 363–364.
Belief in God or gods	Dogs have proto-superstitions (bark at odd movement of parasol on lawn); may look on masters as God-like.	Fuegians do not: they only have superstitions	p.95
Caprice	Animals are capricious in their sexual preferences, e.g. stallions, pheasants.		pp.420, 524 (see Ch. 6).
Conscience	Dogs show shame, so have 'something very like a conscience'	Not in some criminals	pp.116, 103, 71, 107
Cooperation	Baboons cooperate to turn over large rocks and share insects from underneath; animals collectively hunt, defend each other etc.		pp.101–102 & passim

[34] Darwin, *Descent*, p.79: 'It has been asserted that man alone is capable of progressive improvement; that he alone makes use of tools or fire, domesticates other animals, or possesses property; that no animal has the power of abstraction, or of forming general concepts, is self-conscious and comprehends itself; that no animal employs language; that man alone has a sense of beauty, is liable to caprice, has the feeling of gratitude, mystery, &c.; believes in God, or is endowed with a conscience.'

[35] Darwin, *Descent*, pp.79, 74–75, 619.

[36] All page numbers refer to Darwin, *Descent*, unless noted otherwise.

Table 7.1. *Continued*

Attribute	Evidence animals have this?	All humans show this?	Where discussed
Courage, heroism	Dogs defend their masters; monkey protects keeper from baboon; baboons attack eagle to save troop-member		pp.69, 102–103
Domestication of other species	Ants domesticate aphids; make slaves		pp.147–148
Emotions	Yes (main topic of *Expression*), with the exception of weeping and, particularly, blushing	Yes (but young children and some insane people do not blush)	p.79; see Chs 5–6
Feelings of gratitude	Claimed, but no evidence given.		p.79
Feelings of mystery			Listed but not discussed
Insanity	Yes (cites 1871 article)[37]; baboons and monkeys get drunk on alcohol.		pp.7, 79. 'Notebook M,' p.13.[38]
Language	Animals capable of articulating a variety of meaningful sounds to conspecifics (hens, monkeys); monkeys understand some words.	Not in infants	pp.85–87
Memory	Baboons recognize humans after months; ants recognize nest-companions after four months; bees remember the position of 'each clump of flowers in a garden'; see 'Mental individuality.'		pp.74, 84; *Cross and Self-Fertilisation of Plants*, pp.424–426[39]
Mental individuality	Possible to wake a train of old associations in the mind of a dog after an interval of five years, 'although every atom of his brain had probably undergone change more than once.'		p.84

(continued)

---

[37] Lauder Lindsay, 'Madness in Animals,' *Journal of Mental Science* 17, (1871), pp.181–206.
[38] Charles Darwin, 'Notebook M,' (Cambridge: Darwin Online, 1838).
[39] Darwin, *Cross Fertilisation*.

Table 7.1. *Continued*

Attribute	Evidence animals have this?	All humans show this?	Where discussed
Moral sense	Dogs have moral qualities (see Conscience); Darwin does not claim animals have a 'moral sense' as such; but see 'Conscience.'		pp.80, 103
Obedience	Dogs and elephants obey their masters and mahouts; baboons raiding gardens obey their leaders		pp.74, 103–104
Possess property	Monkeys protect their tools; dogs their bones; birds and dogs defend territories (or nests).	Fuegians do not	pp.82, 133
Power of abstraction, or forming general concepts	Dogs categorize distant animals, e.g. dogs, solely as 'dog'—only identifying them as friend, enemy, or stranger when they get closer. Likewise, a command (e.g. Walk! Find it!) alerts a dog to a class of events—of which the specifics are determined later.	Not present in very young children	pp.83–84, 89
Power of imagination	Dogs, cats, and horses have vivid dreams, shown by their movements and utterances when asleep. Dogs bay at moon.		pp.74–75
Progressive improvement	Older animals harder to trap than young; island animals learn wariness of humans; dogs have progressed in 'moral qualities.'	Doubtful in Fuegians	pp.79–81; *Journal* p.216[40]
Reasoning, intelligence	Difficult to tell from instinct; elephants and bears improvise means to retrieve objects; monkeys solve novel problem after just one error; dogs invent solution to novel problem.		pp.75–78

[40] Charles Darwin, *Journal of Researches into the Natural History and Geology of the Countries Visited During the Voyage of H.M.S. Beagle Round the World, under the Command of Capt. Fitz Roy R.N.*, Journal of Researches (London: John Murray, 2006).

Table 7.1. *Continued*

Attribute	Evidence animals have this?	All humans show this?	Where discussed
Religious devotion	Homologous to 'deep love of a dog for his master, associated with complete submission, some fear, and perhaps other feelings'		p.96
Self-command	See Obedience		pp.103–104
Self-comprehension			Listed but not discussed
Self-consciousness	Maybe a dog 'with an excellent memory and some power of imagination … reflects on his past pleasures or pains in the chase.'	Maybe not in 'the hard-worked wife of a degraded Australian savage'	p.83
Self-sacrifice	See Courage, heroism		pp.129–130
Sense of beauty	Female animals have aesthetic sense to appreciate beauty of males' sexual ornaments and song (especially birds).	'Barbarians' and 'uneducated persons' cannot enjoy 'the heavens at night, a beautiful landscape, or refined music'	pp.92ff & Part II.
Sense of humour	Dog teasing owner by retrieving ball, but staying just out of reach rather than give it back.		p.71
Sympathy	Baboons rescuing young from eagle; dog licks sick cat; pelicans feed blind companion		pp.102–104
Tool-making	A baboon mixes mud and water to make a weapon.	Not primeval man	pp.69, 82
Use of fire			Listed but not discussed
Use of tools	Chimps (& monkeys) crack nuts with stones; use sticks as levers. Elephants use branches as fly-swats; etc.		pp.81–82

Descent is economical in proving many of the points summarized in Table 7.1, and at times, almost perfunctory. Darwin was apparently clearing the ground for a more urgent project: his need to prove that even the most *un*-animal-like, quintessentially human, psychical faculties could have evolved according to the laws of natural and or sexual selection. Here, he gave pride of place to conscience, but he also dealt at length with language, and various forms of cooperation and self-sacrifice. Two points were fundamental to his explanation of these 'high' capacities. One has to do with the way the human brain works. The other draws on his hypothesis of group selection.

7.3.1 Growing cerebral integration

The greater the gap between the animal roots of a capacity and its full flowering in humans, the greater Darwin's need for a theory to bridge the two. Hence Darwin's argument that evolution had advanced human 'intellectual powers' by progressive integration of the brains of hominids—making them less and less *modular* (in today's parlance) than those of lower animals.[41] Increasing 'intercommunication' between the various parts of the brain would mean human mentality was less and less siloed than that of pre-hominids, wrote Darwin, leading to higher intelligence, more inventiveness, and greater plasticity of response than in animals.[42] For example, while the dreaming of dogs and birds might show they could respond in a rudimentary way to images rather than reality, the most brilliant products of human imagination depended on both the clarity and variety of mental imagery, a will and ability voluntarily to combine images in novel ways, and hence, 'judgment and taste in selecting or rejecting' the invented combinations.[43] For Darwin, human imagination was not one siloed faculty. Rather it was the result of combining many different capacities to bear on a chosen task—will, clear-mindedness, perception, inventiveness, perseverance, judgement, and taste.

The evolving intercommunication between parts of the human brain underlines that Darwin envisaged evolution moving mental life from fixity in simple animals to a far greater flexibility in humans. This coloured his take on instinct. *Descent* sometimes says humans *do not* have instincts, having lost any which our early progenitors may have possessed.[44] And it sometimes says humans *do have* instincts, *social* instincts. However, humans' social instincts are far more frequently described as if

---

[41] Cf. Jerry Fodor, *Modularity of Mind: An Essay on Faculty Psychology* (Cambridge MA: MIT Press, 1983); Dan Sperber, 'In Defense of Massive Modularity,' in *Language, Brain, and Cognitive Development: Essays in Honour of Jacques Mehler*, ed. Emmanuel Dupoux (Cambridge MA: MIT Press, 2001), pp.47–57 .

[42] Darwin, *Descent*, p.68.

[43] Darwin, *Descent*, p.74.

[44] Darwin, *Descent*, p.109. It very occasionally says the reverse: 'Man has also some few instincts in common, as that of self-preservation, sexual love, the love of the mother for her new-born offspring, the desire possessed by the latter to suck, and so forth,' p.66; see too p.79.

they were *forms of subjectivity*—'a wish and readiness'—not stereotyped habits, as in bees and birds.[45] In this guise, social instincts are said to *provide the foundation* for the relevant forms of subjectivity: the *desires, impulses*, and *motives* energizing human sociability—words that Darwin rarely (desire) or never (impulse, motive) applied to non-humans.[46] Motives are what move us. 'Motives are units in the universe,' Darwin had written in his twenties, and motives have evolutionary roots.[47] This was important because: 'Reason, as reason, can never be a motive to action.'[48] Motives constitute stigmata of our animal ancestry, the biological and conceptual isthmus between humans' refined social agency and its bestial, instinctive beginnings.

Accordingly, Part I of *Descent* goes straight from denying humans have instincts, to saying that there was no reason why we should not have retained from long ago some degree of 'instinct*ive*' love and sympathy for our fellows. Evidence here came from introspection: 'we are indeed all conscious that we do possess such sympathetic feelings.' But introspection proved ambiguous, for it did not tell us whether these feelings had a phylogenetic history, 'having originated long ago in the same manner as with the lower animals,' or an ontogenetic history, having been 'acquired by each of us during our early years.'[49] Hence the value of *Descent*'s copious comparative evidence that animals have social instincts. According to Bacon's principle that nature does not make jumps (*natura non facit saltum*), this proved our social feelings had evolutionary origins.[50]

Finally, as Chapter 8 will elaborate, *Descent* went on to argue that—once it had evolved—human agency could be and had been put to uses which had little or nothing to do with the advantages for fitness which first led to its natural selection. These new uses could be non-adaptive or even maladaptive, as with some peoples' customs and superstitions, which were 'in complete opposition to the true welfare and happiness of mankind' (see 8.2 Relative Autonomy).[51]

7.3.2 Moral economy of groups

To explain the most distinctive kinds of human action, *Descent* invokes the hypothesis of group or 'community selection'—in tandem with the increased plasticity of

---

[45] Darwin, *Descent*, p.98.

[46] Darwin, *Descent*, p.611; e.g. 'the social instincts, which must have been acquired by man in a very rude state, and probably even by his early ape-like progenitors, still give the impulse to some of his best actions,' p.109.

[47] Charles Darwin, 'Old & Useless Notes,' (Cambridge: Darwin Online, 1838-1840), p.25.

[48] Mackintosh, *Progress of Ethical Philosophy*, p.152, or, in Darwin's paraphrase: 'Reason can never lead to action'; Darwin, 'Old & Useless Notes About the Moral Sense & Some Metaphysical Points,' p.10v.

[49] All quotes in paragraph from Darwin, *Descent*, p.109, my italics.

[50] Darwin, *Origin 1876*, pp.156, 166 & passim.

[51] Darwin, *Descent*, pp.121–122.

human agency consequent upon brain-integration.[52] Three crucial chapters in Part I (Chs 3–5) seek to show how primeval versions of language, self-consciousness, conscience, self-sacrifice, fidelity, honesty, courage, obedience, cooperative hunting, communal self-defence—and even more purely intellectual powers like 'sagacity,' reasoning, foresight, and inventive problem-solving—could have promoted the cohesion and fitness of its exponent's tribe.[53] Tribes attract Part I's focus because Darwin held that all the human virtues just listed, 'affect the welfare of *the tribe*,—not that of the species, nor that of an individual member of the tribe.'[54]

So—if we ask how any, distinctively human, moral, or intellectual quality evolved?—the *common form* of Darwin's answer is to say: It evolved because it promoted group survival. To wit:

> When two tribes of primeval man, living in the same country, came into competition, if (other circumstances being equal) the one tribe included a great number of courageous, sympathetic and faithful members, who were always ready to warn each other of danger, to aid and defend each other, this tribe would succeed better and conquer the other ... A tribe rich in the above qualities would spread and be victorious over other tribes: but in the course of time it would, judging from all past history, be in its turn overcome by some other tribe still more highly endowed. Thus the social and moral qualities would tend slowly to advance and be diffused throughout the world.[55]

This hypothesis conformed to all that was known about 'savages,' or might be inferred from their traditions, and from 'old monuments,' namely: 'that from the remotest times successful tribes have supplanted other tribes.'[56] The nub of Darwin's argument was that the victorious tribes in early human battles would be the most *cohesive*. Hence the viability of warlike groups required a moral economy which promoted unselfishness: 'selfish and contentious people will not cohere, and without coherence nothing can be effected.'[57] No tribe could succeed, if 'murder, robbery, treachery, &c., were common.' Consequently such crimes were 'branded with everlasting infamy' within the limits of a single tribe, though they excited 'no such sentiment beyond these limits.'[58] Yet fidelity, self-sacrificing heroism, and generosity would scarcely have profited the individuals strongest in these features. Just the reverse. It was 'extremely doubtful' that the offspring of the most prosocial

---

[52] Referred to as 'community selection' in Robert Richards, *Darwin and the Emergence of Evolutionary Theories of Mind and Behaviour* (Chicago: University of Chicago Press, 1987), passim.
[53] Darwin, *Descent* pp.98–100, 129–131.
[54] Darwin, *Descent*, p.119, my italics.
[55] Darwin, *Descent*, p.131.
[56] Darwin, *Descent*, p.99, 132: 'and this would be natural selection.'
[57] Darwin, *Descent*, p.117.
[58] Darwin, *Descent*, p.130; quoting Walter Bagehot, 'Physics and Politics No.2: The Age of Conflict,' *Fortnightly Review* 3, (1868), p.457.

parents, or of those who were 'the most faithful to their comrades,' would be reared in greater numbers than the children of 'selfish and treacherous' parents belonging to the same tribe: 'He who was ready to sacrifice his life, as many a savage has been, rather than betray his comrades, would often leave no offspring to inherit his noble nature.'[59]

7.4 Language and Conscience

Descent invokes *both* group coherence *and* cerebral plasticity in explaining what it holds to be the most human forms of agency: most particularly, conscientiousness. But Darwin's favoured form of argument is most easily outlined from *Descent's* account of the origin of language.

7.4.1 Language

While Darwin traced animal sound-making to courtship, his explanation for human language invoked not only sexual display, but group-dynamics. A human progenitor would have produced 'true musical cadences,' in singing, just like 'gibbon-apes at the present day.' Such singing would have been especially salient during courtship, when expressing 'various emotions, such as love, jealousy, triumph.' From this base, a proclivity for intelligent imitation would have taken over, assuming our ancestors already lived in cooperative groups. 'May not some unusually wise ape-like animal have imitated the growl of a beast of prey, and thus told his fellow-monkeys the nature of the expected danger?' The first word-like signs would have been imitations of natural events, giving us the 'first step in the formation of a language.'[60]

Language would have improved a group's capacities to coordinate its cooperative ventures (mutual defence, warning against predators, strategic hunting, etc.). So the *continued* evolution of language would have been favoured by natural selection—because it would have strengthened group-solidarity. Simultaneously, development of vocal organs would have been sped by the inherited effects of their increased use in speech. Language-use would also have reflexively 'reacted on' development of the intellect: 'by enabling and encouraging it to carry on long trains of thought,' thus promoting the formation of self-consciousness, and general concepts.[61] Furthermore, speech would enable easier and more exact expression of communal mores, so crystallizing 'the common opinion how each

---

[59] Darwin, *Descent*, p.130.
[60] Darwin, *Descent*, pp.86–87 for quotes in this paragraph.
[61] Darwin, *Descent*, pp.87–88.

member ought to act for the public good.' With increasing clarity, these group-specified—and plastic, because variable from group to group—moral standards would 'naturally' become each member's 'guide to action.' Hence language would enable groups to grow beyond the bounds of animal companionship, and tribalism, because moral standards would ultimately pave the way to supra-tribal civilization (see Ch. 8).[62]

7.4.2 Conscience

Descent's most sustained and complicated argument about human agency concerns conscience. Darwin sometimes drew a distinction between humans' *moral sense*, which gives us the generic capacity and desire to distinguish right from wrong, and *conscience*, which tells us in particular when we ourselves have erred[63]—though he sometimes treated moral sense and conscience as synonyms.[64] And, while he held that the moral sense was 'aboriginally derived from the social instincts'[65]—his treatment of conscience drew on the full gamut of his psychological thinking about: our doubled self; the selfish urgency of strong desire; our growing intellectual power; the use of language; the compulsion to fulfil the desires of valued others (see Chs 4–6); and the importance for human agency of social institutions (see Ch. 8).

One way of portraying *Descent's* account of conscience is as an interaction between conflicting roles or characters, as on the stage. The place of conscience in *Hamlet* provides an illustration, when we take Hamlet's task to have been the transcendence of his own temporizing and indecision, following his vow to take revenge for the murder of his father by his 'remorseless, treacherous, lecherous, kindless,' uncle.[66] Read like this, the early acts of Shakespeare's play depict Hamlet as what Brian Cummings calls, a 'fragmentary repository of alternative selves.'[67] These operate like an unruly troupe of actors—who undertake to evoke conscience by enacting a drama.[68]

I take Cummings' cue, in what follows, to draw out the different characters which animate *Descent's* staging of moral agency.

---

[62] Darwin, *Descent*, p.99.

[63] Darwin, *Descent*, p.116.

[64] Darwin, *Descent*, pp.97–98 & passim. Alexander Bain, *The Emotions and the Will*, 2nd ed. (London: Longmans Green, 1865) did likewise: e.g. p.255.

[65] Darwin, *Descent*, p.119.

[66] William Shakespeare, *Shakespeare's Hamlet* (New York: Henry Holt, 1914), Act 2, scene 2.

[67] Brian Cummings, *Mortal Thoughts: Religion, Secularity, & Identity in Shakespeare and Early Modern Culture* (Oxford: Oxford University Press, 2013), p.180.

[68] Adam Phillips, 'Against Self-Criticism,' *London Review of Books* 37, 5 March 2015. Shakespeare, *Hamlet*, Act 2, scene 2: 'The play's the thing/ Wherein I'll catch the conscience of the king.'

7.4.2.1 Self-gratifier

The first character in Darwin's portrait of conscience is one who selfishly gratifies their own most pressing desires. Darwin held that the performance of any instinct intrinsically gives pleasure, and its prevention, distress. The stronger the instinct, the greater the agent's pleasure or pain. The more important an instinct for survival, the more potent natural selection will have rendered it, and the stronger the desires it will arouse in humans.[69] The strongest instinctual desires would therefore be those of self-preservation: 'hunger, lust, vengeance ... danger shunned at other men's cost.' The actions arising from such instincts have a distinctive profile in time, however: 'the desire to satisfy hunger, or any passion such as vengeance, is in its nature temporary, and can for a time be fully satisfied.' These passions prove hard to recall 'with complete vividness' when not current: 'The instinct of self-preservation is not felt except in the presence of danger; and many a coward has thought himself brave until he has met his enemy face to face.'[70]

The first character of conscience *Descent* describes, therefore, is one who, 'at the moment of action,' is 'apt to follow the stronger impulse,' which will most commonly 'lead him to gratify his own desires at the expense of other men.'[71] Here we should recall the first type of action captured by what the twenty-nine-year-old Darwin hypothesized was our double self: what I have dubbed Train One of the self, entailing a dynamic which unreflectively commits deeds that are easily forgotten, like 'the instincts of animals' (see 5.2.1 The dual self).[72]

7.4.2.2 Praise-seeker

As blushing and courtship prove, humans often feel an involuntary imperative to comply with the wishes of others whose opinions they value. The second character in Darwin's pantheon of conscience is one whose actions are to a high degree determined by 'the expressed wishes and judgment' of fellow-humans.[73] This character grows in the soil of sympathy, which grounds a person's capacity to worry about what others may feel—as in blushing. Sympathy among group-members leads them to value the approbation of their fellows, a form of concern motivated both by the love of others' praise, 'the strong feeling of glory, and the still stronger horror of scorn and infamy.' When, in my mind's eye, I see a friend recoiling from what I have done, or extolling it, I cannot help feeling and judging my actions as does he or she. Consequently humans show influence in the highest degree by 'the wishes, approbation, and blame' of their familiar companions.

---

[69] Darwin, *Descent*, pp.107–108.
[70] Foregoing quotes all from Darwin, *Descent*, pp.112–113.
[71] Darwin, *Descent*, p.114.
[72] Darwin, 'Notebook M,' p.81.
[73] All quotes in this paragraph from Darwin, *Descent*, pp.109–110.

The impulse to act in accordance with the 'approbation and disapprobation' of one's fellow group-members will have amounted to a *compulsion* for our ancestors, wrote Darwin, serving them 'as a rude rule of right and wrong.'[74] Hence we get *Descent*'s second character of conscience: the praise-seeker.

7.4.2.3 Arbitrary monitor

The counterpart of our drive to conform to the dictates of our peers is how we imagine the dictators. Hence, a third feature in *Descent*'s account of conscience is the struggle between short-term selfish motives and the eyes or voice of an 'inward monitor' which gives us the ever-enduring sense of surveillance by tribal judgement.[75]

After a self-serving act—its imperious rationale fast fading—an agent will recast what they have done through a 'deep regard for the good opinion of his fellows,' and 'retribution will surely come.' A man:

> ... will then feel remorse, repentance, regret, or shame; this latter feeling, however, relates almost exclusively to the judgment of others. He will consequently resolve more or less firmly to act differently for the future; and this is conscience; for conscience looks backwards, and serves as a guide for the future.[76]

It is here, in the tussle between self-gratifying instinctual action and complying with community standards, that we see the kind of conflict which requires humans to possess what the young Darwin had called humans' *double* self or double consciousness. Remorse in the waking mind pits two different strains of thought or action against each other. Train One acts more or less unconsciously, producing self-serving deeds. Train Two Darwin called 'reflective consciousness.' It exploits the human capacity for vivid imagination, making links between present, past, and future consequences.[77] Without this second kind of consciousness, and the deliberation it imposes, a creature could not be called a moral being. And so animal behaviour, however brave and self-sacrificing—as when 'a Newfoundland dog drags a child out of the water'—could not be called moral.[78]

Embedded in Train Two processing, the social exemplars by which we judge ourselves prove largely imaginary. Moreover, like the standards of beauty that govern human ornamentation in different tribes (see 6.4.6 Cultural plotlines), moral standards are arbitrary. The *inward monitor*—who dictates that we would have been better to have followed 'the one impulse rather than the other'—operates with

---

[74] Darwin, *Descent*, p.124.
[75] Darwin, *Descent*, p.100.
[76] All quotes from Darwin, *Descent*, pp.113–114.
[77] E.g. Darwin, *Descent*, pp.611–612.
[78] Darwin, *Descent*, p.111.

the caprice of tribal fashions in dress, being as likely to promote 'baneful customs and superstitions' as an agent's genuine welfare. *Descent* gives examples: human sacrifice; murderous persecution of so-called witches; irrational dietary habits; infanticide; suicide; slavery; cruelty to animals; and the 'senseless practice of celibacy'.[79] To all these dictates, however toxic and arbitrary: 'disobedience becomes a crime, *and even abject submission is looked at as a sacred virtue*'.[80]

7.4.2.4 Supreme judge

There is an obvious corollary to *Descent*'s argument that, 'with the less civilised nations reason often errs, and many bad customs and base superstitions come within the same scope, and are then esteemed as high virtues, and their breach as heavy crimes'. Humans have the capacity to transcend the standards of their home-tribe.[81] 'I cannot and will not cut my conscience to fit this year's fashions,' protested playwright and activist Lillian Hellman (1905–1984).[82] So we find Darwin arguing that 'as love, sympathy and self-command become strengthened by habit, and as the power of reasoning becomes clearer,' we become more able to value objectively the judgments of our social group.[83] This capacity for just judgement in the face of tribal demands forms the fourth character in Darwin's take on conscience.

Borne by the advance of reasoning powers and consequently of 'a just public opinion,' which Victorians like Darwin equated with the upward arc of European civilization, reason could win an individual escape from the tyranny of their fellows' judgement. He or she would then have no need to accept the praise or blame of others as a guide to right action.[84] Such independent characters would be impelled by their own power of thinking, 'apart from any transitory pleasure or pain,' to certain lines of conduct. Such a reasonable person might then declare, 'not that any barbarian or uncultivated man could thus think—I am the supreme judge of my own conduct, and in the words of Kant, I will not in my own person violate the dignity of humanity'.[85]

Darwin's own standard of a genuinely humane morality was the cessation of cruelty to slaves and to animals—and here, his actions spoke loudly.[86] Throughout his life, Darwin had avoided political limelight. His one overt political act was the publication in 1863 of 'An Appeal' to the public to outlaw the use of steel traps

[79] Darwin, *Descent*, pp.96, 100, 124, 118–119.

[80] Darwin, *Descent*, p.118, my italics.

[81] Darwin, *Descent*, p.611.

[82] Timothy Adams, *Telling Lies in Modern American Autobiography* (Chapel Hill NC: University of North Carolina Press, 1990), p.161.

[83] Darwin, *Descent*, p.110.

[84] Darwin, *Descent*, p.112.

[85] Darwin, *Descent*, p.110.

[86] Darwin equated the psychology of slave-owners to the derogation of animals: 'Animals whom we have made our slaves we do not like to consider our equals. — Do not slave holders wish to make the black man other kind — animals with affections, imitation, fear of death, pain, sorrow for the dead. — respect': Charles Darwin, 'Notebook B,' (1837–1838), p.231.

(a)   (b)

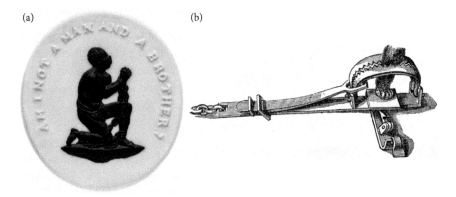

Figure 7.3 Wedgwood slave medallion, circa 1787 (a); the Darwins' Appeal, 1863: 'Some women may never have seen a trap, and therefore I give a wood-cut of one' (b).

Source: (a) AF Fotografie / Alamy Stock Photo; (b) 'Darwin, C. (1863). *An Appeal*. Reproduced by kind permission of the Master and Fellows of Christ's College, Cambridge.

for catching 'vermin.'[87] This disturbing pamphlet was co-written with, and largely distributed by, Emma Darwin.[88] It argued that, despite 'the general increase of humanity,' Victorian traps consigned 'thousands of animals to acute agony, probably of eight or ten hours duration, before it is ended by death.' The pamphlet made great efforts to enlist readers' sympathy and 'compassion.'

A picture was included of an animal's leg caught in jagged steel jaws, the trap itself chained to the ground—just as the slave had been chained in the anti-slavery Medallion distributed by Josiah Wedgwood, grandfather to both Emma and Charles Darwin (Figure 7.3). Under the picture we read: 'The iron teeth shut together with so strong a spring, that a pencil which I inserted was cracked and deeply-indented by the violence of the blow ... when a cat or a rabbit is caught, the limb is cut to the bone and crushed.'[89]

---

[87] Versions of the pamphlet appeared as a letter—without the picture of the trap—in the *Gardeners' Chronicle* (29th August 1863), and in a local newspaper, the *Bromley Record* (1st September 1863). Only the letter published Darwin's initials: Charles Darwin, *An Appeal*, (Cambridge: Darwin Online, 1863), https://www.darwinproject.ac.uk/topics/life-sciences/darwin-and-vivisection/appeal-against-animal-cruelty.

[88] Emma was partly attracted to Charles by his compassion for animals. A few days after he proposed to her on 11th November 1838, when she was aged 30, she wrote to her favourite aunt Jessie Sismondi: 'He is the most open, transparent man I ever saw, and every word expresses his real thoughts. He is particularly affectionate and very nice to his father and sisters, and perfectly sweet tempered, and possesses some minor qualities that add particularly to one's happiness, such as not being fastidious, and being humane to animals ...'; Emma Darwin, *Emma Darwin: A Century of Family Letters*, 2 vols., vol. 2nd (New York: Appleton, 1915), p.6.

[89] All quotes from Darwin, *An Appeal*.

The Darwins pointed out that not all game-keepers were humane. Some would 'have grown callous to the suffering constantly passing under their eyes,' and had been known 'by an eye-witness to leave the traps unvisited for twenty-four or even thirty-six hours.' A verbatim eye-witness account detailed the pathetic response of a rabbit to its torture in a vermin-trap: 'They sit drawn up into a little heap, as if collecting all their force of endurance to support the agony; some sit in a half torpid state induced by intense suffering ... as you approach, [they] start up, struggle violently to escape, and shriek pitiably.'[90]

7.4.2.5 Impulsive hero

To begin with, it might seem that *Descent's* discussion of conscience and agency would equate moral perfection with the subordination of an agent's selfish instinct-like yearnings to the consideration of tribal standards:

> Man prompted by his conscience, will through long habit acquire such perfect self-command, that his desires and passions will at last yield instantly and without a struggle to his social sympathies and instincts, including his feeling for the judgment of his fellows. The still hungry, or the still revengeful man will not think of stealing food, or of wreaking his vengeance.[91]

Then, when we meet the idea of reason as the supreme judge of human action, we might assume that rational conduct represented the pinnacle of moral achievement. Paradoxically then, *Descent* does not give the laurels for moral perfection to actions done deliberately, 'after a victory over opposing desires, or when prompted by some exalted motive,' but to a class of *impulsive* actions. 'Many a civilized man, or even boy,' wrote Darwin—apparently blind to the possibility of female heroism (see Ch. 3, Figure 3.10)—who never before 'risked his life for another, but full of courage and sympathy, has disregarded the instinct of self-preservation, and plunged at once into a torrent to save a drowning man, though a stranger.'[92] In this

---

[90] Twelve years later Darwin was caught up in a nationwide controversy over a public display at the annual meeting of the British Medical Association in 1874. Two unsedated dogs were to be injected with absinthe to produce epileptic seizures for a crowded auditorium. Outrage! Dog two was freed by the scandalized audience. Very soon Frances Cobbe was fronting a move to regulate if not outlaw vivisection. She tried to enlist Darwin publicly to back the 'Henniker' Bill she was submitting to the House of Lords. Darwin refused. Meanwhile, behind the scenes he was working with animal physiologists like Thomas Huxley to prepare an alternative piece of legislation that would not impede the progress of science but which would outlaw all unnecessary cruelty to animals. Many commentators have assumed that Darwin's 'Playfair Bill' was *pro*-vivisection. But David Feller has shown that Darwin's bill did even more to limit animal suffering than Cobbe's. In particular, unlike Cobbe, Darwin was determined to outlaw vivisection that was solely for instructional purposes. He also proposed harsher penalties than she for illegal cruelty: David Feller, 'Dog Fight: Darwin as Animal Advocate in the Antivivisection Controversy of 1875,' *Studies in History and Philosophy of Biological and Biomedical Sciences* 40, (2009), pp.265–271.

[91] Darwin, *Descent*, p.115.

[92] Darwin, *Descent*, pp.110–111.

case, Darwin wrote, the agent was impelled by the same instinctive motive that impelled the heroism in Darwin's favourite monkey:

> Several years ago a keeper at the Zoological Gardens shewed me some deep and
> scarcely healed wounds on the nape of his own neck, inflicted on him, whilst
> kneeling on the floor, by a fierce baboon. The little American monkey, who was a
> warm friend of this keeper, lived in the same large compartment, and was dread-
> fully afraid of the great baboon. Nevertheless, as soon as he saw his friend in peril,
> he rushed to the rescue, and by screams and bites so distracted the baboon that
> the man was able to escape, after, as the surgeon thought, running great risk of his
> life.[93]

This monkey was *Descent*'s epitome of moral perfection, as against praise-seekers
of the tribal type. Thus, echoing Huxley's famous put-down of Bishop Wilberforce
in 1860,[94] the book's penultimate paragraph attests that its author:

> would as soon be descended from that heroic little monkey, who braved his
> dreaded enemy in order to save the life of his keeper ... as from a savage who
> delights to torture his enemies, offers up bloody sacrifices, practises infanticide
> without remorse, treats his wives like slaves, knows no decency, and is haunted by
> the grossest superstitions.[95]

Darwin's praise for this little monkey underlines the fact that *Descent* unfolds a
surprisingly critical take on the arbitrary and often-baneful standards to which a
group-governed conscience truckles, and the praise-seekers it spawns. Apparently,
the conformist character of conscience can diminish moral valour and make
'cowards of us all.'[96] *Descent*'s elevation of Darwin's monkey hero over the tribal
rule-follower suggests that, beyond the inveiglements of group dynamics and our
motives for conscientious compliance, and beyond reason, there stands another
form of morality, one inspired *by* desires, especially sympathetic desires—rather
than by the 'perfect self-command' that subordinates all desires and passions to an
inward monitor, or by a controlled rationality.

---

[93] Darwin, *Descent*, p.103.

[94] The Bishop, in animated opposition to the evolutionary theses recently published in *Origin*, had
supposedly taunted Huxley by asking whether he was descended from a gorilla on his father's or his
mother's side? Huxley purportedly responded by saying he would sooner claim kindred with an ape
than with a man like the Bishop, who used all his great powers to strangle an important scientific debate.

[95] Darwin, *Descent*, p.619

[96] Shakespeare, *Hamlet*, Act 3, Scene 1.

7.5 Resonances

7.5.1 Consciousness and conscientiousness

Conscience stood front and central in Darwin's staging of human agency. This contrasts with psychology today. It is hard to find any sustained discussion of moral struggle in contemporary English-language courses on psychology—or in the textbooks that introduce students to the subject—unless, of course, psychoanalysis remains within the fold. Which touches on a deeper point. Increasing fascination with artificial intelligence and cognitive neuroscience has led many of today's psychologists to invest heavily in research on what they call consciousness. But they equate consciousness to a 'neuronal workspace' within which 'external' information influences the organism's subsequent activity.[97]

When the word consciousness occurs in *Descent*, however, it nearly always refers to a reflexive process of *self*-consciousness, equating to Train Two agency in Darwin's schema of the double self (see 5.2.1 The dual self).[98] The meanings of the words *conscience* and *consciousness* overlapped for Victorians.[99] So, given the group origins *Descent* finds for the higher psychical processes, it is no surprise that the main exercise of consciousness for Darwin was in reflexive assessment of one's actions with regard to moral standards.

One strand of recent research on consciousness converges with Darwin's thinking, namely, that showing conscious agents must *know* and *care* about their knowledge.[100] In this view, consciousness includes something more than a 'global workspace' for highlighting bits of external information, namely: an emotionally engaged capacity for *self-monitoring*. This inclusion reconnects consciousness with self-criticism, as in Hannah Arendt's equation of Adolf Eichmann's 'disastrous failure of what we commonly call conscience'—in masterminding the transportation of hundreds of thousands of Jews to Nazi extermination camps during the Second World War—with his inability 'to think.'[101] The make-up of the word

---

[97] Stanislas Dehaene and Lionel Naccache, 'Towards a Cognitive Neuroscience of Consciousness: Basic Evidence and a Workspace Framework,' *Cognition* 79, (2001), pp.1–37; Daniel Dennett, 'Are We Explaining Consciousness Yet?,' *Cognition* 79, (2001), pp.221–237.

[98] As in James Ferrier, 'An Introduction to the Philosophy of Consciousness, Part I,' *Blackwood's Edinburgh Magazine* 43, (1838). See 4.2 A drama of surfaces. The word consciousness occurs 18 times in *Descent*: 5 times in quotes from others; 10 times in the hyphenated word, *self-consciousness*; and twice when invoking self-consciousness.

[99] Phillips, 'Against Self-Criticism.'

[100] Axel Cleeremans, 'The Radical Plasticity Thesis: How the Brain Learns to Be Conscious,' *Frontiers in Psychology* 2, (2011), https://doi.org/10.3389/fpsyg.2011.00086.; Stanislas Dehaene, Hakwan Lau, and Sid Kouider, 'What Is Consciousness, and Could Machines Have It?,' *Science* 358, (2017), pp.486–492.

[101] Hannah Arendt, 'Thinking and Moral Considerations,' *Social Research* 38, (1971), p.418.

con-science underlines this connection, observes Arendt, 'insofar as it means "to know *with and by* myself."'[102]

The characters of conscience described in *Descent* also chime with philosophical discussions about liberty and nobility. *Descent* frowns at some aspects of conscientious conduct, because the inward monitor browbeats praise-seekers into actions both 'abject' and 'absurd.'[103] Reason is our first safeguard against such conformism. Beyond this, *Descent* lauds the unthinking impulsiveness which produces the highest forms of sympathetic nobility and self-sacrifice. Darwin's critique of praise-seekers shows some parallels with Nietzsche's advocacy of escape from the transitional 'slave morality' of European society, *en route* to something that Darwin also envisaged: a future and higher state, 'as we may hope, even than the Caucasian' (in Nietzsche: 'the superman').[104]

7.5.2 Individualism or psychosociality?

Descent gives consciousness a moral form by deriving it from the fact that human progenitors lived in groups. This runs counter to the work of a lively band of Darwin commentators, past and present, who have found Darwin's work to be individualistic, in keeping with his privileged, Whig-loving, class-position. He was 'bourgeois', a 'rampant Malthusian individualist', a 'wealthy squire' or 'squarson' (squire-cum-parson), living on inherited funds in a decommissioned rectory, 'upholding the paternalist order', and craving 'Broughamite respectability'.[105] Friedrich Engels took this line as early as 1875:

> All that the Darwinian theory of the struggle for existence boils down to is an extrapolation from society to animate nature of Hobbes' theory of the *bellum omnium contra omnes* [war of all against all] and of the bourgeois-economic theory of competition together with the Malthusian theory of population.[106]

Historical commentaries which diagnose individualism in Darwin may trace it to influence by apologists for rampant capitalism like Malthus, Adam Smith, and

---

[102] Hannah Arendt, *The Life of the Mind: Vol.1 Thinking* (London: Secker & Warburg, 1978), p.5, my italics.

[103] Darwin, *Descent*, pp.118, 122.

[104] Darwin, *Descent*, p.156; Friedrich Nietzsche, *Thus Spoke Zarathustra* (Harmondsworth: Penguin Books, 1975).

[105] Adrian Desmond, *The Politics of Evolution: Morphology, Medicine, and Reform in Radical London* (Chicago: University of Chicago Press, 1989), p.408 & passim; Adrian Desmond and James Moore, *Darwin's Sacred Cause: Race, Slavery and the Quest for Human Origins* (London: Penguin, 2009), p.370; Adrian Desmond and James Moore, *Darwin* (Harmondsworth: Penguin, 1991), pp.657–658.

[106] Frederick Engels, 'Letter to Pyotr Lavrov, 12–17 November 1875', in *Karl Marx and Frederick Engels. Works* (Moscow: Politizdat, 1964).

John Ramsay McCulloch.[107] Biologists in thrall to the twentieth century's Modern Synthesis also make Darwin into an individualist, claiming as his central idea the belief that the only entity to benefit from natural selection is the individual, or, more precisely, the individual's 'selfish' genes.[108]

Chapter 6 showed that Darwin did sometimes truckle to the imperatives of Victorian propriety. Yet, in its natural history of human agency, *Descent*'s proposals do not reduce to individualism. Interdependence had long taken centre-stage in Darwin's understanding of nature, a point missed by many contemporary commentators.[109] This emphasis underpins *Descent*'s crucial theoretical distinction between the fates of 'social' versus 'non-social' animals in evolution, a distinction that occupied Darwin's thinking from the 1830s through the 1870s. As its result, Darwin assumes no firm line separating the psychological processes of individuals and the dynamics of groups (see 3.2.4 'What is an individual?').

Here, readers must confront what Johanna Motzkau has called 'the paradox of the psychosocial.'[110] Darwin wrote before psychology split from sociology. By 1910, however, the new psychology took as its main focus the study of individuals, leaving the study of society to sociology.[111] This rift has made *the psychosocial*—those human characteristics which make us both 'one and many'—increasingly unthinkable over the twentieth century. Because each discipline specialized on only one side of the individual–society divide, both psychology and sociology grew up in ways which obscured how we are all members, one of another.[112] Instead, denizens of both disciplines ultimately ended up 'thinking the psychic and the social in separation.' Consequently, a variety of crucial phenomena have become marginal to the human sciences: suggestibility; infant inter-subjectivity; transference and countertransference; sympathy; synchronicity; and 'group mind' or groupness.[113] Darwin's group-based understanding of conscience and the higher psychical faculties sits squarely in this psychosocial space.

---

[107] Silvan Schweber, 'Darwin and the Political Economists: Divergence of Character,' *Journal of the History of Biology* 13, (1980), pp.195–289; Elliott Sober, 'Darwin on Natural Selection: A Philosophical Perspective,' in *The Darwinian Heritage*, ed. David Kohn (Princeton NJ: Princeton University Press, 1985), pp.867–900.; Robert Young, 'Darwinism *Is* Social,' in *The Darwinian Heritage*, ed. David Kohn (Princeton: Princton University Press, 1985), pp.609–639. For a critique of Young et al.'s 'contextualism,' see David Hull, 'Deconstructing Darwin: Evolutionary Theory in Context,' *Journal of the History of Biology* 38, (2005), pp.137–152.

[108] 'The selfish gene theory *is* Darwin's theory,' writes Richard Dawkins, *The Selfish Gene: 30th Anniversary Edition* (Oxford: Oxford University Press, 2006), p.xv, my emphasis.

[109] E.g. Kenneth Gergen, *Relational Being: Beyond Self and Community* (Oxford: Oxford University Press, 2009), pp.381ff.

[110] Johanna Motzkau, 'Exploring the Transdisciplinary Trajectory of Suggestibility,' *Subjectivity* 27, (2009), pp.174–177.

[111] Alan Touraine, 'Is Sociology Still the Study of Society?,' *Thesis Eleven* 23, (1989), pp.5–34.

[112] Lisa Blackman, 'Affect, Relationality and the Problem of Personality,' *Theory, Culture and Society* 25, (2008), pp.81–100; Motzkau, 'Suggestibility,' p.175.

[113] But see e.g. Colwyn Trevarthen and Kenneth Aitken, 'Infant Intersubjectivity: Research, Theory, and Clinical Applications,' *Journal of Child Psychology and Psychiatry* 42, (2001), pp.3–48.; Motzkau, 'Suggestibility'; Blackman, 'Affect, Relationality'; Ben Bradley, 'An Approach to Synchronicity: From

Some might argue Darwin is therefore a social psychologist.[114] *Social psychology* certainly grew from attempts to colonize the no-man's-land created by the split between psychology and sociology—though, obviously, from the side of psychology.[115] Inevitably therefore, these attempts, various as they at first were, have increasingly collapsed the psychosocial events we live and breathe—conflict, cooperation, intimacy, conformity, social influence, social movements, aggression, peace, power, oppression, revolution—back into the heads of individuals, transmuting real-world psychosocial phenomena into the study of your and my so-called 'social cognitions.'[116]

In contrast to the focusing of psychology on individual cognition, psychoanalysis became a practice and theory genuinely rooted in the psychosocial. This should come as no surprise, because *Descent* was an important catalyst in the formation of Freud's take on culture. Freud's central myth in *Totem and Taboo* hailed from one of Darwin's several offbeat and contradictory riffs on primeval promiscuity, namely: that a jealous dominant male would have stamped out promiscuity by expelling his sons from his clan. Such a father, says *Descent*, would:

> not have been a social animal, and yet have lived with several wives, like the gorilla; for all the natives "agree that but one adult male is seen in a band; when the young male grows up, a contest takes place for mastery, and the strongest, by killing and driving out the others, establishes himself as the head of the community." The younger males, being thus expelled and wandering about, would, when at last successful in finding a partner, prevent too close interbreeding within the limits of the same family.[117]

Synchrony to Synchronization,' *International Journal of Critical Psychology* 1, (2001), pp.119–144; Jane Selby and Ben Bradley, 'Infants in Groups: A Paradigm for the Study of Early Social Experience,' *Human Development* 46, (2003), pp.197–221.

[114] E.g. Robert Farr, 'On Reading Darwin and Discovering Social Psychology,' in *The Development of Social Psychology*, ed. Robert Gilmour and Steven Duck (London: Academic, 1980), pp.111–136.

[115] Just as Gabriel Tarde and Émile Durkheim arguably colonized the same no-man's-land *from the side of sociology*: Lisa Blackman, 'Reinventing Psychological Matters: The Importance of the Suggestive Realm of Tarde's Ontology,' *Economy and Society* 36, (2007), pp.574–596; Kenneth Gergen, 'On the Very Idea of Social Psychology,' *Social Psychology Quarterly* 71, (2008), pp.331–337; Émile Durkheim, *Suicide: A Study in Sociology* (London: Routledge & Kegan Paul, 1952), pp.306–320; Katherine Pandora, *Rebels within the Ranks: Psychologists' Critique of Scientific Authority and Democratic Realities in New Deal America* (Cambridge: Cambridge University Press, 1997), pp.90, 99–107, & passim.

[116] Julian Henriques et al., *Changing the Subject: Psychology, Social Regulation and Subjectivity* (London: Routledge, 1998), pp.11–25 & passim; Kenneth Gergen, 'Social Psychology and the Wrong Revolution,' *European Journal of Social Psychology* 19, (1989), pp.465 & passim; Rainer Greifeneder, Herbert Bless, and Klaus Fiedler, *Social Cognition: How Individuals Construct Social Reality*, 2nd ed. (London: Routledge, 2018).

[117] Darwin, *Descent*, p.591. Note, gorillas were not social animals, according to Darwin. Humans were. So the relevance of this passage to human evolution is dubious.

Totem and Taboo extrapolated this into what Freud called a Just-So Story, a scientific myth portraying tribal sons revolting against their tyrannical father, and, like Oedipus, killing him—but also, *eating* him.[118] Such parricide would have created 'a band of brothers,' wherein remorse and its repression would be inescapable.[119] Thereafter, a symbolized father would come to be represented by the tribe's idealized totem animal, whom all members were forbidden to kill—a ban accompanied by a ban on incest (in the dead patriarch's honour). Totemism thus laid 'the beginnings of religion, morality, and social organization,' said Freud—and in particular, the army and the church.[120] The super-ego was ultimate heir to this complex set of repressions.

More generally, Freud's late thought grew from the central insight of *Descent* Part I: that '*the psychology of the group is the oldest human psychology.*'[121] The psychology of the individual *develops from* group psychology—or 'the primal horde'— said Freud, the first step being taken by a hypothetical man in the distant past who 'freed himself from the group and took over the father's part.' This he would have done symbolically and imaginatively, 'disguising the truth' (of parricide) by inventing a meaning-giving origin myth of the horde's heroic self-creation. This legendary first leader would thus have been 'the first epic poet'—harking back to the 'wise ape-like animal' whom Darwin pictured as having invented human language.[122] This legend not only underpinned Freud's understanding of civilization and its discontents, it also informed his account of the growing individual's subjective structuration by his or her Oedipus complex. Indeed, homologous dynamics of individuation-versus-group-demands arguably triangulate our every word.[123]

Freud's division of the human psyche into id, ego, and super-ego resonates with Darwin's psychology in more than one way. It implies human conduct is the product of interaction between alternative selves. Or, in more contemporary lingo, it implies 'multi-subjectivity.'[124] And it merges 'individual' subjectivity with the psychology of the group—or society—in which an individual has their being.[125]

---

[118] Sigmund Freud, *Group Psychology and Analysis of the Ego* (London: Hogarth Press, 1949), p.90.

[119] Sigmund Freud, *Totem and Taboo: Resemblances between the Psychic Lives of Savages and Neurotics* (New York: Moffat Yard & Co., 1918), pp.254–255.

[120] Freud, *Totem and Taboo*, p.258; Freud, *Group Psychology*, p.54.

[121] Freud, *Group Psychology*, p.92, my italics.

[122] Freud, *Totem and Taboo*, pp.54, 65; Sean Dyde tells me that Freud's formulation echoes an older idea, traceable back through Coleridge, Schlegel, and Schelling, to Edmund Burke, who held that the best 'school of moral sentiments' for 'the swinish multitude' was the theatre and its 'poets who ... apply themselves to the moral constitution of the heart': Edmund Burke, *Reflections on the Revolution in France* (London: Dent, 1910), pp.76–78.

[123] Jacques Lacan, *Speech and Language in Psychoanalysis*, ed. Anthony Wilden (Baltimore: Johns Hopkins University Press, 1981).

[124] Jane Selby, 'Feminine Identity and Contradiction: Women Research Students at Cambridge University' (Doctoral thesis, Cambridge University, 1985); Henriques et al., *Changing the Subject*.

[125] Sigmund Freud, *Civilization and Its Discontents* (London: Hogarth Press, 1930).

The idea that the individual is fundamentally groupish has developed into a significant tradition in psychoanalysis, of which early highlights were the works of Wilfred Bion and Isabel Menzies Lyth.[126]

A possible conduit between Darwin and Freud was the polymath William Clifford (1845–1879), whose essays 'On the Scientific Basis of Morals,' 'The Ethics of Belief,' and 'Body and Mind' attracted the eyes of Hughlings Jackson, and the ire of William James.[127] Clifford's explicitly drew his multi-subjective psychology of morals from *Descent*. He proposed humans had multiple selves, the first of which, 'the conscious subject, *das Ich*, [comprised] the whole stream of feelings which make up consciousness regarded as bound together by association and memory.' Within this, Clifford divided off an 'ethical' self: 'a selected aggregate of feelings and of objects related to them which hang together as a conception by virtue of long and repeated association. My self does not include all my feelings, because I habitually separate off some of them, say they do not properly belong to me, and treat them as my enemies.'[128] Beyond or beneath these selves lies 'the tribal self':

> The actual pains or pleasures which come from the woe or weal of the tribe, and which were the sources of this conception [the tribal self], drop out of consciousness and are remembered no more; the symbol which has replaced them becomes a centre and goal of immediate desires, powerful enough in many cases to override the strongest suggestions of individual pleasure or pain.[129]

The self that follows unconscious tribal mores Clifford calls 'pious.' Finally, when a member of a tribe does something to harm it, 'the tribal self wakes up,' and says: 'In the name of the tribe, I do not like this thing that I, as an individual, have done.' This self-judgement 'in the name of the tribe is called Conscience.'

As Clifford highlighted, Darwin's schema of moral development referred to human collectives more than individuals. It is culture, civilization, and collective intelligence that leads 'the standard of morality' to 'rise higher and higher.' In modern psychology, on the other hand, moral development has usually been understood individualistically. Lawrence Kohlberg's famous six-stage progression from the toddler's 'punishment-and-obedience orientation' to the seldom-reached 'universal-ethical-principle orientation' sounds superficially similar to Darwin's account. But it is described solely in terms of *individual* development: the step-wise

---

[126] Wilfred Bion, *Experiences in Groups and Other Papers* (London: Tavistock, 1961); Isabel Menzies-Lyth, 'The Functioning of Social Systems as a Defence against Anxiety,' *Human Relations* 13, (1959), pp.95–121.

[127] William Clifford, 'On the Scientific Basis of Morals: A Discussion,' *Contemporary Review* 26, (1875), pp.650–660; Austin Duncan-Jones, 'The Notion of Conscience,' *Philosophy* 30, (1955), p.136; David Hollinger, 'James, Clifford and Scientific Conscience,' in *The Cambridge Companion to William James*, ed. Ruth Putnam (Cambridge: Cambridge University Press, 1997)

[128] Clifford, 'Basis of Morals,' pp.652–653.

[129] Clifford, 'Basis of Morals,' p.655.

growth of moral understanding in 'the child' running parallel to Jean Piaget's ladder of cognitive stages. Furthermore, the universality of Kohlberg's highest stage is confined in its application to human rights and does not include animal rights—contrary to Darwin.[130]

7.5.3 Group selection?

Perhaps the most controversial resonance between Darwin's group-based human psychology and modern evolutionary thinking is traceable to the Modern Synthesis, and concerns the possibility of group-level evolution, or *group selection*. This controversy dates back at least fifty years, to George Williams' iconoclastic book, *Adaptation and Natural Selection* (1966). Williams' 'gene's eye' account of evolution gained traction a decade later when popularized by Richard Dawkins' best-seller, *The Selfish Gene*. Both men argued natural selection could never promote 'the good of the group,' in a way that did not profit *the genes of* the individual members of those groups. Hence Williams spent several chapters arguing that 'group-related adaptations do not, in fact, exist'—which gave Dawkins his starting point for *The Selfish Gene*.[131]

Their minds focused by the likes of Williams and Dawkins, biologists and Evolutionary Psychologists continue to debate the pros and cons of group selection—though seldom informed by what Darwin actually wrote. One contention surrounds the place of phenotypes in natural selection. Do they play passive or active roles? Williams, Dawkins, and their allies today assume the passivity of phenotypes: organisms are the effects of struggles between genes.[132] However, a number of biologists are now advocating a 'multilevel' approach to understanding macro-evolutionary change—levels including selection at molecular, genetic, and meiotic levels right up to the phenotypic level (the agentive, 'plastic,' organism), the 'trait' group, and even an entire species—although the plastic phenotype leads the parade.[133] Moreover, as discussed in Chapter 3, cooperation and mutual aid play

---

[130] Darwin, *Descent* p.125; Lawrence Kohlberg, 'The Claim to Moral Adequacy of a Highest Stage of Moral Judgment,' *Journal of Philosophy* 70, (1973), pp.630–646.

[131] George Williams, *Adaptation and Natural Selection* (Princeton NJ: Princeton University Press, 1966), pp.93ff; Dawkins, *Selfish Gene*.

[132] E.g. Steven Pinker, 'The False Allure of Group Selection,' in *Handbook of Evolutionary Psychology*, ed. David Buss (Hoboken NJ: Wiley, 2016), pp.867–879.

[133] Mary Jane West-Eberhard, *Developmental Plasticity and Evolution* (New York: Oxford University Press, 2003), p.616; Elliott and Wilson Sober, David, *Unto Others: The Evolution and Psychology of Unselfish Behaviour* (Cambridge MA: Harvard University Press, 1998), p.92; David Wilson and Edward Wilson, 'Rethinking the Foundation of Sociobiology,' *Quarterly Review of Biology* 82, (2007), pp.327–348; Martin Nowak, Corina Tarnita, and Edward Wilson, 'The Evolution of Eusociality,' *Nature* 466, (2010), pp.1057–1062. For a critique of the last paper, see Patrick Abbott et al., 'Inclusive Fitness Theory and Eusociality,' *Nature* 466, (2011), pp.E1–E4.

an increasing role in contemporary biologists' revisions of the Modern Synthesis, across a wide range of species, and at many different levels of selection.[134]

7.5.4 The brain

A rather different version of evolutionary psychology from that based in Santa Barbara proves more syntonic with Darwin's arguments in Part I of *Descent*. This version champions the 'social brain': the hypothesis that the cognitive demands of living in complexly bonded social groups have long shaped human evolution— particularly of the neo-cortex.[135] According to this hypothesis, one need assume just one good reason for our ancestors to have lived in groups: better protection; easier hunting; more successful breeding; language use; whatever. Thenceforth, the need to manage the demands of group-life would be the primary force shaping human evolution. Once we accept this logic, then we can see how all the social virtues *Descent* discusses would thereafter be selected, *given that our ancestors already lived in groups*, and that such groups needed to be as cohesive as possible, the better to survive and the better for their members to multiply.[136]

Contemporary understandings of the neo-cortex strengthen *Descent*'s explanation for human intelligence as arising from increased intercommunication between parts of the brain. Some Evolutionary Psychologists favour an opposite view, that the human brain exhibits *massive modularity*, such that the skull houses a Stone Age collection of fossilized, domain-specific adaptations, each the product of natural selection, making our minds like glorified Swiss army knives (for more on this see 8.4.2 Cerebral modularity in Evolutionary Psychology today).[137] Evidence from today's neurobiologists militates against massive modularity, suggesting that cerebral development shows an enormous inherent plasticity, and 'contextual interactivity'—meaning the individual brain's immediate environment has a crucial role in shaping later function.[138]

---

[134] Martin Nowak, 'Evolving Cooperation,' *Journal of Theoretical Biology* 299, (2012), pp.1–188; Nowak, 'Why We Help.'

[135] Robin Dunbar, 'The Social Brain: Mind, Language, and Society in Evolutionary Perspective,' *Annual Review of Anthropology* 32, (2003), pp.163–181; Robin Dunbar, ed. *Oxford Handbook of Evolutionary Psychology* (Oxford: Oxford University Press, 2007).

[136] William Wcislo, 'Behavioural Environments and Evolutionary Change,' *Annual Review of Ecology and Systematics* 20, (1989), pp.156–157.

[137] Leda Cosmides and John Tooby, *Evolutionary Psychology: A Primer*, (Santa Barbara CA: University of California, 1997), https://www.cep.ucsb.edu/primer.html; Sperber, 'In Defense of Massive Modularity.'

[138] Clark Barrett, 'Modularity,' in *Evolutionary Perspectives on Social Psychology*, ed. Virgil Zeigler-Hill, Lisa Welling, and Todd Shackelford (Cham: Springer, 2015), p.43; David Buller and Valerie Hardcastle, 'Evolutionary Psychology, Meet Developmental Neurobiology: Against Promiscuous Modularity,' *Brain and Mind* 1, (2000), pp.307–325; Jaak Panksepp and Jules Panksepp, 'The Seven Sins of Evolutionary Psychology,' *Evolution and Cognition* 6, (2000), pp.108–131; Johan Bolhuis et al.,

7.5.5 Return to observation

Finally, what are we to make of the evidence Darwin assembled in *Descent*? Darwin's observational approach to animals was dismissed by the first modern psychologists. This stigma has stuck. A century after James Angell laughed in *Psychological Review* at the 'simple-mindedness' of Darwin's observations of animals, Henry Plotkin's history, *Evolutionary Thought in Psychology*, rubbishes the evidence Darwin presented in *Descent* as 'a catalogue of outrageous anecdotes gained from his correspondence,' to which 'no credence at all can be given.'[139] Plotkin overlooks the fact that Darwin himself painstakingly recorded many of the observations in Part I of *Descent*, particularly on apes, babies, and dogs. Moreover, and doubtless to Plotkin's puzzlement, many of the observations of animals reported in *Descent* have proven robust, being borne out by the latest research (see also Ch. 9 on Darwin's path-breaking observations of infants).

The rising credit of Darwin's animal observations partly stems from a swing towards *evolutionary cognition* in today's zoology.[140] Drawing on both comparative psychology and ethology, the last decade has seen a steep rise in the level of mental sophistication attributable to animals. Experimenters now recognize that different species have been primed by their evolutionary history to be experts in different kinds of task. Designing species-friendly experiments, which draw on what a species does well in the wild, renders scientists a plethora of findings that back, or even exceed, many of Darwin's less rigorous observations about the cleverness of ape, bird, bee, and dog.[141] Rooks make tools.[142] Pelican flocks foraging on a lake can increase their food-intake if all flock-members closely coordinate the moment at which they dip their bills in the water: the better the synchronization, the more likely each is to catch a fish.[143] Bumblebees can perform a novel task better (moving a tiny bee-sized 'football' from one place to another)—and better than

---

'Darwin in Mind: New Opportunities for Evolutionary Psychology,' *PLoS Biology* 9, (2011), https://journals.plos.org/plosbiology/article/file?id=10.1371/journal.pbio.1001109. Bolhuis et al. imply *Descent*'s view of 'cognitive evolution' underpins the one elaborated by Tooby and Cosmides which assumes massive modularity. They then refute this view by advocating the plasticity of the human brain.

[139] James Angell, 'The Influence of Darwin on Psychology,' *Psychological Review* 16, (1909), p.158; Henry Plotkin, *Evolutionary Thought in Psychology: A Brief History* (Oxford: Blackwell, 2004), pp.36–37.

[140] Danielle Sulikowski and Darren Burke, 'From the Lab to the World: The Paradigmatic Assumption and the Functional Cognition of Avian Foraging,' *Current Zoology* 61, (2015), pp.328–340.

[141] Frans De Waal, *Are We Smart Enough to Know How Smart Animals Are?* (New York: Norton, 2016).

[142] Christopher Bird and Nathan Emery, 'Insightful Problem-Solving and Creative Tool-Modification by Captive Non-Tool-Using Rooks,' *Proceedings of the National Academy of Sciences of the United States of America* 106, (2009), http://www.pnas.org/content/106/25/10370.long. See also Nathan Emery, *Bird Brain: An Exploration of Avian Intelligence* (Princeton NJ: Princeton University Press, 2016); Gisela Kaplan, *Bird Minds: Cognition and Behaviour of Australian Native Birds* (Clayton South VIC: CSIRO, 2015).

[143] Blair McMahon and Roger Evans, 'Foraging Strategies of American White Pelicans,' *Behaviour* 120, (1992), pp.69–89.

their model—having observed another bee perform the required behaviour.[144] And the peaceable, sociable, group-living, highly-sexed, 'pygmy' chimpanzee or bonobo is able to reason, maintain a culture, plan, make tools, laugh, interpret signs by context, cooperate (against bullies for example), sympathize, be compassionate, altruistic, and more.[145]

In particular, Darwin's stress on the intellectual refinements of dogs—or *dognition*—has gained strong support from studies showing that the domestication of dogs has resulted in a heightened capacity to understand human body-language, meaning that dogs understand better than most other animals what human experimenters intend by their intelligence tests, and so perform 'better' (more in accord with human standards).[146] This research resonates strongly with *Descent*.

Take, for example, what dog-owners call the 'guilty look.' Dog-lover Darwin reported: 'when a dog rushes after a hare, is rebuked, pauses, hesitates, pursues again, [he] returns ashamed to his master.'[147] Or again: 'Dogs ... feel shame, when doing anything which is wrong.— as eating meat, doing their dirt, running home ... Squib at Maer, used to betray himself by looking ashamed *before it was known* he had been on the table, — guilty conscience.'[148]

A recent study tries to disentangle the psychosocial dynamics going on here—by enlisting dog-owners in a *pretended* test of the obedience of dogs in the absence of their owners.[149] The study was filmed. First, each owner was instructed to order their dog not to eat a treat, after which the owner left the room. Half the dogs then received the forbidden treat from the experimenter; the other half got no treat. When the owner returned, he or she was told whether or not their dog had obeyed their command, having been instructed either to greet or to scold their dog accordingly. *But the experimenter lied.* 'Guilty looks' were then counted.

Findings showed sinful dogs did not differ significantly from the virtuous in the frequency of guilty looks shown. The key difference was between dogs whose owners *thought* they had sinned, and scolded them, versus owners who heard their dogs had obeyed, and so greeted them approvingly. Dogs acted guiltiest when scolded—*even if they had not been fed the treat*. In fact they acted *more* guiltily if they were scolded unjustly (when they had *not* had a treat), than if they were

---

[144] Olli Loukola et al., 'Bumblebees Show Cognitive Flexibility by Improving on an Observed Complex Behaviour,' *Science* 355, (2017), pp.833–836.

[145] Frans De Waal and Frans Lanting, *Bonobo: The Forgotten Ape* (San Francisco CA: University of California Press, 1997).

[146] Brian Hare and Vanessa Woods, *The Genius of Dogs* (London: Oneworld, 2013). Interestingly, the behaviour and sensitivities of the albino Wistar (or Norway) rats—used by psychology's behaviourists from the early decades of the 1900s on—have been similarly altered during their domestication: Bonnie Clause, 'The Wistar Rat as a Right Choice: Establishing Mammalian Standards and the Ideal of a Standardized Mammal,' *Journal of the History of Biology* 26, (1993), pp.329–349.

[147] Darwin, *Descent*, p.107.

[148] Darwin, 'Notebook M,' pp.23–24, my italics.

[149] Alexandra Horowitz, 'Disambiguating the "Guilty Look": Salient Prompts to a Familiar Dog Behaviour,' *Behavioural Processes* 81, (2009), pp.447–452.

scolded justly! This shows that dogs respond sensitively to the disapproval and approval of their pack 'leader' (owner)—something Darwin claimed was essential to human moral conduct, and to blushing (cf. Praise-seeker, p. 249–250; see also Ch. 5).[150]

7.6 Conclusion

The instincts peculiar to social animals struck the twenty-nine-year-old Darwin as his best foundation for explaining moral conduct, the domain of agency he was to deem 'highest' in humans by the time he finished the second edition of Descent—higher even than intellectual powers.[151] Yet working out in detail *how* sociality could spawn conscience required decades more work, both regarding the processes producing its natural selection, and how to depict the group dynamics likely to have dominated early tribal life.

The fruit of this work forms three key chapters in Part I of Descent—the part of the book Darwin wrote last. These chapters portray human social life as governed by several conflicting vectors of impulse and self-consciousness. Just as we saw Descent Part III argue that the standards of sexual beauty diverge in different human tribes (see Ch. 6), so Part I describes the human moral sense as being focused on different kinds of moral standard in different social groups, whether tribal or 'civilized.' Group-dynamics do not completely explain moral agency, however. Reason may elevate an act above the demands of one's immediate fellows to embrace all sentient creatures. The highest form of moral action transcends even reason, being an impulsive form of self-sacrifice bestowed on any other creature—whether from one's own species or another.

True to the reflexiveness of any form of agency (see Ch. 3), group-living transforms how human beings think and act, and is thence, itself transformed. Moral understanding advances and cultures diverge. Which raises the question of whether, and to what extent, Darwin's understanding of, and psychological explanations for, human agency were entirely evolutionary—particularly in more complex human societies. Chapter 8 examines this question.

---

[150] This study does not show that dogs do not feel guilt, because it is impossible for any observation to prove *the lack* of a capacity. Also: eating an unsolicited treat proffered by an adult experimenter was an ambiguous test of an owner's prohibition—especially when the owner was absent. Perhaps the tempted dogs felt that the experimenters were 'in loco parentis,' so it would have been impolite to refuse the treats!

[151] Darwin, *Descent*, p.611.

8

Culture

To His true believers, God is a one-stop shop for explanations, is He not? God created everything. God effects everything. God knows everything ...? Well, not entirely. Because there is always free will. Without free will there would be no sin, no
moral struggle, and hence no practice of virtue. Free will means ordinary, sinful
human mental life, and ordinary human action, cannot have a divine origin. Hence
any psychology of the common-or-garden here-and-now must appeal to secular,
or diabolical, explanations.

What about evolution? Is evolution a one-stop shop? Has evolution—or, better
perhaps, natural selection—created all organisms and all human beings, made our
minds, pushed us to do what we do, created our ideas, made the truth of science? Does
evolutionary explanation have a limit? For some scientists, it does not—even where
humans are concerned. Natural selection not only provides the explanatory framework for studying all the organisms that inhabit planet Earth, but natural selection
and processes homologous to natural selection give the best explanatory framework
for studying everything in the domain which includes the human mind and brain-
function ('neural Darwinism'), emotional relationships, the progress of science, and—
most relevantly here—the aspects of human agency we call culture (sometimes called
'the *noösphere*': see Figure 8.1).[1] Proponents of a truly general theory of evolution
have further extended Darwin's template downwards and outwards, to explain the
origin of every kind of atom in the physical *cosmosphere*—which embraces the whole
universe—and the entire suite of molecules in the chemically-conceived *astrosphere*
(or *planetosphere*: Figure 8.1). To its true believers, Darwinism is universal.[2]

---

[1] *Noösphere* was a term introduced by Teilhard De Chardin, *The Phenomenon of Man*
(London: Collins, 1959); Gerald Edelman, *Neural Darwinism: The Theory of Neuronal Group Selection*
(New York: Basic Books, 1987).

[2] For illustrations of universal selectionism/Darwinism, see: Julian Huxley, *Evolution in Action*
(New York: Harper & Row, 1953), pp.2ff; Richard Dawkins, 'Universal Darwinism,' in *The Nature of
Life: Classical and Contemporary Perspectives from Philosophy and Science*, ed. Mark Bedau and Carol
Cleland (Cambridge: Cambridge University Press, 2010); the idea of 'universal acid' in Daniel Dennett,
Darwin's Dangerous Idea: Evolution and the Meanings of Life (New York: Simon & Schuster, 1995);
Alex Mesoudi, *Cultural Evolution: How Darwinian Theory Can Explain Human Culture and Synthesize
the Social Sciences* (Chicago: University of Chicago Press, 2011); Joseph Carroll, Dan McAdams, and
Edward Wilson, eds., *Darwin's Bridge: Uniting the Humanities and Sciences* (Oxford: Oxford University
Press, 2016); Matt Ridley, *The Evolution of Everything* (New York: Harper, 2016). For discussion, see
Betty Smocovitis, 'Unifying Biology: The Evolutionary Synthesis and Evolutionary Biology,' *Journal
of the History of Biology* 25, (1992), pp.1–65; Steven Rose, *Lifelines: Biology, Freedom, Determinism*
(Harmondsworth: Penguin Books, 1997), pp.174–208 & passim.

Darwin's Psychology. Ben Bradley, Oxford University Press (2020). © Oxford University Press.
DOI: 10.1093/oso/9780198708216.001.0001

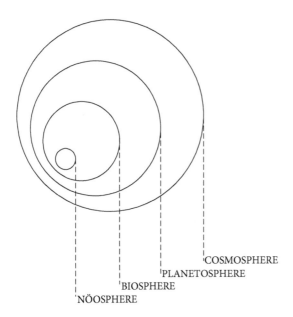

COSMOSPHERE
PLANETOSPHERE
BIOSPHERE
NÖOSPHERE

Figure 8.1 According to the most general theory of evolution, it has four spheres: 'Each is characterised by time and space scales of decreasing magnitude from cosmic to psycho-social; each involves units of increasing magnitude; each achieves products of increasing complexity.'

Source: Peter Sylvester-Bradley, 'An Evolutionary Model for the Origin of Life', in *Understanding the Earth*, ed. Ian Gass, Peter Smith, and Richard Wilson (Horsham: Open University Press, 1971), pp.123–124 (including Figure) & passim.

Which poses a question about Darwin. Was his psychology evolutionary through and through? Did he think evolution explained every facet of human agency? Or, do Darwin's writings about agency suggest there are limits beyond which evolutionary explanations cannot go? If so, what are those limits? And how right was he in his views? These questions set up the argument of this chapter, namely, that while Darwin's two-axis treatment of the living world *does* encompass the cultural aspects of human agency—some aspects of culture *do* exceed the reach of evolutionary explanation, just as do some phenotypic excesses of animal agency (see 3.4.5 Phenotype-led evolution).

For Darwin, the theatre of agency generates variation. *Some* of the struggles and interdependencies which constitute that theatre also feed the second, long-term axis of the natural world: evolutionary change. Accordingly, an explanation of 'cultural evolution' should not need *adding* to Darwin's explanation for the origin of the human species, as is often assumed by authors building on the Modern

Synthesis.[3] Logically, whatever 'culture' feeds into the life of human groups that has an impact on survival is already covered by Darwin's larger theory, culture being but one player in the theatre of agency. Following this logic, we would expect to find *Descent* elaborating a *non-evolutionary* view of culture.

Accordingly, this chapter asks whether and how the connections between culture and selection are treated in *Descent*. I begin by examining how the book understands what Victorians called 'civilization' and 'culture'—two words that were sometimes, but not always, treated as synonyms. I then show that *Descent* does indeed frame human culture as a set of processes and institutions which function in *relative autonomy* from natural selection. I illustrate such relative autonomy by surveying *Descent's* discussions of language, religion, virtue, and aesthetic refinement.

8.1 Civilization and Culture in *Descent*

Several Victorian preoccupations fed and confused *Descent* Part III's discussion of human sexuality (see 6.3.1 Nodal concerns). Similar concerns are at work in Darwin's discussion of *culture* and *civilization*. For Darwin, as for Victorian ethnologists like Edward Tylor, these two terms never occurred in the plural—and were often treated as synonyms.[4] When synonymous, civilization was a single, cumulative continuum, connecting 'races high and low in the scale of culture' in an ascending series of *progressive* stages—with European men of Tylor's and Darwin's class standing on top.[5]

Against this self-congratulatory progressivism, social critics like William Blake and Thomas Paine indicted the savagery of the conditions to which the new satanic mills of urban industry consigned Britain's workers.[6] But the writings of Darwin

---

[3] E.g. Kevin Laland, *Darwin's Unfinished Symphony: How Culture Made the Human Mind* (Princeton: Princeton University Press, 2017), pp.13–14, justifies the argument of his recent book—that Darwin did not recognize 'the central role played by culture in the origins of mind'—by appropriating Darwin's disingenuously modest description ('imperfect and fragmentary') of just *one* of the chapters (Ch. 5) in Charles Darwin, *The Descent of Man, and Selection in Relation to Sex* (London: Murray, 1874), p.127. Laland repurposes this remark to deny the existence of Darwin's entire account of psychological matters. Hence, Laland continues, 'comprehending the evolution of the human mind is Darwin's unfinished symphony'—the symphony Laland's book undertakes to complete.

[4] In 1871, Edward Tylor produced a famous definition, making the two terms refer to the same entity: 'Culture or Civilisation, taken in its wide ethnographic sense, is that complex whole which includes knowledge, belief, art, morals, law, custom, and any other capabilities and habits acquired by man as a member of society': Edward Tylor, *Primitive Culture: Researches into the Development of Mythology, Philosophy, Religion, Language, Art, and Custom* (London: Murray, 1971), p.1. The relations between these two words had a complex and polyglot history, however: see George Stocking Jr, *Victorian Anthropology* (New York: Free Press, 1987), pp.144–185.

[5] Tylor, *Primitive Culture*, p.114.

[6] Thomas Paine, *The Rights of Man Being an Answer to Mr Burke's Attack on the French Revolution* (London: Sherwin, 1817), p.19; William Blake, 'Milton, a Poem in 2 Books,' in *Blake: Complete Writings*, ed. Geoffrey Keynes (Oxford: Oxford University Press, 1969), p.481.

and his wealthy contemporaries scanted what he called 'the inferior members of society'—the impoverished—who attracted description as 'poor and reckless,' 'intemperate, profligate, and criminal.'[7] Darwin belonged to a 'superior class.' And it was to this class of men, not their 'inferiors'—nor women—that Darwin's writings referred when they proposed that 'the more efficient causes of progress seem to consist of a good education during youth whilst the brain is impressible, and of a high standard of excellence, inculcated by the ablest and best men.'[8]

Progress impassioned Victorians. Fed by the speed of social transformation we now call the Industrial Revolution, and the spread of the British empire, the rapid technological sophistication of British society was epitomized for men like Darwin by the success of railways, and the building of the Crystal Palace to house the Great Exhibition of 1851 (Figure 8.2).[9] In keeping, *Descent* uses the words '*civilisation*' and '*culture*' to both evoke progress and assert British culture's pre-eminence.

But was this an example of Darwin's text 'truckling' to the presumed expectations of his audience?[10] Because his optimistic statements about civilization clash with darker comments in his books, explicitly refuting the inevitability of progress—as we shall see.[11] Yet he still liked sometimes to assert the 'more cheerful' view, that 'progress has been much more general than retrogression; that man has risen, though by slow and interrupted steps, from a lowly condition to the highest standard as yet attained by him in knowledge, morals and religion.'[12]

Twinned with the creed of progress is the disparaging contrast *Descent* constantly draws—not only with the 'inferior' orders of Victorian society—but even more starkly and outrageously, with 'uncivilized' or 'barbarian' peoples: 'negroes'; the denizens of Tierra del Fuego; and tribes of the Australian outback.[13] Throughout *Descent*, and hence throughout this chapter, we see 'savages' derided

---

[7] Darwin, *Descent*, pp.138–140, & passim.

[8] Darwin, *Descent*, p.143.

[9] Faith in progress may also have helped give meaning to a world in which the Bible and belief in divine guidance were losing force, thanks to rapid social change and the rising authority of science: see James Secord, *Visions of Science: Books and Readers at the Dawn of the Victorian Age* (Chicago: University of Chicago Press, 2014).

[10] Re. 'truckling': Darwin privately confessed to Hooker, 'I have long regretted that I truckled to public opinion & used Pentateuchal term of creation, by which I really meant "appeared" by some wholly unknown process.' In *Origin*'s third edition (1861), he removed some of the contested terms. Darwin's truckling continued, however, and not just to the foibles of his Christian public. In 1866, he admitted to Lyell that an insertion about climatic cooling was 'a mere piece of truckling' to a botanical quibble from Hooker. (This was removed in the fifth and sixth editions.): Charles Darwin, 'Letter to Hooker, 29th March,' *The Correspondence of Charles Darwin* (Cambridge: Darwin Correspondence Project, 1863), Charles Darwin, 'Letter to Lyell, 15th February,' *The Correspondence of Charles Darwin* (Cambridge: Darwin Correspondence Project, 1866), pp.374, 484, 490, Charles Darwin, *On the Origin of Species by Means of Natural Selection or the Preservation of Favoured Races in the Struggle for Life* (London: Murray, 1859).

[11] Darwin, *Descent*, pp.132–133, 140–141.

[12] Darwin, *Descent*, p.145. For similar optimistic assertions, see Charles Darwin, *The Origin of Species by Means of Natural Selection, or the Preservation of Favoured Races in the Struggle for Life*, 6th ed. (London: Murray, 1876), pp.61, 428.

[13] Darwin, *Descent*, pp.65, 674 & passim.

Figure 8.2 The Crystal Palace (a) was built to house the Great Exhibition of 1851. This celebrated the many technological and industrial advances made by Victorian Britons—thereby establishing Britain's pre-eminent progress in this regard (b).

Source: (a) Philip Brannan, *View of the south side from near the Princes Gate, looking West,* 1851. Copyright © Victoria and Albert Museum, London; (b) Machinery Hall, Crystal Palace Exhibition, London, 1851. The Print Collector / Alamy Stock Photo.

in repellent terms. So it becomes important to state that the aim of this chapter is *not* to give credence to Darwin's Victorian prejudices, but to see how he uses statements about culture and civilization—however deplorable—in the explanation of human agency. In this, it is important to underline the well-established multivalence of Darwin's texts, and the dangers of reducing books like *Origin* and *Descent* to a single message.[14]

8.1.1 Culture as the cultivation of individuals

While the equation between civilization and culture championed by Victorian ethnologists and *Descent* projected a one-track ascent to the sophisticated marvels of present-day European society, a different sense of the word culture sometimes overlapped with or might even be *opposed* to the inexorable advance of civilization. This invoked development or (self-) cultivation in *an individual*.[15] For example, *Descent* describes 'men of a gentle nature,' as those 'given to meditation or culture of the mind.'[16] A like sense exists today: a winemaker may debate the best soil for the culture of a new vine. In this sense, not just in grapes, but in all creatures—plant or animal, unicellular or complex, embryonic or mature—observable characters in individual organisms are produced by their *culture* or *manner of rearing*.

Culture *as cultivation* always draws on environmental information, and so helps create and cement the interdependencies synonymous with the economy of nature. Twenty-first century theories of the phenotype give a helpful way of understanding this meaning of culture (see 3.4.5 Phenotype-led evolution). The Modern Synthesis posed random genetic events—mutation and recombination—as the creators of individual variation, thereby averting any need to consider a relatively autonomous process of individual development. Which made the environmental factors that determined the fates of 'random' genetic variation a parallel lottery of haphazard events. Once we equip ourselves with a theory of the phenotype, however, there is no need to pit 'genes versus environment.'[17] Rather, both genes and habitat play essential, regulatory roles in the development of phenotypic variations and, hence, of adaptation. At whatever age, it is the state of the organism itself which guides development. It is the organism which is structured to make particular use of 'external' substances and 'internal' information in specific ways—for example, by entering into specific symbioses, or by means of fine-tuned sensory processes.

---

[14] Gillian Beer, *Darwin's Plots: Evolutionary Narrative in Darwin, George Eliot and Nineteenth-Century Fiction*, 2nd ed. (Cambridge: Cambridge University Press, 2000).

[15] Tony Bennett, Lawrence Grossberg, and Meaghan Morris, *New Keywords: A Revised Vocabulary of Culture and Society* (Hoboken: Wiley, 2005): aligning somewhat with what Burke had called 'the spirit of a gentleman,' Germans conceived *culture* as the culmination of personal self-cultivation ('*Bildung*'), achieved by studying the highest moral and aesthetic manifestations of the human spirit. Against this, civilization was positioned as an external process of social organization and material advance. A similar notion of 'high' culture is to be found in Matthew Arnold, *Culture and Anarchy* (Cambridge: Cambridge University Press, 1932).

[16] Darwin, *Descent*, p.141.

[17] Mary Jane West-Eberhard, *Developmental Plasticity and Evolution* (New York: Oxford University Press, 2003), p.108.

Unsurprisingly, therefore, a structuring, developmental role for environmental input can be shown at all taxonomic grades of life. Parental exposure to climate (how much lizards bask in the sun) or diet (sugar and fat intake in fruit flies) can alter the physiology and fitness of offspring.[18] Docility in some rodents is increased by handling—a docility which then transmits to their offspring, or to genetically dissimilar foster offspring, as in wild pacas (*Agouti paca*).[19] Not that the growth-effects of a habitat are always beneficial. A pregnant mother's proclivity for gin may imperil her unborn child (foetal alcohol syndrome).

8.2 Relative Autonomy

Darwin looked on the living world as a theatre of agency comprising an organism's engagement with its habitat, including interdependencies with: conspecifics; other species; and inanimate surrounds. In this vein, we might predict—to the extent that the habitat of any member of the quintessentially social species, *Homo sapiens*, comprises the customs, social behaviour, products, and way of life of his or her particular community (i.e. culture)—that this habitat should cue the way each human develops, just as carbon dioxide levels cue the development of plants. We can test this prediction by examining the ways *Descent* discusses the relation of human social worlds to evolution.

Descent generally accepts that a stairway of stages in moral culture rises from a low in 'savages' to Europeans—or, more precisely, the English—at the top.[20] *However*, Darwin saw both the lower and the higher stages of this advance as having had little to do with natural selection. At most, the differing achievements of 'barbarian' and 'civilized' cultures were rooted in the psychical faculties which all humans owe to natural selection. *Descent* sandwiches mental processes between natural selection on the one hand, and, on the other, barbaric or civilized cultures and their more-or-less laudable customs and social institutions. 'With civilised nations,' the book states, 'as far as an advanced standard of morality, and an increased number of fairly good men are concerned, *natural selection apparently effects but little;* though the fundamental social instincts were originally

---

[18] E.g. Lisa Schwanz, 'Parental Thermal Environment Alters Offspring Sex Ratio and Fitness in an Oviparous Lizard,' *Journal of Experimental Biology* 219, (2016), pp.2349–2357; Carmen Emborski and Alexander Mikheyev, 'Ancestral Diet Transgenerationally Influences Offspring in a Parent-of-Origin and Sex-Specific Manner,' *Philosophical Transactions of the Royal Society of London* 374, (2018), https://royalsocietypublishing.org/doi/pdf/10.1098/rstb.2018.0181.
[19] West-Eberhard, *Developmental Plasticity and Evolution*, pp.114–115.
[20] Darwin, *Descent*, pp.66, 141–142: Darwin's paragons, 'the highest men of the highest races,' were all English: Shakespeare and Newton, plus '[prison-reformer John] Howard or [abolitionist Thomas] Clarkson'—Howard and Clarkson being proverbial heroes of Victorian philanthropy: see David Roberts, *The Social Conscience of Early Victorians* (Stanford CA: Stanford University Press, 2002), pp.244–245. Eurocentrism was a staple of Victorian ethnology in the 1860s and 1870s: Stocking Jr, *Victorian Anthropology*.

thus gained.'[21] In fact, as we shall see, *Descent* shows more concern for the ways in which social life *interferes with* natural selection, than with totting up the products of natural selection in contemporary humans.

The importance of cultural provision to Darwin's psychology bears contrast with the views of his colleague Alfred Wallace—because the views of Wallace overlap with modern ideas. Just as Modern Synthesizers like Dawkins exalt humans above the rest of the living world—as the only species on earth who can 'rebel against the tyranny' of our selfish genes—Wallace held that *only humans* were 'social and sympathetic.'[22] Both thus split off the way natural selection affects humans from its effects in animals.

In animals, natural selection changed 'physical form and structure,' observed Wallace. In humans, due to their sociality—their sympathy, division of labour, and 'mutual assistance'— the impact of natural selection on human bodies would halt. Humans look after each other, he said, so 'the weaker, the dwarfish, those of less active limbs, or less piercing eyesight, do not suffer the extreme penalty which falls upon animals so defective.' In humans, therefore, natural selection solely concerned the 'intellectual and moral faculties.' Animal species adapted to changes in climate and their physical surroundings by alteration in bodily structure. Facing similar changes, hominids would not have required 'longer nails or teeth, greater bodily strength or swiftness.' Instead, Wallace said, our species would make 'sharper spears, or a better bow,' construct 'a cunning pitfall,' or combine in a hunting party to surround new prey. The mental and moral capacities which enabled humanoids to invent new technologies and forms of agency were what had been gradually modified by natural selection, while bodily form and structure had remained static.[23]

According to Wallace, this new domain of psycho-cultural natural selection could explain everything that was most human—from the actions of the members of simple hunting tribes to the creations of a Beethoven or a Goethe.[24] From the moment that the evolution of the human body ceased, the mind would become subject 'to those very influences from which his body had escaped,' wrote Wallace. 'Every slight variation' in 'mental and moral nature,' which might enable humans better 'to guard against adverse circumstances, and combine for mutual

---

[21] Darwin, *Descent*, pp.137–138, my italics.

[22] Alfred Wallace, 'The Origin of Human Races and the Antiquity of Man Deduced from the Theory of "Natural Selection"; *Journal of the Anthropological Society of London* 2, (1864), p.clxii; Richard Dawkins, *The Selfish Gene: 30th Anniversary Edition* (Oxford: Oxford University Press, 2006), p.201. Rose, *Lifelines*, pp.213ff, underlines the catastrophic implications of Dawkins' concession for the rest of his argument.

[23] Wallace, 'Origin of Human Races,' p.clxiii.

[24] To Darwin's dismay, Wallace reversed this opinion in 1869, arguing that creation of the higher human faculties must have required intervention from 'the unseen universe of Spirit': Alfred Wallace, *Darwinism: An Exposition of the Theory of Natural Selection, with Some of Its Applications* (London: Macmillan, 1889), p.478.

comfort and protection,' would be preserved and accumulated. Over the millennia, the 'better and higher' humans would increase and spread. The 'lower and more brutal' would give way and die out. And a rapid advancement of mental organization would result.[25]

Wallace's hypothesis assumes a far more parcellated, modular human brain than that pictured in *Descent*. As we have seen, Darwin held that the human brain had become increasingly integrated as our intellectual powers advanced (see 7.3.1 Growing cerebral integration). As a result, our evolved psychical faculties underpin and mediate, but do not specify, any given act. It was when a person's actions were cultivated by 'foul customs,' that these faculties afforded cannibalism or slavery.[26] Alternatively, with a good British education, evolved faculties could inform the highest acts of moral culture: the fight for prison-reform; or the abolition of slavery.[27]

Wallace found no place in his theory for psychical processes enabling a culture to specify an action. He saw no opportunity for one evolved faculty to afford a wide range of contrasting deeds. Just as modern synthesizers hypothesize 'genes for' an enormous range of complex human traits and actions, Wallace assumed that *every slight variation* in humans' mental and moral life was inherited and hence naturally selected as such.[28] Just as a marginally longer bill or slightly sharper eye in a Galapagos finch would be preserved and accumulated if it increased its bearer's fitness, so even tiny variations in mental function could supposedly be inherited, and thus fall under the law of natural selection.

Darwin agreed with Wallace that, during the more recent phases of human evolution, 'the mind will have been modified more than the body.'[29] Unlike Wallace, he argued that, once the foundations of a broad faculty like sympathy, or imitation, or reason, had been explained by natural selection, that faculty would produce *specific* acts which *could not* be explained by natural selection. An example we have already met is the way sympathy grounds human sensitivity to approbation and disapprobation, the foundation of our sense of moral right and wrong, according to *Descent* (see Ch. 7). What different peoples treat as right and wrong pans out very

---

[25] Wallace, 'Origin of Human Races,' p.clxiv.

[26] Darwin, *Descent*, pp.46, 117, 182.

[27] Darwin, *Descent*, p.66.

[28] Wallace's reference to 'every slight variation' recalls the celebrated definition from *Origin*: 'I have called this principle, by which each slight variation, if useful, is preserved, by the term of Natural Selection'; Darwin, *Origin 1859*, p.61. Against the view that there are 'genes for' complex human traits, see: West-Eberhard, *Developmental Plasticity and Evolution*, pp.18–20 & passim; Richard Lewontin, 'Gene, Organism and Environment,' in *Evolution from Molecules to Men*, ed. Derek Bendall (Cambridge: Cambridge University Press, 1983), pp.273–285; Steven Rose, Leon Kamin, and Richard Lewontin, *Not in Our Genes: Biology, Ideology and Human Nature* (Harmondsworth: Penguin Books, 1984).

[29] Charles Darwin, 'Letter to Wallace, May 28th,' *Correspondence of Charles Darwin* (Cambridge: Darwin Correspondence Project, 1864), https://www.darwinproject.ac.uk/letter/?docId=letters/DCP-LETT-4510.

differently in different tribes and types of society. Furthermore, the more social humans became, the more their sociality would have *rebounded* on their agency. As our evolutionary ancestors grew more social—'and this probably occurred at a very early period'—the principles of imitation, and reason, and the capacity to learn from experience would have increased, 'and much modified the intellectual powers in a way, of which we see only traces in the lower animals.'[30]

I emphasize the *rebounding effect* of an increased facility for a certain kind of action on the growth of human agency. This theme constantly recurs in Darwin's psychology. Witness his accounts of blushing, language, conscience, and reason. His account of human agency rebounding follows his account of the rebounding of agency in creatures like mistletoe and earthworms. As I argued in Chapter 3, such rebounding effects or *reagency* (which overlaps with *cultural drive* in today's parlance),[31] is partly what gave rise to the sociability of social animals, according to Darwin's theory. Over the very long term, the 'rebounding' effects of agency *on survival* amount to what Darwin called natural selection. The reverberations of agency are far more pronounced in humans than in other animals. But not all the reflexive effects of human action have long-term evolutionary consequences.

The most obvious way human sociality rebounds on human agency is through social institutions. Amongst these, *Descent* highlights schooling. Wallace had proposed that the social aspect of agency in humans largely depended on 'simple imitation,' not reason: the houses of 'most savage tribes' were 'each as invariable as the nest of a species of bird.'[32] In contrast, *Descent* sees no parallel between human learning and the one-shot mimicry of birds like parrots: 'man cannot, on his first trial, make, for instance, a stone hatchet or a canoe, through his power of imitation. He has to learn his work *by practice*.'[33] Darwin's concept of practice made education, and religious instruction, central to cultural advance.

Which had a corollary. Were an advanced society's institutions to change or fail, moral culture might succumb. Progress towards civilization was neither inevitable

---

[30] Darwin, *Descent*, p.129.

[31] Cultural drive: 'With each spread of a new habit through the population . . . natural selection would favour improvements in a species' capacity to copy the discoveries made by others, resulting in bigger brains. New habits would also generate selection for changes to the animal's anatomy better suited to the behaviour, leading to the fixation of novel mutations. Each increment in brain size . . . would enhance the species' ability to generate and propagate new habits, making the spread of further innovations and the fixation of additional mutations even more likely. This runaway process [has] driven brain evolution in a multitude of animals, particularly primates, but [has] climaxed in humanity—the brainiest, most creative, and most culturally reliant species of all': Laland, *Darwin's Unfinished Symphony*, pp.123ff; Michael Muthukrishna et al., 'The Cultural Brain Hypothesis: How Culture Drives Brain Expansion, Sociality, and Life History,' *PLoS Computational Biology* 14, (2018), https://doi.org/10.1371/journal.pcbi.1006504.

[32] Alfred Wallace, *Contributions to the Theory of Natural Selection* (London: Macmillan, 1870), pp.212ff, 226 Wallace's talk of 'unreasoning imitation' does not cover all forms of imitation, of course. Some kinds of imitation require constant practice, particularly those promulgated in the Christian tradition. See for example: Thomas A Kempis, *The Imitation of Christ* (New York: Dover, 2003).

[33] Darwin, *Descent*, p.68, my italics.

nor permanent. 'We are apt to look at progress as normal in human society,' wrote Darwin, 'but history refutes this.'[34] The ancient Greeks and the Spanish had inexplicably 'retrograded' from the heights of their civilizations. Equally perplexing was 'the awakening of the nations of Europe' from the 'dark ages': the Renaissance. Progress seemed to depend on a concatenation of favourable conditions, 'far too complex to be followed out,' Darwin wrote. A cool climate was an advantage to industry and the arts. Cool but not too severe: the Eskimos and the Fuegians, who lived at the freezing ends of the earth, were not paragons of enlightenment. Perhaps this was not just because of the weather, but because they lacked some other crucial element of social organization. 'Whilst observing the barbarous inhabitants of Tierra del Fuego, it struck me that the possession of some property, a fixed abode, and the union of many families under a chief, were the indispensable requisites for civilisation,' Darwin confided, concluding: 'any form of government is better than none.'[35]

All this puts culture above and beyond the law of effects known as natural selection. *Descent* formally underlines the high degree of autonomy that specific acculturated acts have from their evolutionary foundations by affirming a remark from Darwin's American ally Chauncey Wright (1830–1875). There were 'many consequences' of the law of natural selection, wrote Wright: evolution of one 'useful power' would bring with it 'many resulting advantages as well as *limiting disadvantages*,' disadvantages which might undermine fitness, and which the law of natural selection 'may not have comprehended in its action.' Darwin agreed, confirming that *Descent* had tried to show that Wright's argument had 'an important bearing on the acquisition by man of some of his mental characteristics.'[36] Such mental characteristics included what have more recently been called the 'non-adaptive sequelae' of communal living, for example: maladaptive customs and superstitions; and mating choices on grounds of 'mere wealth or rank'— rather than fitness.[37]

Reframed in Darwin's terms, Wright's principle highlights a crucial consequence of our brains' plasticity during development. Cerebral impressibility had given humans their most efficient spring of progress, by enabling the effects of good education.[38] Yet this same plasticity had fostered many dismaying *disadvantages* so far as natural selection was concerned—as a sizeable section of *Descent* attests. For there

---

[34] Darwin, *Descent*, p.132.

[35] Darwin, *Descent*, pp.130–133, 141.

[36] Darwin, *Descent*, p.571; Chauncey Wright, 'Review [of Contributions to the Theory of Natural Selection. A Series of Essays by Alfred Russell Wallace],' *North American Review* 111, (1870), p.293.

[37] Darwin, *Descent*, pp.121–122, 617. Regarding 'non-adaptive sequelae,' see Stephen Jay Gould, 'Challenges to Neo-Darwinism and Their Meaning for a Revised View of Human Consciousness,' *Tanner Lectures on Human Values* (1984), http://tannerlectures.utah.edu/_documents/a-to-z/g/gould85.pdf.

[38] Darwin, *Descent*, p.143.

were copious ways in which the culture of more civilized societies impeded natural selection:[39]

> We civilised men ... do our utmost to check the process of elimination; we build asylums for the imbecile, the maimed, and the sick; we institute poor-laws; and our medical men exert their utmost skill to save the life of every one to the last moment. There is reason to believe that vaccination has preserved thousands, who from a weak constitution would formerly have succumbed to small-pox. Thus the weak members of civilised societies propagate their kind ... It is surprising how soon a want of care, or care wrongly directed, leads to the degeneration of a domestic race; but excepting in the case of man himself, hardly any one is so ignorant as to allow his worst animals to breed.[40]

There follows a long litany of the follies of civilized culture, albeit weighed against countervailing advantages, as viewed from the perspective of natural selection. Yet the main thrust of this section underlines that, ultimately, natural selection is not the decisive issue when considering the formation or character of human agency.

Descent voices many gripes. Primogeniture was an 'evil' because the advantages it gave firstborn sons, especially the means to marry, did not reflect worth. 'The finest young men' were conscripted into standing armies, being thus exposed to 'early death during war', besides being 'often tempted into vice', and prevented from marrying during the prime of life. Meanwhile 'shorter and feebler men, with poor constitutions' stayed at home to marry and breed.[41]

Against this, however, *Descent* lauds 'the presence of a body of well-instructed men', who, by inheriting wealth (like Darwin), 'have not to labour for their daily bread', because 'all high intellectual work is carried on by them, and on such work, material progress of all kinds mainly depends, not to mention other and higher advantages'. And while Darwin went on to observe that 'the most eminent men did not necessarily leave more offspring than their inferiors'—this was not ultimately decisive, because: 'Great lawgivers, the founders of beneficent religions, great philosophers and discoverers in science, aid the progress of mankind in a far higher degree by their works than by leaving a numerous progeny.' Darwin did not deny that the fittest humans—'men of a superior class ... the careful and frugal, who are generally otherwise virtuous'—bred more slowly than 'the very poor and reckless, who are often degraded by vice'.[42] But, he reassured his readers, such evils

---

[39] Darwin, *Descent*, pp.133–143. The section is called 'Natural Selection as affecting Civilised Nations.'

[40] Darwin, *Descent*, pp.133–134.

[41] Darwin, *Descent*, pp.134–135.

[42] Darwin, *Descent*, pp.135–138.

were outweighed by the fact that, 'with highly civilized nations continued progress depends in a subordinate degree on natural selection.'

So, while taken in isolation, some of Darwin's remarks might suggest he deplored care of the weak—comparing it to a bad farmer breeding his cattle from runts—*Descent* instances such apparent follies rhetorically. Darwin's book does list the many ways in which European societies flout what might, by others, be called the requirements for optimal human evolution. But the purpose of these lists is to underline *the minor role played by natural selection in shaping adult agency*. To the extent that the British made provision for the poor, the maimed, the halt, and the blind, they mitigated these selective disadvantages: superior morals undermined the processes that led to natural selection. But Darwin put a premium on moral culture. So, taken overall, *Descent*'s argument *favours* moral benevolence, because 'if we were intentionally to neglect the weak and helpless,' we would be abandoning 'the noblest part of our nature,' our moral sympathy—and this would be to advocate an 'overwhelming present evil,' destroying the very grounds upon which European societies claimed to be called highly civilized.[43]

8.3 Culture Specifies

Evolution proposes; culture disposes. According to *Descent*, civilization specifies, supplements, or completes the generic roots of agency which all humans can trace to their common descent with animals. A vivid example came from the twenty-two year old Darwin's acquaintance with the three Fuegians—nicknamed Fuegia Basket, Jemmy Button, and York Minster—who shared space with him on *HMS Beagle* from 1831 to 1832. These three hostages had been punitively kidnapped in 1830 and taken back to England—along with a fourth Fuegian, named Boat Memory, who died of smallpox in captivity—by Darwin's captain, Robert Fitzroy (1805–1865), on the previous voyage of the *Beagle*.[44] All three had then been schooled in English and instructed in religion. The *Beagle* repatriated them in December 1832.[45] Fitzroy hoped that, having imbibed Christianity in England, the returned hostages would work alongside a young English missionary Fitzroy had on board to bring all Fuegians to Christ.

---

[43] Darwin, *Descent*, p.134.

[44] Who suspected some of their fellow-natives of stealing one of his boats: Charles Darwin, *Journal of Researches into the Natural History and Geology of the Countries Visited During the Voyage of H.M.S. Beagle Round the World, under the Command of Capt. Fitz Roy R.N.*, Journal of Researches (London: John Murray, 2006), pp.206–207.

[45] In fact Jemmy Button had not been one of the three initial hostages. He had later been bought by Fitzroy, for a pearl button. Hence his new name.

Figure 8.3 Two views of Jemmy Button. The left-hand picture shows him in 1834, after he had reverted to his former ways. The right-hand picture shows him in 1832, prior to his return to South America.

Source: Fitz-Roy, R. (1839). *Narrative of the Surveying Voyages of his Majesty's Ships Adventure and Beagle, between the years 1826 and 1836 describing their examination of the southern shores of South America and the Beagle's circumnavigation of the globe*, Vol. II. (London: Henry Colburn, Great Marlborough Street), p.324.

The young Darwin reports being 'continually struck with surprise' by how closely the returning Fuegians 'resembled us in disposition and in most of our mental faculties.'[46] Jemmy Button, for example, had become 'vain'—always wearing gloves, his hair neatly cut, fond of admiring himself in a looking-glass, and getting distressed 'if his well-polished shoes were dirtied.'[47] This refuted those who said savages could not be civilized, Darwin noted: 'in contradiction of what has often been stated, 3 years has been sufficient to change savages, into, as far as habits go, complete & voluntary Europeans' (Figure 8.3).[48]

So impressed was Darwin with this cultural transformation, that he was sure his three shipmates would maintain their civilized ways on being returned to home soil: 'York, who was a full grown man & with a strong violent mind, will I am certain in every respect live as far as his means go, like an Englishman.'[49] Such confidence may strike us as comical, bespeaking a lack of familiarity with faraway cultures during Victorian times. In any case, Darwin's prediction failed. Once landed back in their homeland, the three Fuegians reverted to their former ways. Jemmy, Darwin's favourite, had within weeks gone back to living 'as if he had never

---

[46] Darwin, *Descent*, p.65.

[47] Charles Darwin, *Journal of Researches into the Natural History and Geology of the Countries Visited During the Voyage of H.M.S. Beagle Round the World, under the Command of Capt. Fitz Roy R.N.* (London: John Murray, 1890), p.207.

[48] Charles Darwin, *Diary of the Voyage of H.M.S. Beagle*, ed. Paul Barrett and Richard Freeman, vol. I, The Works of Charles Darwin (London: William Pickering, 1986), pp.142–143.

[49] Darwin, *Diary of the Voyage of H.M.S. Beagle*, I, p.143.

left his country.'[50] This only underlined, however, that just as European civilization could be successfully added to even 'the lowest barbarians' by means of education, it could just as easily fall away.[51]

As I now go on to show across several domains, *Descent* typically argues that the cultural specification of the flexible psychical foundations *laid through* natural selection did not itself fall under the law of natural selection. Chapter 7 has already documented how the moral sense is shared among all peoples, but each different tribe or society specifies different moral standards. Chapter 6 drew a similar picture with regard to the diverse standards of sexual attractiveness in different human cultures. Earlier in this chapter, we saw how 'civilized nations' principally owed their progress to their school system, according to *Descent*.[52] Likewise with the superior resilience of European societies. Darwin observed that, in his day, 'civilised nations' were 'everywhere supplanting barbarous nations, excepting where the climate opposes a deadly barrier.' And they principally vanquished others through what Darwin called their 'arts,' which were 'products of the intellect.'[53]

The word *art* refers in *Descent* to an acquired skill, specified by knowledge gained through cultural experience. Thus, language—found in all humans—was 'half-art, half-instinct.'[54] Darwin compared speech to bird-song. Infants cooed and babbled just as nestlings squawked and chirped. But gradually birds learnt to 'sing their song round' by practicing the dialect of the songs they heard in their locality. Humans had an analogous 'instinctive tendency to speak.' Nevertheless, 'every language has to be learnt.'[55]

This was significant because it was language that had largely made humans 'the most dominant animal that has ever appeared on this earth.' Articulated speech was *the* 'great stride in the development of the intellect,' because the effects of language-use would have *rebounded* on the development of other mental faculties 'such as those of ratiocination, abstraction, self-consciousness, &c.' Language was a 'wonderful engine which affixes signs to all sorts of objects and qualities, and excites trains of thought which would never arise from the mere impression of the senses, or if they did arise could not be followed out.' Hence, language-use had been the chief cause of 'the largeness of the brain in man relatively to his body, compared with the lower animals.' The continued use of language 'will have reacted on the brain and produced an inherited effect; and this again will have reacted on the improvement of language.'[56]

---

[50] Darwin, *Diary of the Voyage of H.M.S. Beagle*, I, p.227.
[51] Darwin, *Descent*, p.65.
[52] Darwin, *Descent*, p.143.
[53] Darwin, *Descent*, p.128. NB Darwin's belief that acquired characters could become inherited blurs the line many now draw between culture and biology.
[54] Darwin, *Descent*, p.126.
[55] Darwin, *Descent*, p.86.
[56] Darwin, *Descent*, p.610.

Another half-art produced humans' wildly different forms of fear and veneration for the unseen. This was to be found at work in all levels of human society, including Darwin's family, where his wife Emma faithfully practiced Christianity. *Descent* finds belief in 'unseen or spiritual agencies' to be universal in the 'less civilised races,' and therefore, by extrapolation, in our primitive ancestors. How might such beliefs have originated? 'As soon as the important faculties of the imagination, wonder, and curiosity, together with some power of reasoning, had become partially developed,' they would have rebounded: 'man would naturally crave to understand what was passing around him, and would have vaguely speculated on his own existence.'[57]

Darwin affirmed Edward Tylor's suggestion that dreams had spawned respect for the unseen, because 'savages do not readily distinguish between subjective and objective impressions.' However, the *very* earliest beliefs would have resulted from a different process, the *projection* of human faculties into anything non-human which exhibited signs of agency: 'anything which manifests power or movement is thought to be endowed with some form of life, and with mental faculties analogous to our own.' He even found this form of belief in his own dog, who barked at a parasol twitching on the lawn when jostled by the breeze.[58]

Projection would have made it easy to step from belief in unseen spiritual agencies to belief in gods: 'for savages would naturally attribute to spirits the same passions, the same love of vengeance or simplest form of justice, and the same affections which they themselves feel.' As long as human reasoning remained poorly developed, it would encourage 'strange superstitions and customs,' many of which were 'terrible to think of—such as the sacrifice of human beings to a blood-loving god; the trial of innocent persons by the ordeal of poison or fire; witchcraft etc.' Escape from such benightedness resulted solely from improvements in reason, and the advance of science, that is, from 'our accumulated knowledge.'[59]

Eventually, cultural advances would have produced religious ideas and practices that rebounded on the cultures producing them, leading to yet further progress. 'The highest form of religion—the grand idea of God hating sin and loving righteousness,' was a human achievement which could not have arisen 'in the mind of man, until he ha[d] been elevated by long-continued culture.'[60] 'To do good in return for evil, to love your enemy, is a height of morality to which it may be doubted whether the social instincts would, by themselves, have ever led us.' It was only after social instincts had been highly 'cultivated and extended by the aid of reason, instruction, and the love or fear of God,' that the 'golden rule' would ever be thought of or obeyed.[61] The golden rule was, 'As ye would that men should do to you, do ye

---

[57] Darwin, *Descent*, pp.93–95.
[58] Darwin, *Descent*, p.95. See 3.1.1 Agency itself.
[59] Darwin, *Descent*, pp.95–96.
[60] Darwin, *Descent*, pp.144, 612.
[61] Darwin, *Descent*, p.112.

to them likewise'—which Darwin called 'the foundation of morality.'[62] Hence the ideals of Christian action, which resulted *from* culture, subsequently rebounded back *onto* culture.

Descent equates high civilization with a refinement of virtue. Any form of human society exhibits so-called *social virtues*, we hear: namely, those kinds of action that attract approbation in a particular social group. But social virtues reach a new level in civilized nations, promoting actions that no longer patently relate to communal advantage. *Descent* calls these 'self-regarding' virtues, most notably: temperance; chastity; and a hatred of indecency. All of these required a strengthening of the will, resulting in 'self-command'—meaning none were 'esteemed by savages' (cf. Ch. 7.4.2.4 'Supreme judge').[63] *Descent* here invokes a distinction drawn by John Stuart Mill between self- and other-regarding, or 'social' acts. Mill rated social above self-regarding virtues.[64] But *Descent*'s argument overturns Mill's distinction.

Social virtues are often displayed for selfish reasons, *Descent* notes—not least, to avoid the 'agony' of self-recrimination.[65] However, self-regarding virtues are not asocial, said Darwin, because they *do* affect general social welfare, whether or not we recognize their beneficent consequences. For example, temperance was a social good, intemperance being 'highly destructive.'[66] Self-regarding virtues therefore came to be 'highly esteemed or even held sacred,' once they entered the scope of public opinion.[67] They then received praise, 'and their opposites blame,' just as did other-regarding actions.[68] Which meant their effects rebounded, making them 'social virtues' too! In time, therefore, when living in groups, benefits to the self may prove impossible to disentangle from benefits to others. Thus *Descent* calls courage a social virtue, yet it requires self-command—to such an extent that some of Darwin's peers classed it as self-regarding.[69]

A refined sense of beauty was something else that educated Europeans owed to culture, said Darwin. This showed in two ways. As we saw in Chapter 6: 'The taste for the beautiful, at least as far as female beauty is concerned, is not of a special nature in the human mind; for it differs widely in the different races of man, and is not quite the same even in the different nations of the same race.'[70] At the same time, *Descent* describes the music and ornaments admired by 'most savages' as 'hideous'—thereby underlining, wittingly or not, the justice of Darwin's point that taste is culturally relative. This train of thought was capped by the class-based,

---

[62] Darwin, *Descent*, p.126.
[63] Darwin, *Descent*, pp.118–119.
[64] John Mill, *On Liberty* (London: Longman & Green, 1867), pp.44ff.
[65] Darwin, *Descent*, p.121.
[66] Darwin, *Descent*, p.137.
[67] Darwin, *Descent*, pp.119, 132.
[68] Darwin, *Descent*, p.611.
[69] Darwin, *Descent*, p.129; Henry Sidgwick, *The Methods of Ethics*, 7 ed. (London: Macmillan, 1907), pp.332–334; Gabriele Taylor and Sybil Wolfram, 'The Self-Regarding and Other-Regarding Virtues,' *Philosophical Quarterly* 18, (1968), pp.238–248.
[70] Darwin, *Descent*, p.93.

Figure 8.4 *'White dove feeling the universe.'*

(Blake, 'Heaven and Hell,' p.150: 'How do you know but ev'ry Bird that cuts the airy way/ Is an immense world of delight, clos'd by your senses five?')

Source: Brett Whiteley, *White dove feeling the universe*, 1985–92. Oil on plywood. Gift of the Josephine Ulrick and Win Schubert Foundation for the Arts through the Queensland Art Gallery Foundation 2012. Donated through the Australian Government's Cultural Gifts Program. Collection: Queensland Art Gallery | Gallery of Modern Art. © Wendy Whiteley. Photograph: Natasha Harth, QAGOMA.

ethnocentric—if not, anthropocentric (see Figure 8.4)—comment: 'Obviously no animal would be capable of admiring such scenes as the heavens at night, a beautiful landscape, or refined music; but such high tastes are acquired through culture, and depend on complex associations; they are not enjoyed by barbarians or by uneducated persons.'[71]

8.4 Resonances

Scientists have long sought a theory of everything. Search for such a theory featured as much in Darwin's day as ours. Even anti-evolutionary apologists for religion, such as philosopher-scientist William Whewell (1794–1866)—who had

[71] Darwin, *Descent*, p.93.

mentored Darwin in his twenties—entered the chase. For Whewell, with the rise of science and the strengthening Victorian belief that natural laws had formed all phenomena—everything from nebular dust through human psychology to the highest artistic achievement should come under one explanatory rubric.[72]

The twentieth century saw a series of evolutionary Theories of Everything, also called 'general theories of evolution.' The Jesuit palaeontologist Teilhard de Chardin proposed one and Julian Huxley a second (and my father a third)—each theory drawing variously on Darwin and other early evolutionists.[73] Would Darwin have countenanced a *general* theory of evolution? As shown in this chapter, he would not. Because most of the higher achievements of European civilization were argued in *Descent* to be products of non-evolutionary *culturally*-specified processes, albeit processes *developed from* evolved foundations.

Such a conclusion must today face down three countervailing arguments:

1. a historiographical argument that Darwin was a 'Social Darwinist': his supposedly progressive 'biological' theory made the social order of Victorian capitalism an evolutionary product, and so justified socio-cultural privilege and oppression;
2. a psychological argument that culture is generated by cerebral modules which evolved 'biologically' in the Stone Age;
3. an anthropological argument that cultures evolve through a parallel kind of natural selection—not of genes—but of ideas, behavioural traits, and or *memes*.

I will consider each in turn.

---

[72] 'And thus we are led by a close and natural connexion, through a series of causes, from those which regulate the imperceptible changes of the remotest nebulae in the heavens, to those which determine the diversities of language, the mutations of art, and even the progress of civilisation, polity, and literature': William Whewell, *The Philosophy of the Inductive Sciences*, 2 vols., vol. 2 (London: Parker, 1840), p.115—this was a book Darwin had read: see, Charles Darwin, *The Variation of Animals and Plants under Domestication*, 2 vols., vol. 2 (London: John Murray, 1875), p.350. See also Walter Cannon, 'The Problem of Miracles in the 1830's,' *Victorian Studies* 4, (1960), pp.4–32. Auguste Comte, the anonymous author of *Vestiges of Creation* (Robert Chambers), and Herbert Spencer all proposed accounts with similar, universal ambitions—Chambers' and Spencer's theories both being evolutionary. The idea was almost a commonplace, being stated and restated in books such as John Atkinson's and Harriet Martineau's joint *Letters on the Laws of Man's Nature and Development*, published in 1851: 'Every thought and every act is part of the development of nature, and of the history of the world, and of the universe,—of the eternal and undeviating law of laws,—of the fundamental but incomprehensible origin of things': John Atkinson and Harriet Martineau, *Letters on the Laws of Man's Nature and Development* (Boston: Mendum, 1851), p.135.

[73] De Chardin, *Phenomenon*; Huxley, *Evolution in Action*; Sylvester-Bradley, 'Evolutionary Model'; Peter Sylvester-Bradley, *Evolution and the Destiny of Man*, Syracuse University: Geology Contributions 6, (1979); John Tooby and Leda Cosmides, 'The Psychological Foundations of Culture,' in *The Adapted Mind: Evolutionary Psychology and the Generation of Culture*, ed. Jerome Barkow, Leda Cosmides, and John Tooby (New York: Oxford University Press, 1992), p.24.

8.4.1 Social Darwinism

Social Darwinism is a phrase coined in the 1890s. It refers to the view that, in order to make progress, a human society should promote the survival of the fittest, and hence, encourage unrestricted competition between individuals, groups, and nations—such that the strong displace the weak. These views belong more to Herbert Spencer than Darwin, because Spencer overtly promoted competitive individualism as a means of social progress.[74] The importance of Social Darwinism as a concept, and a trope in American politics, was boosted in 1944 by Richard Hofstadter's *Social Darwinism in American Thought*.[75]

Today, commentators divide over whether Darwin was a Social Darwinist. Writers like Robert Young, James Moore, and Adrian Desmond forthrightly declare that, to anyone who actually reads it, *Descent* is inescapably a Social Darwinist tract: 'a piece of self-congratulatory science for the Liberal *nouveau riche* and emerging agnostic patriarchs of middle England.'[76] To adherents of the Modern Synthesis, on the other hand, Social Darwinism is 'an odious misapplication of Darwinian thinking'—presumably, because Darwin never explicitly *said* that the survival of the fittest was a means of promoting social progress.[77] (More circumspect scholars draw a more complex picture.)[78]

Read carefully, however, *Descent* does not advance a single view about whether and how to 'apply to moral and social questions analogous views to those which I have used in regard to the modification of species.'[79] It advances three views.

First, like many of his peers (Wallace, for example), Darwin knew that the expansionist imperial powers of the mid-1800s were displacing and often exterminating native peoples: 'the civilised races have extended, and are now everywhere extending their range, so as to take the place of the lower races.'[80] This he framed as a form of natural selection among groups. On the other hand, *within* European nations, vaccination and like social programmes were *checking* 'in many ways the action of natural selection,' by allowing 'the weak members of civilised societies' to

---

[74] Raymond Williams, 'Social Darwinism,' in *Herbert Spencer: Critical Assessments*, ed. John Offer (London: Routledge, 2000).

[75] Richard Hofstadter, *Social Darwinism in American Thought* (Philadelphia PA: University of Pennsylvania Press, 1944).

[76] James Moore and Adrian Desmond, 'Introduction,' in *Charles Darwin: The Descent of Man and Selection in Relation to Sex*, ed. James Moore and Adrian Desmond (Harmondsworth: Penguin Books, 2004), pp.lv–liv; Dennett, *Darwin's Dangerous Idea*, p.393; Robert Young, 'Darwinism *Is* Social,' in *The Darwinian Heritage*, ed. David Kohn (Princeton: Princeton University Press, 1985), pp.609–639.

[77] Dennett, *Darwin's Dangerous Idea*, p.393.

[78] E.g. Diane Paul, 'Darwinism, Social Darwinism and Eugenics,' in *The Cambridge Companion to Darwin*, ed. Jonathan Hodge and Gregory Radick (Cambridge: Cambridge University Press, 2003), pp.219–245.

[79] Charles Darwin, 'Letter to Thiel, February 25th,' *Correspondence of Charles Darwin* (Cambridge: Darwin Correspondence Project, 1869), https://www.darwinproject.ac.uk/letter/?docId=letters/DCP-LETT-6634.

[80] Darwin, *Descent*, p.169.

'propagate their kind'—something he called 'highly injurious to the race of man,' threatening it with 'degeneration.'[81]

Writing in this vein, Darwin sounds as if he might deplore care of the weak—attracting the conclusion he was a Social Darwinist. Yet Darwin *does not advocate* neglect of the weak. On the contrary, *Descent* argues that to abandon the poor and needy would be to undo what was best about Europeans: their civilization and humanity. This is his second argument. Hence, if we are to maintain high moral standards, 'we must bear without complaining the undoubtedly bad effects' of the weak 'surviving and propagating their kind.'[82]

Finally, there is no need to despair at the mitigation of the struggle for existence in humane societies, says *Descent*, because natural selection is of little importance in promoting the march of human progress, the influence of social institutions being far more powerful. This appeal to a superordinate, specificatory culture connects directly to what Chapter 3 called *phenotypic excess*. The theatre of agency inevitably exceeds what is grist for evolution's mill. Phenotypes exceed genotypes. Hence, not all phenotypic action results from (or feeds) the diachronic axis of phylogeny. The generic capacities for action, endowed to humans by their common descent with animals, *may afford* cultural specifications which yield great advantages, morally, intellectually, or aesthetically— specifications which may or may not be *biologically* transmissible. By the same token, many of the ways in which culture specifies these evolved capacities may prove adaptively neutral, or even maladaptive, as we saw previously—which is why Darwin advocated the arguments of Chauncey Wright.

8.4.2 Cerebral modularity in Evolutionary Psychology today

How does Darwin's perspective on culture compare with claims by twenty-first century Darwinists? The evolutionary explanation of cultural phenomena has long been a goal of the Santa Barbara school of Evolutionary Psychology (see 3.4.4 Psychology and the Modern Synthesis).[83] According to this school, natural selection long ago endowed human beings with 'functionally specialized … evolved information-processing mechanisms instantiated in the human nervous system.'[84] Evolved to fit humans to the Stone Age world, these cognitive modules are not generic, or intercommunicative, as in *Descent*. They 'produce behaviour that solves *particular* [Palaeolithic] problems, such as mate selection, language acquisition, family relations, and cooperation.' In so doing, they 'generate some of

---

[81] Darwin, *Descent*, pp.134–136.

[82] Darwin, *Descent*, p.134.

[83] Tooby and Cosmides, 'The Psychological Foundations of Culture.'

[84] John Tooby and Leda Cosmides, 'The Theoretical Foundations of Evolutionary Psychology,' in *The Handbook of Evolutionary Psychology*, ed. David Buss (Hoboken NJ: Wiley, 2016), p.24.

the *particular* content of human culture, including certain behaviours, artefacts, and linguistically transmitted representations.'[85] Evolutionary Psychology consists in 'reverse engineering'—that is, working backwards, using knowledge of 'design by natural selection' as key—from current behaviour to work out what the panhuman, Palaeolithic modules are, which currently generate our behaviour, and our cultures.[86]

This perspective arouses controversy—particularly regarding its reliance on the Modern Synthetic assumption that genes are the only means by which 'functional design features replicate themselves from parent to offspring.'[87] Here I consider its degree of alignment with the theses of *Descent* (regarding its alignment with *Origin*, see 3.4.4 Psychology and the Modern Synthesis). Most notably, and unlike Santa Barbara Evolutionary Psychology, *Descent* does not argue that natural selection is the principal mould of 'civilized' human action. This is partly because it holds that cerebral domains are interconnected and flexible in application, *not* modules custom-made to answer *'definite and inherited'* problems (see 7.3.1 Growing cerebral integration).[88]

Beyond this, evolved generic capacities are not, according to Darwin, *specified* by the culture of the Stone Age. They are specified—*for good or ill*—by the particular cultures and institutions that influence the habitat of their development and operation in given populations. The causal arrow for the Santa Barbara school goes from genes to the brain to culture, such that genes generate cerebral modules, and cerebral modules generate culture.[89] For Darwin, a significant causal arrow goes

---

[85] Tooby and Cosmides, 'The Psychological Foundations of Culture,' p.24, my italics.

[86] Tooby and Cosmides, 'Theoretical Foundations,' pp.19ff. See also John Dupré, *Human Nature and the Limits of Science* (Oxford: Clarendon Press, 2001), pp.77–81, regarding perceived weaknesses of reverse-engineering.

[87] Tooby and Cosmides, 'Theoretical Foundations,' p.23. Cf. Jaak Panksepp and Jules Panksepp, 'The Seven Sins of Evolutionary Psychology,' *Evolution and Cognition* 6, (2000), The Seven Sins of Evolutionary Psychology,' *Evolution and Cognition* 6, (2000), pp.108–131; Hilary Rose and Steven Rose, eds., *Alas, Poor Darwin: Arguments against Evolutionary Psychology* (London: Jonathan Cape, 2000); Dupré, *Human Nature*; John Dupré, 'Against Maladaptationism: Or What's Wrong with Evolutionary Psychology?,' in *Knowledge as Social Order: Rethinking the Sociology of Barry Barnes*, ed. Massimo Mazzoti (London: Routledge, 2008), pp.165–180.

[88] Darwin, *Descent*, p.68.

[89] Caveats are required here. First, in more recent writings, Santa Barbara theorists John Tooby and Leda Cosmides make a concession that 'many properties of organisms are not adaptations' and hence 'most "parts" of an organism are not functional.' This is partly because, à la Darwin, such parts are 'by-products' of adaptations, which have no necessary evolutionary utility. Writing gives them their example (just as it gave Darwin *his* example of a facet of language which was *not* based on 'an instinctive tendency to speak': Darwin, *Descent*, p.86). 'All brain-intact persons learn to speak (or sign) the language of their surrounding community without explicit instruction,' they write, 'whereas reading and writing require explicit schooling, are not mastered by some individuals, and are entirely absent from some cultures.' Tooby and Cosmides make this show two things. First, people may fail to read and write through mutational 'noise,' as in dyslexia. Yet, secondly, even the successful only succeed through 'laborious study and schooling.' Hence, a neural-programme that was once naturally selected to ensure humans acquire their mother tongue when young, can 'be deployed in activities that are unrelated to its original function'—as Wright and Darwin long ago argued: Tooby and Cosmides, 'Theoretical Foundations,' pp.28–30.

the other way, from culture and society to the specification of human action— and ultimately to what may be inherited, and so can evolve.[90]

8.4.3 Theories of cultural evolution

The Santa Barbara approach to culture makes it hard to explain aspects of human agency not obviously coded by genes—like writing and other facets of cultural change. Theorists of cultural evolution step in to fill this gap. However, as argued in Chapter 3, such theories are only needed for as long as we assume that evolution by natural selection is gene-led, and the phenotype is a passive follower, whose character is determined by DNA. (For, if evolution is phenotype-led, human culture fits under the rubric of the same evolutionary theory that explains insect-orchid cooperation, or a duck's webbed feet.)

Cultural evolutionists owe more than one debt to the Modern Synthesis, their theories mimicking several key features of the gene-centric take on 'biological' evolution (see Ch. 3).[91] Amongst these, the most remarkable is their definition of culture *as information*—mental states, acquired by social learning, stored in individual brains—just as genes are equated with information, stored in an individual's coils of DNA.[92] Which is a departure from the usual ways of looking at culture. Thus the Oxford English Dictionary defines culture as 'the distinctive ideas, customs, social behaviour, products, or way of life of a particular nation, society, people, or period. Hence: a society or group characterized by such customs, etc.'[93]

According to *this* definition, culture is not solely made up of invisible, intangible, individualized bits of information buried in your or my grey matter. It does include such information, insofar as human action does. But it also includes material objects, like cathedrals, and embodied practices, like football games, and 'society' as well, with its historically entrenched locale, borders, roads, cities, and institutions. Which raises questions. Is the 'information' cultural evolutionists theorize *co-extensive* with the observable *material* components of a culture—of which ideas may or may not be formed? Can my idea of the cathedral of Notre Dame in Paris— or anyone's idea—equate to the cathedral itself, before it burnt down? Is a gun, say

---

[90] West-Eberhard, *Developmental Plasticity and Evolution*.

[91] Peter Richerson and Robert Boyd, *Not by Genes Alone: How Culture Transformed Human Evolution* (Chicago: University of Chicago Press, 2005) make this mimicry absolutely explicit. For example, quoting foundational modern synthesizer Theodosius Dobzhansky, as saying, 'Nothing about biology makes sense, except in the light of evolution,' they write their book's concluding chapter under the heading, 'Nothing about culture makes sense except in the light of evolution' (pp.237–257).

[92] Richerson and Boyd, *Not by Genes Alone*, p.5; Mesoudi, *Cultural Evolution*, pp.2–3; Tim Lewens, *Cultural Evolution: Conceptual Challenges* (Oxford: Oxford University Press, 2015), pp.44–60. Cf. John Maynard-Smith, 'The Concept of Information in Biology,' *Philosophy of Science* 67, (2000), pp.177–194.

[93] 'Culture, noun, 7a,' in Anon, *Oxford English Dictionary Online*, (Oxford: Oxford University Press, 2018), https://www.lexico.com.

Figure 8.5 The Maxim gun, Idea and Reality: diagram of first prototype, 1881 (a); in use by the Red Army, 1943 (b).
Source: (a) Sir Hiram S. Maxim. (1915). *My Life*; (b) World History Archive / Alamy Stock Photo.

a Maxim gun, and the details of the mayhem it has afforded since its invention—veiled from its inventor, and absent from general knowledge—the same thing as, say, its inventor's *idea* of the gun and its first uptake (Figure 8.5)? Even if it were, is information the sufficient unidirectional cause of cultural phenomena: actions; suffering; customs; wars; roads; synagogues; train timetables; laptops; lockdowns; the internet; books; guns; political watersheds? Or does an (often unintended) causal arrow also go in the opposite direction? As when the ringing of a mobile phone causes a car crash?

Whatever the answers to these questions, the idea of *cultural evolution* takes us away from the main tenets of Darwin's psychology—which neither proposes that culture evolves, nor that culture occupies a domain separate from but parallel to that of biological evolution.

The nearest *Descent* comes to countenancing cultural evolution is in three paragraphs which draw an analogy between biological natural selection and the natural selection of 'words and grammatical forms,' leading to 'the formation of different languages.'[94] *Descent* notes several points of similarity between the genealogy of languages and the evolution of species: the existence of rudimentary (or redundant) letters and sounds; the hierarchical classification of languages; extinction of some languages by other, 'dominant languages and dialects'; correlated growth; homologies; and analogies. But Darwin did not extend this ingenious argument beyond its brief compass, to culture in general. Because he was using it for a specific purpose: to rebut a Christian idea that modern European languages had *degenerated* from more perfect, Eden-like, beginnings, as evidenced in 'the perfectly regular and wonderfully complex construction of the languages of many barbarous nations,' which hence, owed 'their origin to a special act of [divine] creation.'[95]

Finally, like Wallace, theories of cultural evolution make a disjunction between the processes of evolution in physical versus cultural domains. Darwin's take on human evolution was more complex and more integrated—*all* evolution being mediated by the characters of the organism, or phenotype. So: what do the facts show? *Are* humans still *physically* evolving—contrary to Wallace's claim? Apparently we are—and *in response to cultural changes.*

Examples come from studies of digestion. In populations from places which have seen the rise of cow-keeping and milk-based diets—North Europe and parts of Africa—there is evidence of increased persistence after infancy of the enzyme responsible for absorbing lactose (the sugar in milk) through the gut wall. This change is less than ten thousand years old.[96] Likewise, when peoples from places that have traditionally farmed grains (like Japan, Europe, and USA), and so have a starchy diet, are compared with low-starch-eating populations (like the Mbuti and Biaka in Africa)—we find a much higher presence of the gene coding starch-digesting enzymes (salivary amylase) in the starchy areas.[97] Observations like these—that changes in agriculture and hence diet lead to inherited biochemical

---

[94] Darwin, *Descent*, pp.90–92.

[95] Darwin, *Descent*, pp.90–92, argued that, if languages had been changed over time by a process akin to natural selection, they could not be seen as products of intentional design: as 'elaborately and methodically formed.' Furthermore, if they were like products of biological evolution, 'the most symmetrical and complex ought not to be ranked above irregular, abbreviated, and bastardized languages.' After all: 'a naturalist does not consider an [echinoderm which] ... consists of no less than 150,000 pieces of shell, all arranged with perfect symmetry in radiating lines ... as more perfect than a bilateral one with comparatively few parts, and with none of these parts alike, excepting on the opposite sides of the body.' In fact, the more *bastardized* a language and the more its symmetry was *spoiled* by 'borrowed expressive words and useful forms of construction from various conquering, conquered, or immigrant races'—like English—then the more it showed 'differentiation and specialisation,' so the more perfect it arguably was. See also Gregory Radick, *The Simian Tongue: The Long Debate About Animal Language* (Chicago: University of Chicago Press, 2007), pp.36–37.

[96] E.g. Sarah Tishkoff et al., 'Convergent Adaptation of Human Lactase Persistence in Africa and Europe,' *Nature Genetics* 39, (2007), pp.31–40.

[97] E.g. George Perry et al., 'Diet and the Evolution of Human Amylase Gene Copy Number Variation,' *Nature Genetics* 39, (2007), pp.1256–1260.

changes —confirm that natural selection in humans is culturally conditioned and so phenotype-led, in accordance with the argument of *Descent*.

A further powerful case for the importance of phenotype-led evolution of the human brain via 'transitional habits' (see 3.3.4 Agency directs selection), has been advanced by Terrence Deacon.[98] Deacon elaborates at length the argument made by Darwin that use of language precipitated *the* 'great stride' in the development of the human intellect and evolution of the large human brain.[99] This would have been a two-way process, or what Deacon calls a *co-evolution* of language and the brain. In Darwin's words, continued use of language will have 'reacted on the brain and produced an inherited effect,' while growth of the brain, will have led to improvements in language.[100]

8.5 Conclusion

Many crave a scientific insight that can illuminate all human experience and behaviour.[101] Hence the appeal of Dawkins' iconoclastic claim that, because all genes are selfish, we are all 'born selfish.'[102] Hence too the appeal of the Santa Barbara school of Evolutionary Psychology, and its leading idea that the default explanation for any human action—including language skills, selfishness, and male–female rape—makes it result from some still-operative, evolved, brain-mechanism that adapted our distant ancestors to their Stone Age world.[103]

Would Darwin have accepted such uses of his theory? Not according to the evidence in this chapter. Yes, he undoubtedly advanced an 'evolutionary' psychology, insofar as he understood that the foundations for the actions of his fellow men and women, including not only their most abhorrent but their noblest qualities—of sympathy 'for the most debased,' of benevolence which extended to 'the humblest living creature,' and of 'god-like intellect which has penetrated into the movements and constitution of the solar system'—were ultimately traceable to bestial beginnings.[104] Yet, civilized agency could not be explained entirely, or even predominantly, in terms of natural selection and animal origins. The cultivating institutions of modern societies played the main role in how he and his contemporaries acted—particularly, organised education, the advance of science, and the golden rule of moral instruction.

---

[98] Terrence Deacon, *The Symbolic Species: The Co-Evolution of Language and the Brain* (New York: Norton, 1997), pp.335–345 & passim.

[99] NB This is an example of convergent thinking, as Deacon seems unaware of Darwin's argument about the crucial reagency of language-use in evolution of the human brain.

[100] Darwin, *Descent*, p.610.

[101] Darwin, *Descent*, p.94.

[102] Dawkins, *Selfish Gene*, p.3.

[103] Michael Gard and Ben Bradley, 'Getting Away with Rape: Erasure of the Psyche in Evolutionary Psychology,' *Psychology, Evolution & Gender* 2, (2000), pp.313–319.

[104] Darwin, *Descent*, p.619.

9

The Critical Test

Falsifiability is the fail-safe. Any theory which cannot survive the challenge of empirical disproof must lose the seal of science. How does this touchstone judge Darwin's psychology?[1] Is there an observational test—an *experimentum crucis*—by the results of which the approach described in these pages would stand or fall? With this question, the two streams of scholarship sustaining this book join force: Darwin's writings on agency; and the requirements of a biologically informed twenty-first century psychology. For the needed test must ask a question hitherto unanswered by contemporary scientists, though crucial to their conception of the human. Are you and I born preadapted to group-life?

Darwin has been hailed, alongside Herbert Spencer and Alexander Bain, for trail-blazing the study of babies in psychology (see 4.6 Resonances).[2] Unlike his peers, however, Darwin based his statements about infancy on a meticulous, years-long observational study that he and his wife Emma had conducted—mostly but not entirely concentrated on their oldest child, William Erasmus (1839–1914), nicknamed 'Doddy,' who was born when Charles was thirty and Emma thirty-one. Darwin called this a 'natural history of babies.'[3]

Doddy was a particularly good subject for Darwin's purposes because, being first-born, he 'could not have learnt anything by associating with other children.'[4] The bulk of the notebooks Emma and Charles filled are in Darwin's hand, and are concerned with the first year of Doddy's development.[5] Darwin's correspondence shows he also solicited specific kinds of observation of babies from family, friends, and acquaintances, especially from young mothers: 'A dear young lady near here,

---

[1] The rule that all scientific conjectures must be refutable—or falsifiable—was argued by Karl Popper, *Objective Knowledge: An Evolutionary Approach* (Oxford: Clarendon Press, 1972); cf. Michael Ghiselin, *The Triumph of the Darwinian Method* (Berkeley CA: University of California Press, 1969).
[2] Thomas Dixon, *From Passions to Emotions: The Creation of a Secular Psychological Category* (Cambridge: Cambridge University Press, 2003), pp.143–144.
[3] Charles Darwin, 'Notebook M,' (Cambridge: Darwin Online, 1838), pp.142, back cover; Charles Darwin and Emma Darwin, *Notebook of Observations on the Darwin Children*, (Cambridge: Darwin Online, 1839–1856), http://darwin-online.org.uk/content/frameset?itemID=CUL-DAR210.11.37.
[4] Charles Darwin, *The Expression of the Emotions in Man and Animals*, 2nd ed. (London: Murray, 1890), p.379.
[5] The notes go from December 1839 (Doddy was born on the 27th Dec. 1839) to July 1856. Emma Darwin's observations become more copious towards the end of the notebook, and particularly concern what her children *said*.

Darwin's Psychology. Ben Bradley, Oxford University Press (2020). © Oxford University Press.
DOI: 10.1093/oso/9780198708216.001.0001

plagued a very young child for my sake, till it cried, & saw the eyebrows for a second or two beautifully oblique, just before the torrent of tears began.'[6]

Darwin's early observations gained consequence in his mature publications on human agency: for three reasons. First, as we saw in Chapter 4, Darwin held that, if an infant exhibited a recognizable action it was likely innate, and therefore had an evolutionary base. Secondly, infants—in being both human but, by definition, unable to speak and, hence, largely unaffected by culture—provided a useful bridge between civilized adults and their animal forbears. The springs of human agency, instinctive or not, were at their most obvious in the early months of life. For example, *Expression* argues the observation of infants shows expressive movements in a 'pure and simple' form which they lose over time.[7] Similarly, *Descent*'s claim that '*Reason* stands at the summit' of human faculties found support from 'a daily record of the actions of one of my infants':

> ... when he was about eleven months old, and before he could speak a single word, I was continually struck with the greater quickness, with which all sorts of objects and sounds were associated together in his mind, compared with that of the most intelligent dogs I ever knew.[8]

Third, Darwin drew *an analogy* between infant development and the evolution of agency. In both, there was a gradual and natural transition from lesser to greater complexity. The analogy could be used in two ways. One was rhetorical. A lack of obvious evidence for the evolution of human capacities did not mean such evolution had not occurred: 'At what age does the new-born infant possess the power of abstraction, or become self-conscious, and reflect on its own existence? We cannot answer; nor can we answer in regard to the ascending organic scale.' Conversely, when in doubt about 'the probable steps and means by which the several mental and moral faculties of man have been gradually evolved,' we could study infant development as an analogous, because equally gradual and natural, process: 'That such evolution is at least possible, ought not to be denied, for we daily see these faculties developing in every infant.'[9]

---

[6] Charles Darwin, 'Letter to Huxley, 30th January,' *Correspondence of Charles Darwin* (Cambridge: Darwin Correspondence Project, 1868), https://www.darwinproject.ac.uk/letter/?docId=letters/DCP-LETT-5817. Cf. pp.122–135, Samantha Evans, ed. *Darwin and Women: A Selection of Letters* (Cambridge: Cambridge University Press, 2017); Marjorie Leach and Paula Hellal, 'Darwin's 'Natural History of Babies', *Journal of the History of the Neurosciences* 19, (2010), pp.140–157.

[7] Darwin, *Expression*, p.14.

[8] Charles Darwin, *The Descent of Man, and Selection in Relation to Sex* (London: Murray, 1874), pp.75–77.

[9] Darwin, *Descent*, pp.126–127. Note that here I advisedly use the words 'analogy' and 'analogous,' rather than 'homology' and 'homologous.' Darwin made no argument in *Descent* or elsewhere that infant development (ontogeny) was *homologous* to the evolution of human agency (phylogeny). In fact, his response to Taine's article in 1877 ('The Sketch') can be seen as a rebuttal of even a mild homology, in that Taine believed: 'the child presents in a passing state the mental characteristics that are found in a fixed state in primitive civilisations, very much as the human embryo presents in a passing state

An example here was the heated Victorian argument over the origin of human language.[10] Having suggested that the hominin voice-box perhaps evolved through gibbon-like singing, *Descent* nominates imitation backed by advanced mental powers as the chief drivers in the human species' acquisition of speech.[11] This suggestion ran foul of the eminent philologist and anti-evolutionist, Max Müller (1823–1900).[12] Müller pooh-poohed the 'bow-wow' theory that language had started from semi-human ancestors imitating the sounds around them. For him, language and thought were 'inseparable,' and no animal could attain to the level of thought intrinsic to speech. Hence language could not have emerged gradually from animal beginnings: it was created from nothing by God. Darwin refuted Müller by pointing out that, long before infants link sounds with ideas, let alone speak, they demonstrate a capacity to form general concepts.[13] For example, Darwin held that preverbal infants (and dogs) form both general and specific concepts that guide their recognition of others. At nine weeks, Doddy would orientate to his mother when she first approached as he would to any other adult—showing he had a 'general concept' of a person.[14] But when she came close enough, his recognition of her was signalled by a change 'in the little noises he was uttering.'[15] Likewise, at seven months, he associated an idea of his nurse with her name, 'so that if I called it out he would look round for her.'[16]

9.1 Publications on Infancy

While *Descent* lays down a crucial rationale for observing infants as proxy for proving the evolution of human agency, and uses Darwin's own observations to make occasional points, the most extensive discussions of his infant observations occur in *Expression*. These largely concern crying, screaming, weeping, and

the physical characteristics that are found in a fixed state in the classes of inferior animal.' (Even here, note, Taine only proposes an *analogy*—'very much as'—between *cultural* and embryological or *physical* recapitulation.): Hippolyte Taine, 'The Acquisition of Language by Children,' *Mind* 2, (1877), p.259; Charles Darwin, 'A Biographical Sketch of an Infant,' *Mind: A Quarterly Review of Psychology and Philosophy* 2, (1877). Cf. John Morss, *The Biologising of Childhood: Developmental Psychology and the Darwinian Myth* (Hove UK: Lawrence Erlbaum, 1990).

[10] Gregory Radick, *The Simian Tongue: The Long Debate About Animal Language* (Chicago: University of Chicago Press, 2007), pp.1–83.
[11] Darwin, *Descent*, pp.568–573, 84–92.
[12] This debate paralleled the Victorian ethnology–anthropology debate over an Edenic versus an evolutionary origin for the human species: see Ch. 2, 'Natural histories of man.'
[13] Darwin, *Descent*, pp.88–89; Darwin, 'Biographical Sketch,' pp.289–291.
[14] For the same argument, applied to dogs, see Darwin, *Descent*, p.83: 'when a dog sees another dog at a distance, it is often clear that he perceives that it is a dog in the abstract; for when he gets nearer his whole manner suddenly changes, if the other dog be a friend.'
[15] Darwin and Darwin, *Notebook of Observations on the Darwin Children*, p.5r.
[16] Darwin, 'Biographical Sketch,' p.290.

sobbing—which infants do 'with extraordinary force' (see Ch. 4).[17] *Expression* also draws inferences from the *lack* of blushing in infancy: namely, that blushing implies self-consciousness, something which children take time to develop. But *Expression*'s most significant exploitation of Darwin's baby observations target infants' readings, or 'recognition,' of *others*' expressions.

As we saw in Chapter 4, a distinctive element in Darwin's understanding of non-verbal gestures was that the significances of expressive movements were best approached via their readings by others, as mediated by the circumstances which provided their specific occasions. This led Darwin to hypothesize that, as expressive movements became instinctive, 'their recognition would likewise have become instinctive.'[18] To emphasize the viability of this hypothesis, *Expression* describes four observations from Darwin's study of Doddy.

First, Darwin reported Doddy 'understood a smile and received pleasure from seeing one, answering it by another,' at much too early an age 'to have learnt anything by experience.' Second, new father Darwin had experimented by making 'many odd noises and strange grimaces, and tried to look savage' at the four-month-old Doddy. These noises (if not too loud) and grimaces, 'were all taken as good jokes' by his son, a fact that Darwin attributed to their being 'preceded or accompanied by smiles'—proving the baby's reading of Darwin's experimental vocalizations and scary facial movements was context-sensitive. Third, at around five months, Doddy 'seemed to understand a compassionate expression and tone of voice.' Fourth, at six months, when Doddy's nurse pretended to cry, his face 'instantly assumed a melancholy expression, with the corners of the mouth strongly depressed,' suggesting that 'an innate feeling' had told him that 'the pretended crying of his nurse expressed grief,' and this, 'through the instinct of sympathy,' excited grief in Doddy.[19]

Five years after *Expression* appeared, Darwin published an article 'A Biographical Sketch of an Infant' in the recently launched serial *Mind*, a journal dedicated to settling whether psychology deserved to be called a science. Darwin's Sketch was framed in response to an article by the French philosopher Hippolyte Taine (1828–1893) in the previous issue of *Mind*, called: 'The Acquisition of Language by Children.' Taine's essay recounted a series of observations of a little girl from birth onwards, related to her use and understanding of language and language-like sounds.

Darwin's attention would have been pricked by Taine's article because his own decades-old observations of Doddy had been informed by a long-standing dispute over language acquisition. This concerned the origins of the 'natural language' of emotions which Scottish 'common sense' philosophers like Thomas Reid

---

[17] Darwin, *Expression*, p.14.
[18] Darwin, *Expression*, p.378.
[19] Darwin, *Expression*, p.379.

(1710–1796) and Dugald Stewart (1753–1828) had argued to provide the foundations for a child's learning of the 'artificial language' of words.[20] Consequently, the
slant of Darwin's article differed from Taine's in significant ways. Most obviously,
Darwin's focus was primarily on non-vocal, non-verbal expressions during the first
year of life. Taine's comments largely addressed the vocalizations of a toddler.

Darwin's observations were more precise and science-like than Taine's. Darwin
often dated to the day the age his child had been when a particular observation
had been made. Taine never did this. Darwin revelled in precise descriptions of his
son's movements and the circumstances which provoked them. Taine described
his subject's actions more globally and impressionistically: her vocalizations at one
year of age were 'the twittering of a bird.' Moreover, Taine's interpretations were
seldom justified by reasoning or experimental test. For instance he baldly concludes, without further comment, that, even at age one, 'she attaches no meaning
to the sounds she utters. She has learned only the materials of language.'[21] Taine
had no evolutionary element in the psychology he projected onto his girl: she apparently learnt solely through trial and error.[22] Doddy on the other hand, was born
with reflexes and expressions ready-to-go, abetted by an experimentally demonstrated capacity to read the expressions of others, and a burgeoning ability to
reason.

The principal drive in Darwin's early note-taking on Doddy had been to gather
evidence of the form of early human expressions and for the existence and type of
any innate responses. Hence the copious use of these observations in *Expression*.
The 'Biographical Sketch of an Infant' was intended to relate Darwin's *other* findings about Doddy, those not directly bearing on expression. Nonetheless, the article finishes up by stressing the same four critical observations that *Expression* had
also exploited, and in almost the same words: to show that, 'an infant understands
to a certain extent, and as I believe at a very early period, the meaning or feelings of
those who tend him, by the expression of their features.'[23]

Leaving aside this crucial conclusion, the Sketch reads as a recapitulation in
small of several key themes from Darwin's treatment of the most distinctive qualities of human agency in *Descent*. Whilst the article does not address *Descent*'s

---

[20] A section of Erasmus Darwin, *Zoonomia, or, the Laws of Organic Life, Vol.1* (London: Johnson,
1794) is called 'Of Instinct.' This Charles Darwin had read closely, probably shortly before Doddy's
birth, as his marginalia attest. The elder Darwin had used an associationist framing of observational
claims about infants' expressions—especially weeping—to refute Reid's argument that humans are born
possessing a God-given 'natural language.' Although *Expression* neither refers to this debate, nor to any
of its protagonists, it nevertheless covertly offers a point-by-point refutation of Erasmus Darwin's work,
whilst recasting the natural-theological conclusions of Reid and Stewart in an evolutionary form. See
Ben Bradley, 'Darwin's Intertextual Baby: Erasmus Darwin as Precursor in Child Psychology,' *Human
Development* 37, (1994), pp.86–102.
[21] Taine, 'Acquisition of Language'pp.252–253.
[22] Though she did recapitulate the stages of mental activity in 'primitive civilisations,' that is *cultural*
evolution, according to Taine, 'Acquisition of Language,' p.259.
[23] Darwin, 'Biographical Sketch,' pp.293–294.

group-hypothesis directly, several of the Sketch's observations relate to Darwin's theses about human groups. Most obvious is the section on 'moral sense', which also discusses conscience.[24] Doddy's first sign of feeling abashed at others' disapprobation, and a resultant need for reconciliation, occurred at 13 months when his father said, 'Doddy won't give poor papa a kiss,—naughty Doddy'! By two and a quarter years of age, Doddy was congratulating himself on his own generosity to his younger sister (giving her his last bit of gingerbread), with 'high self-approbation', exclaiming, 'Oh kind Doddy, kind Doddy.' Two months later he became sensitive to ridicule, imagining that when others were laughing, they were laughing at him. And so on.[25]

Beyond this, it is obvious that many of Doddy's and his infant siblings' acts had been observed when more than one other person was present, people with whom the infant was also interacting.[26] Witness for example Doddy's jealousy: 'Jealousy was plainly exhibited when I fondled a large doll, and when I weighed his infant sister, he being then 15½ months old.' Or his pride: 'When pleased after performing some new accomplishment, being then almost a year old, he evidently studied the expression of those around him.' This latter observation linked directly with Descent's emphasis on the human concern with the approbation and disapprobation of one's acts by others.

An instructive contrast can be drawn between Taine's and Darwin's understandings of infant agency. For Taine (rather like Bain) there was no evolved structure of agency built into the new-born. The little girl he observed is said 'from the first hour' to cry incessantly and to 'kick about', moving all her limbs 'and perhaps all her muscles.' Taine brackets these movements as 'probably reflex action'—by which he appears to mean, the baby had not specifically directed reflexes, but merely a non-specific bodily reactiveness. Even by three months, there was no aim: 'she touches and moves at random; she tries the movements of her arms and the tactile and muscular sensations which follow them; nothing more.' Slowly, out of the 'enormous number' of random movements and sensations, 'there will be evolved by gradual selection the intentional movements having an object and attaining it.'[27]

Taine applies the same approach to explaining the acquisition of language. Sounds are first emitted at random. Hence the little girl's meaningless 'twittering' during her first twelve months: 'Its flexibility is surprising; I am persuaded that all the shades of emotion, wonder, joy, wilfulness and sadness are expressed by differences of tone; in this she equals or even surpasses a grown up person.' But over time, through trial and error—or 'gropings and constant attempts'—'the child

---

[24] Darwin, 'Biographical Sketch', pp.291–292.

[25] Darwin, 'Biographical Sketch', pp.291–292.

[26] Darwin's household included not just his wife and children but the baby's beloved nurse, and up to twenty other domestic servants, not to mention visiting family-members or other guests.

[27] Taine, 'Acquisition of Language', p.252. NB Taine's use of evolutionary language: showing its wide use in Victorian times—far beyond the bounds of Darwin's particular theory.

learns to emit such or such a sound.' A small role is given to imitation. But otherwise, trial and error reign supreme.[28]

By today's understanding of developmental plasticity, Taine's views would be subsumed under *somatic selection*, that is, adjustment involving an overproduction of movements, with a resulting selection of favourable combinations or 'happy hits.'[29] Darwin perceived the plasticity of infant action differently, as being shaped by specific processes. Simplest of all were reflexes with specific functions—although these were often the fruit of plasticity. Thus, during Doddy's first week: 'it seemed clear to me that a warm soft hand applied to his face excited a wish to suck. This must be considered as a reflex or an instinctive action, for it is impossible to believe that experience and association with the touch of his mother's breast could so soon have come into play.'[30] Crying, some fearful and angry reactions, and the differential use of intonation to mark interrogation and exclamation were *instinctive*—actions which involved 'a little dose of judgement' and hence showed more variability than reflexes.[31] Beyond this stood Doddy's more plastic capacities: for purposive movement of his hand to his mouth (evident before six weeks of age); for reason; and for intentional action (both first seen at sixteen weeks).

Here also fits the infant's capacity for understanding 'the meaning or feelings of those who tend him, by the expression of their features'—which Darwin claimed as evidence of innate sympathy, the lynchpin of his group psychology.[32] Whilst the Sketch does not reiterate this point, the final pages of *Expression* propose that the infant's inbuilt capacity to read others' expressive movements proves crucial to social development. Not only does it provide 'the first means of communication between the mother and her infant.' It also underpins the child's first appreciation of group-standards: 'she smiles approval, and thus encourages her child on the right path, or frowns disapproval.'[33]

9.2 Resonances

9.2.1 Trail-blazing descriptions

Darwin's descriptions of infants have proven impressively robust. While not every one of his observations of Doddy's capacities generalizes to other babies, the *kinds*

---

[28] Taine, 'Acquisition of Language,' p.253.

[29] According to James Mark Baldwin's coinages, this would be called 'functional selection' or 'organic accommodation'; Mary Jane West-Eberhard, *Developmental Plasticity and Evolution* (New York: Oxford University Press, 2003), pp.37ff.

[30] Darwin, 'Biographical Sketch,' pp.285–286.

[31] Charles Darwin, *The Origin of Species by Means of Natural Selection, or the Preservation of Favoured Races in the Struggle for Life*, 6th ed. (London: Murray, 1876), p.205.

[32] Darwin, 'Biographical Sketch,' pp.293–294.

[33] Darwin, *Expression*, pp.385–386.

of observation he made have repeatedly proven prescient of later psychologists' concerns, many being a century or more ahead of their time.[34] Take for example the note, made in 1840, that the eleven-month-old Doddy: 'looks very much pleased, after the performance [of] any of these accomplishments. He evidently studies expression of those around him, especially if anything new is done before him'.[35] Psychologists would now recognize this inspection of others' faces when anything new or ambiguous fills a baby's attention as *social referencing*. Yet social referencing was not 'discovered' until the late 1970s.[36] And acknowledgement of the pleasure infants take in amusing other people has had to await the observational work of Vasu Reddy.[37]

Likewise, Darwin's Sketch acutely contrasts apes' reactions to their mirror-images with those of babies. Doddy was 'like all infants', in that he 'much enjoyed' looking at his reflection, 'and in less than two months perfectly understood that it was an image'. By contrast, the apes Darwin tested did not recognize their image as an image but 'placed their hands behind the glass', and, 'far from taking pleasure in looking at themselves they got angry and would look no more'.[38] The way infants' prolonged jubilation at their own mirror-images contrasts with the swift exhaustion of other primates' interest in their reflections, gave rise to a great deal of discussion about the origins of self-consciousness in the 1950s, and again in the 1980s.[39]

A third example is Darwin's observation of Doddy's jealousy when his father pretended to show affection for a large doll.[40] Jealousy is something that has only been consistently attributed to infants by psychologists from the late 1990s. This is largely due to a series of experiments by Sybil Hart and her associates, who devised

---

[34] For example, we now know that even neonates can visually locate a sound, something Darwin did not observe Doddy to do until he was 18 weeks old: Darwin, 'Biographical Sketch', p.286; Marcella Castillo and George Butterworth, 'Neonatal Localisation of a Sound in Visual Space', *Perception* 10, (1981), pp.331–338.

[35] Darwin and Darwin, *Notebook of Observations on the Darwin Children*, p.26r.

[36] Saul Feinman, 'Social Referencing in Infancy', *Merrill-Palmer Quarterly* 28, (1982), pp.445–470

[37] Vasu Reddy, 'Infant Clowns: The Interpersonal Creation of Humour in Infancy', *Enfance* 53, (2003), pp.247–256; Vasu Reddy, 'On Being the Object of Attention: Implications for Self–Other Consciousness', *Trends in Cognitive Sciences* 7, (2003), pp.397–402; Vasu Reddy, 'Before the 'Third Element': Understanding Attention to Self', in *Joint Attention: Communication and Other Minds*, ed. Naomi Eilan, et al. (Oxford: Clarendon Press, 2005), pp.85–109.

[38] Darwin, 'Biographical Sketch', pp.289–290.

[39] See the famous talk given in 1949 by Jacques Lacan, 'The Mirror-Stage as Formative of the Function of the I as Revealed in Psychoanalytic Experience', in *Ecrits: A Selection*, ed. Alan Sheridan (London: Tavistock, 1977), pp.1–8. This was taken up by Donald Winnicott, 'Mirror-Role of Mother and Family in Child Development', in *Playing and Reality* (Harmondsworth: Penguin Books, 1967), pp.149–159, and Ben Bradley and Colwyn Trevarthen, 'Babytalk as an Adaptation to the Infant's Communication', in *The Development of Communication*, ed. Natalie Waterson and Catherine Snow (London: Wiley, 1978), pp.75–92. Experimentalists rediscovered this phenomenon in the 1980s: e.g. Michael Lewis and Jeanne Brooks-Gunn, 'The Development of Early Visual Self-Recognition', *Developmental Review* 4, (1984), pp.215–219.

[40] Darwin, 'Biographical Sketch', p.289.

an experiment in which mothers directed conversation away from their baby to 'an infant-size doll.'[41]

Perhaps most resonant are the observations that led Darwin to conclude that infants read meaning into their caregivers' facial expressions, and so can communicate, in the first weeks of life.[42] There had long been no place for this insight in the surge of research on infancy and language acquisition by psychologists and psycho-linguists, which began in the 1960s. Piaget and his followers, who saw infant development as a cognitive process, largely pictured the infant as a being who had slowly to figure out what other people were, and how to be social. Attachment theory also made no place for a young baby communicating with, or understanding, adults. However, in the early 1970s, the Edinburgh-based psycho-biologist Colwyn Trevarthen, who has long admired Darwin's approach to agency, began to present a series of minutely analysed descriptions of filmed infant–adult 'conversations' that forced him to conclude babies only two months old were furnished with innate intersubjectivity—the capacity to coordinate their actions (and mental states) with others.[43]

Trevarthen's observations of infant–mother intersubjectivity connected his research with, and so strengthened, two significant strands of developmental theory newly emerging in the 1970s. One proposed that infants needed to have a pragmatic or *communicative* competence prior to using what Noam Chomsky's called their *grammatical* competence in an appropriately conversational way.[44] The other strand took up the work of Lev Vygotsky (1896–1934), who had proposed that every aspect of a child's social or cultural abilities appeared, not once, but twice in development: first, embedded in *social* agency, 'between people' (*inter*mentally), and later, after a process of *internalization*, in individual agency (*intra*mentally).[45]

[41]   E.g. Sybil Hart, Tiffany Field, and Claudia De Valle, 'Infants Protest Their Mothers' Attending to an Infant-Size Doll,' *Social Development* 7, (1998), pp.54–61.

[42]   Darwin, 'Biographical Sketch,' p.294.

[43]   Colwyn Trevarthen, 'Conversations with a Two-Month-Old,' *New Scientist* 62, (1974), pp.230–235; Colwyn Trevarthen, 'Communication and Cooperation in Early Infancy: A Description of Primary Intersubjectivity,' in *Before Speech: The Beginnings of Human Communication*, ed. Margaret Bullowa (Cambridge: Cambridge University Press, 1979), pp.321–347.

[44]   Jurgen Habermas, 'Towards a Theory of Communicative Competence,' in *Recent Sociology No.2*, ed. Hans Dreitzel (London: Macmillan, 1970), pp.360–375; Robin Campbell and Roger Wales, 'The Study of Language Acquisition,' in *New Horizons in Linguistics*, ed. John Lyons (Harmondsworth: Penguin Books, 1970), pp.1–13; Lois Bloom, *Language Development: Form and Function in Emerging Grammars* (Cambridge, MA: M.I.T. Press, 1970); Joanna Ryan, 'Early Language Development: Towards a Communicational Analysis,' in *The Integration of a Child into a Social World*, ed. Martin Richards (Cambridge: Cambridge University Press, 1974), pp.185–213; Jerome Bruner, 'The Ontogenesis of Speech Acts,' *Journal of Child Language* 2, (1975), pp.1–19.

[45]   Lev Vygotsky, *Mind in Society: The Development of Higher Psychological Processes* (Cambridge MA: Harvard University Press, 1978), pp.55–57. Vygotsky died in 1934, so it may seem odd to call his theory 'newly emerging.' However, his influence in the English-speaking world only took off after his 1934 book Lev Vygotsky, *Thought and Language* (Cambridge MA: MIT Press, 1962) was translated into English. His 1930 book, Vygotsky, *Mind in Society*, was not translated into English until 1978. Arguably, Vygotsky partly drew his theory from Darwin's writings: see Anna Stetsenko, 'Darwin and Vygotsky on Development: An Exegesis on Human Nature,' in *Children, Development and Education: Cultural, Historical, Anthropological Perspectives*, ed. Michalis Kontopodis, Christoph Wulf, and Bernd Fichtner (Dordrecht: Springer, 2011), pp.24–40.

This chimes directly with *Descent's* argument that, while infants may owe their *capacity* for moral discrimination to some sort of (part) instinct, the actual moral standards by which they will come to judge themselves are specified by the customs and institutions of the social world in which they grow up (Chs 7–8).[46]

9.2.2 Baldwin and plasticity

The fact that Darwin's Sketch put such stress on his infant children's capacities to reason and learn underlines that there is no incompatibility between human agency having an evolutionary basis and being highly flexible: 'What a contrast does the mind of an infant present to that of the pike,' exclaims the Sketch—pikes being paragons of mental inflexibility.[47] To the extent that humans were, in Darwin's terms, the *frontier instance* of intelligence, and therefore of developmental plasticity, his observations of infancy chime with and can inform any biology which promotes a phenotype-led vision of evolution (see 4.3.5 Comparative evidence).[48]

The first psychologist to underline the importance of developmental plasticity, and hence of infancy, to an evolutionary understanding of human agency was someone who explicitly drew his theory from Darwin: James Mark Baldwin (see 4.6.1 Wundt, Baldwin, Mead). Baldwin argued that, as mammalian species 'advance upward,' they show 'decreasing instinctive endowment and increasing plasticity, accompanied by increasing mental capacity and educability.' Thus human infants are 'poorest in instinctive endowment, most helpless at birth, but most teachable and most highly equipped with brain and mind.' Plasticity 'is a real character,' he went on, 'the opposite of fixity.' It gives an organism genuine alternatives; 'genuine novelties of adjustment are possible. And consciousness, intelligence, is also a real character, correlated with plasticity.'[49]

Baldwin's theoretical and empirical elaborations of his evolutionary understanding of the structure and subsequent trajectory of infant plasticity directly extended Darwin's groupness hypothesis. For Baldwin, the fundamental psychological unit was the *socius*—a more or less socialized individual—integrated into a group of like socii, thereby constituting a *social situation*.[50] His work

---

[46] See also Paul Ricoeur, *Freud and Philosophy: An Essay on Interpretation* (New Haven CT: Yale University Press, 1970), pp.506–514, for a different but fascinating and Darwin-compatible account of internalization.

[47] Darwin, 'Biographical Sketch,' pp.290–291, where Darwin cites the case of a pike, who, 'during three whole months dashed and stunned himself against a glass partition which separated him from some minnows; and when, after at last learning that he could not attack them with impunity, he was placed in the aquarium with these same minnows, then in a persistent and senseless manner he would not attack them!'

[48] Scott Gilbert and David Epel, *Ecological Developmental Biology: The Environmental Regulation of Development, Health, and Evolution*, 2nd ed. (Sunderland, MA: Sinauer Ass., 2015), p.xiii.

[49] James Baldwin, *Darwin and the Humanities* (Baltimore: Review Publishing, 1909), pp.23–25.

[50] Baldwin, *Darwin*, pp.44–45.

significantly influenced Freud's and, particularly, Piaget's understandings of child development.[51]

While Piaget drew several of the key terms in his vocabulary from Baldwin (accommodation, assimilation, circular reaction), he later disavowed Baldwin's influence, and, in particular, Baldwin's idea that development comprised a social–individual dialectic. He explained this renunciation by pointing out that the term (social) cooperation could be rewritten 'co'-'operation.' This, he said, allowed him to avoid putting the 'whole burden' of development 'on society.' He justified this significant little piece of word-play by saying it would be an 'incredible mistake' to forget that, 'without the nervous system, nothing at all would happen.' Consequently, since no one would *dare to say that the nervous system was a social product*, it was 'necessary to recognize its existence prior to society and social forces.'[52] Yet, both Baldwin—and Baldwin's muse, Darwin—*did* dare to suggest that the nervous system was (in part) a social product. As do many of today's evolutionary psychologists and brain scientists (see 7.5 Resonances).[53]

In contrast to Piaget, Baldwin's elaboration of plasticity went out of its way to emphasize how Darwin's theory could account for the evolution of social cooperation, or what Baldwin called *joint action*. He asked how nature had provided 'not simply a correlation of characters within a single organism, but *as between two or more different individuals*'?[54] Plasticity was the answer, or, in Darwin's language, *transitional habit* backed up by natural selection—which is how *Origin* accounts for the joint acquisition of mammary glands in women and the instinct of sucking in infants, amongst other things (see 3.3.4 Agency directs selection).[55]

The slow cementing of habits by natural selection was what Baldwin called *organic selection* (and what we call the *Baldwin effect*).[56] Explaining the biological basis for social cooperation required an assumption that actual 'life activities'

---

[51] Jean Piaget and Jacques Vonèche, 'Reflections on Baldwin,' in *The Cognitive-Developmental Psychology of James Mark Baldwin: Current Theory and Research in Genetic Epistemology*, ed. John Broughton and John Freeman-Moir (Norwood NJ: Ablex, 1982), pp.80–86. See Jean Piaget, *The Moral Judgement of the Child* (London: Routledge & Kegan Paul, 1932), which shows considerable engagement with Baldwin.

[52] Piaget and Vonèche, 'Reflections,' pp.84–85, my italics.

[53] E.g. John Cacioppo, Gary Berntson, and Jean Decety, 'Social Neuroscience and Its Relation to Social Psychology,' *Social Cognition* 28, (2010), pp.675–685; Robin Dunbar, 'The Social Brain: Mind, Language, and Society in Evolutionary Perspective,' *Annual Review of Anthropology* 32, (2003), pp.163–181; Chris Frith, 'The Social Brain?,' *Philosophical Transactions of the Royal Society B* 362, (2007), http://doi.org/10.1098/rstb.2006.2003.

[54] Baldwin, *Darwin*, p.30, his italics.

[55] Darwin, *Origin 1876*, pp.138ff, 190.

[56] James Baldwin, 'A New Factor in Evolution,' *American Naturalist* 30, (1896), pp.441–451; James Baldwin, 'A New Factor in Evolution, Continued,' *American Naturalist* 30, (1896), pp.536–553; James Baldwin, *Development and Evolution: Including Psychophysical Evolution, Evolution by Orthoplasy, and the Theory of Genetic Modes* (New York: Macmillan, 1902); George Simpson, 'The Baldwin Effect,' *Evolution* 7, (1953), pp.110–117; Jacy Young, 'The Baldwin Effect and the Persistent Problem of Preformation Versus Epigenesis,' *New Ideas in Psychology* 31, (2013), pp.355–362; West-Eberhard, *Developmental Plasticity and Evolution*, pp.151ff.

require individuals to live '*in associated pairs, groups* etc.'[57] Living in pairs or groups, argued Baldwin, individuals would learn to coordinate their actions, which would yield them immediate benefits. Over time, 'the individual actions required' would be gradually 'moulded by variation from generation to generation and fitted together for the performance of the joint function.' In each generation, 'groups of individuals best' fitted for the joint action would be formed.' Thus individuals would evolve according to the extent of their ability 'to play each their part "for the benefit of the community." '[58] Meanwhile the whole group would prove more resilient than less-coordinated groups due to the advantages of their improved cooperation.[59]

9.2.3 The groupness hypothesis

Recently, psychologists have returned to the problem of how joint action might have evolved.[60] That such evolution has occurred forms the focus of the evidence compiled by advocates of the social brain hypothesis—which proposes that humans owe their big brains to the evolutionary demands of group-living (see 7.5 Resonances).[61] Apart from this, the idea that young babies might be able actively to participate in groups—as Darwin's theory predicts—had until recently made few inroads into psychological thinking. Yet Darwin's publications set a clear precedent for using infancy to test evolutionary claims about humans' group-living.

Suggestive as Darwin's books and articles on infancy are, they do not directly aim to provide evidence for his central psychological hypothesis—that the most human of human characters draw from an inbuilt 'groupness': humans' capacity to interact at the same time with more than one other group-member, and so participate in group-dynamics. This hypothesis gets its clearest formulation in an aside to *Descent*: 'It seems possible that the connection between the related members of the same barbarous tribe, exposed to all sorts of danger, might be so much more important, owing to the need of mutual protection and aid, than that between the mother and her child.'[62] Yet this critical distinction between a 'dyadic' or two-person sociability and a supra-dyadic *group*-sociability has hardly figured in the last half-century's explosion of infancy research—for which one catalyst was

---

[57] Baldwin, *Darwin*, p.30, his italics.
[58] Baldwin's quote here refers to Darwin, *Origin 1876*, p.67.
[59] Baldwin, *Darwin*, pp.30–31.
[60] For example, in David Lee's theory of 'tau coupling': David Lee, 'Guiding Movement by Coupling Taus,' *Ecological Psychology* 10, (1998), http://www.tandfonline.com/doi/abs/10.1080/10407413.1998.9652683. On the developmental and psycho-cultural significance of 'joint action,' see the work of the late John Shotter, e.g. Tim Corcoran and John Cromby, eds., *Joint Action: Essays in Honour of John Shotter* (London: Routledge, 2016).
[61] E.g. Dunbar, 'Social Brain.'
[62] Darwin, *Descent*, p.589.

the advent of mass-produced video-recorders in the late 1960s.[63] This explosion ignited a different debate, prompted by doubts about the claims of cognitive scientists who, influenced by Jean Piaget, still assumed that 'at birth an infant is essentially an asocial being.'[64]

By 1970, a growing number of psychologists claimed babies were *born social*.[65] Many championed 'attachment theory,' based on the work of John Bowlby (1907–1990).[66] Contrary to Darwin's statement about the primacy of intra-tribal connectedness for babies, Bowlby had imagined that the dangers of the Stone Age would mean new-borns needed an evolved set of *fixed action patterns* to promote proximity to the baby's mother: smiling; crying; cooing; following; mutual gaze; clinging; and sucking. Over the first nine months of life, these attachment behaviours would become locked onto one person—usually but not always the baby's mother—and the first attachment bond would form. This bond became the prototype for all a baby's subsequent social relationships, which by implication, would also be dyadic.

Attachment theory still provides psychology with its dominant conceptions of infants' social well-being and development. As a result, the last several decades of research into the first year of social life have focussed almost exclusively on twosomes.[67] Critics of attachment theory have pointed out that the 'fixed action patterns' supposedly seen in new-borns are actually very flexible. The same human baby typically produces many forms of smile, for example.[68] Also, attachment theory has little to say about what might go on between mother and baby, *once proximity has been gained*—which is much of the time, according to Bowlby.[69] Yet critics of attachment theory still typically focus almost exclusively on a two-person, infant–adult scenario.[70] This focus swiftly became entrenched by widespread

[63] Kenneth Kaye, *The Mental and Social Lives of Babies: How Parents Create Persons* (Chicago: University of Chicago Press, 1982); Ben Bradley, *Visions of Infancy: A Critical Introduction to Child Psychology* (Cambridge: Polity Press, 1989); Ben Bradley, 'Jealousy in Infant-Peer Trios: From Narcissism to Culture,' in *Handbook of Jealousy: Theories, Principles and Multidisciplinary Approaches*, ed. Sybil Hart & Marie Legerstee (Hoboken NJ: Wiley-Blackwell, 2010), pp.192–234.

[64] Rudolph Schaffer, *The Growth of Sociability* (Harmondsworth: Penguin, 1971), p.1.

[65] E.g. Mary Ainsworth, 'Object Relations, Dependency, and Attachment: A Theoretical Review of the Infant-Mother Relationship,' *Child Development* 40, (1969), pp.969–1025.

[66] John Bowlby, *Attachment and Loss, Vol.1: Attachment*, 2nd ed., 2 vols. (London: Hogarth Press, 1982).

[67] The 1920s and 1930s saw some research on peer–peer interaction, but this was largely focused on aggression and was conceived dyadically. For a brief review, see Jane Selby and Ben Bradley, 'Infants in Groups: A Paradigm for the Study of Early Social Experience,' *Human Development* 46, (2003), p.203

[68] Colwyn Trevarthen, 'Descriptive Analyses of Infant Communicative Behaviour,' in *Studies in Mother-Infant Interaction*, ed. Rudolph Schaffer (London: Academic Press, 1977), pp.227–270.; Ben Bradley, 'A Study of Young Infants as Social Beings' (University of Edinburgh, 1980).

[69] Trevarthen, 'Primary Intersubjectivity.'

[70] Kaye, *Mental and Social Lives of Babies*. For reviews, see: Selby and Bradley, 'Infants in Groups'; Bradley, 'Jealousy.' Trevarthen's research on infants' 'communicative musicality' provides an exception, e.g. Stephen Malloch and Colwyn Trevarthen, eds., *Handbook of Communicative Musicality* (Oxford: Blackwell-Wiley, 2009).

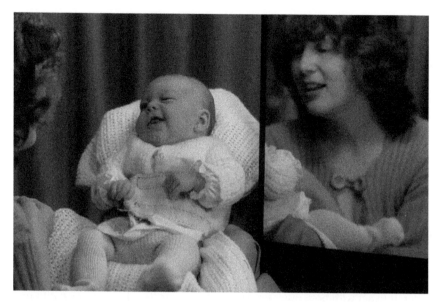

Figure 9.1 The *en face* recording paradigm for research on infant sociability, pioneered by Trevarthen and his colleagues. In this case, a single video-camera and a carefully-positioned mirror (at right) record the expressive movements of both baby and mother (who is sitting at the left of the picture).

E.g. Trevarthen, 'Conversations'. Picture kindly supplied by Professor Colwyn Trevarthen.

adoption of the so-called *en face* recording paradigm: where infant and adult are sat face-to-face, so expressions of both can be easily video-taped—using one or two tripod-mounted cameras—thereby providing optimal data for the micro-analysis of their non-verbal behaviour (Figure 9.1).[71]

The grip of two-person theories of sociability means most psychologies predict that infants, and even school-age children, will not participate in group-dynamics—human sociability being generated by a 'dyadic program'.[72]

This prediction, or assumption, remained untested for several decades. Recently, however, researchers in Switzerland, France, and Australia have begun to examine whether preverbal babies show groupness. European studies have focused on trios which include both babies and adults.[73] In Australia, we have focused on the more

[71] Bradley, 'Jealousy'; Kaye, *Mental and Social Lives of Babies*.

[72] Bowlby, *Attachment*, p.378; Dale Hay, Marlene Caplan, and Alison Nash, 'The Beginnings of Peer Relations,' in *Handbook of Peer Interactions, Relationships, and Groups*, ed. Kenneth Rubin, William Bukowski, and Brett Laursen (New York: Guilford Press, 2011), pp.121–142.

[73] Elisabeth Fivaz-Depeursinge and Antoinette Corboz-Warnery, *The Primary Triangle: A Developmental Systems View of Mothers, Fathers and Infants* (New York: Basic Books, 1999); Jacqueline Nadel and Hélène Tremblay-Leveau, 'Early Perception of Social Contingencies and Interpersonal Intentionality: Dyadic and Triadic Paradigms,' in *Early Social Cognition: Understanding Others in the First Months of Life*, ed. Phillipe Rochat (Mahwah NJ: Lawrence Erlbaum, 1999), pp.189–212; Elisabeth

symmetrical interactions in baby-only groups. We reason that, if groupness were to be shown in groups that included both infants and adults, then sceptics could argue the adults had, like puppet-masters, 'scaffolded' the interaction to make the babies appear more group-capable than they really are.[74] We found, even without adults present, meticulous frame-by-frame analysis of baby-only trios shows several kinds of interaction which could not be generated by a dyadic programme.

The method we have devised for our studies is to ask three or four sets of parents to bring to our film-studio babies aged between six and nine months—having already ensured the babies have not met. Once welcomed, each baby is secured in an immobilized push-chair by their parent. Two video-cameras are already rolling to capture what goes on (Figure 9.2). As soon as the babies are sitting comfortably, all adults vacate the studio to watch what unfolds through closed-circuit television from a nearby room. Now left alone, the babies interact for as long as they like, up to 25 minutes in some cases. As soon as they start to wail or grumble, we finish the session. Then we analyse the films.

All sorts of different dramas develop in infant-peer groups. In trios, babies will often interact with both other group-members at once—playing footsie with one while scowling at another, for example.[75] Or two group members will try sympathetically to comfort a third member who is unhappy, by touching for example—and, to our surprise, their acts of comforting often succeed.[76] In all such scenarios, infants use several different channels of communication at the same time: touch; facial expression; gaze; rhythm; and sound.[77]

In each group, non-verbal 'conversation' can generate idiosyncratic shared meanings over time. Imitation proves a great resource. For example, in one trio, Paula (8 months old) was animated and playful while her mother put her into her chair. However, as Paula watched her mother leave the room, her face fell. Mother

Fivaz-Depeursinge et al., 'Four-Month-Olds Make Triangular Bids to Father and Mother During Trilogue Play with Still-Face,' *Social Development* 14, (2005), pp.361–378.

[74] This had been a criticism of Trevarthen's conclusion from films made using the *en face* paradigm that babies could share mental states with their mothers, thus manifesting what he called their 'innate' or 'primary intersubjectivity': Trevarthen, 'Primary Intersubjectivity.' See e.g. John Shotter and Susan Gregory, 'On First Gaining the Idea of Oneself as a Person,' in *Life Sentences: Aspects of the Social Role of Language*, ed. Rom Harré (New York: Wiley, 1976), pp.3–9.; John Newson, 'The Growth of Shared Understandings between Infant and Caregiver,' in *Before Speech: The Beginnings of Human Communication*, ed. Margaret Bullowa (Cambridge: Cambridge University Press, 1979), pp.207–222; Kaye, *Mental and Social Lives of Babies.*

[75] Selby and Bradley, 'Infants in Groups,' pp.212–216.

[76] Mitzi-Jane Liddle, Ben Bradley, and Andrew McGrath, 'Baby Empathy and Peer Prosocial Responses,' *Infant Mental Health Journal* 36, (2015), pp.446–458. NB While we used current psychological vocabulary when describing infants' responses to a group-member's distress as 'empathy' in this paper, Darwin would undoubtedly have seen our observations as proving the early existence of *sympathy.*

[77] On rhythm, see: Ben Bradley, 'Early Trios: Patterns of Sound and Movement in the Genesis of Meaning between Infants,' in *Handbook of Communicative Musicality*, ed. Stephen Malloch and Colwyn Trevarthen (Oxford: Blackwell-Wiley, 2009), pp.263–280.

(a)   (b)

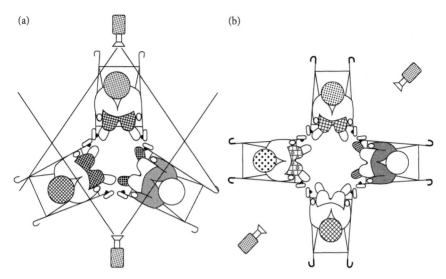

Figure 9.2 Arrangement of babies and cameras for experiments on infant-only trios and quartets.
Image credit: Picture kindly supplied by Professor Colwyn Trevarthen.

gone, she grabbed her toe—which she continued to do, off and on, for about half the ensuing five-minute session. This comforted her—as was shown by her greater curiosity about (i.e. looks at) the two other babies in her group when holding her toe than when not.

The two other girls in Paula's all-girl trio, Esther (7 months) and Ethel (6 months), showed no dismay when their mothers left them, looking around the studio with interest. Soon Esther noticed Paula holding her toe, and copied her, then looked at Ethel. Ethel did not follow suit. So Esther let her foot drop. A little while later, Esther tries again: holding her toe and looking at Paula. Paula looks away. Two tries later, Esther succeeds. She grabs her toe and looks at Paula—who is not holding her toe at this point. Paula copies Esther (Figure 9.3). Thereafter, Paula and Esther start playing a toe-holding game that lasts for several minutes. In the process, Paula's toe-holding has been transformed from an effort to contain her anxiety into a turn in a game.[78]

Psychologists have traditionally found it hard to credit narrative evidence of the kind produced by descriptive analyses of babies in groups.[79] Photographs notwithstanding, might perceptions of group-level infant interaction exist solely in the

[78] Selby and Bradley, 'Infants in Groups,' pp.209–211.
[79] Such disbelief—not to say ridicule—was not uncommon in response to early publications of Trevarthen's pioneering non-experimental, descriptive research on early infancy. Compare, for example, Trevarthen, 'Conversations,' with, Edward Start, 'Is Politeness Innate?,' *New Scientist* 71, (1976), pp.586–587.

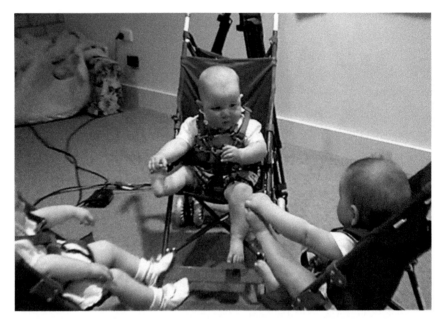

Figure 9.3 Paula (centre) following Esther's (right) imitative initiation of a toe-holding game while Ethel looks on with interest.
Selby and Bradley, 'Infants in Groups', p.211.

mind of the beholder? Evidence is sometimes deemed more valid if it can *mathematically* be shown to have very little chance of having resulted from a random play of events. Could anyone stage a convincing mathematically analysed, chance-proofed, test that confirms infant groupness?

An obvious barrier to devising such a test stems from the idiosyncratic actions and meanings generated in different infant-only groups. Statistical testing requires we compare the numbers of frequently repeated events *of the same kind* occurring under given conditions. The most frequent act performed when infants socialize is alteration in their direction of looking. So, change in gaze-direction becomes the obvious target-behaviour for any quantitative test of infant groupness.

Well-focused (macular, *foveal*) vision has a very narrow field of view in humans (around 13° in width). This means visually directed infant acts—pointing, smiling, frowning—cannot help but have a one-to-one (dyadic) interpersonal form, because an agent can only focus on one person at once, especially when in close proximity (Figures 9.2; 9.3). Reading or responding to others can be far more inclusive and groupy, because we can easily pick up what others are doing 'out of the corners of our eyes,' that is, through our ambient or *peripheral* vision, which has a wide field of view (approaching 180°).[80] So, rather than asking, 'Can infants direct their

---

[80] Colwyn Trevarthen, 'Two Mechanisms of Vision in Primates', *Psychologische Forschung* 31, (1968), pp.299–337.

actions toward two or more other babies at once?'—we would do better to examine the *responsiveness* of infants in groups, because the wide field detectable through their peripheral vision should allow them to respond to what more than one other group-member is doing at the same time. The key question then becomes: Can a baby's behaviour at Time B be predicted from what *two or more* other group-members were simultaneously doing at a previous Time A?

So far, only one study has been published addressing this question. The answer is 'Yes': infants *can and do* respond to what two or more others are simultaneously doing, and with a frequency so high that it has a likelihood of less than one in a thousand of occurring by chance.[81] Using data from a larger study,[82] Michael Smithson and I analysed every change of gaze-direction during five-minute-long interactions in two baby-only quartets (average age 8 months). Studying quartets rather than trios has the advantage of ensuring that one-to-one, 'mutual' gaze between two members does not *necessarily* entail the exclusion or isolation of the residual member(s), as it does in trios (cf. Figures 9.3 and 9.4).

One intriguing finding was that the start of mutual gaze between two group members strongly attracted the attention of one or both their companions—rather than the two companions pairing off in parallel (Figure 9.4). The paucity of parallel mutual gaze in our data directly counters the assumption that infant sociability is (always) generated by a dyadic programme. This study *does* prove that young, pre-verbal infants *are capable of* groupness—which is what Darwin's group hypothesis predicts. Further research will help elaborate the extent to which groupness plays a part in babies' everyday lives.

For instance, several national curricula for the group-care of infants have made the concept of *belonging* a hallmark of quality.[83] As a result, a 'sense of belonging' becomes something that should be observable and assessable in the group settings where more and more infants and toddlers spend their days. Much is said about belonging that aligns it with foundational, pan-human requirements for normal psychological development—like attachment, grammatical competence, and theory of mind:[84] it is held to apply across 'different settings'; and it 'shapes who children are and who they can become.'[85] However, notwithstanding the good intentions

[81] Ben Bradley and Michael Smithson, 'Groupness in Preverbal Infants: Proof of Concept,' *Frontiers of Psychology* 8, (2017), https://doi.org/10.3389/fpsyg.2017.00385.

[82] Ben Bradley, Jane Selby, and Cathy Urwin, 'Group Life in Babies: Opening up Perceptions and Possibilities,' in *Infant Observation and Research: Emotional Processes in Everyday Lives*, ed. Cathy Urwin and Janine Sternberg (Hove: Routledge, 2012), pp.275–295.

[83] See for example the national curricula of Australia, New Zealand, and less overtly, Greece, the UK, and, Italy: Theodora Papathedorou, 'Being, Belonging and Becoming: Some Worldviews of Early Childhood in Contemporary Curricula,' *Forum on Public Policy Online* 2, (2010),

[84] Though, on theory of mind, see Ben Bradley, 'A Serpent's Guide to Children's "Theories of Mind",' *Theory & Psychology* 3, (1993), pp.497–521.

[85] Department of Education Employment and Workforce Relations, *Belonging, Being, Becoming: The Early Years Learning Framework for Australia* (Canberra: Commonwealth of Australia, 2009), pp.7, 20 & passim.

Figure 9.4 Mutual gaze between the two closest babies in a quartet attracts the attention of their companions.

of its promoters, 'belonging' in infants has proven impossible to observe.[86] So an alternative is needed. Which has led some to suggest that the comfortable inter-dependency between infants and their companions, which advocates of 'belonging' commend, would be better viewed through the Darwin-inspired prism of infants' precocious capacity to participate in groups.[87]

9.2.4 Founding father?

All of which brings us to a final conundrum. Darwin has often been hailed as fore-father in developmental psychology, and his 'Biographical Sketch' of Doddy, its first significant publication.[88] Yet, as I have shown in this chapter, Darwin's central

---

[86] This is because belonging is most easily conceived *negatively*, being observable only in the breach: a baby who is observably *excluded from* a group, *can* be said *not* to 'belong.'

[87] Jane Selby et al., 'Is Infant Belonging Observable? A Path through the Maze,' *Contemporary Issues in Early Childhood* 19, (2018), pp.404–416 argue that 'belonging' has effectively been taken up as a proxy for the concept of groupness—prior to the empirical demonstration of groupness in young infants. As such, belonging can now be dispensed with in discussions of the 'needs' of infants and toddlers in early childhood education and care.

[88] E.g. Bradley, *Visions of Infancy*; Bradley, 'Darwin's Intertextual Baby'; James Russell, *The Acquisition of Knowledge* (London: Macmillan, 1978); Alan Costall, 'Specious Origins? Darwinism

thesis about human agency—its evolutionary grounding in the group life of our ancestors—has, for a century, had zero impact. Behind this conundrum hides another one: concerning the relationship Darwin thought to obtain between *ontogeny* (an individual human child's development) and *phylogeny* (the evolution of *Homo sapiens* from earlier species).

According to Darwin's German colleague, Ernst Haeckel (1834–1919), phylogeny *causes* ontogeny, or, put differently, the child's development from conception onwards *recapitulates* the stages of human evolution from single-cell organisms. Did Darwin subscribe to Haeckel's recapitulatory or, in Haeckel's coinage, 'biogenetic' law? If so, he has good claim to be a founding father for child psychology, because the traditional model of normal child development does have a biogenetic imprint. Whatever Darwin himself wrote, Victorians typically pictured evolution as a triumphant step-wise ascent, crowned by the emergence of civilized Man.[89] In parallel fashion, the most celebrated pioneers of child psychology assumed mental development to: have a single track; be progressive, in that a child got better and better as time passed; advance up a ladder of stages, the stages being tied to certain ages; and point in the direction of a desirable end-point (e.g. mature 'genital' sexuality; true scientific thinking). The recapitulationist Freud's theory of psychosexual stages and the recapitulationist Piaget's theory of cognitive stages are the two best-known examples.[90]

Did Darwin blaze this trail? At least one commentator reads the principal theme in Darwin's Sketch as being that: 'his son's early developmental stages *recapitulated* the behaviour and intelligence of lower animals, advancing into those of his savage ancestors while still very young.'[91] This reading apparently rests on one sentence

and Developmental Theory,' in *Evolution and Developmental Psychology*, ed. George Butterworth, Julie Rutkowska, and Michael Scaife (Brighton: Harvester, 1985), pp.30–41; George Butterworth and Margaret Harris, *Principles of Developmental Psychology* (London: Psychology Press, 2014), p.6.

[89] Peter Bowler, *The Non-Darwinian Revolution: Reinterpreting a Historical Myth* (Baltimore: Johns Hopkins University Press, 1992).

[90] For discussion, and many more examples, see Morss, *Biologising of Childhood*, pp.173ff & passim. Piaget was a cultural recapitulationist, like Taine.

[91] Evelleen Richards, *Darwin and the Making of Sexual Selection* (Chicago: Chicago University Press, 2017), p.286, my italics. Claims that Darwin 'was' a recapitulationist are also to be found in Robert Richards, *The Meaning of Evolution: The Morphological Construction and Ideological Reconstruction of Darwin's Theory* (Chicago: University of Chicago Press, 1992); Lynn Nyhart, 'Embryology and Morphology,' in *The Cambridge Companion to the Origin of Species*, ed. Michael Ruse and Robert Richards (Cambridge: Cambridge University Press, 2009), pp.194–215; and Jonathon Hodge, 'Darwin's Book: *On the Origin of Species,' Science & Education* 22, (2013), pp.2267–2294. E. Richards is convinced the same embryological argument, 'given Haeckel's authoritative stamp,' underlay Darwin's entire conception of 'the evolution of human mental, social, and ethical faculties,' all of which 'were recapitulated in individual ontogeny': Richards, *Darwin*, p.448. However, Darwin never cited or approved Haeckel's biogenetic theory—though he had ample opportunity to do so. (E.g. he never used Haeckel's word *ontogeny*). Privately, he held a low opinion of Haeckel's work: see Ch. 1, 'Darwin and modernism.' Darwin actually championed the work of *another* embryologist, Karl Ernst von Baer, who was *not* a recapitulationist: Dov Ospovat, *The Development of Darwin's Theory: Natural History, Natural Theology,*

in the Sketch. After noting that Doddy, aged two and a quarter years, was 'much alarmed' by the various larger animals whom he had seen in cages at the London Zoo, Darwin added: 'May we not suspect that the vague but very real fears of children, which are quite independent of experience, are the inherited effects of real dangers and abject superstitions during ancient savage times?'

In order to make this brief speculative digression a serious espousal of Haeckel's biogenetic law, we would have to interpolate several proposals into the Sketch which it nowhere makes. For example: that the fears of Doddy's savage adult forebears were not just inherited by Doddy, but had been 'pushed back to an earlier juvenile stage by subsequent and more advanced inherited ancestral beliefs and behaviours'—as Haeckel's theory of embryological recapitulation requires.[92] This proposal is not only missing from the Sketch, but seems alien to Darwin's published thinking, as can be seen from two observations in *Expression*. One is its argument that, reversing Haeckel, ontogeny may become phylogeny over large stretches of time, as with the contraction of the eyebrows in distress (see 4.5 'Kinks and Muddles'). The other is Darwin's unstoppable instinctive recoil from the strike of a caged puff adder at the London Zoo (see 4.3 Methods of Study). Darwin commented that his will and reason were powerless against 'the imagination of a danger which had never been experienced.'[93] Instinctive dread of snakes dated back much further than the dawn of humanity (if that is what was meant by the phrase 'ancient savage times'), being seen in Old World monkeys—as shown by experiments Darwin reported in *Descent*.[94] This adder-induced flinch must make us doubt Darwin's adherence to Haeckel's biogenetic law.[95] For clearly the adult Darwin had no more transcended his fear of snakes than had the monkeys he teased with a stuffed serpent in London Zoo.

Leaving aside the question of whether or not Darwin subscribed to Haeckel's law, it may still seem far-fetched to claim Darwin as founding father for today's child psychology. Darwin's research on children comprised detailed observations of single cases, whereas developmental psychologists nowadays largely analyse aggregate data, homogenized from scaled measures of many children.[96] Nine months

---

and Natural Selection, 1838–1859 (Cambridge: Cambridge University Press, 1981), p.96; Darwin, *Origin 1876*, pp.97, 308–309, 387–388; Darwin, *Descent*, pp.9, 164, 205.

[92] Richards, *Darwin*, p.286; see Stephen Jay Gould, *Ontogeny and Phylogeny* (Cambridge, MA: Harvard University Press, 1977).

[93] Darwin, *Expression*, pp.39–40.

[94] Darwin, *Descent*, pp.71–72.

[95] Unless, of course, in the tradition of Ptolemy, one makes his flinch an 'exception.' Even if it were an exception, Haeckel's law still fails to address all the *other* observations reported in the Sketch.

[96] Not that this makes Darwin's research impotent or unscientific, see Bent Flyvberg, 'Five Misunderstandings of Case-Study Research,' *Qualitative Inquiry* 12, (2006), pp.219–245.

before he died, Darwin did give his imprimatur to the analysis of aggregate data, however.

In 1881, the Committee on Education of the American Social Science Association launched a circular exhorting mothers to fill in an enclosed set of questions about 'mental development in their infant children,' in a manner 'that might be at some future time invaluable to the psychologist.' Emily Talbot, the enterprising secretary of the Committee, sent the circular to Darwin. He enthused back to her, noting that isolated observations would 'add but little to our knowledge,' whereas, 'tabulated results from a very large number of observations systematically made,' would likely throw 'much light on the sequence and period of development of the several faculties.' For example, he wrote that 'it would be desirable to *test statistically*,' the common view that 'coloured children at first learn as quickly as white children,' but afterwards 'fall off in progress.'[97]

As Darwin's reply to Talbot suggests, statistical tests lend themselves primarily to comparing groups and testing *existing* generalizations. Case studies serve better to illuminate general developmental processes, reveal new theoretical possibilities, and so prompt *new* generalizations (see Ch. 2). Case-work is thus eminently scientific.[98]

9.3 Conclusion

The idea of an *experimentum crucis* or 'critical experiment' goes back to Isaac Newton at the dawn of science. Unless a theory can be subjected to an observational test—the failure of which amounts to the theory's refutation—we do not deem the theory scientific.[99] If there were no good evidence to support Darwin's contention that the most distinctive features of human agency have evolved through the pressures and demands of group-living, scientists might justifiably reject or ignore his psychology. Darwin himself suggested one such test, a natural history of babies. His own observations of infancy have proved both prescient and remarkably durable. But they do not add up to a test of his groupness hypothesis.

A growing number of contemporary studies throw light on the capacities of babies to participate in groups—in a way that goes beyond their ability to conduct

---

[97] Charles Darwin, 'Letter to Mrs. Emily Talbot on the Mental and Bodily Development of Infants,' (1881), http://darwin-online.org.uk/content/frameset?itemID=F1995&viewtype=text&pageseq=1., my italics.

[98] Flyvberg, 'Case-Study Research,' passim.

[99] Popper, *Objective Knowledge*, passim. NB The very possibility of a crucial experiment—at least, in physics—is not without challengers, however. See e.g. Pierre Duhem, *The Aim and Structure of Physical Theory*, 2nd ed. (Princeton NJ: Princeton University Press, 1954).

one-on-one interactions with individual group-members. Perhaps the most telling data comprise detailed, video-recorded, case-studies of communication involving infants in trios and quartets. The first direct, statistical test of infants' capacity for groupness has also yielded positive results. Thus, though we are in the early stages of empirically researching Darwin's group-based understanding of human agency, the foundations for such an understanding steadily firm.

10

Conclusion

From the age of eight, Darwin revelled in the infinitely complex and diversified face of the natural world. His children were to make a joke of this enthusiasm, comparing his description of barnacle larvae in *The Origin of Species*—with their 'six pairs of beautifully constructed natatory legs, a pair of magnificent compound eyes, and extremely complex antennae'—to a commercial advertisement.[1] This indefatigable delight, and the minute and laborious attentions to life-forms it fuelled throughout his life, stood apart from and prior to his theorizing—both biographically and logically.

Holed up by storms in Plymouth Sound on board *HMS Beagle*, poised to start its voyage round Earth, the twenty-two-year-old Darwin confided in his journal that he looked forward to the 'fine scenery' the coming years held with more enthusiasm than any other part of the voyage.[2] A few months later, his first walk through the Brazilian rainforest became an epiphany: 'if the eye attempts to follow the flight of a gaudy butter-fly, it is arrested by some strange tree or fruit; if watching an insect one forgets in the strange flower it is crawling over; if turning to admire the splendour of the scenery, the individual character of the foreground fixes the attention. The mind is a chaos of delight.'[3] Move forward twenty-six years to a Thursday in spring, and Darwin is in Surrey, on a break away from home and the burden of writing his big book on natural selection. After a morning of hydrotherapy, he picks up a pen to write to his dearest Emma:

> Yesterday after writing to you I strolled a little beyond the glade for an hour & half & enjoyed myself—the fresh yet dark green of the grand Scotch Firs, the brown of the catkins of the old Birches with their white stems & a fringe of distant green from the larches, made an excessively pretty view.— At last I fell fast asleep on the grass & awoke with a chorus of birds singing around me, & squirrels running up

---

[1] Charles Darwin, *The Origin of Species by Means of Natural Selection, or the Preservation of Favoured Races in the Struggle for Life*, 6th ed. (London: Murray, 1876), p.389; Francis Darwin, 'Reminiscences of My Father's Everyday Life,' in *The Life and Letters of Charles Darwin*, ed. Francis Darwin (London: John Murray, 1887), p.155.

[2] Charles Darwin, *Diary of the Voyage of H.M.S. Beagle*, ed. Paul Barrett and Richard Freeman, vol. I, The Works of Charles Darwin (London: William Pickering, 1986), p.12.

[3] Darwin, *Diary of the Voyage of H.M.S. Beagle*, I, pp.41–42; Ben Bradley, 'Darwin's Sublime: The Contest between Reason and Imagination in *On the Origin of Species*,' *Journal of the History of Biology* 44, (2011), pp.205–232.

Darwin's Psychology. Ben Bradley, Oxford University Press (2020). © Oxford University Press.
DOI: 10.1093/oso/9780198708216.001.0001

the trees & some Woodpeckers laughing, & it was as pleasant a rural scene as ever I saw, & I did not care one penny how any of the beasts or birds had been formed.[4]

Here lies Darwin, immersed once again in his beloved theatre of agency, his bed, 'clothed with many plants of many kinds'—around him 'birds singing on the bushes, with various insects flitting about, and with worms crawling through the damp earth ... forms, so different from each other, and dependent upon each other in so complex a manner'[5]—a theatre which, for him, came before and stood independently of any question about origins or evolution.

Darwin did not see evolution by natural selection as something, or some mechanism, *added to* the theatre of agency. It *resulted from* the theatre of agency. Natural selection was, for him, a law like the law of gravity: it named the long-term effects of what *Origin* calls *the struggle for life*—which goes on within any interdependent habitat of organisms—and the *variations* between creatures of the same species, which partly result from that struggle. In his theory, just like that of his geological mentor Charles Lyell, the present holds the key to the past.[6]

Perhaps because he was not in thrall to the as-yet-uninvented idea of the gene, Darwin recognized that the production of individual differences was consequent upon an individual's *development* in a given habitat. The first question he asked himself after he opened his post-*Beagle* 'transmutation' notebooks, aged twenty-eight, was 'Why is life short'? After noting that 'we *know* world subject to cycle of change,' he observed that the immature were malleable while they developed, whereas the old were fixed in their ways and their physiological organization. The cycle of generations which resulted from the brevity of life had a profound consequence, therefore: 'to adapt & alter the race to *changing* world.' We see the young of living beings 'become permanently changed or subject to variety, according to circumstances.'[7] And, just as the variations in a given species were produced through the reproduction, development, and the flexible habits of entangled organisms participating in the economy of nature, so the interconnections between conspecifics, other organisms, and inanimate aspects of their habitat winnowed the better-fitted variants from the less. Consequently evolution by natural selection—and by sexual selection—resulted from the here-and-now dramas of agency.

[4] Charles Darwin, 'Letter to Emma, April 29th,' *Correspondence of Charles Darwin* (Cambridge: Darwin Correspondence Project, 1858), https://www.darwinproject.ac.uk/letter/?docId=letters/DCP-LETT-2261.
[5] Darwin, *Origin 1876*, p.429.
[6] Charles Darwin, *The Autobiography of Charles Darwin, 1809-1882: With Original Omissions Restored*, ed. Nora Barlow (London: Collins, 1958), p.87: 'Everything in nature is the result of fixed laws.' See also Douglas Erwin, 'Evolutionary Uniformitarianism,' *Developmental Biology* 357, (2011), pp.27–34. Contrast John Tooby and Leda Cosmides, 'The Past Explains the Present: Emotional Adaptations and the Structure of Ancestral Environments' *Ethology and Sociobiology* 11, (1990), pp.375–424.
[7] Charles Darwin, 'Notebook B,' (1837–1838), pp.2–7, Darwin's emphases.

Natural selection was not Darwin's only deduction from the ubiquity of agency. Witness his publications on the form and movements of gemmules, colonial invertebrates, insectivorous and climbing plants, the fertilization of orchids, and the habits of earthworms (see 3.1 Contours of Agency). Observation of the simpler things in nature presents a clear image of the complex web of relations that binds together all plants and animals with their physical conditions. It also brings out the key dimensions of Darwin's understanding of agency: the immanent dynamism in all living forms; the intrinsic link between purposive movement and the possession of sense-organs (see 3.1 Contours of Agency); its relation to an organism's development, specific form and needs; the flexibility of means by which an agent attains its ends; and intelligence, the transcendence of random trial and error.

Every action prompts a reaction. The agency of living creatures inevitably has effects which change the habitat in which the agent lives, whether these effects be purposed or not. Which is to say, each act *rebounds* upon its agent. Gulliver-like, an organism's agency therefore ties it to the world in which it lives with countless ligatures of interdependency. Such reagency, and the mutual relations it forms, become particularly intricate when they take place between animate creatures, and yet more so when rebounding between creatures of the same species. Even between plants and animals, the results can be wonderfully complex—as between orchids and moths. But among conspecifics, the results include all those features which distinguished what Darwin called *social animals* from others. The distinctiveness of social animals—and the rebounding of action and reaction within groups of conspecifics—gives the key to Darwin's treatment of human agency: the way non-verbal expressions signify; the dynamics of blushing; as well as his take on conscience, and culture. And it informs his take on the dramas of sexual desire throughout the animal kingdom.

Darwin's approach to human agency is most manifest in *Expression*. *Expression* focuses at length on the musculature and physiology of the surface-movements we read as emotional. It argues that these movements are purposeless, and so, not *designed* to express anything. Even the word 'expression' comes into question, as, while facial movements often *reveal* the thoughts and intentions of others, this result was neither intended nor expected when the movement first occurred.[8] Darwin's invention of what we now call 'judgement tests' illustrates his point of view. The significance of an expression is best approached through *its reading by others*. In fact, and counter-balancing the intrinsic purposelessness of 'expressive' movements, Darwin argued the *recognition or reading* of meaning into such movements had become instinctive, appearing in the earliest months of life (see Ch. 9). Such reading, even in babies occurs *contextually*, according to the circumstances

---

[8] Charles Darwin, *The Expression of the Emotions in Man and Animals*, 2nd ed. (London: Murray, 1890), pp.377, 386.

accompanying the movements read. When the innate human capacity for reading others rebounds, it can cause blushing.

10.1 A *Psychologist*?

Taken together, Darwin's publications describe and explain many different kinds of agency in many different species—in ways that today retain considerable currency, in both biology and psychology. Taking human agency alone, Darwin presented complex, pioneering accounts of infancy, expressions, blushing, courtship, conscience, and culture. Yet historians seldom class Darwin as a psychologist. Rick Rylance's book, *Victorian Psychology and British Culture, 1850–1880*, entirely bypasses Darwin's psychological work, focusing instead on Herbert Spencer, Alexander Bain, and George Henry Lewes.[9] George Stocking, the founding editor of the trail-blazing *Journal of the History of the Behavioural Sciences*, reported an 'absence' in place of 'an adequate treatment of human psychology by Darwin'.[10] And historian of science, Robert Young, described Darwin's psychological work as dated and 'unoriginal'.[11]

Leaving aside how closely such commentators may have read *Expression* and *Descent*, this paradox highlights a crucial point. Darwin deliberately placed his writings on agency at a distance from the tradition of scholarship which began to style itself 'psychology' in the mid-1800s, the pioneers of which today's psychologists hail as grounding their discipline. And, unlike Spencer, Bain, and Lewes, he never published a plump tome with *Psychology* on its spine. A reason for this reticence emerges from a memorial notice published in the scientific journal *Nature* in 1882, shortly after Darwin died—penned by Darwin's protégé George Romanes, an aspiring psychologist.

Romanes' essay, 'Psychology', was the last of a five-part tribute evaluating what Darwin had contributed to science. It begins: 'The effects upon Psychology of Mr. Darwin's writings have been so immense, that we shall not over-state them by saying that they are fully comparable with those we have previously considered as

---

[9] Rick Rylance, *Victorian Psychology and British Culture, 1850–1880* (Oxford: Oxford University Press, 2000), pp.227–228 & passim. Rylance's book does refer to the uptake of Darwin's 'biological' work, especially by Lewes—but states that Darwin 'did not attempt' to develop a psychology. Rylance, *Victorian Psychology* makes no reference to *Expression*, and dismisses Charles Darwin, *The Descent of Man, and Selection in Relation to Sex* (London: Murray, 1874) as containing but a single psychological chapter. For an opposing argument, see e.g. John Durant, 'The Ascent of Nature in Darwin's the *Descent of Man*', in *The Darwinian Heritage*, ed. David Kohn (Princeton NJ: Princeton University Press, 1985), pp.283–306.

[10] George Stocking Jr, *Victorian Anthropology* (New York: Free Press, 1987), p.142. For the significance of this journal for psychology, see: Robert Young, 'Scholarship and the History of the Behavioural Sciences,' *History of Science* 5, (1966), pp.1–51.

[11] Robert Young, *Darwin's Metaphor: Nature's Place in Victorian Culture* (Cambridge: Cambridge University Press, 1985), pp.56–78

having been exerted by the same writings on geology, botany, and zoology.'[12] Yet, almost immediately, Romanes appears to contradict himself, observing:

> Mr. Darwin was not only not himself a psychologist, but had little aptitude for, and perhaps less sympathy with, the technique of psychological method. The whole constitution of his mind was opposed to the subtlety of the distinctions and the mysticism of the conceptions which this technique so frequently involves; and therefore he was accustomed to regard the problems of mind in the same broad and general light that he regarded all the other problems of nature.[13]

It is not far to gather what Romanes meant by 'the subtlety of the distinctions and the mysticism of the conceptions' typical of Victorian psychology. Romanes' own attempts to gain acceptance as a psychologist give illustration. Romanes' two most psychological books—*Mental Evolution in Animals* (1883) and *Mental Evolution in Man* (1888), particularly the latter—are typical of the preoccupation with terminology that seemed universal in the psychology of his day.[14] For instance, Romanes' regretted that his abstruse term *recept* never gained traction in psychology—'recept' supposedly denoting mental products intermediate between 'percepts' and 'concepts.'[15]

No Victorian psychology would pass muster which did not carefully distinguish and pedantically define a whole suite of technical terms referring to mental operations. For example, St George Mivart criticized Darwin's *Descent* as psychologically naïve—having a 'lack of appreciation of technical terms'—because it did not maintain clear distinctions between 'six kinds of action to which the nervous system ministers': reflexes; sensations; sensible perceptions; associations; self-consciousness; and, reason. 'These two latter kinds of action,'—self-consciousness and reason—Mivart went on:

> are deliberate operations, performed, ... by means of representative ideas implying the use of a reflective representative faculty. Such actions distinguish the

---

[12] George Romanes, 'Charles Darwin, [Part] V. [Psychology],' *Nature* 26, (1882), p.169.

[13] Romanes, 'Darwin V', p.169.

[14] George Romanes, *Mental Evolution in Animals, with a Posthumous Essay on Instinct by Charles Darwin* (London: Kegan Paul Trench & Co, 1883); George Romanes, *Mental Evolution in Man: Origin of Human Faculty* (London: Kegan Paul & Trench, 1888). Characteristically, in a letter from 1878, having summed up a book he had just read on psychology by Delboeuf [Joseph Delboeuf, *La Psychologie Comme Science Naturelle: Son Présent Et Son Avenir* (Brussels: Librairie Européene C. Muquardt, 1876)] as 'rather like Herbert Spencer,' Darwin went on gently to suggest an *observational* project to Romanes: 'Have you ever thought of keeping a young monkey, so as to observe its mind...?': Charles Darwin, 'Letter to Romanes, September 2nd,' *Correspondence of Charles Darwin* (Cambridge: Darwin Correspondence Project, 1878), https://www.darwinproject.ac.uk/letter/?docId=letters/DCP-LETT-11684.

[15] George Romanes and Ethel Romanes, *The Life and Letters of George John Romanes*, 2nd ed. (London: Longmans, Green and Co., 1896), p.201; Romanes, *Mental Evolution in Man*, pp.199–201. For an example of psychologists' rejection of Romanes terms 'recept' and 'receptual,' see C Lloyd Morgan, 'Mind in Man and Brute,' *Nature* 39, (1889), pp.313–315.

intellect or rational faculty. Now, we assert that possession in perfection of all the first four (presentative) kinds of action by no means implies the possession of the last two (representative) kinds.[16]

Thus equipped with the requisite metaphysical niceties, Mivart had the means to insulate the special, God-given status of the human mind against Darwin's arguments, by denying animals the top two levels of mentation. Mivart's approach was entirely deductive and definitional. He required no observations to refute Darwin.[17]

Darwin had no patience with psychologists' terminological obsessions, which produced 'mere verbiage.'[18] In this connection he approved Leslie Stephen's damnation of psychologies which assumed that 'because you can give two things different names, they must therefore have different natures.'[19] Further evidence that Darwin had little time for Victorian psychology is to be found in his attitude to Bain and Spencer (see 4.6 Resonances).[20] He was particularly severe on Spencer, who was for Darwin the archetypal psychologist. Spencer was as adept as anyone at imperiously asserting finical metaphysical distinctions between invisible mental operations. See for example his eightfold division between presentative, presentative-representative, representative, and re-representative cognitions and feelings, published in 1860.[21]

For all their enormous extent, Darwin's major publications mention 'psychology' just twice. Both references were to Spencer's psychology.[22] In private, shortly after the *Origin* came out, Darwin had briefly taken up with Lyell the place of 'man' and 'psychology' in the case for evolution. With customary self-deprecation, he told Lyell that, 'psychologically I have done scarcely anything. Unless indeed expression

---

[16] George St. Mivart, '[Review of] the Descent of Man, and Selection in Relation to Sex,' *Quarterly Review* 131, (1871), pp.67–68.

[17] In like vein, Romanes, *Mental Evolution in Man*, p.199, mildly chastised Darwin, *Descent*, p.83, for saying the whimpers and twitches of a dreaming dog show that he 'reflects on his past pleasures or pains in the chase.' Romanes commented: 'Of course a psychologist may take technical exception to the word "reflects" in this passage,' but added, in Darwin's defence, 'this type of receptual reflection *does* take place in dogs.'

[18] Darwin, *Descent*, p.78.

[19] Darwin, *Descent*, p.78.

[20] Bain certainly treated Darwin as a psychologist, claiming that Darwin's book on *Expression* (1872) had purloined one of Bain's own theses and adding a critical appendix to a new edition of one of his books: Alexander Bain, 'Review of "Darwin on Expression": Being a Postscript to the Senses and the Intellect,' in *The Senses and the Intellect* (London: Longmans Green & Co, 1873), pp.697–715.

[21] Herbert Spencer, 'Bain on the Emotions and the Will,' in *Essays: Scientific, Political, and Speculative (Second Series)* (London: Williams and Norgate, 1863), pp.139–142.

[22] Darwin's only other published reference to psychology—other than in borrowed quotations and book titles—was to Erasmus Darwin's work: Charles Darwin, 'Preliminary Notice,' in *Life of Erasmus Darwin*, ed. Ernst Krause (London: John Murray, 1879), p.104. NB Neither of Darwin's two most psychological books, *Expression* and *Descent*, use the word 'psychology.' By contrast, Alexander Bain, *The Senses and the Intellect*, 3rd ed. (London: Longmans Green, 1868), and Herbert Spencer, *The Principles of Psychology*, 2 vols., vol. 1 (London: Williams & Norgate, 1870) use the word 39 and 38 times, respectively.

of countenance can be included.' He then added, 'I *suspect* (for I have never read it) that "Spencer's Psychology" has a bearing on Psychology, as *we* should look at it.'[23]

In later editions of the *Origin*, Darwin added the two aforementioned references to Spencer's *Psychology*, most notably, in the sixth and final edition: 'Psychology will be securely based on the foundation *already well laid by Mr. Herbert Spencer*, that of the necessary acquirement of each mental power and capacity by gradation.'[24] This was a characteristic bit of diplomatic flattery, as is shown by the consistently scathing remarks—'wildly abominably speculative' and 'a thinking pump' are just two—Darwin made about Spencer in his *Autobiography* and private correspondence.[25]

What most distinguished Darwin's approach to agency from the approach which was trail-blazed by Spencer and Bain concerns observation. Darwin invested lifelong labours and considerable ingenuity and intelligence in absorbing and developing the long-standing tradition of natural history (see Ch. 2). When he and his friend, the botanist and Director of London's Kew Gardens Joseph Hooker (1817–1911), reviled Spencer, what they most deplored was Spencer's 'deductive manner' of treating every subject, as if 'his speculative conclusions' were 'realized facts.'[26] What Darwin most regretted was that Spencer had not 'trained himself to observe more.'[27] Avoiding the labour of observation, Spencer's generalizations partook more 'of the nature of definitions than of laws of nature,' did not aid one 'in

---

[23] Charles Darwin, 'Letter to Lyell, Jan 10th,' *The Correspondence of Charles Darwin* (Cambridge: Darwin Correspondence Project, 1860), https://www.darwinproject.ac.uk/letter/?docId=letters/DCP-LETT-2647, Darwin's emphasis. NB the implied difference between the 'bearing' of *Spencer*'s view on psychology, and psychology 'as *we* should look at it.'

[24] Darwin, *Origin 1876*, p.428, my italics. Editions one and two of Charles Darwin, *On the Origin of Species by Means of Natural Selection or the Preservation of Favoured Races in the Struggle for Life* (London: Murray, 1859), p.488, had also referred to psychology, but without referring to Spencer: 'Psychology will be based on a new foundation, that of the necessary acquirement of each mental power and capacity by gradation.' Editions three to five (1863–1869) had added 'An Historical Sketch of the Recent Progress of Opinion on the Origin of Species' which contained a sentence about Spencer whose Herbert Spencer, *The Principles of Psychology* (London: Longman, Brown, Green, and Longmans, 1855) had 'also treated Psychology on the principle of the necessary acquirement of each mental power and capacity by gradation.' In the final edition (1872), Darwin referred readers to Spencer in the body of the book (see the italicized seven words). Arguably, Darwin's 1859 formulation of the 'new foundation' evolutionary theory laid for 'Psychology'—namely, 'that of the necessary acquirement of each mental power and capacity by gradation'—he was *always* referring to the views of Spencer, given that he used precisely these words to summarize Spencer's approach in his 'Historical Sketch.'

[25] Charles Darwin, 'Letter to Hooker, April 9th,' *Correspondence of Charles Darwin* (Cambridge: Darwin Correspondence Project, 1866), https://www.darwinproject.ac.uk/letter/?docId=letters/DCP-LETT-5051; Joseph Hooker, 'Letter to Darwin, December 14th,' *Correspondence of Charles Darwin* (Cambridge: Darwin Correspondence Project, 1866), https://www.darwinproject.ac.uk/letter/?docId=letters/DCP-LETT-5305; Charles Darwin, 'Letter to Hooker, December 24th,' *Correspondence of Charles Darwin* (Cambridge: Darwin Correspondence Project, 1866), https://www.darwinproject.ac.uk/letter/?docId=letters/DCP-LETT-5321.

[26] Darwin, *Autobiography*, pp.108–109; Joseph Hooker, 'Letter to Darwin, January 24th,' *Correspondence of Charles Darwin* (Cambridge: Darwin Cooorespondence Project, 1864), https://www.darwinproject.ac.uk/letter/?docId=letters/DCP-LETT-4396.xml.

[27] Charles Darwin, 'Letter to Hooker, 10th December,' *Darwin Correspondence Project* (Cambridge 1866), https://www.darwinproject.ac.uk/letter/?docId=letters/DCP-LETT-5300.xml.

predicting what will happen in any particular case,' and so were not of any 'strictly scientific use.'[28]

Unlike Spencer and Bain, Darwin was no armchair philosopher, nor a superficial looker-on at creatures who happened to cross his path. By the time of Darwin's apprenticeship in the 1820s, natural history had come to embody a complex and demanding suite of practical skills, involving an array of sophisticated instruments, which took years to master. Darwin's reputation among Victorians as an eminent, award-winning man of science largely hung on his achievements as a naturalist, not as an evolutionist. And many of his observational discoveries retain relevance today—in geology, plant physiology, invertebrate zoology, ecology, as well as in psychology.

10.2 The Cost of Suspicion

Naturalists like Darwin were well aware of the pitfalls of observation. Hence their need for rigorous training, and the many safeguards built into their practice. Observable realities were enormously—Darwin would say 'infinitely'—complex, and required years of labour to grasp. This was as true of human action as of leaf-pulling by earthworms or the fertilization of orchids by insects. Human facial expressions were not to be made sense of by introspection into an inner world of mental states—as in Bain and Spencer. They were first and foremost *bodily movements*, primarily explicable in terms of muscle and nerve. Surface movements acquired semiotic significance, that is, became expressive, *a posteriori*, through their recognition by others. Such recognition thus served as the main avenue to their interpretation in Darwin's research (see Ch. 4).

Twentieth-century psychology took a contrary path. The kinds of observation undertaken by Darwin, and other Victorian naturalists, were deemed suspect and peremptorily dismissed.[29] The only observations that could be trusted should take place in a laboratory, under tightly controlled conditions, decontaminated from bias. Likewise, ordinary language could not be trusted. Terms had often to be invented, or at least, newly defined, preferably to conform to a hypothesis that could be tested by the activities of experimenters.

Such distrust is a famous mark of modernist thinking, dubbed the 'hermeneutic of suspicion' by French philosopher Paul Ricoeur (1913–2005)[30]: a dynamic of interpretation which not only sanctifies the laboratory experiment in psychology, but

[28] Darwin, *Autobiography*, p.109.
[29] E.g. James Angell, 'The Influence of Darwin on Psychology,' *Psychological Review* 16, (1909). Cf. Kurt Danziger, *Constructing the Subject: Historical Origins of Psychological Research* (Cambridge: Cambridge University Press, 1990). See Ch. 1.
[30] Paul Ricoeur, *Freud and Philosophy: An Essay on Interpretation* (New Haven CT: Yale University Press, 1970), passim.

structures its explanations—unobservable mental events providing the determinants of observable action.[31] In the laboratory, researchers do have human action available for first-hand observation, however brief and choreographed it may be.[32] Yet they seek to explain such action by virtue of a *non-observable* psychological state, condition, or mechanism.[33] This theoretical manoeuvre prevails today. Surface realities get explained by invisible processes which dwell in the depths, and to which qualified experts alone have reliable access.[34] For example:

- Whatever the manifest content of the dream I remember from last night, its meaning is ravelled up in unconscious dream-thoughts which cannot be directly known.
- I like my own company and find large groups exhausting? Then I must have an introverted personality, or be 'on the spectrum.'[35]
- My son is constantly chattering and won't sit still? Maybe he is hyperactive.[36]
- I doubt my abilities and worry my partner finds me too needy? My attachment template was damaged by my mother when I was a baby.[37]
- I think I am a kind-hearted, benevolent fellow? Not true: unless I have found some (non-genetic) way to free myself from the tyranny of my selfish genes.[38]

There is nothing wrong with having a theory of course. The problem is that, when psychology jettisons natural history, it inverts the roles that observation and description best play in scientific method. Naturalists like Darwin took seriously the complex relationship between observation and supposition (see 2.6.2 Three takes on observation). That is why they took such pains over their descriptions: the biases and blinkers fashioned by adherence to a pet theory would succumb to dogged immersion in real-life details, repeatedly observed—abetted in Darwin's case by a

---

[31] Roy Baumeister, Kathleen Vohs, and David Funder, 'Psychology as the Science of Self-Reports and Finger Movements: Whatever Happened to Actual Behavior?', *Perspectives on Psychological Science* 2, (2007), pp.396–403; Elizabeth Stokoe, *Talk: The Science of Conversation* (London: Little Brown, 2018), pp.138–139 & passim.

[32] Something that is not true of the (online or mailed-out) survey, or self-administered questionnaire.

[33] Kenneth Gergen, 'The Limits of Language as the Limits of Psychological Explanation', *Theory & Psychology* 28, (2018), p.700; Baumeister, Vohs, and Funder, 'Psychology as the Science of Self-Reports.' See also the critique by Hannah Arendt, *The Life of the Mind: Vol.1 Thinking* (London: Secker & Warburg, 1978), pp.25ff, of: 'modern science's relentless search for the base underneath mere appearances.'

[34] Alfred Whitehead, *Process and Reality: An Essay in Cosmology* (New York: Free Press, 1978), p.18, refuted this manoeuvre by asserting: 'There is no going behind actual entities to find anything more real.' Cf. Paul Stenner, 'A.N. Whitehead and Subjectivity', *Subjectivity* 22, (2008), p.98.

[35] That is, the spectrum containing Asperger's Syndrome.

[36] See e.g. American Psychiatric Association, *Diagnostic and Statistical Manual of Mental Disorders*, 5th ed. (Arlington VA: American Psychiatric Publishing, 2013).

[37] Nancy Collins and Stephen Read, 'Adult Attachment, Working Models, and Relationship Quality in Dating Couples', *Journal of Personality and Social Psychology* 58, (1990), pp.644–663.

[38] Richard Dawkins, *The Selfish Gene: 30th Anniversary Edition* (Oxford: Oxford University Press, 2006), pp.200–201; Evelyn Keller, *The Century of the Gene* (Cambridge MA: Harvard University Press, 2000).

genius for noting exceptions.[39] Much more detail will strike a naturalist, observing, listening, and describing the face of the living world over months or years, than can be recorded over minutes in the simplified conditions of a laboratory. Not surprisingly therefore, more robust discoveries arise from the type of intense observation fuelling case-studies of a particularly organism or social group *in vivo* than stem from statistics extracted from pre-defined measures applied to large anonymous samples, studied *in vitro*, which slight real-life complexities (see 2.6 Resonances).[40]

In championing laboratory experiments, twentieth-century psychologists championed *hypothetico-deductive* research. This can have value, but diverts psychologists from independent observation. It constrains *what* they observe— and limits their range of insights by pre-defining a small number of variables that need to be counted in different experimental conditions so as to test a given theory. Such research diminishes the living world. For once the validity of a theory of 'deep' causation has been established *in vitro*, its advocates resurface with a mission to elaborate the 'reality' their theory affords them.

This dynamic gives rise to the claim that the sensible world is impoverished when compared to the invisible realm which underlies it.[41] Witness Chomsky's argument about 'the poverty of the stimulus.' This contrasts the riches of our visual perceptions with the under-determined ambiguity of the visual stimuli which strike our senses. Regarding language, Chomsky imputed a 'qualitative difference' between the intricate if invisible linguistic structures he called 'deep' (grammatical competence), and the 'impoverished and unstructured' aural environment of 'surface' speech accessible to babies acquiring language—a claim not based on observational evidence, which, when gathered, convincingly refutes it.[42]

My book *Visions of Infancy* illustrates this dynamic at length. It documents how competing theories of infancy license ways of seeing babies which are so various,

[39] See his son Francis's comment, in Darwin, 'Reminiscences,' pp.148–149: 'There was one quality of mind which seemed to be of special and extreme advantage in leading him to make discoveries. It was the power of never letting exceptions pass unnoticed. Everybody notices a fact as an exception when it is striking or frequent, but he had a special instinct for arresting an exception. A point apparently slight and unconnected with his present work is passed over by many a man almost unconsciously with some half-considered explanation, which is in fact no explanation. It was just these things that he seized on to make a start from. In a certain sense there is nothing special in this procedure, many discoveries being made by means of it.' See also Ch. 5, 'The dual self.'

[40] Bent Flyvberg, 'Five Misunderstandings of Case-Study Research,' *Qualitative Inquiry* 12, (2006), pp.234–241.

[41] Elliott Sober, 'Language and Psychological Reality: Some Reflections on Chomsky's *Rules and Representations*,' *Linguistics and Philosophy* 3, (1980), pp.395–405.

[42] Noam Chomsky, *Rules and Representations* (New York: Columbia University Press, 1980), pp.34–38 & passim; Noam Chomsky, 'A Review of B.F. Skinner's *Verbal Behaviour*,' *Language* 35, (1959), pp.26–58. Cf. Catherine Snow, 'Mothers' Speech to Children Learning Language,' *Child Development* 43, (1972), pp.549–565; Jerome Bruner, 'From Communication to Language—a Psychological Perspective,' *Cognition* 3, (1974), pp.255–287; Ben Bradley and Colwyn Trevarthen, 'Babytalk as an Adaptation to the Infant's Communication,' in *The Development of Communication*, ed. Natalie Waterson and Catherine Snow (London: Wiley, 1978), pp.75–92.

yet so specific and narrowly based, that they prove observationally incommensurable. For example:[43]

- Freud's baby is a boy who has erections, flushes with erotic pleasure when sucking at his mother's breast, and hallucinates what he wishes;
- Piaget's baby is an epistemologist in embryo, intent on figuring out whether objects continue to exist when out of sight, how to coordinate hand and eye, and the logic of intentional action—no place here for genitals hardening or nipples mouthed;
- Chomsky's baby is possessed of a deep and universal capacity for grammar, using which she or he processes and reproduces the language which bathes him or her: pointing observers to collect rule-governed errors in toddlers' speech (e.g. 'mouses' for 'mice'), and bypass completely markers of epistemological assumption, or sensuality;[44]
- Bowlby's baby is born with a set of seven reflexive attachment behaviours which ensure he or she is kept in close proximity to his or her mother, especially when stressed: smiling; looking; crying; following; clinging; and so on (see 9.2.3 The groupness hypothesis).[45]

This hermeneutic of suspicion is most prominent in self-styled *critical* psychology, a newer tradition which challenges the validity of science as ground for social understanding.[46] Here, not only is the face of the ordinary world dubious as a source of 'natural history' knowledge. Observation in the laboratory is equally suspect: it neither reflects nor engages with independent 'reality'. This makes the 'facts' of any matter 'constructions': observers' biases have created what they claim to have seen (see Ch. 2).[47] So observation in science becomes a farce. Instead, say

---

[43] Arguably, it is this dynamic that helps produce the *crisis of replication* in today's experimental psychology: Scott Maxwell, Michael Lau, and George Howard, 'Is Psychology Suffering from a Replication Crisis?: What Does "Failure to Replicate" Really Mean?', *American Psychologist* 70, (2015), pp.487–498.; Marcus Munafo and John Sutton, 'There's This Conspiracy of Silence around How Science Really Works', *Psychologist* 30, (2017), pp.36–39.

[44] E.g. David McNeill, *The Acquisition of Language: The Study of Developmental Psycholinguistics* (New York: Harper & Row, 1970).

[45] For several more examples, see Ben Bradley, *Visions of Infancy: A Critical Introduction to Child Psychology* (Cambridge: Polity Press, 1989). For parallel evidence from a different science (genetics), see: Jan Sapp, 'The Nine Lives of Gregor Mendel,' in *Experimental Inquiries: Historical, Philosophical and Social Studies of Experimentation in Science*, ed. Homer Le Grand (Dordrecht Netherlands: Kluwer Academic Publishers, 1990), pp.137–166.

[46] E.g. Ian Parker, *The Crisis in Modern Psychology and How to End It* (London: Taylor & Francis, 2013); Erica Burman, *Deconstructing Developmental Psychology*, 3rd ed. (London: Routledge, 2017).

[47] Mary Gergen, 'Induction and Construction: Teetering between Two Worlds,' *European Journal of Social Psychology* 19, (1989), pp.432–433, 436, her inverted commas around 'reality' and 'facts'; see also Kenneth Gergen, 'The Social Constructionist Movement in Modern Psychology,' *American Psychologist* 40, (1985), pp.266–275.

the critics, the knowledge purveyed by empirical psychologies reflects—and so naturalizes—imbalances in social power.[48]

When critical psychology 'deconstructs' the possibility of access to reality, it prioritizes instead the reality of *discourse*, pictured as 'a multiplicity of polyvalent versions, perspectives, language games, meanings, narratives.'[49] In so doing, deconstructions invoke yet another unobservable reality, 'a system of signs that are in place prior to observation.'[50] The assumed pre-eminence of signs in human affairs means that, regardless of observation, knowledge claims must always 'fall victim to the system of language.'[51] This means that critical psychologists have license to *make up* 'facts' to suit their arguments—provided her or his political aims are emancipatory and so above reproach—because all facts are deemed to be discursively constructed anyway.[52]

Prioritizing discourse entails a commitment to a starkly dualistic world-view which puts humans at odds with their evolutionary history—ironically reinstituting an un-Darwin-like anthropocentrism that assumes the inferiority, or irrelevance, of the animal order to the human.[53] Ironically, because this worldview means 'post-modern' critical psychology aligns perfectly with the dualistic take on culture generated by the *Modern* Synthesis (Chs 3 and 8).[54] Both critical psychology and gene-centric biology make human organisms impossible phenomena: 'hybrids of nature and culture.'[55] Should human scientists take Darwin's path, however—restoring priority to observation and, thereby, to pan-organic agency—the breach between nature and culture becomes susceptible of reconciliation, complex though such reconciliation must be.

10.3 Darwin Afresh

Few events of the kind that shape, spice, or poison, people's lives are liable to take place in a laboratory: aesthetic transport; longed-for achievement; fateful

[48] Michel Foucault, *Power-Knowledge: Selected Interviews and Other Writings, 1972–1977* (New York: Pantheon, 1980). E.g. Gergen, 'Induction and Construction'; Erica Burman, 'Child as Method: Implications for Decolonising Educational Research,' *International Studies in Sociology of Education* 28, (2019), pp.1–23.

[49] Paul Stenner, 'On the Actualities and Possibilities of Constructionism: Towards Deep Empiricism,' *Human Affairs* 19, (2009), p.203. Cf. Gergen, 'Social Constructionist Movement'; Martha Augoustinos and Cristian Tileagă, 'Twenty Five Years of Discursive Psychology,' *British Journal for Social Psychology* 51, (2012), pp.405–412.

[50] Michel Foucault, 'The Order of Discourse,' in *Unifying the Text: A Post-Structuralist Reader*, ed. Robert Young (London: Routledge & Kegan Paul, 1981), pp.48–78; Gergen, 'Limits of Language,' p.698.

[51] Gergen, 'Limits of Language,' p.698.

[52] E.g. see the epistemological defence of made-up claims by Erica Burman, 'Limits of Deconstruction, Deconstructing Limits,' *Feminism & Psychology* 25, (2015), p.413, in response to Ben Bradley, 'Deconstruction Reassigned? 'The Child,' Antipsychology and the Fate of the Empirical,' *Feminism & Psychology* 25, (2015), pp.293–297.

[53] Jacques Derrida, *The Animal That Therefore I Am* (New York: Fordham University Press, 2008), p.136.

[54] Bruno Latour, *We Have Never Been Modern* (Cambridge MA: Harvard University Press, 1993), p.104: '*the very notion of culture is an artefact created by bracketing Nature off,*' Latour's italics.

[55] Latour, *Never Been Modern*, p.10, & passim

accident; puberty; violence; sexual passion; criminality; romance; chronic neglect; or the peace which passeth understanding. On these grounds, some cast
the experimental, hypothetico-deductive method adopted by twentieth-century
psychologists, as conditioning a flight from (or defence against) reality, and from
the phenomena most likely to challenge a patient observer of humanity and in-
humanity.[56] This book argues for a return to the accessible, the visible, the audible
in the study of agency—to the description of what occurs on the knowable surface
of the living world, to a respect for the irreducible richness of this most immediate
source of meaning.

Especially when illuminated by today's phenotype-led theory of evolution (see
Ch. 3), Darwin's work shows that a return to the surface is not an *alternative* to effective theory-building, but its precondition. Given the currency of the icons of
modernism, both in today's psychology and educated common sense—the laboratory, the 'underlying' condition or mental state, the deep structure, the gene, the
meme, the brain, nature versus culture, the evils of anthropomorphism, the privileged access to reality we cede to experts—a genuine reappraisal of the approach
to evidence and argument found in Darwin's publications requires extreme icono-
clasm.[57] Only when we have loosened the grip of modernist icons on our thinking
will we see that Darwin's work has long stood at the destination upon which various
prophets of renewal in psychology, social science, and biology, today converge.

Witness the late John Shotter's summary of the 'profound conceptual shift'
which 'has been, and currently still is, taking place in social theory.' Drawing on
Wittgenstein, and particularly on the 'agential realism' promoted by Karen Barad,
Shotter proffers what he calls a re-visioning of psychology.[58] This, he says, will require us to re-situate ourselves as 'spontaneously responsive, moving, embodied
living beings,' within a reality in which nothing exists 'in separation from anything else,' a reality within which we are immersed 'as participant agencies and to
which we also owe significant aspects of our own natures.'[59] Such a positioning will
require us to rethink observation, says Shotter. Rather than seeking deliberately
to observe 'expectations arising out of our theories,' our enquiries should be focused more 'on what we can *notice* as occurring, spontaneously, within the flow of
people's activities within a particular circumstance.'[60] Shotter's advocacy of a shift
from observing to noticing in psychological research underlines that, when we

[56] George Devereux, *From Anxiety to Method in the Behavioural Sciences* (The Hague: Mouton,
1968), passim; Arendt, *Life of the Mind*, p.4: 'standardized codes of expression and conduct have the
recognized social function of protecting us against reality, that is, against the claim on our thinking attention that all events and facts make by virtue of their existence'; Ian Shapiro, *The Flight from Reality in
the Human Sciences* (Princeton NJ: Princeton University Press, 2005).
[57] Ricoeur, *Freud and Philosophy*, p.27.
[58] John Shotter, 'Agential Realism, Social Constructionism, and Our Living Relations to Our
Surroundings: Sensing Similarities Rather Than Seeing Patterns,' *Theory & Psychology* 24, (2014), p.307;
Karen Barad, *Meeting the Universe Halfway: Quantum Physics and the Entanglement of Matter and
Meaning* (Durham, North Carolina: Duke University Press, 2007).
[59] Shotter, 'Agential Realism,' pp.306–307.
[60] Shotter, 'Agential Realism,' p.311, his italics.

observe, 'we respond to differences that make a difference,' or *matter* to us in some way, so valuing and changing how we 'relate ourselves to our surroundings.'[61]

A like point is gained from a different direction in the forward-looking writings of Paul Stenner, which advocate a new kind of empiricism in psychology ('deep' empiricism). Stenner frames his argument in terms drawn from the cosmology of Alfred Whitehead (1861–1947).[62] For Shotter, we are essentially agents, and no agent exists in separation from its circumstances: entities only come into existence through 'our performing actions ... in relation to our different acquired ways of relating ourselves to our surroundings.'[63] Stenner puts like stresses on 'activity' and 'essential relatedness': 'Human beings are not separate from, but *part of* the wider universe, and metaphysics must articulate this general "togetherness." '[64] Shotter finds it 'difficult to differentiate' within any occurring phenomenon, which aspects are subjective, and which objective—whether in a laboratory or elsewhere.[65] Likewise, Stenner bases his new empiricism on the notion that 'subjective and objective aspects are fused together in each occasion of actuality.' Any described event thus exemplifies the care or *concern* an observer has for the objects she or he describes.[66]

Regarding subjectivity and objectivity, compare Darwin (see 2.4 Empirical Orientation; 2.6.2 Three takes on observation). A vision of the living world as a theatre of agency would be nothing without an *audience*, including Darwin and us, his readers.[67] Darwin did not spend hours, weeks, years observing other organisms for no reason. He went, as any curious naturalist would go, looking for answers to questions. This is inevitable ... and advisable: all 'the best scientific theories are self-projections,' wrote Nobel prize-winner Werner Heisenberg (1901–1976).[68] Just as we project ourselves into the characters we see enacted on a stage off Broadway—and so learn about our world and ourselves—so Darwin projected the possibility of what he wanted to discover into what he observed, the better to test its reality. If this sounds like a recipe for anthropomorphism, it is. Darwin's observations were often anthropomorphic.

However, as I argued in Chapter 6, recent years have restored value to anthropomorphism as a route to discovery. Anthropomorphism often helps scientists predict animal behaviour more effectively than would be possible from a description

[61] Shotter, 'Agential Realism,' pp.308, 320–321. Cf. Donna Haraway, 'Situated Knowledges: The Science Question in Feminism and the Privilege of Partial Perspective,' *Feminist Studies* 14, (1988), pp.575–599.
[62] Stenner, 'Actualities and Possibilities,' p.196.
[63] Shotter, 'Agential Realism,' p.319.
[64] Stenner, 'A.N. Whitehead,' p.97; Stenner, 'Actualities and Possibilities,' p.206, his italics.
[65] Shotter, 'Agential Realism,' p.308.
[66] Stenner, 'A.N. Whitehead,' p.98.
[67] Arendt, *Life of the Mind*, pp.21–22.
[68] Quoted in Marilyn Gaull, 'From Wordsworth to Darwin: "On the Fields of Praise",' *Wordsworth Circle* 10, (1979), p.34.

framed in mechanical terms. This is not surprising. Animals 'are made of flesh and blood, have limbic systems, and share thousands upon thousands of features with us that are absent in computers and robots.'[69] Hence, the granting of intelligence, purpose, design, and human attributes to nonhuman animals can enrich the range of hypotheses a scientist may consider. It also changes our view of ourselves: the more human-like we permit animals to become the more animal-like we become in the process—a change central to the thrust of Darwin's comparative project (see Ch. 4).

Whitehead's philosophy derives all events from activities or processes characteristic of what he called *organisms*. Even rock he construed under the rubric of organism, albeit, organism on a very simple level. The concept of organism figures large in Stenner's deep empiricism, therefore, just as it does in Darwin's psychology, and in the phenotype-led theory of evolution promulgated in today's biology (see Ch. 3). While this allows serious consideration of the 'bewildering complexity and uniqueness' proper to human agency, it also affirms a certain commonality between human beings, animals, plants and even rocks, but 'without "flattening out" or otherwise denying' the important differences between the kinds of event characteristic of, 'a mountain, a penguin, an organ, and a human being.'[70]

Emphasis on *organisms* leads yet another contemporary renovation of psychology into the ambit of Darwin's arguments and evidence about agency, namely, efforts to formulate a more biologically informed 'evolutionary psychology' than what comes out of Santa Barbara (see 3.4 Resonances).[71] Whilst the champions of this renovation typically draw on a large but relatively narrow body of evidence— when compared to the transdisciplinary breadth of the resonances informing and evoked by Darwin's psychology—they structure this evidence using themes familiar from the forgoing chapters.

For example, Karola Stotz envisions a rebooted evolutionary psychology which does not focus on cerebral modules, but organisms, actively engaged in the business of being alive: 'acting on their own behalf.' Such agents are not to be viewed as abstract, symbol-processing machines, or passive gene-controlled vehicles, but as bodily engaging with the environment in which they are embedded—and plastically adapting to it through their development. Such development should not posit a pre-existing subject and object, because subject and object *emerge from* the organism's engagement with its world. Agency is thus seen as 'a driving force in evolution,' and adaptation to the environment is reframed as the result of 'an interactive construction of self and the environment.' Consequently, says Stotz, the significance of an organism's actions or habits is less attributable to what is 'inside'

[69] Francis De Waal, 'Anthropomorphism and Anthropodenial: Consistency in Our Thinking About Humans and Other Animals ' *Philosophical Topics* 27, (1999), p.270.

[70] Stenner, 'A.N. Whitehead,' p.94.

[71] Danielle Sulikowski, 'Evolutionary Psychology: Fringe or Central to Psychological Science?,' *Frontiers in Psychology* 7, (2016), https://www.frontiersin.org/articles/10.3389/fpsyg.2016.00777/full.

the organism, or to its brain, than to the world that the organism's habits (and brain) are *inside of*—a viewpoint Stotz advances under the aegis of 'extended cognition,' and which chimes tunefully with the take on emotion in *Expression* (see 4.6 Resonances).[72]

New biologies also converge on Darwin's psychology. For example 'processual' biology—recently announced by twenty authors in a collection of essays edited by Dan Nicholson and John Dupré—criticizes the notion of biological individuality, and stresses inter-specific interrelations and agency across a hierarchy of organic forms, but especially at the level of the organism, just as Darwin did.[73] And, as argued from the first chapter of this book on, the 'extended evolutionary synthesis,' 'developmental systems theory,' plus theories of 'niche construction' and of the phenotype, are all in the business of rediscovering the importance to evolutionary theory of themes I have drawn out from Darwin's writings: development (not just transmission) as a crucial aspect of inheritance; the plasticity of growing forms; the agentive production of variability by living creatures; hierarchies of organization; the interdependencies interfusing habits with habitat; mutual aid; and co-adaptation among different species.[74]

Of course, these twenty-first century biologies have an incalculable amount of detailed substance which adds to, draws out, challenges, and refines Darwin's original coverage of the natural world. All the same, if we see Darwin's publications on agency as a small-scale map of a domain requiring future elaboration, the magnifications of this map produced by a century and a half of subsequent research show a moth-eaten patchwork, not a full or even coverage of the whole domain.

One reason for this incompleteness was the break-up of intellectual endeavours into what we now call 'disciplines,'—a break-up taking place throughout Darwin's lifetime. One sign of change was the founding of a series of new societies in London: the Linnaean Society, in 1788, specializing in natural history and especially botany; the Geological Society, in 1807; the Zoological Society, in 1825; the British Association for the Advancement of Science, in 1831; the Royal Statistical Society, in 1834; the Chemical Society, in 1841; the Ethnological Society, in 1843;

---

[72] Karola Stotz, 'Extended Evolutionary Psychology: The Importance of Transgenerational Developmental Plasticity,' *Frontiers in Psychology* 5, (2014), https://www.frontiersin.org/articles/10.3389/fpsyg.2014.00908/full., pp.10–12. Cf. Susan Oyama, *Evolution's Eye: A Systems View of the Biology-Culture Divide* (Durham NC: Duke University Press, 2000), pp.166–191

[73] Daniel Nicholson and John Dupré, eds., *Everything Flows: Towards a Processual Philosophy of Biology* (Oxford: Oxford University Press, 2018).

[74] See for example: Massimo Pigliucci and Gerd Muller, eds., *Evolution: The Extended Synthesis* (Cambridge MA: MIT Press, 2010); Susan Oyama, Paul Griffiths, and Russell Gray, *Cycles of Contingency: Developmental Systems and Evolution* (Cambridge, MA: MIT Press, 2001); John Odling-Smee, Kevin Laland, and Marcus Feldman, *Niche Construction: The Neglected Process in Evolution* (Princeton NJ: Princeton University Press, 2003); Mary Jane West-Eberhard, *Developmental Plasticity and Evolution* (New York: Oxford University Press, 2003); Ben Bradley, Darwin 1.0: Is the EES Playing Catch-Up?, 2018, http://extendedevolutionarysynthesis.com/darwin-1-0-is-the-ees-playing-catch-up/.

and the Anthropological Society, in 1863. Victorians also saw the philosophy of science emerge as a recognizable discipline.[75] Psychology was a late-comer, in this regard. The American Psychological Association began in 1892; the British Psychological Society not until 1901.

Working at arm's length from institutions, Darwin withstood the fragmentation of the academy. He did serve for three onerous years as a secretary to the Geological Society (1838–1841), but made very slight commitments to other specialist societies.[76] And he did sometimes refer to himself as a geologist.[77] But his principal scientific identity was as a naturalist. Thus *Descent* tells us that the distinguishing mark of Darwin's study of human conscience was that no one else had approached it 'exclusively from the side of natural history.'[78]

10.4 The Restoration of Meaning

As we saw in Chapter 2, the mark of a naturalist was work across many branches of science, not just one. Thomas Huxley elaborated: a naturalist might begin with 'the investigation of habits' as a descriptive exercise, but would also need to consider physiology, psychology, geography, and geological distribution. Likewise, an investigator of 'the relations of forms,' would be impelled into systematics, zoology, botany, development, morphology, and philosophical anatomy. Any of these specialisms would itself furnish 'sufficient occupation for a lifetime,' wrote Huxley: 'but in their aggregate only, are they the equivalent of the science of natural history: and the title of naturalist, in the modern sense, is deserved only by one who has mastered the principles of all.'[79]

Anyone who grapples with Darwin's work, published and unpublished, soon gathers that he was an exemplary naturalist in Huxley's sense. Indeed, his scientific needs took him even farther afield than Huxley's lists imply. Horticulture, agriculture, travelogues, essays on aesthetics, histories, tracts on natural theology, moral philosophies, and poetry, all had impact in his writings.[80] Perhaps due to this very transdisciplinarity, certain concepts central in his work have slipped between the

---

[75] Sandra Herbert, 'The Place of Man in the Development of Darwin's Theory of Transmutation. Part 2,' *Journal of the History of Biology* 10, (1977), p.163.

[76] Martin Rudwick, 'Charles Darwin in London: The Integration of Public and Private Science,' *Isis* 73, (1982), p.190.

[77] E.g. Charles Darwin, 'Notebook M,' (Cambridge: Darwin Online, 1838), p.40.

[78] Darwin, *Descent*, p.97.

[79] Thomas Huxley, 'On Natural History, as Knowledge, Discipline, and Power,' *Proceedings of the Royal Institution* 6, (1856), http://aleph0.clarku.edu/huxley/SM1/NatHis.html., p.306.

[80] Today's parlance makes Darwin's treatment of the living world *trans*disciplinary. See Paul Stenner, *Liminality and Experience: A Transdisciplinary Approach to the Psychosocial* (London: Palgrave Macmillan, 2017), p.2: this concept 'has been used in efforts to describe integrative activity, reflection and practice that addresses, crosses and goes through and beyond the limits of established disciplinary borders, in order to address complex problems that escape conventional definition and intervention.'

divisions of twentieth-century scholarship—notably, his concepts of *social animals* and of *expression*.

Worse, when constructed from *within* any one modern discipline, the extent and interest of Darwin's vision of life-forms have become seriously diminished. Richard Dawkins explains. Whilst it might seem a truism to say that 'all biologists nowadays believe in Darwin's theory,' he observes, this does not mean that, 'each biologist has, graven in their brain, an identical copy of the exact words of Charles Darwin himself.' Because nowadays biologists learn about what Darwin said, 'not from Darwin's writings, but from more recent authors.'[81] Namely, biologists like Dawkins—who have also not read Darwin.[82] Over the decades, an erosion of Darwin's original propositions takes place, as in the game of Chinese whispers— until only a trace of what he actually wrote may be detected in the pronouncements of contemporary Darwinists.

A further rift opened in twentieth-century academies and exacerbated this erosion: splitting the humanities from the sciences.[83] Darwin's writings had traditionally been deemed bereft of literary quality. 'Scientifically, the *Origin* is a classic ... but verbally it is a rag-and-bone shop,' wrote Walter Cannon in 1968, echoing a view which went back to George Eliot's snap-judgement in 1859 that, whilst epoch-making scientifically, *Origin* was 'ill-written.'[84] Darwin's self-deprecation feeds into this view. He unfailingly presented himself as a plain man without literary skill, 'artless' but candid, 'courteous, trustworthy and friendly,' a true gentleman-scientist, with great sympathy for the presumed doubts and confusions aroused in his readers as they waded through his clumsy texts.[85] However, literary scholars have recently shown how Darwin's work is deeply imaginative. And they remind Darwinists of an old maxim: the highest art is to conceal art ('*Art est celare artem*').[86] Which means, paradoxically, the measure of Darwin's artistry reflects the degree to which his books have *not* been seen as imaginative or rhetorical, but as the work of a literary innocent: 'although the art of rhetoric may make a

---

[81] Dawkins, *Selfish Gene*, p.195.

[82] For example, Richard Dawkins, *The Blind Watchmaker: Why the Evidence of Evolution Reveals a Universe without Design* (Harlow: Longman, 1986), p.290, states that while Darwin believed in the 'Lamarckian' principle that acquired characters were heritable, this belief 'was not a part of his theory of evolution.' For evidence to the contrary, read Darwin, *Expression* and Ch. 3 of this book.

[83] Witness, for example, the faculty-structures of contemporary universities, and the contrasting levels at which research in the arts/humanities and in the sciences is funded by governments. Cf. Charles Snow, 'The Two Cultures,' *Leonardo* 23, (1990), pp.169–173.

[84] Walter Cannon, 'Darwin's Vision in *On the Origin of Species*,' in *The Art of Victorian Prose*, ed. George and Madden Levine, William (New York: Oxford University Press, 1968), p.172; George Eliot, *The George Eliot Letters 1859–1861*, 9 vols., 3rd vol. (New Haven CT: Yale University Press, 1954), p.227. Cf. Bradley, 'Darwin's Sublime.'

[85] Janet Browne, *Charles Darwin: The Power of Place* (London: Jonathan Cape, 2002), pp.54–55.

[86] A saying attributed to the Roman poet Ovid, who lived at the time of Christ.

speech or book striking, if its artistry is detected, that very fact may be advanced as reason for rejecting it.'[87]

Origin's author impersonates a lucky ship's naturalist who just happened to have been 'struck with certain facts.'[88] Yet evolution is impossible to observe *in vivo*, advancing too slowly to be seen.[89] While Darwin's books go to great lengths to base their argument in observation, their chief aim is to get readers to adopt a vision of nature embracing aeons. To this end, *Descent* and *Origin* require us to *imagine* scenarios from the past to help explain some fact of agency, or problem of theory. Thus the only diagram in *Origin* is fictional, comprising a long commentary on the made-up family tree of 'the species of a genus large in its own country' (Figure 3.9).[90] Despite this imaginary status, however, Darwin's scientific success depended on his readers believing that he 'did not *invent* laws. He *described* them. Indeed, it was essential to his project that it should be accepted not as invention, but as description.'[91]

The art and craft of describing the world—natural or human—will stand for little if not graced by imagination.[92] Given that description is fundamental to science, and particularly to psychology, as this book argues, then rich description—capable of capturing what Darwin called the 'infinite complexity' of freely occurring events—is the essential complement to its much-bruited antithesis: the use of bespoke definitional rules and a severely restricted style to describe events manufactured in a laboratory.[93] As Stengers and Prigogine urge, even exponents of the exact sciences:

> need to get out of the laboratories where they have little by little learned the need to resist the fascination of a quest for the general truth of nature. They now know that idealized situations will not give them a universal key; therefore, they must

---

[87] John Campbell, 'Charles Darwin: Rhetorician of Science,' in *The Rhetoric of the Human Sciences: Language and Argument in Scholarship and Public Affairs*, ed. John Nelson, Megill, Allan and McCloskey, Donald (Madison: University of Wisconsin Press, 1987), p.72; Ilse Bulhof, *The Language of Science: A Study of the Relationship between Literature and Science in the Perspective of a Hermeneutical Ontology, with a Case Study of Darwin's 'the Origin of Species.'* (Leiden: E.J. Brill., 1992), pp.11–24.

[88] Darwin, *Origin 1859*, p.1.

[89] Gillian Beer, *Darwin's Plots: Evolutionary Narrative in Darwin, George Eliot and Nineteenth-Century Fiction*, 2nd ed. (Cambridge: Cambridge University Press, 2000), p.8; Isabelle Stengers, 'Who Is the Author?,' in *Power and Invention: Situating Science* (Minneapolis: University of Minnesota Press, 1997), pp.153–175.

[90] E.g. Darwin, *Descent*, pp.152, 378, 609; Darwin, *Origin 1859*, p.159.

[91] Beer, *Darwin's Plots*, p.46, her italics.

[92] Ricoeur, *Freud and Philosophy*, p.551.

[93] Robert Madigan, Susan Johnson, and Patricia Linton, 'The Language of Psychology: APA Style as Epistemology,' *American Psychologist* 50, (1995), pp.428–436. Cf. John Constable (1776–1837), quoted in Charles Tomlinson, 'A Meditation on John Constable,' in *The Penguin Book of Contemporary Verse*, ed. Kenneth Allott (Harmondsworth: Penguin Books, 1962), p.364 NB Darwin was well-connected to the Pre-Raphaelite painters (e.g. Thomas Woolner sculpted a bust of him in 1870), who pioneered painting 'en plein air,' long before its more famed exponents, the Impressionists (Monet, Renoir, Cezanne et al.): Lang. Cecil, *The Pre-Raphaelites and Their Circle*, 2nd ed. (Chicago: University of Chicago Press, 1975), p.xiii.

finally become again 'sciences of nature,' confronted with the manifold richness that they have for so long given themselves the right to forget.[94]

Convergently, unscripted 'infinite complexity' is the quarry of all those social scientists, who have argued that, the richer or 'thicker' an initial description of human drama, the greater its value in grounding an investigation.[95] Thus many of psychology's more inventive teachers kick off a course by getting their students to absorb a well-written biography, read a novel, or see a film, as a means of opening a window on the real-life complexity of the phenomena they study.[96]

Contemporary psychology still presents itself as an *empirical* science, implying it is descriptive at some level. Meanwhile research grows increasingly introspective, in realignment with the work of Spencer and Bain. Today's students read that the most important thing about human agency, especially as manifest in social behaviour, is 'the inner processes that mediate and produce' it.[97] Why? Because the vast majority of recent research into social life now focuses on *self-reports by isolated experimental subjects*, asked to fill in questionnaires, at a table, or on a computer. Which means that finger movements, as in keystrokes and pencil marks, constitute the vast majority of visible social action psychologists examine.[98]

Surely, Roy Baumeister and his colleagues protest wryly, 'some important behaviour involves *standing up*?' And shouldn't language be treated as a topic of research, in its own right—not just as a window onto 'inner processes'—that is: people actually talking to other people, something that goes well beyond 'getting instructions about how to sign a consent form and activate the computer program?'[99] For language, when framed as conversational *speech acts*, plays a leading part in the theatre of human agency.[100]

---

[94] Isabelle Stengers and Ilya Prigogine, 'The Reenchantment of the World,' in *Power and Invention: Situating Science* (Minneapolis: University of Minnesota Press, 1997), p.45.

[95] E.g. Edwin Ardener, 'Some Outstanding Problems in the Analysis of Events,' in *The Voice of Prophecy and Other Essays* (Oxford: Basil Blackwell, 1989); Clifford Geertz, *The Interpretation of Cultures: Selected Essays* (New York: Basic Books, 1973); Flyvberg, 'Case-Study Research,' p.237; John Shotter, 'The 'Poetry' Prior to Science,' *New Ideas in Psychology* 11, (1993), pp.415–417; Shotter, 'Agential Realism,' p.321.

[96] E.g. May Sarton, *The House by the Sea: A Journal* (New York: Norton, 1981), pp.236–237. In the 1990s, Alan Parker and Gerald Scarfe, 'The Wall,' (Beverly Hills CA: Metro-Goldwyn-Mayer, 1982) served as a fruitful introduction to 'Developmental Psychology' at James Cook University in Australia—perhaps a comment on student culture at that place and time.

[97] Baumeister, Vohs, and Funder, 'Psychology as the Science of Self-Reports,' p.401; Kenneth Gergen, 'Social Psychology and the Wrong Revolution,' *European Journal of Social Psychology* 19, (1989), pp.463–484.

[98] Baumeister, Vohs, and Funder, 'Psychology as the Science of Self-Reports,' p.397.

[99] Baumeister, Vohs, and Funder, 'Psychology as the Science of Self-Reports,' p.399, my italics.

[100] John Austin, *How to Do Things with Words* (Oxford: Clarendon Press, 1962); Shotter, 'Agential Realism,' passim; Stokoe, *Talk*. Even 'thought' can be framed in this way, as dialogic 'inner speech': Lev Vygotsky, *Thought and Language* (Cambridge MA: MIT Press, 1962); Vladimir Voloshinov, *Marxism and the Philosophy of Language* (New York: Seminar, 1973); Raymond Williams, 'On Dramatic Monologue and Dialogue (Particularly in Shakespeare),' in *Writing in Society* (London: Verso, 1991), pp.31–64.

Psychology's flight from *in vivo* study of the world which embraces us—human and non-human—invites explanation in terms of money and time, not truth and method. Observational engagement of the kind Darwin undertook is slow, diffi-cult, and costly. Advance in research psychologists' careers requires a quick-fire flow of articles, the quality of which is measured by the status of the journals which publish them. The highest ranking journals in most avenues of psychology put greater value on swift and easy sit-down questionnaire studies—which target 'inner states'—than on burdensome inquiries into meaningful, worldly action. Hence, little incentive exists for real-life observational research. So little gets done.[101]

Darwin's publications show another psychology is possible.[102] My exposition of his work is no invite to a return of the repressed. Such invitation would imply there existed some real memory of Darwin's work, inscribed in the discipline, or the minds of its practitioners.[103] Neither in its reception by Victorians, nor in the works of the protagonists of modern Darwinism, however, is there good evidence that Darwin's observations on and conceptualizations about what we now call be-haviour have ever been well assimilated into psychology.[104]

Rather Darwin's writings open out a route to investigation that does not idealize the laboratory, a route which restores meaning to the visible, tangible universe of agency in which we move and breathe and have our being. No longer need we solely choreograph the world into laboratory studies which test *pre-existing* already-held hypotheses.[105] We can look forward. Read diligently in the book of nature—wherever in our vegetable, animal, human surroundings we find descrip-tive purchase—and we create new possibilities for sign and symbol, with signifi-cances that stretch forwards in time, *towards* the answers we seek. By projecting ourselves, via its self-checking rigours, into disciplined observation, we can unveil, to those of us with ears to hear and eyes to see, new substance in the dramas of our life-world, both personal and scientific.[106] Darwin's psychology proffers a re-enchantment of our world.[107]

---

[101] Baumeister, Vohs, and Funder, 'Psychology as the Science of Self-Reports,' pp.399ff.

[102] Isabelle Stengers, *Another Science Is Possible: A Manifesto for Slow Science* (Cambridge: Polity Press, 2018).

[103] Ricoeur, *Freud and Philosophy*, p.537.

[104] Peter Bowler, *The Non-Darwinian Revolution: Reinterpreting a Historical Myth* (Baltimore: Johns Hopkins University Press, 1992); Julian Huxley, *Evolution: The Modern Synthesis* (London: George Allen & Unwin, 1942). There are, of course, exceptional scholars who *have* assimilated Darwin more completely: Thomas Clifford and James Mark Baldwin prominent amongst them, and, as representative of this century: West-Eberhard, *Developmental Plasticity and Evolution.*

[105] The contrast between regressive and progressive forms of interpretation comes from Ricoeur, *Freud and Philosophy*, pp.494–551.

[106] Shotter, 'Agential Realism,' p.317; Kenneth Gergen, 'Social Psychology as History,' *Journal of Personality and Social Psychology* 26, (1973), pp.309–320.

[107] E.g. Ricoeur, *Freud and Philosophy*, passim; Stengers and Prigogine, 'Reenchantment.'

10.5 Final Word

This book gives readers a point of access to Darwin's writings about psychological matters. Darwin's work comprises a largely-unmined trove of insight and argument, across a multitude of species and circumstances, establishing considerable resources for the elaboration of a socialized, agentic account of how we and our fellow creatures live. Reading the writings I discuss proves Darwin a scientist who advanced a challenging and often-ennobling vision of the world. His take on human agency went beyond the evolutionary, entailing an integrated understanding of the relations between biology, psychology, and culture.

Whatever one makes of Darwin's writings, and the unseemly positions they sometimes voice in discussions about sex, race, and class, his work proves mostly thoughtful and compassionate. At the very least, how he cast the relationship between human agency and our descent from animals, poses questions for anyone considering the project of psychology in the twenty-first century—a century which shows increasing appetite for broader debate about the topics Darwin's psychology addresses. Beyond this, an engagement with the theatre of agency Darwin documented cannot help but enhance and deepen our grasp of the variety and richness of the observable world, and of the lives of its inhabitants. I speak from experience.

Glossary

NB Italicized words are defined elsewhere in this Glossary.

Accommodation Change of old *habits* to adapt to new conditions.

Acquired characters These are characteristics acquired during an individual's lifetime, or *ontogeny*.

Agency, agent An entity's capacity to act, with purpose, or move of its own accord. An agent is an entity possessed of agency.

Anthropology, anthropologicals The branch of science that studies human beings. In the mid-nineteenth century, the Anthropological Society assumed a progressive, quasi-evolutionary view of human civilization, which was often taken to imply human races were different species (see *polygenism*). The so-called 'anthropologicals' differed from the 'ethnologicals' on this (see *ethnology*).

Assimilation Modes of action in which old *habits* are imposed on new content.

Attitude For Darwin, an attitude was an observable physical posture—often an exaggerated whole-body orientation suggesting a readiness to respond to current circumstances in a certain way.[1] For example, a Bengalese man accused of theft, 'listened silently and scornfully to the accusation; his attitude erect, chest expanded, mouth closed, lips protruding, eyes firmly set and penetrating. He then defiantly maintained his innocence, with upraised and clenched hands, his head being now pushed forwards, with the eyes widely open and eyebrows raised.'[2] Darwin's understanding differs from contemporary psychological definitions of attitude, which largely refer to invisible cognitive constructs.[3]

Baldwin Effect The *plasticity* of living organisms, including their *agency*, enables them to adapt in novel ways to changing circumstances during their lifetime (or *ontogeny*). Such adaptations, *phenotypic accommodations*, or *acquired characters*, will create new selection pressures: conditions under which some genetic changes will be favoured over others. For example, adoption of a milky diet may lead to selection for increased persistence of the enzyme responsible for absorbing lactose (the sugar in milk) through the gut wall. In this way *ontogeny* leads *phylogeny*.[4] Baldwin called this effect 'organic selection,' a

---

[1] This physical understanding of the word was, and still is, common in the arts, and in aeronautical engineering. Thus John Keats, *Keats: Poetical Works* (London: Oxford University Press, 1956), p.210, hails the 'silent form' of a Grecian urn: 'Fair attitude!'

[2] Charles Darwin, *The Expression of the Emotions in Man and Animals*, 2nd ed. (London: Murray, 1890), p.259.

[3] For further discussion, see Robert Farr, 'On Reading Darwin and Discovering Social Psychology,' in *The Development of Social Psychology*, ed. Robert Gilmour and Steven Duck (London: Academic, 1980), pp.121–123.

[4] James Baldwin, 'A New Factor in Evolution,' *American Naturalist* 30, (1896), pp.441–451; Mary Jane West-Eberhard, *Developmental Plasticity and Evolution* (New York: Oxford University Press, 2003), pp.151–154.

phenomenon *Origin* discusses under *transitional habits*. In the 1950s, it was renamed 'the Baldwin Effect'.[5]

Behaviourism A school of experimental psychology which advocates analysis of observable behaviour as the true scientific method. Usually associated with the view that all behaviour can be explained in terms of rewards (pleasures) and punishments (pains) previously offered by the environment during an individual's lifetime.

Biogenetic law Ernst Haeckel's term for the hypothesis that *ontogeny* recapitulates *phylogeny*: the stages of growth of an individual, from fertilized egg to maturity—but especially when an embryo—represents a condensed re-run of the main stages of the evolutionary history of the species to which that individual belongs. The biogenetic laws thus holds that phylogeny *causes* ontogeny.[6]

Character Distinguishable feature of an organism; phenotypic trait.

Circumnutation The movement of a part of a plant, typically the tip of stem or *radicle*, which 'bends successively to all points of the compass, so that the tip revolves'.[7]

Civilization Civilization was typically equated with *culture* in Darwin's day. The word referred to 'that complex whole which includes knowledge, belief, art, morals, law, custom, and any other capabilities and habits acquired by man as a member of society'.[8] It was typically referred to as a single ladder of stages, rising up from a low, in *savages* or barbarians, to a high in Europeans. The high-point of civilization for Darwin, was the refinement of the *moral sense*, and in particular, the humane treatment of all human beings—and animals.

Clinometer An instrument used by geologists to measure the gradient and bearing of slopes.

Co-evolution The idea that different species evolve synergistically, producing mutual interdependence or symbiosis. See for example, Darwin's study of the co-evolution of orchids and insects.[9] Co-evolution is sometimes held to be a late twentieth-century discovery.[10]

Common descent, community of descent Terms Darwin used to refer to what we call *evolution*.

Community selection Another term for *group selection*.

---

[5] George Simpson, 'The Baldwin Effect', *Evolution* 7, (1953), pp.110–117.

[6] For a critique, see Stephen Jay Gould, *Ontogeny and Phylogeny* (Cambridge, MA: Harvard University Press, 1977).

[7] Charles Darwin, *The Power of Movement in Plants* (London: John Murray, 1880), p.1.

[8] Edward Tylor, *Primitive Culture: Researches into the Development of Mythology, Philosophy, Religion, Language, Art, and Custom* (London: Murray, 1971), p.1.

[9] Charles Darwin, *The Various Contrivances by Which Orchids Are Fertilised by Insects*, 2nd revised ed. (London: John Murray, 1882).

[10] E.g. Kenneth Gergen, *Relational Being: Beyond Self and Community* (Oxford: Oxford University Press, 2009), pp.381–382; Gilles Deleuze and Felix Guattari, *A Thousand Plateaus: Capitalism and Schizophrenia* (Minneapolis: University of Minnesota Press, 1987), pp.10–12—who claim that emphasis on the kind of relationship that obtains between 'orchid and wasp' introduces a *new* kind of *rhizomatic* ontology to philosophy—a dimension which they hold to be unknown to Darwin, despite his book on *Orchids*. This, they go on, will force *arborescent* 'evolutionary schemas' to 'abandon the old model of the tree and descent'.

Component analysis Analysis of facial expressions in terms of the muscle-movements that produce them.

Conditions of life The organic and inorganic circumstances that comprise a creature's *habitat*.

Conscience The culturally informed standards of right and wrong by which an individual judges the probity of their own actions. A cultural specification of the *moral sense*.

Correlated growth One of the principles or processes that shapes evolutionary outcomes, according to Darwin. Because creatures have a particular physical form—the knee bone is connected to the thigh bone, the thigh bone is connected to the hip bone, the hip bone is connected to the back bone, and so on—an adaptive change to one part of a creature's anatomy is likely to have knock-on effects for other parts of its anatomy. One of Darwin's examples was that white cats with blue eyes are often deaf.

Creature In this book: a synonym for organism—microbe, plant, or animal.

Cubism An important pictorial movement arising out of the rejection of traditional Western single-viewpoint perspective in or around the year 1910. Its first 'analytical' stages were characterized by simple geometric forms which soon gave way to further complexes of interlocking semi-transparent planes.[11] Pablo Picasso was its most famous exponent.

Cultural drive A contemporary hypothesis which posits a positive feedback loop between cultural advances and brain-growth.[12]

Cultural evolution 'Cultural evolution is the idea that changes in human cultural beliefs, knowledge, customs, skills, and languages can be understood as an evolutionary process.'[13] I argue that a theory of cultural evolution is only necessary for as long as one believes that biological evolution is led by the genes, as per the *Modern Synthesis*. If phenotypic change leads evolution (as evidence reviewed in Ch. 8 shows that it can), then cultural change is subsumed under the theory of the phenotype.

Culture A word with two meanings in Darwin's work. One is synonymous with *civilization*, and applies to the sophistication over human history of civic society. The other refers to the self-development of an individual, usually by way of a good education and immersion in great literature (what German's call 'Bildung').

Darwin industry The branch of the study of the history of science which deals with Darwin—called an 'industry' because it has: an enormous amount of Darwin-material to mine; engaged a great many scholars; been very productive; and gone through various fashions.[14]

---

[11] 'Cubism, *n*.' in Anon, *Oxford English Dictionary Online*, (Oxford: Oxford University Press, 2018), https://www.lexico.com.
[12] E.g. Michael Muthukrishna et al., 'The Cultural Brain Hypothesis: How Culture Drives Brain Expansion, Sociality, and Life History,' *PLoS Computational Biology* 14, (2018), https://doi.org/10.1371/journal.pcbi.1006504.
[13] Kevin Laland, *Darwin's Unfinished Symphony: How Culture Made the Human Mind* (Princeton: Princeton University Press, 2017), p.323
[14] Timothy Lenoir, 'The Darwin Industry,' *Journal of the History of Biology* 20, (1987), pp.115–130; Robert Young, 'Darwinism Is Social,' in *The Darwinian Heritage*, ed. David Kohn (Princeton: Princton University Press, 1985), pp.609–638; Gowan Dawson, 'Darwin Decentred,' *Studies in History and Philosophy of Biological and Biomedical Sciences* 46, (2014), pp.93–96.

Development The unique life-course of an individual (*ontogeny*). One of the two processes subsumed by Darwin under the term *inheritance*—the other being *transmission*. In Victorian times, 'development' was also a synonym of *evolution*.

Differential psychology The study of the ways that individual human beings consistently differ from each other in behaviour, and of the underlying processes which produce those differences.

Direct action One of the processes that produces evolutionary change, according to Darwin, who proposed that the direct action of the physical conditions of life on an organism could produce heritable effects. For example, climate could explain 'differences of colour in the races of man.'[15]

DNA Deoxyribose nucleic acid, two parallel coils of which inhabit the nuclei of cells, and make up the most crucial component of the heritable material in organisms' chromosomes. Portions of DNA code proteins, with the help of other substances in each cell. These portions are called *genes*.

Dyadic Involving two people.

Ecology Ecology was first defined by Ernst Haeckel as the body of knowledge concerning the *economy of nature*: 'the study of all those complex interrelationships referred to by Darwin as the conditions of the struggle for existence.'[16]

Economy of nature Writers like Carl Linnaeus and Charles Lyell conceived nature as a divinely constructed economy in which the division of labour between the interdependent activities and stations of its various inhabitants were 'fitted to produce general ends, and reciprocal uses,' and thus 'ensure the health and well-being of the natural world.'[17]

Entropy The degree of disorder in a physical system. The second law of thermodynamics states that, over time, entropy can never decrease in any closed physical system.

Epigenetic Any heritable phenotypic change relevant to an organism's reproductive success which does not alter its DNA. Such changes often vary gene activity or expression, and may result from environmental factors or normal developmental processes.[18]

Eras 1, 2, and 3 Darwin's time-scale for human pre-history had three phases: a 'most ancient,' 'primeval,' 'very remote epoch'—when our ancestors were 'semi-human' (which I call Era 1); a fully-human era, comprising 'long ages' of savagery (Era 2); that of 'civilised nations,' culminating in Victorian society (Era 3).

Essentialism The view that facial and other human expressions 'externalize' internal essences (emotions)

[15] Charles Darwin, *The Descent of Man, and Selection in Relation to Sex* (London: Murray, 1874), p.196.
[16] Haeckel quoted in Robert Stauffer, 'Ecology in the Long Manuscript Version of Darwin's *Origin of Species* and Linnaeus' *Oeconomy of Nature*,' *Proceedings of the American Philosophical Society* 104, (1960), p.235
[17] Trevor Pearce, '"A Great Complication of Circumstances"–Darwin and the Economy of Nature,' *Journal of the History of Biology* 43, (2010), p.498.
[18] Cf. West-Eberhard, *Developmental Plasticity and Evolution*, p.112.

Ethnography The descriptive branch of *ethnology*; the fieldwork that produced descriptions of (traditionally) 'exotic'—that is 'uncivilized'—peoples and cultures.

Ethnology, ethnologicals The branch of knowledge concerned with human culture and society, and its development. Nowadays this would overlap considerably with sociology and social anthropology. In Victorian times, the Ethnological Society, and its members and adherents—called 'ethnologicals'—differed from *anthropologicals* in assuming that all human races belonged to one species (*monogenism*). They had a degenerative account of racial differences, and a biblical time-scale, such that humans emerged from the Garden of Eden as one race. But some races (guess which!) had since degenerated from the prototype.

Ethology Originally equated with natural history. Today the word identifies the biological study of animal (including human) behaviour—especially from an evolutionary perspective.

Evolution The transformation of organisms over many generations into new forms. This word did not appear in any edition of *Origin* before the final one (1872). Darwin's first published use of 'evolution' is in *Variation* (1868), so far as I can find. Darwin used a variety of terms to refer to evolution, amongst them: 'the theory of descent with modification'; 'the transmutation of species'; *community of descent*; and *common descent*.

Expression Non-verbal sounds and movements by animals that can be read as signifying emotions or states of mind. Not intentional.

Extended Evolutionary Synthesis Refers to post-1980 theoretical additions to the *Modern Synthesis* in evolutionary biology, including: *epigenetic factors*; developmental plasticity; and *niche construction*.

Externalism See *situationism*.

Fixed action pattern A term used in *ethology* to designate an innate or instinctive behavioural sequence that is relatively invariant.

Galvanism The use of electrical current to stimulate muscles.

Gemmule According to Darwin and his peers, the simplest unit of life (a 'living atom'). Also, the unit of heredity, present in all cells, and transmitted from generation to generation.[19]

Gene The basic 'unit' of heredity in an organism now often hypothesized to be positioned in chromosomes and to comprise doubled coils of *DNA* (deoxyribose nucleic acid). Genes code proteins. Chromosomes contain a lot of other stuff (e.g. protein) which affects how the DNA is expressed. Also, genes are not 'units' in any simple sense of the term, in that: they may code different proteins if read 'backwards'; the same protein may be coded by different genes; the same bit of DNA may code different proteins.[20]

Genotype The genetic makeup which distinguishes an individual organism or one of its characters when compared with other similar individuals.

---

[19] See Charles Darwin, *The Variation of Animals and Plants under Domestication*, 2nd ed., 2 vols., vol. 2 (London: John Murray, 1875).

[20] John Dupré, *Processes of Life: Essays in the Philosophy of Biology* (Oxford: Oxford University Press, 2012), pp.135ff.

Gesture Conventional expressions or gestures are, according to Darwin, 'acquired by the individual during early life,' being likely to differ in different cultural groups, 'in the same manner as do their languages.'[21]

Goniometer An instrument Darwin used for measuring angles of the crystal faces in rock.

Gradualism The doctrine that evolution proceeds gradually, and so is not punctuated by major discontinuities.

Group selection The name for natural selection operating at the level of groups, based in the idea that fitter or more cohesive tribes replace or flourish better than less fit, less cohesive tribes.

Habit Pattern of action.

Habitat An organism's place of habitation.

Heritability Loosely speaking, this term refers to characters that are 'heritable,' or can be inherited. But heritability nowadays refers to a statistic which assesses the degree to which the observed variation in a given *character* is not the result of environmental circumstances or chance. This means that the calculation of heritability is determined by the variability of environmental conditions under which observations are made. In a uniform environment, the variations in a character's heritability may approach 100%. In a widely varying environment, heritability of the same variations may approach 0%.

Homology Similarities between organisms that can be traced to their common descent or evolutionary kinship: distinguished from 'analogy,' which refers to similarities which do *not* result from genealogical kinship.

Individual differences This refers to variations in a *character*. Individual differences are products of the process of *variation*.

Individualism The assumption that the primary units in the world are free-standing organisms—whose primary, generic characteristics owe nothing to interdependence with other organisms.[22]

Inheritance The means by which a character passes, through the process of reproduction, from generation to generation. Darwin stressed that two distinct processes contributed to inheritance—the *transmission*, and the *development* of characters. He published several 'laws of inheritance,' both in *Origin* and in *Descent*.

Inheritance of acquired characters Darwin sometimes stated that *characters* acquired during an organism's lifetime could be passed on to its descendants—via the creation and *transmission* of *gemmules*.[23] Such *transmission* was commonly hypothesized in Victorian times, and experimental evidence had been published to support it.[24] It is now called 'Lamarckism,' as it was promoted by Jean Baptiste Lamarck (1744–1829; and before him by Darwin's grandfather, Erasmus Darwin). Darwin was very critical of Lamarck.

---

[21] Darwin, *Expression*, p.16.

[22] Steven Lukes, *Individualism* (Colchester: ECPR, 2006).

[23] E.g. Charles Darwin, 'Inherited Instinct,' *Nature* 7, (1873).

[24] E.g. Charles-Edouard Brown-Séquard, 'Hereditary Transmission of Effects of Certain Injuries to the Nervous System,' *Lancet* 105, (1875), pp.7–8; Gregory Radick, 'Darwin's Puzzling *Expression*,' *Comptes Rendus Biologies* 333, (2010), pp.181–187.

Recently, biologists have suggested that acquired characters *can indeed* be inherited, albeit, not in a Lamarckian fashion.[25]

Instinct An inherited *habit*, or pattern of action. Darwin elaborated: 'An action, which we ourselves require experience to enable us to perform, when performed by an animal, more especially by a very young one, without experience, and when performed by many individuals in the same way, without their knowing for what purpose it is performed, is usually said to be instinctive. But I could show that none of these characters are universal. A little dose of judgment or reason, as Pierre Huber expresses it, often comes into play, even with animals low in the scale of nature.'[26]

Intelligence Use by an agent of variable means to reach a purposed end.

Judgement tests One of Darwin's methods for determining the meaning of human facial movements—photographs of which would be presented to judges, without any explanation. The judges were then asked to read the photographed expression and write down a description.

Kin selection, family selection The idea that a character in an individual could evolve by natural selection when family-members of that individual were advantaged by the character—even if the individual itself died without reproducing. Darwin called this 'family selection.'

Law According to Darwin, a law was a repeated 'sequence of events as ascertained by us.'[27] His example was the law of gravity: smaller objects are always drawn towards much larger bodies. *Natural* and, by implication, *sexual selection* were laws, Darwin argued—not mechanisms, agencies, or causal processes (though he sometimes wrote about them as if they were mechanisms, causal processes, and agencies).

Materialism, materialist A complex term, with many different meanings in the late 1830s—when Darwin used the term a total of four times. (He never discussed it in his books). It could refer to a denial of the existence of God; or to a belief that only physical entities had existence (hence 'mind' could be reduced to matter/did not exist). There were also 'vital' materialists, and 'physical' materialists.[28] Despite it not uncommonly being maintained that Darwin 'was' a materialist, he seems to have dallied with materialism for only a few months in 1838–1839 (see Ch. 3, 'Living as agency').[29]

Meme A term invented by Richard Dawkins to refer to 'a unit of cultural transmission,' analogous to a *gene*. Dawkins' examples of memes include: 'tunes, ideas, catch-phrases,

---

[25] West-Eberhard, *Developmental Plasticity and Evolution*, pp.37, 192.
[26] Charles Darwin, *The Origin of Species by Means of Natural Selection, or the Preservation of Favoured Races in the Struggle for Life*, 6th ed. (London: Murray, 1876), p.205.
[27] Darwin, *Origin 1876*, p.63.
[28] See Edward Manier, *The Young Darwin and His Cultural Circle. A Study of Influences Which Helped Shape the Language and Logic of the First Drafts of the Theory of Natural Selection* (Dordrecht: Reidel, 1978), Ch. 4.
[29] E.g. Howard Gruber, *Darwin on Man: A Psychological Study of Scientific Creativity* (Chicago: University of Chicago Press, 1981), p.30.; Jonathan Hodge, 'Chance and Chances in Darwin's Early Theorizing and in Darwinian Theory Today,' in *Chance in Evolution*, ed. Grant Ramsey and Charles Pence (Chicago: Chicago University Press, 2016), p.50.

clothes fashions, ways of making pots or of building arches.'[30] Mimicking the *Modern Synthesis*, Dawkins equates meme-inheritance solely with *transmission*, ignoring *development,* and so bypassing any need to invoke the *agency* of the people who absorb, adapt, transform, improve, or criticize cultural materials. *Transmission* of memes supposedly occurs by 'imitation,' a process which Dawkins equates with a meme 'leaping from brain to brain'—an image which removes *agency* from people, awarding it instead to memes themselves.

Metaphysics Loosely speaking, philosophy. More precisely, the consideration of the fundamental 'first principles' required for understanding the world around us. The etymology of metaphysics—meta-physics—implies a 'higher level' of physics.

Modern Synthesis The synthesis, made in the 1930s, of discoveries in genetics and population genetics with a particular understanding of *natural selection* as a mechanism (not a law). Sometimes called the gene-centric or gene's-eye view of evolution—the Modern Synthesis has lately been augmented to produce the *Extended Evolutionary Synthesis*. The Modern Synthesis is widely criticized for ignoring both *agency*, and also the place of *development* in the process of *inheritance*. It typically equates *inheritance* solely with the *transmission* of DNA.

Modernism A movement, most notably in art and literature, that began in or around 1910. Elements of modernism persisted after the 2nd World War. *Cubism* was an early example of modernist painting. Modernist movements often involved the production of manifestoes which critiqued classical forms—typically on the grounds that the traditional forms of art, science, or culture, no longer reflected the needs, demands, and sensibilities of the 'modern' industrial world.

Modularity, modular traits West-Eberhard defines modularity as 'the properties of discreteness and dissociability of parts and integration within parts'—or more precisely, she defines modular traits as: 'subunits of the phenotype that are determined by the switches or decision points that organise development, whether of morphology, physiology, or behaviour.'[31] Her long, sophisticated and richly evidence-based discussion of anatomical and physiological modularity, in particular, contrasts with speculation in Evolutionary Psychology, where modularity is a central if contentious term. For example, there is little if any *direct* evidence that: 'the mind is organized into modules or mental organs, each with a specialized design that makes it an expert in one arena of interaction with the world. The modules' basic logic is specified by our genetic program. Their operation was shaped by natural selection to solve the problems of the hunting and gathering life led by our ancestors in most of our evolutionary history.'[32]

Monogenism The view that the human races are all members of one species, *Homo sapiens.*

[30] Richard Dawkins, *The Selfish Gene: 30th Anniversary Edition* (Oxford: Oxford University Press, 2006), pp.192ff.

[31] West-Eberhard, *Developmental Plasticity and Evolution*, Ch.4.

[32] Steven Pinker, *How the Mind Works* (New York: Norton & Co, 1997), p.21; David Buller and Valerie Hardcastle, 'Evolutionary Psychology, Meet Developmental Neurobiology: Against Promiscuous Modularity,' *Brain and Mind* 1, (2000), pp.307–325; Johan Bolhuis et al., 'Darwin in Mind: New Opportunities for Evolutionary Psychology,' *PLoS Biology* 9, (2011), https://journals.plos.org/plosbiology/article/file?id=10.1371/journal.pbio.1001109.

Moral sense The inborn capacity to distinguish between right and wrong.

Multi-level selection The idea that natural selection operates at several levels, having several different 'units': genes; phenotypes; individuals; families; social groups; species; ecosystems.

Mutation An abrupt random change in the *character* of an *organism*, nowadays traced to a change of DNA sequence.

Natura non facit saltum A principle stated by Francis Bacon in 1620, meaning 'nature does not make jumps,' or that all changes in nature are gradual. This was one of Darwin's key maxims in *Origin*, and the credo of his evolutionary *gradualism*. In the late twentieth century, Darwin's gradualism was challenged by proponents of the idea that evolutionary history comprised long periods of little change, alternating with short periods of fast change, and so comprised 'punctuated equilibria.'[33] However, careful empirical tests have shown more variability in the rates of evolution than can easily be accommodated by the idea of equilibria or plateaus of stasis.[34]

Natural history A tradition of scientific enquiry that goes back centuries. In the 1700s and early 1800s, natural history was viewed as the foundation for each of what we today call the 'natural sciences.' It focused on the observation and description of the natural world, inanimate and animate—especially living creatures and their life-cycles. *Ethology* was first proposed as a branch of natural history.

Natural selection For Darwin, the *law* or sequence of events betokening the preservation of favourable *variations* and rejection of injurious *variations* in the struggle for life.

Naturalist One who practices *natural history*. Early exponents of *ethology* called themselves naturalists.

Niche The role an *organism* plays in its ecosystems, hence the expression 'ecological niche.' Niche is nowadays sometimes defined as including *organisms'* 'niche-constructing acts, and selection from [the] sources' available in their environment—that is their *agency*.[35]

Niche construction An organism's alteration of the *habitat* (or ecological *niche*) in which it lives. Niche construction has been proposed as a factor in evolution—in that it can alter the selection pressures bearing on the niche-constructor. See *agency* and *transitional habits*.

Ontogeny Haeckel's term for the stage-progressive, *phylogeny*-mimicking, development of an individual from fertilized egg to sexual maturity. (Darwin never used this term.)

Operationalization In *psychology*, a procedure intended to make the meaning of a theoretical concept—referring to an hypothesized but invisible process—susceptible to

[33] Niles Eldredge and Stephen Jay Gould, 'Punctuated Equilibria: An Alternative to Phyletic Gradualism,' in *Models in Paleobiology*, ed. Tom Schopf (San Fransisco: Freeman Cooper, 1972), pp.82–115.
[34] E.g. Andrew Johnson and Christopher Lennon, 'Evolution of Gryphaeate Oysters in the Mid-Jurassic of Western Europe,' *Paleontology* 33, (1990), pp.453–485.
[35] John Odling-Smee, Kevin Laland, and Marcus Feldman, *Niche Construction: The Neglected Process in Evolution* (Princeton NJ: Princeton University Press, 2003), p.419.

observational examination by rendering it synonymous with the activities of the experimenter, or 'operations', pre-assigned to measure it.

Organism A living being.

Parental investment theory This proposes that animals in which the relative investment per offspring varies between males and females will show a concomitant dimorphism in mating strategy. A large investment will be protected more diligently than a small investment. Hence females, who typically invest more per gamete than males, will be choosier about their mates, than males.

Phenotype A single character of an organism, or all its characters, other than those that describe the chemistry of its *genes*. Phenotypes include: 'enzyme products of genes, behaviours, metabolic pathways, morphologies, nervous tics, remembered phone numbers, and spots on the lung following a bout of flu.'[36]

Phenotypic accommodation When *organisms* develop functional *phenotypes* through 'adaptive mutual adjustment among variable parts during development, without genetic change.'[37]

Phenotypic excess Aspects of an organism's agency not relevant to—and even less—a direct result of, *evolution*. Non-adaptive *plasticity*.

Phylogeny Haeckel's term for the evolutionary genealogy or 'line of descent' of an organic being, or the lines of all organic beings.[38]

Physiognomy The study of the features of the face, or of the form of the body generally, as being supposedly indicative of character.[39]

Plasticity The quality of what Darwin called 'impressibility', or of being 'readily capable of change'.[40] Mary Jane West-Eberhard equates it with 'responsiveness' and 'flexibility', defining it as: 'the ability of an organism to react to an external or internal environmental input with a change of form, state, movement, or rate of activity.'[41]

Polygenism The Victorian view that there were several human species alive in the nineteenth century (contrary to the thesis of *monogenism*). Polygenism gave scientific justification to slavery, as black slaves were deemed come from a different ('lower') species than their white masters.

Progressivism The belief that *evolution* or *civilization* inevitably progresses.

Psychology A branch of study that has been defined in many different ways, for example, as: 'the science of behaviour' (by *behaviourists*); and 'the science of mental life' (by William James). Each definition implies a very different *metaphysics*, whether or not its advocates claim to eschew metaphysics.[42]

---

[36] West-Eberhard, *Developmental Plasticity and Evolution*, p.31.

[37] West-Eberhard, *Developmental Plasticity and Evolution*, p.51.

[38] Darwin, *Origin 1876*, p.381. I think this was the only mention of 'phylogeny' in Darwin's writings, published and unpublished.

[39] 'Physiognomy, n,.' in Anon, *Oxford English Dictionary Online*.

[40] Darwin, *Origin 1876*, p.438.

[41] West-Eberhard, *Developmental Plasticity and Evolution*, p.33.

[42] Paul Stenner, 'On the Actualities and Possibilities of Constructionism: Towards Deep Empiricism,' *Human Affairs* 19, (2009), pp.194–210.

Psychometrics The branch of *psychology* that aims objectively to measure invisible psychological *characters*.

Psycho-social An adjective applying to phenomena that are both subjective and social. For example, the suggestibility of children and adults; the intersubjectivity of infants and adults; inner speech; groupness.[43]

Radicle The first root to sprout from a seed.

Reagency The way in which actions directly or indirectly rebound, have effects on, or react back upon an agent (see also Reflexive effects).

Recapitulation A word used as shorthand for the processes producing the *biogenetic law*.

Reciprocal altruism The hypothesis that altruism can evolve if altruists, or their kin, are later 'repaid' for acts of sacrifice by the beneficiaries of those acts. In effect, this hypothesis means that, if altruism has evolved, it was not altruism.

Recombination The production of genetic variation by the jumbling up of *DNA* sequences through cell-division or during the combination of two parents' genes in sexual reproduction.

Reflexive effects The rebounding of the effects of *agency* upon *agents* (see also *Reagency*).[44]

Replicators Dawkins' term for *genes*, which he extends to *memes*, as both supposedly make copies of themselves (replicate) with great fidelity.

Savages The term Victorians gave to peoples who lived in relatively small groups or 'tribes,' and gained their food by hunting and gathering. In Darwin's work, examples included the inhabitants of Tierra del Fuego, and Australian Aborigines. The term referred to peoples supposedly *lacking* the features which identified more 'advanced' *civilization* but *present* in upper class Victorian Englishmen, and their European (and American) peers.

Sexual selection A law which covered two kinds of sequence of events, namely, the agentic and anatomical results of: inter-sexual courtship itself, and; intra-sexual competition for mates—as by peahens, or male game-birds, posturing and fighting to subdue each other in order to 'gain access to' court fertile members of the opposite sex.

Situationism Views which attribute emotion according to how expressive actions interface with their external circumstances—what seems to prompt them and what effects they have. Also known as *externalism*.

Social animals Animals which habitually live in groups (not just in families), even in the harshest conditions.

Somatic selection The large-scale production of diverse ('random') variant characters and movements which are then either winnowed away, or preserved and reinforced, in a

[43] Johanna Motzkau, 'Exploring the Transdisciplinary Trajectory of Suggestibility,' *Subjectivity* 27, (2009), pp.172–194; Wendy Hollway and Tony Jefferson, *Doing Qualitative Research Differently: A Psychosocial Approach*, 2nd ed. (London: Sage, 2013).

[44] The generic form of what Ulrich Beck has described as the source of the *specific* ailments of today's advanced 'risk society,' namely 'reflexive modernization.' See: Ben Bradley and John Morss, 'Social Construction in a World at Risk: Towards a Psychology of Experience,' *Theory & Psychology* 12, (2002), pp.512–515.

trial-and-error fashion. The result is establishment of a functional pattern without a central coordination of elements—as in the rotational *circumnutation* of plants.

Species According to Darwin, a label of convenience for a group or taxonomic class of organisms having certain permanent *characters* in common which clearly distinguish it from other such groups.

Speech acts Linguists have traditionally viewed language as a complex of vocabulary and grammar. In the middle of the last century, a different view emerged: that what we attend to in everyday conversation is not so much the words that utterances contain but the speech acts that those utterances perform: curses, requests, vows, warnings, invitations, promises, jokes, apologies, predictions, and their like.

States of mind In Darwin, often synonymous with a subjective meaning that can be read into objective changes in *attitude* in human beings (or animals). According to Ferrier, such states cannot be assumed to be 'conscious of themselves.'[45]

Struggle for existence, struggle for life Derived from Malthus and Lyell, Darwin used these terms, 'in a large and metaphorical sense,' to refer to all the activities upon which the life and future survival of an individual organism (or *species*-member) depended, including, both 'dependence of one being on another,' and 'success in leaving progeny.'[46] The 'struggle' could be as much of a plant with other plants, to attract insects, as with climate change, or for an organism to collaborate with conspecifics.

Symbolic interactionism Holds that both the mind and the self are socially produced through interpersonal interaction in a process mediated by symbols, verbal and non-verbal.

Sympatric species Species that inhabit the same place—in contrast to 'allopatric' species, which live in different places.

Systematics Taxonomy; the study of the organization of types in the 'system of nature' or 'natural system' (see Ch. 2, 'Empirical orientation').

Taxonomy The systematic classification of fossil and living organisms. Typically, taxonomic classification is hierarchical, going, in descending order from: kingdoms (e.g. Protista; plants; animals); Phyla (e.g. vertebrates, molluscs); class (e.g. mammals); order (e.g. primates); families (e.g. old world monkeys); genus (e.g. chimpanzees); species (e.g. Bonobo).

Transdisciplinary A concept that has been used in efforts to describe integrative activity, reflection, and practice that addresses, crosses and goes through and beyond the limits of established disciplinary borders, in order to address complex problems that escape conventional definition and intervention.[47]

Transitional habits Darwin's term for habits that create new selection pressures. See *Baldwin Effect, niche construction.*

[45] James Ferrier, 'An Introduction to the Philosophy of Consciousness, Part I,' *Blackwood's Edinburgh Magazine* 43, (1838), passim.

[46] Darwin, *Origin 1876*, p.50.

[47] Paul Stenner, *Liminality and Experience: A Transdisciplinary Approach to the Psychosocial* (London: Palgrave Macmllan, 2017), p.2.

Transmission With *development*, a process that forms one part of *inheritance*; the passing of heritable genetic material (*gemmules* or *genes*)—which codes or helps determine phenotypic characters—from generation to generation. In more complex organisms, transmission typically involves sexual reproduction.

Triangulation In social science triangulation means the combination of data or methods so that contrasting viewpoints cast light upon a topic. Triangulation is argued to strengthen and validate the results of research.

Use and disuse of parts An *agentic* process which is ambiguous in subsuming both, *transitional habits*, and *the inheritance of acquired characters*. Its effects contribute to evolutionary change, according to Darwin. For example: 'as the larger ground-feeding birds seldom take flight except to escape danger, it is probable that the nearly wingless condition of several birds, now inhabiting or which lately inhabited several oceanic islands, tenanted by no beast of prey, has been caused by disuse.'[48]

Variation, variations *Variation* is one of two processes producing the sequences of effects Darwin called *natural selection* (also necessary for *sexual selection*)—the other being the *struggle for life*. Variation was, according to Darwin, largely a result of the *plasticity*, and particularly, the *agency*, of living *organisms*, when struggling for life. Variation produces *variations*, or what Darwin called *individual differences*. Darwin's use of the word 'variation' can be ambiguous. In most cases *variation* refers to a process. But that process produces effects called individual differences, each of which may also be referred to as a 'variation.'

Web of affinities The evolved homologies between related taxonomic groups.

Wright's principle Chauncey Wright argued that the law of natural selection might result in one 'useful power' which would bring with it 'many resulting advantages as well as limiting disadvantages,' disadvantages which might **undermine** fitness, and which the law of natural selection 'may not have comprehended in its action.'[49] Darwin wrote that *Descent* had tried to show that Wright's principle had 'an important bearing on the acquisition by man of some of his mental characteristics.'[50]

Zoophyte An animal which resembles a plant, including by Darwin's day: sponges, corals, sea anemones, and bryozoans (like *Flustra*, see Ch. 2).

---

[48] Darwin, *Origin 1876*, p.108.
[49] Chauncey Wright, 'Review [of Contributions to the Theory of Natural Selection. A Series of Essays by Alfred Russell Wallace],' *North American Review* 111, (1870), p.293.
[50] Darwin, *Descent*, p.571.

Bibliography

Abbott, Patrick, Jun Abe, John Alcock, et al. 'Inclusive Fitness Theory and Eusociality.' *Nature* 466 (2011): pp.1057–1062.

Abercrombie, John. *Inquiries Concerning the Intellectual Powers and the Investigation of Truth.* 8th ed. London: John Murray, 1838.

Adams, Timothy. *Telling Lies in Modern American Autobiography.* Chapel Hill NC: University of North Carolina Press, 1990.

Ainsworth, Mary. 'Object Relations, Dependency, and Attachment: A Theoretical Review of the Infant-Mother Relationship.' *Child Development* 40 (1969): pp.969–1025.

Alem, Sylvain, Clint Perry, Xingfu Zhu, Olli Loukola, Thomas Ingraham, Eirik Søvik, and Lars Chittka. 'Associative Mechanisms Allow for Social Learning and Cultural Transmission of String Pulling in an Insect.' *PLoS Biology* 14 (2016): e1002564. https://doi.org/10.1371/journal.pbio.1002564.

Allan, Mea. *Darwin and His Flowers.* New York: Taplinger, 1977.

Alpers, Svetlana. *The Art of Describing: Dutch Art in the Seventeenth Century.* London: Penguin, 1989. First published 1983.

American Psychiatric Association. *Diagnostic and Statistical Manual of Mental Disorders.* 5th ed. Arlington VA: American Psychiatric Publishing, 2013.

Anderson, Richard. 'The Untranslated Content of Wundt's *Grundzüge Der Physiologischen Psychologie.' Journal of the History of the Behavioural Sciences* 11 (1975): pp.381–386.

Andersson, Malte, and Leigh Simmons. 'Sexual Selection and Mate Choice.' *Trends in Ecology and Evolution* 21 (2006): pp.296–302.

Angell, James. 'The Influence of Darwin on Psychology.' *Psychological Review* 16 (1909): pp.152–169.

Angell, James. 'Psychology at the St. Louis Congress.' *The Journal of Philosophy, Psychology and Scientific Methods* 2 (1905): pp.533–546.

Anon. 'Emotion Experiment.' (2019). Cambridge: Darwin Correspondence Project. https://www.darwinproject.ac.uk/commentary/human-nature/expression-emotions/emotion-experiment.

Anon. 'Metaphysics.' In *Encyclopaedia Britannica or, a Dictionary of Arts, Sciences and Miscellaneous Literature*, edited by David Millar. Edinburgh: Archibald Constable & co., 1810.

Anon. 'Metaphysics.' In *The Edinburgh Encyclopaedia*, edited by David Brewster. Edinburgh: Blackwood, 1830.

Anon. *Oxford English Dictionary Online.* Oxford: Oxford University Press, 2018. https://www.lexico.com .

Anon. 'Royal Medal.' In *Wikipedia*, 2018, https://en.wikipedia.org/wiki/Royal_Medal.

Ardener, Edwin. 'Behaviour'—a Social Anthropological Critique.' In *The Voice of Prophecy and Other Essays*, edited by Michael Chapman, pp.104–108. Oxford: Blackwell, 1989.

Ardener, Edwin. 'The New Anthropology and Its Critics.' In *The Voice of Prophecy and Other Essays*, edited by Michael Chapman, pp.45–64. New York: Berghahn, 2018.

Ardener, Edwin. 'Social Anthropology and the Decline of Modernism.' In *The Voice of Prophecy*, pp.191–210. Oxford: Blackwell, 1989.

Ardener, Edwin. 'Some Outstanding Problems in the Analysis of Events.' In *The Voice of Prophecy and Other Essays*, pp.86–104. Oxford: Basil Blackwell, 1989.

Arendt, Hannah. *The Life of the Mind: Vol.1 Thinking*. London: Secker & Warburg, 1978.

Arendt, Hannah. 'Thinking and Moral Considerations.' *Social Research* 38 (1971): pp.417–446.

Arnold, Matthew. *Culture and Anarchy*. Cambridge: Cambridge University Press, 1932. First published 1869.

Atkinson, John, and Harriet Martineau. *Letters on the Laws of Man's Nature and Development*. Boston: Mendum, 1851.

Augoustinos, Martha, and Cristian Tileagă. 'Twenty Five Years of Discursive Psychology.' *British Journal for Social Psychology* 51 (2012): pp.405–412.

Austin, John. *How to Do Things with Words*. Oxford: Clarendon Press, 1962.

Bacon, Francis. *Novum Organum Scientiarum*. 1620. http://www.gutenberg.org/files/45988/45988-h/45988-h.htm.

Bagehot, Walter. 'Physics and Politics No.2: The Age of Conflict.' *Fortnightly Review* 3 (1868): pp.452–471.

Bain, Alexander. *The Emotions and the Will*. 2nd ed. London: Longmans Green, 1865. First published 1859.

Bain, Alexander. 'Phrenology and Psychology.' *Fraser's Magazine* 61 (1860): pp.692–708.

Bain, Alexander. 'The Respective Spheres and Mutual Helps of Introspection and Psychophysical Experiment in Psychology.' *Mind (New Series)* 2 (1893): pp.42–53.

Bain, Alexander. 'Review of "Darwin on Expression": Being a Postscript to the Senses and the Intellect.' In *The Senses and the Intellect*, pp.697–714. London: Longmans Green & Co, 1873.

Bain, Alexander. *The Senses and the Intellect*. 3rd ed. London: Longmans Green, 1868. First published 1855.

Baldwin, James. *Darwin and the Humanities*. Baltimore: Review Publishing, 1909.

Baldwin, James. *Development and Evolution: Including Psychophysical Evolution, Evolution by Orthoplasy, and the Theory of Genetic Modes*. New York: Macmillan, 1902.

Baldwin, James. *Mental Development in the Child and Race: Methods and Processes*. New York: Macmillan, 1895.

Baldwin, James. 'A New Factor in Evolution.' *American Naturalist* 30 (1896): pp.441–451.

Baldwin, James. 'A New Factor in Evolution, Continued.' *American Naturalist* 30 (1896): pp.536–553.

Baldwin, James. *Social and Ethical Interpretations in Mental Development: A Study in Social Psychology*. London: Macmillan, 1897.

Ball, Gregory, and Jacques Balthazart. 'How Useful Is the Appetitive and Consummatory Distinction for Our Understanding of the Neuroendocrine Control of Sexual Behaviour?' *Hormones and Behaviour* 53 (2008): pp.307–311.

Barad, Karen. *Meeting the Universe Halfway: Quantum Physics and the Entanglement of Matter and Meaning*. Durham, North Carolina: Duke University Press, 2007.

Barham, James. 'Normativity, Agency, and Life.' *Studies in History and Philosophy of Biological and Biomedical Sciences* 43 (2012): pp.92–103.

Barrett, Clark. 'Modularity.' In *Evolutionary Perspectives on Social Psychology*, edited by Virgil Zeigler-Hill, Lisa Welling, and Todd Shackelford, pp.39–50. Cham: Springer, 2015.

Barrett, Lisa. 'Was Darwin Wrong about Emotional Expressions?' *Current Directions in Psychological Science* 20 (2011): pp.400–406.

Barrett, Paul, Peter Gautrey, Sandra Herbert, David Kohn, and Sydney Smith, eds. *Charles Darwin's Notebooks, 1836–1844*. Ithaca NY: Cornell University Press, 1987.

Barthes, Roland. 'The Death of the Author.' In *Image-Music-Text*, pp.142–148. Glasgow: Fontana, 1977.

Barthes, Roland. *S/Z*. Malden, MA: Blackwell, 1990. First published 1973.

Baum, David, and Allan Larson. 'Adaptation Reviewed: A Phylogenetic Methodology for Studying Character Macroevolution.' *Systematic Zoology* 40 (1991): pp.1–18.

Baumeister, Roy, Kathleen Vohs, and David Funder. 'Psychology as the Science of Self-Reports and Finger Movements: Whatever Happened to Actual Behavior?' *Perspectives on Psychological Science* 2 (2007): pp.396–403.

Beach, Frank. 'The Snark Was a Boojum.' *American Psychologist* 5 (1950): pp.115–124.

Beckett, Samuel. 'Three Dialogues with Georges Duthuit.' In *Proust*. London: Calder & Boyars, 1970.

Beer, Gillian. *Darwin's Plots: Evolutionary Narrative in Darwin, George Eliot and Nineteenth-Century Fiction*. 2nd ed. Cambridge: Cambridge University Press, 2000. First published 1983.

Beer, Gillian. 'Late Darwin and the Problem of the Human.' (2010). https://nationalhumanitiescenter.org/on-the-human/2010/06/late-darwin-and-the-problem-of-the-human/.

Beer, Gillian. *Open Fields: Science in Cultural Encounter*. Oxford: Clarendon Press, 1996.

Bell, Charles. *The Anatomy and Philosophy of Expression as Connected with the Fine Arts*. 3rd ed. London: John Murray, 1844. First published 1806.

Bell, Charles. *Essays on the Anatomy of Expression in Painting*. London: Longman, Hurst, Rees, and Orme, 1806.

Bennett, Tony, Lawrence Grossberg, and Meaghan Morris. *New Keywords: A Revised Vocabulary of Culture and Society*. Hoboken: Wiley, 2005.

Bergson, Henri. *Creative Evolution*. New York: Henry Holt, 1911. First published 1907.

Bhabha, Homi K. *The Location of Culture*. New York: Routledge, 1994.

Billig, Michael. 'Repopulating Social Psychology: A Revised Version of Events.' In *Reconstructing the Psychological Subject: Bodies, Practices and Technologies*, edited by Betty Bayer and John Shotter, pp.126–152. London: Sage, 1998.

Bion, Wilfred. *Experiences in Groups and Other Papers*. London: Tavistock, 1961.

Bird, Christopher, and Nathan Emery. 'Insightful Problem-Solving and Creative Tool-Modification by Captive Non-Tool-Using Rooks.' *Proceedings of the National Academy of Sciences of the United States of America* 106 (2009) pp.10370–10375.

Birkhead, Tim. 'Studies of West Palearctic Birds. 189. Magpie.' *British Birds* 82 (1989): pp.583–600.

Blackman, Lisa. 'Affect, Relationality and the Problem of Personality.' *Theory, Culture and Society* 25 (2008): pp.23–47.

Blackman, Lisa. 'Habit and Affect: Revitalizing a Forgotten History.' *Body & Society* 19 (2013): pp.186–216.

Blackman, Lisa. 'Reinventing Psychological Matters: The Importance of the Suggestive Realm of Tarde's Ontology.' *Economy and Society* 36 (2007): pp.574–596.

Blake, William. 'The Marriage of Heaven and Hell.' In *Blake: Complete Writings*, edited by Geoffrey Keynes, pp.148–160. Oxford: Oxford University Press, 1969.

Blake, William. 'Milton, a Poem in 2 Books.' In *Blake: Complete Writings*, edited by Geoffrey Keynes, pp.480–535. Oxford: Oxford University Press, 1969.

Bloom, Lois. *Language Development: Form and Function in Emerging Grammars*. Cambridge, MA: MIT. Press, 1970.

Blumenthal, Arthur. 'Wilhelm Wundt and Early American Psychology: A Clash of Cultures.' In *Wilhelm Wundt and the Making of Scientific Psychology*, edited by Robert Rieber, pp.117–135. New York: Plenum, 1980.

Boero, Ferdinando. 'From Darwin's *Origin of Species* toward a Theory of Natural History.' *F1000 Prime Reports* 7 (2015): 1–13. doi:10.12703/P7-49.

Bögels, Susan, Wendy Rijsemus, and Peter de Jong. 'Self-Focused Attention and Social Anxiety: The Effects of Experimentally Heightened Self-Awareness on Fear, Blushing, Cognitions, and Social Skills.' *Cognitive Therapy and Research* 26 (2002): pp.461–472.

Bolhuis, Johan, Gillian Brown, Robert Richardson, and Kevin Laland. 'Darwin in Mind: New Opportunities for Evolutionary Psychology.' *PLoS Biology* 9 (2011): 1–8. https://doi.org/ 10.1371/journal.pbio.1001109.

Boring, Edwin. *A History of Experimental Psychology.* 2nd ed. Englewood-Cliffs NJ: Prentice-Hall, 1950.

Bouchard, Frédéric, and Philippe Huneman, eds. *From Groups to Individuals: Evolution and Emerging Individuality.* Cambridge MA: MIT Press, 2013.

Bowlby, John. *Attachment and Loss, Vol.1: Attachment.* 2nd ed. 2 vols. London: Hogarth Press, 1982. First published 1969.

Bowler, Peter. *Evolution: The History of an Idea.* 2nd ed. Berkeley: University of California Press, 1989. First published 1983.

Bowler, Peter. *The Non-Darwinian Revolution: Reinterpreting a Historical Myth.* Baltimore: Johns Hopkins University Press, 1992.

Bradley, Ben. 'An Approach to Synchronicity: From Synchrony to Synchronization.' *International Journal of Critical Psychology* 1 (2001): pp.119–144.

Bradley, Ben. 'Darwin 1.0: Is the EES Playing Catch-Up?', edited by Kevin Laland, 2018, http://extendedevolutionarysynthesis.com/darwin-1-0-is-the-ees-playing-catch-up/.

Bradley, Ben. 'Darwin's Intertextual Baby: Erasmus Darwin as Precursor in Child Psychology.' *Human Development* 37 (1994): pp.86–102.

Bradley, Ben. 'Darwin's Sublime: The Contest between Reason and Imagination in *On the Origin of Species*.' *Journal of the History of Biology* 44, no. 2 (2011): pp.205–232.

Bradley, Ben. 'Deconstruction Reassigned? "The Child," Antipsychology and the Fate of the Empirical.' *Feminism & Psychology* 25 (2015): pp.284–304.

Bradley, Ben. 'Early Trios: Patterns of Sound and Movement in the Genesis of Meaning between Infants.' In *Handbook of Communicative Musicality*, edited by Stephen Malloch and Colwyn Trevarthen, pp.263–280. Oxford: Blackwell-Wiley, 2009.

Bradley, Ben. 'Experiencing Symbols.' In *Symbolic Transformations: Toward an Interdisciplinary Science of Symbols*, edited by Brady Wagoner, pp.93–119. London: Routledge, 2010.

Bradley, Ben. 'Infancy as Paradise.' *Human Development* 34 (1991): pp.35–54.

Bradley, Ben. 'Jealousy in Infant-Peer Trios: From Narcissism to Culture.' In *Handbook of Jealousy: Theories, Principles and Multidisciplinary Approaches*, edited by Sybil Hart and Marie Legerstee, pp.192–234. Hoboken NJ: Wiley-Blackwell, 2010.

Bradley, Ben. 'Language and the Dissolution of Individuality.' *MOSAIC Monographs* 3 (1988): pp.1–34.

Bradley, Ben. 'Pedagogy.' In *Encyclopedia of Critical Psychology*, edited by Thomas Teo New York: Springer, 2014. https://doi.org/10.1007/978-1-4614-5583-7_213.

Bradley, Ben. *Psychology and Experience.* Cambridge: Cambridge University Press, 2005.

Bradley, Ben. 'Rethinking "Experience" in Professional Practice: Lessons from Clinical Psychology.' In *Understanding and Researching Professional Practice*, edited by Bill Green, pp.65–82. Rotterdam: Sense Publishers, 2009.

Bradley, Ben. 'A Serpent's Guide to Children's "Theories of Mind".' *Theory & Psychology* 3 (1993): pp.497–521.

Bradley, Ben. 'A Study of Young Infants as Social Beings.' PhD thesis. University of Edinburgh, 1980.

Bradley, Ben. *Visions of Infancy: A Critical Introduction to Child Psychology.* Cambridge: Polity Press, 1989.

Bradley, Ben, and Colwyn Trevarthen. 'Babytalk as an Adaptation to the Infant's Communication.' In *The Development of Communication*, edited by Natalie Waterson and Catherine Snow, pp.75–92. London: Wiley, 1978.

Bradley, Ben, and Michael Smithson. 'Groupness in Preverbal Infants: Proof of Concept.' *Frontiers of Psychology 8* (2017). https://doi.org/10.3389/fpsyg.2017.00385.

Bradley, Ben, Jane Selby, and Cathy Urwin. 'Group Life in Babies: Opening up Perceptions and Possibilities.' In *Infant Observation and Research: Emotional Processes in Everyday Lives*, edited by Cathy Urwin and Janine Sternberg, pp.137–148. Hove: Routledge, 2012.

[Brewster, David]. 'Review of Cours de philosophie positive, by Auguste Comte.' *Edinburgh Review 67* (1838): pp.271–308.

Broughton, John, and John Freeman-Moir, eds. *The Cognitive-Developmental Psychology of James Mark Baldwin: Current Theory and Research in Genetic Epistemology.* Norwood NJ: Ablex, 1982.

Brown, Steven, and Paul Stenner. *Psychology without Foundations: History, Philosophy and Psychosocial Theory.* London: Sage, 2009.

Brown-Séquard, Charles-Edouard. 'Hereditary Effects of Transmission of Certain Injuries to the Nervous System.' *Lancet 105* (1875): pp.7–8.

Browne, Janet. 'Darwin's Botanical Arithmetic and the "Principle of Divergence," 1854–1858.' *Journal of the History of Biology 13* (1980): pp.53–89.

Browne, Janet. 'Darwin and the Expression of the Emotions.' In *The Darwinian Heritage*, edited by David Kohn, pp.307–326. Princeton NJ: Princeton University Press, 1985.

Browne, Janet. *Charles Darwin: Voyaging.* 2 vols. Vol. 1. London: Pimlico, 1996. First published 1995.

Browne, Janet. *Charles Darwin: The Power of Place.* 2 vols. Vol. 2. London: Jonathan Cape, 2002.

Bruner, Jerome. 'From Communication to Language—a Psychological Perspective.' *Cognition 3* (1974): pp.255–287.

Bruner, Jerome. 'The Ontogenesis of Speech Acts.' *Journal of Child Language 2* (1975): pp.1–19.

Buckley, Kerry. *Mechanical Man: John Broadus Watson and the Beginnings of Behaviourism.* New York: Guilford Press, 1989.

Bulhof, Ilse. *The Language of Science: A Study of the Relationship between Literature and Science in the Perspective of a Hermeneutical Ontology, with a Case Study of Darwin's 'The Origin of Species.'* Leiden: E.J. Brill., 1992.

Bull, James. 'Sex Determination in Reptiles.' *Quarterly Review of Biology 55* (1980): pp.3–21.

Buller, David, and Valerie Hardcastle. 'Evolutionary Psychology, Meet Developmental Neurobiology: Against Promiscuous Modularity.' *Brain and Mind 1* (2000): pp.307–325.

Burgess, Stuart. 'The Beauty of the Peacock Tail and the Problems with the Theory of Sexual Selection.' *Journal of Creation 15* (2001): pp.94–102.

Burgess, Thomas Henry. *The Physiology or Mechanism of Blushing: Illustrative of the Influence of Mental Emotion on the Capillary Circulation; with a General View of the Sympathies, and the Organic Relations of Those Structures with Which They Seem to Be Connected.* London: John Churchill, 1839.

Burke, Darren. 'Why Isn't Everyone an Evolutionary Psychologist?' *Frontiers in Psychology 5* (2014): 1–8. https://www.frontiersin.org/articles/10.3389/fpsyg.2014.00910/full.

Burke, Edmund. *A Philosophical Inquiry into the Origin of Our Ideas of the Sublime and the Beautiful: With an Introductory Discourse Concerning Taste.* London: Nimmo, 1887. First published 1757. http://www.gutenberg.org/files/15043/15043-h/15043-h.htm.

Burke, Edmund. *Reflections on the Revolution in France*. London: Dent, 1910. First published 1790.

Burke, Sean. *The Death and Return of the Author: Criticism and Subjectivity in Barthes, Foucault, and Derrida*. 3rd ed. Edinburgh: Edinburgh University Press, 2010. First published 1992.

Burkhardt, Frederick. 'England and Scotland: The Learned Societies.' In *The Comparative Reception of Darwinism*, edited by Thomas Glick, pp.32–74. Chicago: University of Chicago Press, 1988.

Burkhardt, Richard. 'Darwin on Animal Behavior and Evolution.' In *The Darwinian Heritage*, edited by David Kohn. Princeton NJ: Princeton University Press, 1985.

Burman, Erica. 'Child as Method: Implications for Decolonising Educational Research.' *International Studies in Sociology of Education* 28 (2019): pp.4–26.

Burman, Erica. *Deconstructing Developmental Psychology*. 3rd ed. London: Routledge, 2017. First published 1994.

Burman, Erica. 'Limits of Deconstruction, Deconstructing Limits.' *Feminism & Psychology* 25 (2015): pp.408–422.

Buss, David, ed. *Handbook of Evolutionary Psychology*. Hoboken NJ: Wiley, 2005.

Buss, David, ed. *Handbook of Evolutionary Psychology*. 2 vols. Hoboken NJ: Wiley, 2016. First published 2005.

Buss, David. 'The Great Struggles of Life: Darwin and the Emergence of Evolutionary Psychology.' *American Psychologist* 64 (2009): pp.140–148.

Butterworth, George, and Margaret Harris. *Principles of Developmental Psychology*. London: Psychology Press, 2014. First published 1994.

Cacioppo, John, Gary Berntson, and Jean Decety. 'Social Neuroscience and Its Relation to Social Psychology.' *Social Cognition* 28 (2010): pp.675–685.

Campbell, John. 'Charles Darwin: Rhetorician of Science.' In *The Rhetoric of the Human Sciences: Language and Argument in Scholarship and Public Affairs*, edited by John Nelson, Allan Megill, and Donald McCloskey, pp.69–86. Madison: University of Wisconsin Press 1987.

Campbell, Robin, and Roger Wales. 'The Study of Language Acquisition.' In *New Horizons in Linguistics*, edited by John Lyons, pp.242–260. Harmondsworth: Penguin Books, 1970.

Campbell, Susan. 'Emotion as an Explanatory Principle in Early Evolution Theory.' *Studies in History and Philosophy of Science* 28 (1997): pp.453–474.

Cannon, Walter. 'Darwin's Vision in *On the Origin of Species*.' In *The Art of Victorian Prose*, edited by George Levine and William Madden, pp.154–176. New York: Oxford University Press, 1968.

Cannon, Walter. 'The Problem of Miracles in the 1830's.' *Victorian Studies* 4 (1960): pp.4–32.

Carey, Daniel. 'Compiling Nature's History: Travellers and Travel Narratives in the Early Royal Society.' *Annals of Science* 54 (1997): pp.269–292.

Carlyle, Thomas. 'Shooting Niagara--and After?' *Macmillan's Magazine* 16 (1867): pp.674-687.

Carpenter, William. *Principles of Mental Physiology, with Their Applications to the Training and Discipline of the Mind, and the Study of Its Morbid Conditions*. 2nd ed. London: Henry S. King, 1875. First published 1874.

Carroll, Joseph, Dan McAdams, and Edward Wilson, eds. *Darwin's Bridge: Uniting the Humanities and Sciences*. Oxford: Oxford University Press, 2016.

Carroll, Lewis. *Through the Looking-Glass and What Alice Found There*. 1871. https://archive.org/details/ThroughTheLookingGlassAndWhatAliceFoundThere/mode/2up.

Casimir, Michael, and Michael Schnegg. 'Shame across Cultures: The Evolution, Ontogeny and Function of a "Moral Emotion".' In *Between Culture and Biology: Perspectives on*

Ontogenetic Development, edited by Heidi Keller, Ype Poortinga, and Axel Schölmerich, pp.270–300. Cambridge: Cambridge University Press, 2002.

Castillo, Marcella, and George Butterworth. 'Neonatal Localisation of a Sound in Visual Space.' *Perception* 10 (1981): pp.331–338.

Chandler, Christopher, Charles Ofria, and Ian Dworkin. 'Runaway Sexual Selection Leads to Good Genes.' *Evolution* 67 (2012): pp.110–119.

Chapman, Anne. *Darwin in Tierra del Fuego*. Buenos Aires: Imago Mundi, 2006.

Charlesworth, Brian, and Deborah Charlesworth. 'Darwin and Genetics.' *Genetics* 183 (2009): pp.757–766.

Chomsky, Noam. 'A Review of B.F. Skinner's *Verbal Behaviour*.' *Language* 35 (1959): pp.26–58.

Chomsky, Noam. *Rules and Representations*. New York: Columbia University Press, 1980.

Clause, Bonnie. 'The Wistar Rat as a Right Choice: Establishing Mammalian Standards and the Ideal of a Standardized Mammal.' *Journal of the History of Biology* 26 (1993): pp.329–349.

Cleeremans, Axel. 'The Radical Plasticity Thesis: How the Brain Learns to Be Conscious.' *Frontiers in Psychology* 2 (2011): 1–12. https://doi.org/10.3389/fpsyg.2011.00086.

Clifford, William. 'On the Scientific Basis of Morals: A Discussion.' *Contemporary Review* 26 (1875): pp.650–660.

Clutton-Brock, Tim. 'Co-Operation between Non-Kin in Animal Societies.' *Nature* 462 (2009): pp.51–57.

Collins, Nancy, and Stephen Read. 'Adult Attachment, Working Models, and Relationship Quality in Dating Couples.' *Journal of Personality and Social Psychology* 58 (1990): pp.644–663.

Cooley, Charles. *Human Nature and the Social Order*. New York: Scribner's, 1902.

Coon, Dennis and John Mitterer. *Introduction to Psychology: Gateways to Mind and Behaviour*. 2nd ed. Belmont CA: Wadsworth, 2010.

Corcoran, Tim and John Cromby, eds. *Joint Action: Essays in Honour of John Shotter*. London: Routledge, 2016.

Coriale, Danielle. 'When Zoophytes Speak: Polyps and Naturalist Fantasy in the Age of Liberalism.' *Nineteenth-Century Contexts* 34 (2012): pp.19–36.

Cosmides, Leda, and John Tooby. *Evolutionary Psychology: A Primer*. Santa Barbara CA: University of California, 1997. https://www.cep.ucsb.edu/primer.html.

Costall, Alan. 'How Lloyd Morgan's Canon Backfired.' *Journal of the History of the Behavioural Sciences* 29 (1993): pp.113–122.

Costall, Alan. 'Specious Origins? Darwinism and Developmental Theory.' In *Evolution and Developmental Psychology*, edited by George Butterworth, Julie Rutkowska, and Michael Scaife, pp.30–41. Brighton: Harvester, 1985.

Crichton-Browne, James. 'Letter to Darwin, 16th April.' In *Correspondence of Charles Darwin* Cambridge: Darwin Correspondence Project, 1871. https://www.darwinproject.ac.uk/letter/?docId=letters/DCP-LETT-7689.xml.

Crivelli, Carlos, and Alan Fridlund. 'Facial Displays Are Tools for Social Influence.' *Trends in Cognitive Sciences* 22 (2018): pp.388–399.

Cromby, John. *Feeling Bodies: Embodying Psychology*. London: Palgrave-Macmillan, 2015.

Crozier, W. Ray and Peter De Jong. 'The Study of the Blush: Darwin and After.' In *The Psychological Significance of the Blush*, edited by W. Ray Crozier and Peter De Jong, pp.1–11. Cambridge: Cambridge University Press, 2012.

Csengei, Ildiko. *Sympathy, Sensibility and the Literature of Feeling in the Eighteenth Century*. Basingstoke: Palgrave Macmillan, 2012.

Cuff, Benjamin, Sarah Brown, Laura Taylor, and Douglas Howit. 'Empathy: A Review of the Concept.' *Emotion Review* 8 (2016): pp.144–153.

Cummings, Brian. *Mortal Thoughts: Religion, Secularity, & Identity in Shakespeare and Early Modern Culture*. Oxford: Oxford University Press, 2013.

Danziger, Kurt. *Constructing the Subject: Historical Origins of Psychological Research*. Cambridge: Cambridge University Press, 1990.

Darwin, Charles. *An Appeal*. Cambridge: Darwin Online. Composed 1863. https://www.darwinproject.ac.uk/topics/life-sciences/darwin-and-vivisection/appeal-against-animal-cruelty.

Darwin, Charles. *The Autobiography of Charles Darwin, 1809–1882: With Original Omissions Restored*. Edited by Nora Barlow London: Collins, 1958.

Darwin, Charles. 'A Biographical Sketch of an Infant.' *Mind: A Quarterly Review of Psychology and Philosophy* 2 (1877): pp.285–294.

Darwin, Charles. 'Books to Be Read' and 'Books Read' Notebook.' In *Darwin Manuscripts*. Cambridge: Darwin Online. Composed 1838–1851. http://darwin-online.org.uk/content/frameset?viewtype=text&itemID=CUL-DAR128.-&pageseq=1.

Darwin, Charles. *Charles Darwin's Natural Selection: Being the Second Part of His Big Species Book Written from 1856 to 1858*. Cambridge: Cambridge University Press, 1975.

Darwin, Charles. 'Darwin's Journal.' *Bulletin of the British Museum (Natural History). Historical Series* 2, no. 1 (1959): pp.1–21.

Darwin, Charles. 'Darwin's Ornithological Notes [1837]. Edited with an Introduction, Notes, and Appendix by Nora Barlow.' *Bulletin of the British Museum (Natural History) Historical Series* 2 (1963): pp.201–278.

Darwin, Charles. *The Descent of Man and Selection in Relation to Sex Vol.1.* London: Murray, 1871.

Darwin, Charles. *The Descent of Man and Selection in Relation to Sex Vol.2.* London: Murray, 1871.

Darwin, Charles. *The Descent of Man, and Selection in Relation to Sex.* 2nd ed. London: Murray, 1874.

Darwin, Charles. *Diary of the Voyage of H.M.S. Beagle Vol. I. The Works of Charles Darwin*. Edited by Paul Barrett and Richard Freeman. London: William Pickering, 1986.

Darwin, Charles. *The Effects of Cross and Self Fertilisation in the Vegetable Kingdom*. 2nd ed. London: John Murray, 1878. First published 1876.

Darwin, Charles. *The Expression of the Emotions in Man and Animals*. 2nd ed. London: Murray, 1890. First published 1872.

Darwin, Charles. *The Formation of Vegetable Mould, through the Action of Worms, with Observations on Their Habits*. London: John Murray, 1882. First published 1881.

Darwin, Charles. *Journal of Researches into the Natural History and Geology of the Countries Visited During the Voyage of H.M.S. Beagle Round the World, under the Command of Capt. Fitz Roy R.N.* London: John Murray, 2006. First published 1860. http://darwin-online.org.uk/content/frameset?itemID=F20&viewtype=text&pageseq=1

Darwin, Charles. *Insectivorous Plants*. 2nd ed. London: John Murray, 1888. First published 1875.

Darwin, Charles. 'Letter.' *Gardeners' Chronicle*, no. 9th February (1861): p.122.

Darwin, Charles. 'Letter to Bain, 9th October.' In *Correspondence of Charles Darwin* Cambridge: Darwin Correspondence Project, 1873. https://www.darwinproject.ac.uk/letter/?docId=letters/DCP-LETT-9092.xml.

Darwin, Charles. 'Letter to Bradlaugh, 6th June.' In *Correspondence of Charles Darwin* Cambridge: Darwin Correspondence Project, 1877. https://www.darwinproject.ac.uk/letter/?docId=letters/DCP-LETT-10988.xml.

Darwin, Charles. 'Letter to Caroline Kennard, 9th January.' In *Correspondence of Charles Darwin* Cambridge: Darwin Correspondence Project, 1882. https://www.darwinproject. ac.uk/letter/?docId=letters/DCP-LETT-13607.xml.

Darwin, Charles. 'Letter to Catherine Darwin, 22nd May to 14th July.' In *Correspondence of Charles Darwin* Cambridge: Darwin Correspondence Project, 1833. https://www. darwinproject.ac.uk/letter/?docId=letters/DCP-LETT-206.xml.

Darwin, Charles. 'Letter to Charles Whitley, 9th September.' In *Correspondence of Charles Darwin* Cambridge: Darwin Correspondence Project, 1831. http://www.darwinproject. ac.uk/entry-121.xml.

Darwin, Charles. 'Letter to Crichton-Browne, 22nd May.' In *Correspondence of Charles Darwin* Cambridge: Darwin Correspondence Project, 1869. https://www.darwinproject. ac.uk/letter/?docId=letters/DCP-LETT-6755.xml.

Darwin, Charles. 'Letter to Crichton-Browne, 8th June.' In *Correspondence of Charles Darwin* Cambridge: Darwin Correspondence Project, 1869. https://www.darwinproject. ac.uk/letter/?docId=letters/DCP-LETT-6779.xml.

Darwin, Charles. 'Letter to Crichton-Browne, 8th June.' In *Correspondence of Charles Darwin* Cambridge: Darwin Correspondence Project, 1870. https://www.darwinproject. ac.uk/letter/?docId=letters/DCP-LETT-7224.xml.

Darwin, Charles. 'Letter to Emma, 29th April.' In *Correspondence of Charles Darwin* Cambridge: Darwin Correspondence Project, 1858. https://www.darwinproject.ac.uk/ letter/?docId=letters/DCP-LETT-2261.xml.

Darwin, Charles. 'Letter to Falconer, 22nd April.' In *Correspondence of Charles Darwin* Cambridge: Darwin Correspondence Project, 1863. https://www.darwinproject.ac.uk/ letter/?docId=letters/DCP-LETT-4121.xml.

Darwin, Charles. 'Letter to Fawcett, 18th September.' In *Correspondence of Charles Darwin* Cambridge: Darwin Correspondence Project, 1861. https://www.darwinproject.ac.uk/ letter/?docId=letters/DCP-LETT-3257.xml.

Darwin, Charles. 'Letter to Fitzroy, 4th or 11th October 1831.' In *Correspondence of Charles Darwin* Cambridge: Darwin Correspondence Project, 1831. https://www.darwinproject. ac.uk/letter/?docId=letters/DCP-LETT-139.xml.

Darwin, Charles. 'Letter to Fox, 23rd May.' In *Correspondence of Charles Darwin* Cambridge: Darwin Correspondence Project, 1833. https://www.darwinproject.ac.uk/ letter/?docId=letters/DCP-LETT-207.xml.

Darwin, Charles. 'Letter to Fritz Müller, 22nd Feb.' In *Correspondence of Charles Darwin* Cambridge: Darwin Correspondence Project, 1867. https://www.darwinproject.ac.uk/ letter/?docId=letters/DCP-LETT-5410.xml.

Darwin, Charles. 'Letter to Gaskell, November 15th.' In *Correspondence of Charles Darwin* Cambridge: Darwin Correspondence Project, 1878. https://www.darwinproject.ac.uk/ letter/?docId=letters/DCP-LETT-11745.xml.

Darwin, Charles. 'Letter to Gray, 24th February.' In *Correspondence of Charles Darwin* Cambridge: Darwin Correspondence Project, 1860. https://www.darwinproject.ac.uk/ letter/?docId=letters/DCP-LETT-2713.xml.

Darwin, Charles. 'Letter to Gray, 22nd May.' In *Correspondence of Charles Darwin* Cambridge: Darwin Correspondence Project, 1860. https://www.darwinproject.ac.uk/ letter/?docId=letters/DCP-LETT-2814.xml.

Darwin, Charles. 'Letter to Gray, 26th November.' In *Correspondence of Charles Darwin* Cambridge: Darwin Correspondence Project, 1860. https://www.darwinproject.ac.uk/ letter/?docId=letters/DCP-LETT-2998.xml.

Darwin, Charles. 'Letter to Gray, 11th April.' In *Correspondence of Charles Darwin* Cambridge: Darwin Correspondence Project, 1861. https://www.darwinproject.ac.uk/letter/?docId=letters/DCP-LETT-3115.xml.

Darwin, Charles. 'Letter to Gray, 23rd–24th July.' In *Correspondence of Charles Darwin* Cambridge: Darwin Correspondence Project, 1862. www.darwinproject.ac.uk/letter/?docId=letters/DCP-LETT-3662.xml.

Darwin, Charles. 'Letter to Gray, 11th May.' In *Correspondence of Charles Darwin* Cambridge: Darwin Correspondence Project, 1863. www.darwinproject.ac.uk/letter/?docId=letters/DCP-LETT-4153.xml.

Darwin, Charles. 'Letter to Gray, 13th September.' In *Correspondence of Charles Darwin* Cambridge: Darwin Correspondence Project, 1864. https://www.darwinproject.ac.uk/letter/?docId=letters/DCP-LETT-4611.xml.

Darwin, Charles. 'Letter to Haeckel, 7th November.' In *Correspondence of Charles Darwin* Cambridge: Darwin Correspondence Project, 1868. https://www.darwinproject.ac.uk/letter/?docId=letters/DCP-LETT-6450.

Darwin, Charles. 'Letter to Henslow, 9th September.' In *Correspondence of Charles Darwin* Cambridge: Darwin Correspondence Project, 1831. https://www.darwinproject.ac.uk/letter/?docId=letters/DCP-LETT-123.xml.

Darwin, Charles. 'Letter to Henslow, 24th July to 7th November.' In *Correspondence of Charles Darwin* Cambridge: Darwin Correspondence Project, 1834. https://www.darwinproject.ac.uk/letter/?docId=letters/DCP-LETT-251.xml.

Darwin, Charles. 'Letter to Hooker, 3rd Feb.' In *Correspondence of Charles Darwin* Cambridge: Darwin Correspondence Project, 1850. https://www.darwinproject.ac.uk/letter/?docId=letters/DCP-LETT-1300.xml.

Darwin, Charles. 'Letter to Hooker, 25th December.' In *Correspondence of Charles Darwin* Cambridge: Darwin Correspondence Project, 1857. https://www.darwinproject.ac.uk/letter/?docId=letters/DCP-LETT-2194.xml.

Darwin, Charles. 'Letter to Hooker, 14th Feb.' In *Correspondence of Charles Darwin* Cambridge: Darwin Correspondence Project, 1860. www.darwinproject.ac.uk/letter/?docId=letters/DCP-LETT-2696.xml.

Darwin, Charles. 'Letter to Hooker, 29th March.' In *Correspondence of Charles Darwin* Cambridge: Darwin Correspondence Project: 1863. https://www.darwinproject.ac.uk/letter/?docId=letters/DCP-LETT-4065.xml.

Darwin, Charles. 'Letter to Hooker, 9th April.' In *Correspondence of Charles Darwin* Cambridge: Darwin Correspondence Project, 1866. https://www.darwinproject.ac.uk/letter/?docId=letters/DCP-LETT-5051.xml.

Darwin, Charles. 'Letter to Hooker, 10th December.' In *Correspondence of Charles Darwin* Cambridge: Darwin Correspondence Project: 1866. https://www.darwinproject.ac.uk/letter/?docId=letters/DCP-LETT-5300.xml.

Darwin, Charles. 'Letter to Hooker, 24th December.' In *Correspondence of Charles Darwin* Cambridge: Darwin Correspondence Project, 1866. https://www.darwinproject.ac.uk/letter/?docId=letters/DCP-LETT-5321.xml.

Darwin, Charles. 'Letter to Huxley, 3rd September.' In *Correspondence of Charles Darwin* Cambridge: Darwin Correspondence Project, 1855. https://www.darwinproject.ac.uk/letter/?docId=letters/DCP-LETT-1759.xml.

Darwin, Charles. 'Letter to Huxley, 22nd December.' In *Correspondence of Charles Darwin* Cambridge: Darwin Correspondence Project, 1866. www.darwinproject.ac.uk/letter/?docId=letters/DCP-LETT-5315.xml.

Darwin, Charles. 'Letter to Huxley, 30th January.' In *Correspondence of Charles Darwin* Cambridge: Darwin Correspondence Project, 1868. https://www.darwinproject.ac.uk/letter/?docId=letters/DCP-LETT-5817.

Darwin, Charles. 'Letter to Huxley, 14th October.' In *Correspondence of Charles Darwin* Cambridge: Darwin Correspondence Project, 1869. www.darwinproject.ac.uk/letter/?docId=letters/DCP-LETT-6936.xml.

Darwin, Charles. 'Letter to Lyell, 8th October.' In *Correspondence of Charles Darwin* Cambridge: Darwin Correspondence Project, 1845. https://www.darwinproject.ac.uk/letter/?docId=letters/DCP-LETT-919.xml.

Darwin, Charles. 'Letter to Lyell, 23rd February.' In *Correspondence of Charles Darwin* Cambridge: Darwin Correspondence Project, 1860. https://www.darwinproject.ac.uk/letter/?docId=letters/DCP-LETT-2707.

Darwin, Charles. 'Letter to Lyell, 10th January.' In *Correspondence of Charles Darwin* Cambridge: Darwin Correspondence Project, 1860. https://www.darwinproject.ac.uk/letter/?docId=letters/DCP-LETT-2647.xml.

Darwin, Charles. 'Letter to Lyell, 15th February.' In *Correspondence of Charles Darwin* Cambridge: Darwin Correspondence Project, 1866. https://www.darwinproject.ac.uk/letter/?docId=letters/DCP-LETT-5007.xml.

Darwin, Charles. 'Letter to Mrs Emily Talbot on the Mental and Bodily Development of Infants.' In *Correspondence of Charles Darwin* Cambridge: Darwin Correspondence Project, 1881. http://darwin-online.org.uk/content/frameset?itemID=F1995&viewtype=text&pageseq=1.

Darwin, Charles. 'Letter to Murray, 24th September.' In *Correspondence of Charles Darwin* Cambridge: Darwin Correspondence Project, 1861. www.darwinproject.ac.uk/letter/?docId=letters/DCP-LETT-3264.xml.

Darwin, Charles. 'Letter to Murray, 13th April.' In *Correspondence of Charles Darwin* Cambridge: Darwin Correspondence Project, 1871. https://www.darwinproject.ac.uk/letter/?docId=letters/DCP-LETT-7680.xml.

Darwin, Charles. 'Letter to Robert Fitzroy, 10th October.' In *Correspondence of Charles Darwin* Cambridge: Darwin Correspondence Project, 1831. https://www.darwinproject.ac.uk/letter/?docId=letters/DCP-LETT-139.xml.

Darwin, Charles. 'Letter to Romanes, 2nd September.' In *Correspondence of Charles Darwin* Cambridge: Darwin Correspondence Project, 1878. https://www.darwinproject.ac.uk/letter/?docId=letters/DCP-LETT-11684.xml.

Darwin, Charles. 'Letter to Sir James Paget, 29th April' In *Correspondence of Charles Darwin* Cambridge: Darwin Correspondence Project, 1869. https://www.darwinproject.ac.uk/letter/?docId=letters/DCP-LETT-6716.xml.

Darwin, Charles. 'Letter to Thiel, 25th February.' In *Correspondence of Charles Darwin* Cambridge: Darwin Correspondence Project, 1869. https://www.darwinproject.ac.uk/letter/?docId=letters/DCP-LETT-6634.xml.

Darwin, Charles. 'Letter to Wallace, 15th June.' In *Correspondence of Charles Darwin* Cambridge: Darwin Correspondence Project, 1864. https://www.darwinproject.ac.uk/letter/?docId=letters/DCP-LETT-4535.xml.

Darwin, Charles. 'Letter to Wallace, 28th May.' In *Correspondence of Charles Darwin* Cambridge: Darwin Correspondence Project, 1864. https://www.darwinproject.ac.uk/letter/?docId=letters/DCP-LETT-4510.xml.

Darwin, Charles. 'Letter to Wallace, 26th February.' In *Correspondence of Charles Darwin* Cambridge: Darwin Correspondence Project, 1867. https://www.darwinproject.ac.uk/letter/?docId=letters/DCP-LETT-5420.xml.

Darwin, Charles. 'Letter to Wallace, 12th–17th March.' In *Correspondence of Charles Darwin* Cambridge: Darwin Correspondence Project, 1867. https://www.darwinproject.ac.uk/letter/?docId=letters/DCP-LETT-5440.xml.

Darwin, Charles. 'Letter to Woolner, 7th April.' In *Correspondence of Charles Darwin* Cambridge: Darwin Correspondence Project, 1871. https://www.darwinproject.ac.uk/letter/?docId=letters/DCP-LETT-7665.xml.

Darwin, Charles. 'Life. Written August - 1838.' (1838): 56–62. http://darwin-online.org.uk/content/frameset?itemID=CUL-DAR91.56-63&viewtype=text&pageseq=1.

Darwin, Charles. *The Movements and Habits of Climbing Plants*. 2nd ed. London: John Murray, 1882. First published 1865.

Darwin, Charles. 'Notebook B.' Cambridge: Darwin Online. Composed 1837–1838, http://darwin-online.org.uk/content/frameset?itemID=CUL-DAR121.-&viewtype=text&pageseq=1.

Darwin, Charles. 'Notebook C. Cambridge: Darwin Online. Composed 1838. http://darwin-online.org.uk/content/frameset?itemID=CUL-DAR122.-&viewtype=text&pageseq=1

Darwin, Charles. 'Notebook D.' Cambridge: Darwin Online. Composed 1838. http://darwin-online.org.uk/content/frameset?viewtype=text&itemID=CUL-DAR123.-&pageseq=1.

Darwin, Charles. 'Notebook E.' Cambridge: Darwin Online. Composed 1838–1839. http://darwin-online.org.uk/content/frameset?viewtype=text&itemID=CUL-DAR124.-&pageseq=1.

Darwin, Charles. 'Notebook M.' Cambridge: Darwin Online. Composed 1838, http://darwin-online.org.uk/content/frameset?itemID=CUL-DAR125.-&viewtype=text&pageseq=1.

Darwin, Charles. 'Notebook N.' Cambridge: Darwin Online. Composed 1838–1839. http://darwin-online.org.uk/content/frameset?keywords=n%20notebook&pageseq=1&itemID=CUL-DAR126.-&viewtype=side.

Darwin, Charles. 'Old & Useless Notes About the Moral Sense & Some Metaphysical Points.' Cambridge: Darwin Online.Composed 1838–1840. http://darwin-online.org.uk/content/frameset?viewtype=side&itemID=CUL-DAR91.4-55&pageseq=1.

Darwin, Charles. *On the Origin of Species by Means of Natural Selection or the Preservation of Favoured Races in the Struggle for Life*. London: Murray, 1859.

Darwin, Charles. *On the Origin of Species*. 4th ed. London: Murray, 1866.

Darwin, Charles. *On the Origin of Species*. 5th ed. London: Murray, 1869.

Darwin, Charles. 'On the Ova of Flustra, or, Early Notebook, Containing Observations Made by CD When He Was at Edinburgh, March 1827.' In *The Collected Papers of Charles Darwin*, edited by Paul Barrett, pp.285–291. Chicago: Chicago University Press, 1977.

Darwin, Charles. 'On the Use of the Microscope on Board Ship.' In *A Manual of Scientific Enquiry; Prepared for the Use of Her Majesty's Navy: And Adapted for Travellers in General*, edited by John Herschel, pp.389–395. London: Murray, 1849.

Darwin, Charles. 'Origin of Species.' *Athenaeum* 1854 (1863): p.617.

Darwin, Charles. *The Origin of Species by Means of Natural Selection, or the Preservation of Favoured Races in the Struggle for Life*. 6th ed. London: Murray, 1876.

Darwin, Charles. *The Power of Movement in Plants*. London: John Murray, 1880.

Darwin, Charles. 'Preliminary Notice.' In *Life of Erasmus Darwin*, edited by Ernst Krause, pp.1–128. London: John Murray, 1879.

Darwin, Charles. 'Rio Notebook.' Cambridge: Darwin Online. Composed 1832. http://darwin-online.org.uk/content/frameset?itemID=EH1.10&viewtype=text&pageseq=1

Darwin, Charles. *The Structure and Distribution of Coral Reefs*. London: Smith, Elder & Co, 1842.

Darwin, Charles. *The Variation of Animals and Plants under Domestication.* 2 vols. London: John Murray, 1875. First published 1868.

Darwin, Charles. *The Various Contrivances by Which Orchids Are Fertilised by Insects.* 2nd revised ed. London: John Murray, 1882. First published 1862.

Darwin, Charles. 'Zoology Notes.' In *Charles Darwin's Zoology Notes & Specimen Lists from H.M.S. Beagle,* edited by Richard Keynes. Cambridge: Cambridge University Press, 2005.

Darwin, Emma. *Emma Darwin: A Century of Family Letters.* 2 vols. Vol. 2nd, New York: Appleton, 1915.

Darwin, Charles, and Emma Darwin. *Notebook of Observations on the Darwin Children.* Cambridge: Darwin Online. Composed 1839–1856. http://darwin-online.org.uk/content/frameset?itemID=CUL-DAR210.11.37.xml.

Darwin, Erasmus. *The Temple of Nature, or, the Origin of Society: A Poem, with Philosophical Notes.* Baltimore: Butler Bonsal & Niles, 1804. First published 1803.

Darwin, Erasmus. *Zoonomia, or, the Laws of Organic Life, Vol.1.* London: Johnson, 1794.

Darwin, Francis. 'Reminiscences of My Father's Everyday Life.' In *The Life and Letters of Charles Darwin,* edited by Francis Darwin, pp.108–162. London: John Murray, 1887.

Darwin, William. 'Letter to Darwin, 25th March.' In *Correspondence of Charles Darwin* Cambridge: Darwin Correspondence Project, 1868. https://www.darwinproject.ac.uk/letter/?docId=letters/DCP-LETT-6069.xml.

Darwin, William. 'Letter to Darwin, 22nd April.' In *Correspondence of Charles Darwin* Cambridge: Darwin Correspondence Project, 1868. https://www.darwinproject.ac.uk/letter/?docId=letters/DCP-LETT-6137.xml.

Daston, Lorraine. 'The Empire of Observation, 1600–1800.' In *Histories of Scientific Observation,* edited by Lorraine Daston and Elizabeth Lunbeck, pp.81–113. Chicago: University of Chicago Press, 2011.

Daston, Lorraine. 'The Theory of Will versus the Science of Mind.' In *The Problematic Science: Psychology in Nineteenth-Century Thought,* edited by William and Ash Woodward, Mitchell, pp.88–115. New York: Praeger, 1982.

Daston, Lorraine, and Gregg Mitman, eds. *Thinking with Animals: New Perspectives on Anthropomorphism.* New York: Columbia University Press, 2005.

Davis, Robert. 'The Brass Age of Psychology.' *Technology and Culture* 11 (1970): pp.604–612.

Dawes, Robyn. *House of Cards: Psychology and Psychotherapy Built on Myth.* New York: Free Press, 1996.

Dawkins, Richard. *The Blind Watchmaker: Why the Evidence of Evolution Reveals a Universe without Design.* Harlow: Longman, 1986.

Dawkins, Richard. *The Extended Phenotype: The Long Reach of the Gene.* Oxford: Oxford University Press, 1982.

Dawkins, Richard. *The Selfish Gene: 30th Anniversary Edition.* Oxford: Oxford University Press, 2006. First published 1976.

Dawkins, Richard. 'Universal Darwinism.' In *The Nature of Life: Classical and Contemporary Perspectives from Philosophy and Science,* edited by Mark Bedau and Carol Cleland, pp.260–273. Cambridge: Cambridge University Press, 2010.

Dawson, Gowan. 'Darwin Decentred.' *Studies in History and Philosophy of Biological and Biomedical Sciences* 46 (2014): pp.93–96.

Dawson, Gowan. *Darwin, Literature and Victorian Respectability.* Cambridge: Cambridge University Press, 2007.

De Chadarevian, Soraya. 'Laboratory Science versus Country-House Experiments. The Controversy between Julius Sachs and Charles Darwin.' *British Journal for the History of Science* 29 (1996): pp.17–41.

De Chardin, Teilhard. *The Phenomenon of Man*. London: Collins, 1959.

De Gelder, Beatrice. *Emotions and the Body*. Oxford: Oxford University Press, 2016.

De Queiroz, Kevin. 'Branches in the Lines of Descent: Charles Darwin and the Evolution of the Species Concept.' *Biological Journal of the Linnean Society* 103 (2011): pp.19–35.

De Santis, Marco, Elena Cesari, Enigma Nobili, Gianluca Straface, Anna Cavaliere, and Alessandro Caruso. 'Radiation Effects on Development.' *Birth Defects Research (Part C)* 81 (2007): pp.177–182.

De Waal, Frans. 'Anthropomorphism and Anthropodenial: Consistency in Our Thinking About Humans and Other Animals.' *Philosophical Topics* 27 (1999): pp.255–280.

De Waal, Frans. 'Are We in Anthropodenial?' *Discover Magazine*, no. July 1st (1997). http://discovermagazine.com/1997/jul/areweinanthropod1180.

De Waal, Frans. *Are We Smart Enough to Know How Smart Animals Are?* New York: Norton, 2016.

De Waal, Frans, and Frans Lanting. *Bonobo: The Forgotten Ape*. San Francisco CA: University of California Press, 1997.

De Waal, Frans. 'Darwin's Legacy and the Study of Primate Visual Communication.' *Annals of the New York Academy of Siences* 1000 (2003): pp.7–31.

Deacon, Terrence. *The Symbolic Species: The Co-Evolution of Language and the Brain*. New York: Norton, 1997.

Deese, James. *Psychology as Science and Art*. New York: Harcourt Brace Jovanovich, 1972.

Dehaene, Stanislas, and Lionel Naccache. 'Towards a Cognitive Neuroscience of Consciousness: Basic Evidence and a Workspace Framework.' *Cognition* 79 (2001): pp.1–37.

Dehaene, Stanislas, Hakwan Lau, and Sid Kouider. 'What Is Consciousness, and Could Machines Have It?' *Science* 358 (2017): pp.486–492.

Delboeuf, Joseph. *La Psychologie Comme Science Naturelle: Son Présent Et Son Avenir*. Brussels: Librairie Européene C. Muquardt, 1876.

Deleuze, Gilles, and Felix Guattari, *A Thousand Plateaus: Capitalism and Schizophrenia*. Minneapolis: University of Minnesota Press, 1987. First published 1980.

Deleuze, Gilles. *Expressionism in Philosophy: Spinoza*. New York: Zone Books, 1990. First published 1968.

Dennett, Daniel. 'Are We Explaining Consciousness Yet?' *Cognition* 79 (2001): pp.221–237.

Dennett, Daniel. *Darwin's Dangerous Idea: Evolution and the Meanings of Life*. New York: Simon & Schuster, 1995.

Dennis, Alex, and Greg Smith. 'Interactionism, Symbolic.' In *International Encyclopedia of the Social & Behavioral Sciences,* edited by James Wright Amsterdam: Elsevier, 2015. https://doi.org/10.1016/B978-0-08-097086-8.32079-7.

Department of Education Employment and Workforce Relations. *Belonging, Being, Becoming: The Early Years Learning Framework for Australia*. Canberra: Commonwealth of Australia, 2009.

Derrida, Jacques. *The Animal That Therefore I Am*. New York: Fordham University Press, 2008.

Derrida, Jacques. *Of Grammatology*. Baltimore, MA: Johns Hopkins University Press, 1976. First published 1967.

Desmond, Adrian. *The Politics of Evolution: Morphology, Medicine, and Reform in Radical London*. Chicago: University of Chicago Press, 1989.

Desmond, Adrian, and James Moore. *Darwin*. Harmondsworth: Penguin, 1991.

Desmond, Adrian, and James Moore. *Darwin's Sacred Cause: Race, Slavery and the Quest for Human Origins*. London: Penguin, 2009.

Dettlebach, Michael. ' "A Kind of Linnaean Being": Forster and Eighteenth-Century Natural History.' In *Observations Made During a Voyage around the World by Johann Reinhold Forster*, edited by Nicholas Thomas, Harriet Guest, and Michael Dettlebach, pp.lv–lxxvi. Honolulu: University of Hawai'i Press, 1996.

Devereux, George. *From Anxiety to Method in the Behavioural Sciences*. The Hague: Mouton, 1968.

Dewey, John. 'The Theory of Emotion (1) Emotional Attitudes.' *Psychological Review* 1 (1894): pp.553–569.

Dijk, Corine, and Peter De Jong. 'Fear of Blushing: No Overestimation of Negative Anticipated Effects but a High Subjective Probability of Blushing.' *Cognitive Therapy and Research* 33 (2009): pp.53–74.

Dixon, Thomas. *From Passions to Emotions: The Creation of a Secular Psychological Category* Cambridge: Cambridge University Press, 2003.

Dobzhansky, Theodosius. *Genetics and the Origin of Species*. 3rd ed. New York: Columbia University Press, 1951. First published 1937.

Donald, Merlin. *A Mind So Rare: The Evolution of Human Consciousness*. New York: Norton, 2001.

Donders, Franciscus. 'On the Action of the Eye-Lids in Determination of Blood from Expiratory Effort.' *Beale's Archives of Medicine* 5 (1870): pp.20–38.

Dreyfus, Hubert. *Being-in-the-World: A Commentary on Heidegger's Being and Time Division I*. Cambridge MA: MIT Press, 1991.

Dror, Otniel. 'Seeing the Blush: Feeling Emotions.' In *Histories of Scientific Observation*, edited by Lorraine Daston and Elizabeth Lunbeck, pp.326–348. Chicago: University of Chicago Press, 2011.

Drummond, Peter, and Nadia Mirco. 'Staring at One Side of the Face Increases Blood Flow on That Side of the Face.' *Psychophysiology* 41 (2004): pp.281–287.

Duchenne, Guillaume-Benjamin. *Mécanisme De La Physionomie Humaine, Ou, Analyse Électro-Physiologique De L'expression Des Passions*. Paris: Jules Renouard 1862.

Duchenne, Guillaume-Benjamin. *The Mechanism of Human Facial Expression*. Cambridge: Cambridge University Press, 1990. First published 1862.

Duhem, Pierre. *The Aim and Structure of Physical Theory*. 2nd ed. Princeton NJ: Princeton University Press, 1954. First published 1914.

Dunbar, Robin, ed. *Oxford Handbook of Evolutionary Psychology*. Oxford: Oxford University Press, 2007.

Dunbar, Robin. 'The Social Brain: Mind, Language, and Society in Evolutionary Perspective.' *Annual Review of Anthropology* 32 (2003): pp.163–181.

Duncan-Jones, Austin. 'The Notion of Conscience.' *Philosophy* 30 (1955): pp.131–140.

Dupré, John. 'Against Maladaptationism: Or What's Wrong with Evolutionary Psychology?' In *Knowledge as Social Order: Rethinking the Sociology of Barry Barnes*, edited by Massimo Mazzoti, pp.165–180. London: Routledge, 2008.

Dupré, John. *Human Nature and the Limits of Science*. Oxford: Clarendon Press, 2001.

Dupré, John. *Processes of Life: Essays in the Philosophy of Biology*. Oxford: Oxford University Press, 2012.

Dupré, John, and Maureen O'Malley. 'Varieties of Living Things: Life at the Intersection of Lineage and Metabolism.' *Philosophy Theory and Practice in Biology* 1 (2009): 1–25. http://dx.doi.org/10.3998/ptb.6959004.0001.003.

Durant, John. 'The Ascent of Nature in Darwin's *the Descent of Man*.' In *The Darwinian Heritage*, edited by David Kohn, pp.283–306. Princeton NJ: Princeton University Press, 1985.

Durkheim, Émile. *Suicide: A Study in Sociology*. London: Routledge & Kegan Paul, 1952. First published 1897.

Dyson, Frank, Arthur Eddington, and Charles Davidson. 'A Determination of the Deflection of Light by the Sun's Gravitational Field, from Observations Made at the Total Eclipse of May 29, 1919.' *Philosophical Transactions of the Royal Society of London* 579 (1920): pp.291–333.

Edelman, Gerald. *Neural Darwinism: The Theory of Neuronal Group Selection.* New York: Basic Books, 1987.

Edelmann, Robert. 'Embarrassment and Blushing.' In *Encyclopaedia of Human Behaviour*, edited by Vilayanur Ramachandran, pp.24–31. London: Academic Press, 2012.

Einstein, Albert. 'On the Electrodynamics of Moving Bodies.' In *The Principle of Relativity.* New York: Dover, 1952. First published 1905.

Einstein, Albert. 'On the Method of Theoretical Physics.' *Philosophy of Science* 1 (1934): pp.163–169.

Ekman, Paul. 'An Argument for Basic Emotions.' *Cognition and Emotion* 6 (1992): pp.169–200.

Ekman, Paul, and Daniel Cordaro. 'What Is Meant by Calling an Emotion Basic.' *Emotion Review* 3 (2011): pp.364–370.

Elfenbein, Hillary, and Nalini Ambady. 'On the Universality and Cultural Specificity of Emotion Recognition: A Meta-Analysis.' *Psychological Bulletin* 128 (2002): pp.203–235.

Eliot, George. *The George Eliot Letters 1859–1861.* 9 vols. Vol. 3. New Haven CT: Yale University Press, 1954.

Ellsworth, Phoebe. 'William James and Emotion: Is a Century of Fame Worth a Century of Misunderstanding?' *Psychological Review* 101 (1994): pp.222–229.

Elms, Alan. 'The Crisis of Confidence in Social Psychology.' *American Psychologist* 30 (1975): pp.967–976.

Emborski, Carmen, and Alexander Mikheyev. 'Ancestral Diet Transgenerationally Influences Offspring in a Parent-of-Origin and Sex-Specific Manner.' *Philosophical Transactions of the Royal Society of London* 374 (2018): 1–11. https://royalsocietypublishing.org/doi/pdf/10.1098/rstb.2018.0181.

Emery, Nathan. *Bird Brain: An Exploration of Avian Intelligence.* Princeton NJ: Princeton University Press, 2016.

Endersby, Jim. 'Darwin on Generation, Pangenesis and Sexual Selection.' In *The Cambridge Companion to Darwin*, edited by Jonathan Hodge and Gregory Radick, pp.73–95. Cambridge: Cambridge University Press, 2009.

Engels, Frederick. 'Letter to Pyotr Lavrov, 12–17 November 1875.' In *Karl Marx and Frederick Engels. Works.* Moscow: Politizdat, 1964.

Erwin, Douglas. 'Evolutionary Uniformitarianism.' *Developmental Biology* 357 (2011): pp.27–34.

Evans, Jonathan. 'Dual-Processing Accounts of Reasoning, Judgement, and Social Cognition.' *Annual Review of Psychology* 59 (2008): pp.255–278.

Evans, Samantha, ed. *Darwin and Women: A Selection of Letters.* Cambridge: Cambridge University Press, 2017.

Evans, Samantha. 'Preface.' In *Darwin and Women: A Selection of Letters*, edited by Samantha Evans, pp.xix–xxvi. Cambridge: Cambridge University Press, 2017.

Farr, Robert. 'On Reading Darwin and Discovering Social Psychology.' In *The Development of Social Psychology*, edited by Robert Gilmour and Steven Duck, pp.111–136. London: Academic, 1980.

Feinman, Saul. 'Social Referencing in Infancy.' *Merrill-Palmer Quarterly* 28 (1982): pp.445–470.

Feller, David. 'Dog Fight: Darwin as Animal Advocate in the Antivivisection Controversy of 1875.' *Studies in History and Philosophy of Biological and Biomedical Sciences* 40 (2009): pp.265–271.

Fernandez-Dols, José-Miguel. 'Facial Expression and Emotion: A Situationist View.' In *The Social Context of Nonverbal Behaviour*, edited by Pierre Philippot, Robert Feldman, and Erik Coats, pp.242–261. Cambridge: Cambridge University Press, 1999.

Fernández-Dols, José-Miguel. 'Natural Facial Expression: A View from Psychological Constructionism and Pragmatics.' In *The Science of Facial Expression*, edited by James Russell and José-Miguel Fernández-Dols, pp.457–475. Oxford: Oxford University Press, 2017.

Ferrier, James. 'An Introduction to the Philosophy of Consciousness, Part I.' *Blackwood's Edinburgh Magazine* 43 (1838): pp.187–201.

Fischer, Clara. 'Feminist Philosophy, Pragmatism, and the "Turn to Affect": A Genealogical Critique.' *Hypatia* 31 (2016): pp.810–826.

FitzRoy, Robert. *Narrative of the Surveying Voyages of His Majesty's Ships Adventure and Beagle, between the Years 1826 and 1836, Describing Their Examination of the Southern Shores of South America and the Beagle's Circumnavigation of the Globe: Proceedings of the Second Expedition 1831-1836.* 3 vols. Vol. 2. London: Henry Colburn, 1839.

Fivaz-Depeursinge, Elisabeth, Nicolas Favez, Chloe Lavanchy, S De Noni, and Keren Frascarolo France. 'Four-Month-Olds Make Triangular Bids to Father and Mother During Trilogue Play with Still-Face.' *Social Development* 14 (2005): pp.361–378.

Fivaz-Depeursinge, Elisabeth and Antoinette Corboz-Warnery. *The Primary Triangle: A Developmental Systems View of Mothers, Fathers and Infants.* New York: Basic Books, 1999.

Flyvberg, Bent. 'Five Misunderstandings of Case-Study Research.' *Qualitative Inquiry* 12 (2006): pp.219–245.

Fodor, Jerry. *Modularity of Mind: An Essay on Faculty Psychology.* Cambridge MA: MIT Press, 1983.

Fothergill, Alastair. 'Ocean World.' In *The Blue Planet*, 50 minutes. London: BBC, 12th September, 2001.

Foucault, Michel. 'The Order of Discourse.' In *Unifying the Text: A Post-Struturalist Reader*, edited by Robert Young, pp.51–78. London: Routledge & Kegan Paul, 1981.

Foucault, Michel. *The Order of Things: An Archaeology of the Human Sciences.* London: Routledge, 2002. First published 1966.

Foucault, Michel. *Power-Knowledge: Selected Interviews and Other Writings, 1972–1977.* New York: Pantheon, 1980.

Freeman, Richard. 'Queries About Expression.' (1977). http://darwin-online.org.uk/ EditorialIntroductions/Freeman_QueriesaboutExpression.html.

Freud, Sigmund. *Civilization and Its Discontents.* London: Hogarth Press, 1930.

Freud, Sigmund. 'Dostoevsky and Parricide.' In *Standard Edition of the Complete Psychological Works of Sigmund Freud, Volume XXI*, edited by James Strachey, pp.173–194. 24 vols. London: Hogarth Press, 1961. First published 1928.

Freud, Sigmund. *Group Psychology and Analysis of the Ego.* London: Hogarth Press, 1949. First published 1922.

Freud, Sigmund. 'Instincts and Their Vicissitudes.' In *The Standard Edition of the Complete Psychological Works of Sigmund Freud, Volume XIV*, edited by James Strachey. 24 vols. London: Hogarth, 1957. First published 1915.

Freud, Sigmund. *The Psychopathology of Everyday Life.* New York: Macmillan, 1914. First published 1905.

Freud, Sigmund. *Totem and Taboo: Resemblances between the Psychic Lives of Savages and Neurotics*. New York: Moffat Yard & Co., 1918. First published 1913.

Freud, Sigmund, and Josef Breuer. *Studies on Hysteria*. Harmondsworth: Penguin Books, 1977. First published 1895.

Fridlund, Alan. 'Darwin's Anti-Darwinism in the *Expression of the Emotions in Man and Animals*.' *International Review of Studies on Emotion* 2 (1992): pp.117–137.

Fridlund, Alan. 'Evolution and Facial Action in Reflex, Social Motive, and Paralanguage.' *Biological Psychology* 32 (1991): pp.3–100.

Fridlund, Alan. *Human Facial Expression: An Evolutionary View*. San Diego: Academic Press, 1994.

Frith, Chris. 'The Social Brain?' *Philosophical Transactions of the Royal Society B* 362 (2007). http://doi.org/10.1098/rstb.2006.2003.

Gard, Michael, and Ben Bradley. 'Getting Away with Rape: Erasure of the Psyche in Evolutionary Psychology.' *Psychology, Evolution & Gender* 2 (2000): pp.313–319.

Gardner, Howard. *The Mind's New Science: A History of the Cognitive Revolution*. New York: Basic Books, 1987.

Garrison, Jim. 'Dewey's Theory of Emotions: The Unity of Thought and Emotion in Naturalistic Functional "Co-Ordination" of Behaviour.' *Transactions of the Charles S. Peirce Society* 39 (2003): pp.405–443.

Garwood, Jeremy. 'Darwin Reloaded.' *Lab Times* 3 (2017): pp.20–23.

Gaull, Marilyn. 'From Wordsworth to Darwin: "On the Fields of Praise".' *Wordsworth Circle* 10 (1979): pp.33–48.

Gaunt, William. *The Aesthetic Adventure*. London: Jonathan Cape, 1945.

Geary, David. 'Evolution of Parental Investment.' In *Handbook of Evolutionary Psychology*, edited by David Buss, pp.524–541. Hoboken NJ: Wiley, 2016.

Geertz, Clifford. *The Interpretation of Cultures: Selected Essays*. New York: Basic Books, 1973.

Gendron, Maria, and Lisa Barrett. 'Facing the Past: A History of the Face in Psychological Research on Emotion Perception.' In *The Science of Facial Expression*, edited by James Russell and José-Miguel Fernández-Dols, pp.15–36. Oxford: Oxford University Press, 2017.

Gergen, Kenneth. 'The Limits of Language as the Limits of Psychological Explanation.' *Theory & Psychology* 28 (2018): pp.697–711.

Gergen, Kenneth. 'On the Very Idea of Social Psychology.' *Social Psychology Quarterly* 71 (2008): pp.331–337.

Gergen, Kenneth. *Realities and Relationships: Soundings in Social Construction*. Cambridge MA: Harvard University Press, 2009.

Gergen, Kenneth. 'The Social Constructionist Movement in Modern Psychology.' *American Psychologist* 40 (1985): pp.266–275.

Gergen, Kenneth. 'Social Psychology as History.' *Journal of Personality and Social Psychology* 26 (1973): pp.309–320.

Gergen, Kenneth. 'Social Psychology and the Wrong Revolution.' *European Journal of Social Psychology* 19 (1989): pp.463–484.

Gergen, Mary. 'Induction and Construction: Teetering between Two Worlds.' *European Journal of Social Psychology* 19 (1989): pp.431–437.

Ghiselin, Michael. 'Darwin and Evolutionary Psychology.' *Science* 179 (1973): pp.964–968.

Ghiselin, Michael. *The Triumph of the Darwinian Method*. Berkeley CA: University of California Press, 1969.

Gibson, Susannah. 'On Being an Animal, or, the Eighteenth Century Zoophyte Controversy in Britain.' *History of Science* 50 (2012): pp.453–476.

Giddens, Anthony. *The Constitution of Society: Outline of the Theory of Structuration.* Cambridge: Polity Press, 1984.

Gil, Diego, and Manfred Gahr. 'The Honesty of Bird Song: Multiple Constraints for Multiple Traits.' *Trends in Ecology and Evolution* 17 (2002): pp.133–141.

Gilbert, Scott, and David Epel. *Ecological Developmental Biology: The Environmental Regulation of Development, Health, and Evolution.* 2nd ed. Sunderland, MA: Sinauer Ass., 2015. First published 2008.

Gilbert, Scott, Bosch, Thomas, and Ledón-Rettig, Cristina. 'Eco-Evo-Devo: Developmental Symbiosis and Developmental Plasticity as Evolutionary Agents.' *Nature Reviews: Genetics* 16 (2015): pp.611–622.

Gilbert, Scott, Jan Sapp, and Alfred Tauber. 'A Symbiotic View of Life: We Have Never Been Individuals.' *Quarterly Review of Biology* 87 (2012): pp.325–341.

Gleizes, Albert, and Jean Metzinger. 'Cubism 1912.' In *Modern Artists on Art: Ten Unabridged Essays*, edited by Robert Herbert, pp.1–18. Englewood-Cliffs, NJ: Prentice-Hall, 1964. First published 1912.

Godfrey-Smith, Peter. 'Darwinian Individuals.' In *From Groups to Individuals: Evolution and Emerging Individuality*, edited by Frédéric Bouchard, and Philippe Huneman, pp.17–36. Cambridge MA: MIT Press, 2013.

Goffman, Erving. 'Embarrassment and Social Organization.' *American Journal of Sociology* (1956): pp.264–271.

Goldney, Jodie. 'The Othering of a Profession.' *Psychreg Journal of Psychology* 2 (2018): pp.56–67.

Gontier, Nathalie. 'Depicting the Tree of Life: The Philosophical and Historical Roots of Evolutionary Tree Diagrams.' *Evolution* 5 (2011): pp.515–538.

Gould, John. *Birds Part 3 of the Zoology of the Voyage of H.M.S. Beagle.* London: Smith Elder and Co., 2006. First published 1841. http://darwin-online.org.uk/content/frameset?page seq=1&itemID=F9.3&viewtype=text.

Gould, John. *Handbook to the Birds of Australia.* 2 vols. Vol. 1. London: Taylor & Francis, 1865.

Gould, Stephen Jay. 'Challenges to Neo-Darwinism and Their Meaning for a Revised View of Human Consciousness.' *Tanner Lectures on Human Values* (1984). http://tannerlectures. utah.edu/_documents/a-to-z/g/gould85.pdf.

Gould, Stephen Jay. *Ontogeny and Phylogeny.* Cambridge, MA: Harvard University Press, 1977.

Graham, Kirsty, Takeshi Furuichi, and Richard Byrne. 'The Gestural Repertoire of the Wild Bonobo (*Pan paniscus*): A Mutually Understood Communication System.' *Animal Cognition* 20 (2017): pp.171–177.

Gray, Asa. 'Letter to Darwin, 2nd–3rd July.' In *Correspondence of Charles Darwin* Cambridge: Darwin Correspondence Project, 1862. www.darwinproject.ac.uk/letter/ ?docId=letters/DCP-LETT-3637.xml.

Gray, Jane. 'Letter to Susan Loring, 28th October to 2nd November.' In *Correspondence of Charles Darwin* Cambridge: Darwin Correspondence Project, 1868 https://www. darwinproject.ac.uk/people/about-darwin/family-life/visiting-darwins.

Greene, John. 'Darwin as a Social Evolutionist.' *Journal of the History of Biology* 10 (1977): pp.1–27.

Greifeneder, Rainer, Herbert Bless, and Klaus Fiedler. *Social Cognition: How Individuals Construct Social Reality.* 2nd ed. London: Routledge, 2018. First published 2003.

Griffiths, Paul. *What Emotions Really Are: The Problem of Psychological Categories.* Chicago: University of Chicago Press, 1997.

Griffiths, Paul, and Andrea Scarantino. 'Emotions in the Wild: The Situated Perspective on Emotion.' In *The Cambridge Handbook of Situated Cognition*, edited by Murat Aydede, and Philip Robbins, pp.437–453. Cambridge: Cambridge University Press, 2009.

Griffiths, Tom. *The Art of Time Travel: Historians and Their Craft*. Carlton VIC: Black Inc., 2016.

Grosz, Elizabeth. *Becoming Undone: Darwinian Reflections on Life, Politics and Art*. Durham N.C.: Duke University Press, 2011.

Gruber, Howard, and Paul Barrett. *Darwin on Man: A Psychological Study of Scientific Creativity, Together with Darwin's Early and Unpublished Notebooks*. London: Wildwood, 1974.

Habermas, Jurgen. 'Towards a Theory of Communicative Competence.' In *Recent Sociology No.2*, edited by Hans Dreitzel, pp.115–148. London: Macmillan, 1970.

Hacking, Ian. 'Double Consciousness in Britain, 1815–1875.' *Dissociation* 4 (1991): pp.134–146.

Haeckel, Ernst. *Natürliche Schöpfungsgeschichte*. Berlin: Georg Reimer, 1868.

Haraway, Donna. *The Companion Species Manifesto: Dogs, People, and Significant Otherness*. Chicago: Prickly Paradigm, 2003.

Haraway, Donna. 'Situated Knowledges: The Science Question in Feminism and the Privilege of Partial Perspective.' *Feminist Studies* 14 (1988): pp.575–599.

Hare, Brian, and Vanessa Woods. *The Genius of Dogs*. London: Oneworld, 2013.

Harré, Rom, and Paul Secord. *The Explanation of Social Behaviour*. Oxford: Basil Blackwell, 1972.

Hart, Sybil, Tiffany Field, and Claudia De Valle. 'Infants Protest Their Mothers' Attending to an Infant-Size Doll.' *Social Development* 7 (1998): pp.54–61.

Hartley, David. *Observations on Man: His Frame, His Duty, and His Expectations*. Gainsville FL: Scholars' Facsimiles and Reprints, 1966. First published 1749.

Hartley, Lucy. *Physiognomy and the Meaning of Expression in Nineteenth-Century Culture*. Cambridge: Cambridge University Press, 2001.

Haught, John. *God after Darwin: A Theology of Evolution*. Boulder: Westview, 2000.

Hay, Dale, Marlene Caplan, and Alison Nash. 'The Beginnings of Peer Relations.' In *Handbook of Peer Interactions, Relationships, and Groups*, edited by Kenneth Rubin, William Bukowski, and Brett Laursen, pp.121–142. New York: Guilford Press, 2011.

Hebb, Donald. 'Emotion in Man and Animal: An Analysis of the Intuitive Processes of Recognition.' *Psychological Review* 53 (1946): pp.88–106.

Henriques, Julian, Wendy Hollway, Cathy Urwin, Couze Venn, and Valerie Walkerdine. *Changing the Subject: Psychology, Social Regulation and Subjectivity*. 2nd ed. London: Routledge, 1998. First published 1984.

Herbert, Sandra. *Charles Darwin, Geologist*. Ithaca, NY: Cornell University Press, 2005.

Herbert, Sandra. 'The Place of Man in the Development of Darwin's Theory of Transmutation. Part 2.' *Journal of the History of Biology* 10, no. 2 (1977): pp.155–227.

Herbert, Sandra, and Paul Barrett. 'Introduction: Old & Useless Notes.' In *Charles Darwin's Notebooks, 1836–1844*, edited by Paul Barrett et al. Ithaca NY: Cornell University Press, 1987.

Herschel, John. *A Preliminary Discourse on the Study of Natural Philosophy*. London: Longman & Co, 1830.

Hinde, Robert. 'Ethological Models and the Concept of "Drive".' *British Journal for the Philosophy of Science* 6 (1956): pp.321–331.

Hodge, Jonathan. 'Chance and Chances in Darwin's Early Theorizing and in Darwinian Theory Today.' In *Chance in Evolution*, edited by Grant Ramsey, and Charles Pence, pp.41–75. Chicago: Chicago University Press, 2016.

Hodge, Jonathon. 'Darwin as a Lifelong Generation Theorist.' In *The Darwinian Heritage*, edited by David Kohn, pp.207–244. Princeton, NJ: Princeton University Press, 1985.

Hodge, Jonathon. 'Darwin's Book: *On the Origin of Species*.' *Science & Education* 22 (2013): pp.2267–2294.

Hodge, Jonathon. 'The Notebook Programmes and Projects of Darwin's London Years.' In *The Cambridge Companion to Darwin*, edited by Jonathon Hodge and Gregory Radick, pp.40–68. Cambridge: Cambridge University Press, 2003.

Hodge, Jonathon. 'The Structure and Strategy of Darwin's "Long Argument".' *British Journal for the History of Science* 10 (1977): pp.237–246.

Hofstadter, Richard. *Social Darwinism in American Thought*. Philadelphia PA: University of Pennsylvania Press, 1944.

Hoggart, Richard. *The Uses of Literacy*. London: Chatto & Windus, 1957.

Hölldobler, Bert, and Edward Wilson. *The Ants*. Berlin: Springer-Verlag, 1990.

Hollinger, David. 'James, Clifford and Scientific Conscience.' In *The Cambridge Companion to William James*, edited by Ruth Putnam, pp.69–83. Cambridge: Cambridge University Press, 1997.

Hooke, Robert. *Micrographia: Or Some Physiological Descriptions of Minute Bodies Made by Magnifying Glasses. With Observations and Inquiries Thereupon*. London: Royal Society, 1664.

Hooker, Joseph. 'Letter to Darwin, 24th January.' In *Correspondence of Charles Darwin* Cambridge: Darwin Correspondence Project, 1864. https://www.darwinproject.ac.uk/letter/?docId=letters/DCP-LETT-4396.xml.

Hooker, Joseph. 'Letter to Darwin, 14th December.' In *Correspondence of Charles Darwin* Cambridge: Darwin Correspondence Project, 1866. https://www.darwinproject.ac.uk/letter/?docId=letters/DCP-LETT-5305.

Horowitz, Alexandra. 'Disambiguating the "Guilty Look": Salient Prompts to a Familiar Dog Behaviour.' *Behavioural Processes* 81 (2009): pp.447–452.

Howitt, Dennis, and Duncan Cramer. *Introduction to Research Methods in Psychology*. 3rd ed. Harlow UK: Pearson, 2011.

Hrdy, Sarah. 'Empathy, Polyandry, and the Myth of the Coy Female.' In *Feminist Approaches to Science*, edited by Ruth Bleier, pp.119–146. New York: Pergamon Press, 1986.

Hrdy, Sarah. 'Raising Darwin's Consciousness: Female Sexuality and the Prehominid Origins of Patriarchy'. *Human Nature* 8 (1997): pp.1–49.

Hull, David. 'Deconstructing Darwin: Evolutionary Theory in Context.' *Journal of the History of Biology* 38 (2005): pp.137–152.

Hume, David. *An Enquiry Concerning Human Understanding*. Chicago: Open Court, 1921. First published 1748.

Hunter, John. *Essays and Observations on Natural History, Anatomy, Physiology, Psychology, and Geology, Vol. I* London: Van Voorst, 1861.

Hurd, Peter, Wachtmeister, Carl-Adam, and Enquist, Magnus. 'Darwin's Principle of Antithesis Revisited: A Role for Perceptual Biases in the Evolution of Intraspecific Signals.' *Proceedings of the Royal Society B* 259 (1995): pp.201–205.

Huxley, Julian. *Evolution in Action*. New York: Harper & Row, 1953.

Huxley, Julian. *Evolution: The Modern Synthesis*. London: George Allen & Unwin, 1942.

Huxley, Julian. 'Introduction to the Second Edition.' In *Evolution: The Modern Synthesis*, pp. xiii–li. London: George Allen & Unwin, 1963.

Huxley, Julian. 'The Present Standing of the Theory of Sexual Selection.' In *Evolution: Essays on Aspects of Evolutionary Biology*, edited by Gavin De Beer, pp.11–42. Oxford: Clarendon Press, 1938.

Huxley, Julian. *Religion without Revelation*. London: Parrish, 1957. First published 1927.

Huxley, Thomas. 'On the Hypothesis That Animals Are Automata, and Its History.' *Fortnightly Review* 16 (1874): pp.555–580.

Huxley, Thomas. 'On Natural History, as Knowledge, Discipline, and Power.' *Proceedings of the Royal Institution* 6 (1856): 305–314. https://mathcs.clarku.edu/huxley/SM1/NatHis.html.

Ingleby, David. 'The Job Psychologists Do.' In *Reconstructing Social Psychology*, edited by Nigel Armistead, pp.314–327. Harmondsworth: Penguin Books, 1974.

Ingold, Tim. 'Bindings against Boundaries: Entanglements of Life in an Open World.' *Environment and Planning A* 40 (2008): pp.1796–1810.

Ives, Kelly. *Cixous, Irigaray, Kristeva: The Jouissance of French Feminism*. Kidderminster: Crescent Moon, 1996.

Jabr, Ferris. 'How Beauty Is Making Scientists Rethink Evolution.' *New York Times Magazine*, January 9th 2019.

Jack, Rachael, Roberto Caldara, and Philippe Schyns. 'Internal Representations Reveal Cultural Diversity in Expectations of Facial Expressions of Emotion.' *Journal of Experimental Psychology (General)* 141 (2012): pp.19–25.

Jack, Rachael, Oliver Garrod, Hui Tu, Roberto Caldara, and Philippe Schyns. 'Facial Expressions of Emotion Are Not Culturally Universal.' *Proceedings of the National Academy of Sciences of the United States of America* 109 (2012): pp.7241–7244.

James, William. 'Are We Automata?' In *Essays in Psychology*, pp.38–61. Cambridge MA: Harvard University Press, 1984. First published 1879.

James, William. 'Brute and Human Intellect.' In *Essays in Psychology*, pp.1–37. Cambridge MA: Harvard University Press, 1984. First published 1878.

James, William. 'Does Consciousness Exist?' In *Essays in Radical Empiricism*, pp.1–38. New York: Longmans, Green, & Co., 1912. First published 1904.

James, William. *The Principles of Psychology*. New York: Holt, 1890.

James, William. 'What Is an Emotion?' *Mind (New Series)* 9 (1884): pp.188–205.

Janicke, Tim, Ines Häderer, Marc Lajeunesse, and Nils Anthes. 'Darwinian Sex Roles Confirmed across the Animal Kingdom.' *Science Advances* 2 (2016): p.e1500983.

Jardine, Boris. 'Between the Beagle and the Barnacle: Darwin's Microscopy, 1837–1854.' *Studies in History and Philosophy of Science* 40 (2009): pp.382–395.

Jay, Martin. 'Historical Explanation and the Event: Reflections on the Limits of Contextualization.' *New Literary History* 42 (2011): pp.557–571.

Jay, Martin. 'The Textual Approach to Intellectual History.' In *Force Fields: Between Intellectual History and Cultural Critique*, pp.158–166. New York: Routledge, 1993.

Jesus, Paulo. 'The Embodied Nature of Emotions: On Charles Darwin and William James' Legacy.' *Revista da Faculdade de Letras—Série de Filosofia* 27-28 (2010-2011).

Johannsen, Wilhelm. *Elemente Der Exakten Erblichkeitslehre Mit Grundzügen Der Biologischen Variationsstatistik*. Jena: Gustav Fischer, 1909.

Johannsen, Wilhelm. 'The Genotype Conception of Heredity.' *American Naturalist* 45 (1911): pp.129–159.

Jones, Ernest. *Sigmund Freud: Life and Work, Vol.2*. London: Hogarth, 1953.

Jones, Greta. 'The Social History of Darwin's *The Descent of Man*.' *Economy and Society* 7 (1978): pp.1–23.

Joynson, Robert. *Psychology and Common Sense*. London: Routledge and Kegan Paul, 1974.

Juengel, Scott. 'Godwin, Lavater, and the Pleasures of Surface.' *Studies in Romanticism* 35 (1996): pp.73–98.

Kahneman, Daniel, and Shane Frederick. 'Representativeness Revisited: Attribute Substitution in Intuitive Judgement.' In *Heuristics and Biases: The Psychology of Intuitive*

Judgement, edited by Thomas Gilovich, Griffin, Dale and Kahneman, Daniel, pp.49–81. New York: Cambridge University Press, 2002.

Kaplan, Gisela. *Bird Minds: Cognition and Behaviour of Australian Native Birds*. Clayton South VIC: CSIRO, 2015.

Kasschau, Richard. *Psychology: Exploring Behaviour*. Englewood Cliffs NJ: Prentice-Hall., 1980.

Kaye, Kenneth. *The Mental and Social Lives of Babies: How Parents Create Persons*. Chicago: University of Chicago Press, 1982.

Keller, Evelyn. *The Century of the Gene*. Cambridge MA: Harvard University Press, 2000.

Kelley, Laura, and John Endler. 'Male Great Bowerbirds Create Forced Perspective Illusions with Consistently Different Individual Quality.' *Proceedings of the National Academy of Sciences of the United States of America* 109 (2012): pp.20980–20985.

Kelly, Kevin. *Out of Control: The New Biology of Machines, Social Systems and the Economic World*. Boston: Addison-Wesley, 1994.

Koch, Sigmund. 'Psychology's Bridgman vs Bridgman's Bridgman.' *Theory & Psychology* 2 (1992): pp.261–290.

Kohlberg, Lawrence. 'The Claim to Moral Adequacy of a Highest Stage of Moral Judgment.' *Journal of Philosophy* 70 (1973): pp.630–646.

Kohn, David. 'Darwin's Ambiguity: The Secularization of Biological Meaning.' *British Journal for the History of Science* 22 (1989): pp.215–239.

Kokko, Hanna, and Michael Jennions. 'Parental Investment, Sexual Selection and Sex Ratios.' *Journal of Evolutionary Biology* 21 (2008): pp.919–948.

Kolenbrander, Paul, Roxanna Andersen, David Blehert, Paul Egland, Jamie Foster, and Robert Jr. Palmer. 'Communication among Oral Bacteria.' *Microbiology and Molecular Biology Reviews* 66 (2002): pp.486–505.

Kottler, Malcolm. 'Charles Darwin and Alfred Russell Wallace: Two Decades of Debate over Natural Selection.' In *The Darwinian Heritage*, edited by David Kohn, pp.367–432. Princeton NJ: Princeton University Press, 1985.

Krueger, Joel. 'Dewey's Rejection of the Emotion/Expression Distinction.' In *Neuroscience, Neurophilosophy and Pragmatism: Understanding Brains at Work in the World*, edited by Tibor Solymosi and John Shook, pp.140–161. London: Palgrave Macmillan, 2014.

Krueger, Joel. 'Varieties of Extended Emotions.' *Phenomenology and the Cognitive Sciences* 13 (2014): pp.533–555.

Kuo, Zing-Yang. 'A Psychology without Heredity.' *Psychological Review* 31 (1924): pp.427–448.

Kutschera, Ulrich, and Karl Niklas. 'Evolutionary Plant Physiology: Charles Darwin's Forgotten Synthesis.' *Naturwissenschaften* 96 (2009): pp.1339–1354.

Lacan, Jacques. 'Aggressivity in Psychoanalysis.' In *Écrits: A Selection*, pp.8–29. London: Tavistock Publications, 1977. First published 1948.

Lacan, Jacques. 'The Mirror-Stage as Formative of the Function of the I as Revealed in Psychoanalytic Experience.' In *Ecrits: A Selection*, edited by Alan Sheridan, pp.1–7. London: Tavistock, 1977. First published 1949.

Lacan, Jacques. *Speech and Language in Psychoanalysis*. Edited by Anthony Wilden. Baltimore: Johns Hopkins University Press, 1981. First published 1953.

Laland, Kevin. *Darwin's Unfinished Symphony: How Culture Made the Human Mind*. Princeton: Princeton University Press, 2017.

Laland, Kevin, John Odling-Smee, and Gilbert Scott. 'Evodevo and Niche Construction: Building Bridges.' *Journal of Experimental Zoology (Mol Dev Evol)* 310B (2008): pp.549–566.

Landecker, Hannah. 'Metabolism, Autonomy, and Individuality.' In *Biological Individuality: Integrating Scientific, Philosophical, and Historical Perspectives*, edited by Scott Lidgard and Lynn Nyhart, pp.225–248. Chicago: Chicago University Press, 2017.

Landow, George. 'Laus Veneris by Sir Edward Coley Burne-Jones.' In *The Victorian Web*, edited by George Landow. Providence RI, 2006, http://www.victorianweb.org/painting/bj/paintings/21.html.

Lang, Cecil. *The Pre-Raphaelites and Their Circle*. 2nd ed. Chicago: University of Chicago Press, 1975. First published 1968.

Laplanche, Jean, and Jean-Bertrand Pontalis. *The Language of Psycho-Analysis*. London: Hogarth Press, 1983. First published 1973.

Larsen, Anne. 'Equipment for the Field.' In *Cultures of Natural History*, edited by Nick Jardine, James Secord, and Emma Spary, pp.358–377. Cambridge: Cambridge University Press, 1996.

Larson, Allan. 'Review of John Alcock's *"The Triumph of Sociobiology"*.' *Isis* 93 (2002): pp.348–349.

Larson, James. 'An Alternative Science: Linnaean Natural History in Germany, 1770–1790.' *Janus* 66 (1979): pp.267–283.

Latour, Bruno. *We Have Never Been Modern*. Cambridge MA: Harvard University Press, 1993.

Latour, Bruno. 'What Is Given in Experience? A Review of Isabelle Stengers, *Penser Avec Whitehead*.' *Boundary* 2 (2005): pp.222–237.

Latour, Bruno. *What Is the Style of Matters of Concern?* Assen: Van Gorcum, 2008.

Lavater, John. *Essays on Physiognomy, Designed to Promote the Knowledge and the Love of Mankind, Vol.I*. London: Murray, Hunter & Holloway, 1789.

Lavine, Thelma. 'Reflections on the Genetic Fallacy.' *Social Research* 29, no. 3 (1962): pp.321–336.

Lawlor, Len. 'The End of Phenomenology: Expressionism in Deleuze and Merleau-Ponty.' *Continental Philosophy Review* 31 (1998): pp.15–34.

Lawrence, Christopher. 'The Nervous System and Society in the Scottish Enlightenment.' In *Natural Order: Historical Studies of Scientific Culture*, edited by Barry Barnes and Steven Shapin, pp.19–40. London: Sage, 1979.

Leach, Marjorie, and Paula Hellal. 'Darwin's "Natural History of Babies".' *Journal of the History of the Neurosciences* 19 (2010): pp.140–157.

Leary, Mark, and Kaitlin Toner. 'Theories of the Blush.' In *The Psychological Significance of the Blush*, edited by W. Ray Crozier and Peter De Jong, pp.63–76. Cambridge: Cambridge University Press, 2012.

Lecoq, Jacques. *The Moving Body: Teaching Creative Theatre*. London: Methuen, 2002.

Lee, David. 'Guiding Movement by Coupling Taus.' *Ecological Psychology* 10 (1998). http://www.tandfonline.com/doi/abs/10.1080/10407413.1998.9652683.

Lenoir, Timothy. 'The Darwin Industry.' *Journal of the History of Biology* 20 (1987): pp.115–130.

Lewens, Tim. *Cultural Evolution: Conceptual Challenges*. Oxford: Oxford University Press, 2015.

Lewens, Tim. *Darwin*. London: Routledge, 2007.

Lewens, Tim. 'Human Nature: The Very Idea.' *Philosophy & Technology* 25 (2012): pp.459–474.

Lewes, George. 'Mr Darwin's Hypotheses, Part I.' *Fortnightly Review* 3 (1868): pp.353–373.

Lewes, George. 'Mr. Darwin's Hypotheses, Part III.' *Fortnightly Review (New Series)* 4 (1868): pp.61–80.

Lewis, Michael, and Jeanne Brooks-Gunn. 'The Development of Early Visual Self-Recognition.' *Developmental Review* 4 (1984): pp.215–239.

Lewontin, Richard. 'Gene, Organism and Environment.' In *Evolution from Molecules to Men*, edited by Derek Bendall, pp.273–285. Cambridge: Cambridge University Press, 1983.

Leys, Ruth. 'The Turn to Affect: A Critique.' *Critical Inquiry* 37 (2011): pp.434–472.

Liddle, Mitzi-Jane, Ben Bradley, and Andrew McGrath. 'Baby Empathy and Peer Prosocial Responses.' *Infant Mental Health Journal* 36 (2015): pp.446–458.

Lidgard, Scott, and Lynn Nyhart, eds. *Biological Individuality: Integrating Scientific, Philosophical, and Historical Perspectives*. Chicago: Chicago University Press, 2017.

Lieber, Francis. 'A Paper on the Vocal Sounds of Laura Bridgeman the Blind Deaf-Mute at Boston; Compared with the Elements of Phonetic Language.' *Contributions to Knowledge* 2 (1851). https://archive.org/details/101299133.nlm.nih.gov/mode/2up.

Lindsay, W. Lauder. 'Madness in Animals.' *Journal of Mental Science* 17 (1871): pp.181–206.

Loukola, Olli, Clint Perry, Louie Coscos, and Lars Chittka. 'Bumblebees Show Cognitive Flexibility by Improving on an Observed Complex Behaviour.' *Science* 355 (2017): pp.833–836.

Love, Rosaleen. 'Darwinism and Feminism: The Woman Question in the Life and Work of Olive Schreiner and Charlotte Perkins Gilman.' In *The Wider Domain of Evolutionary Thought*, edited by David Oldroyd and Ian Langham, pp.113–132. Dordrecht: Reidl, 1983.

Loyau, Adeline, Marion Petrie, Michel Saint Jalme, and Gabriele Sorci. 'Do Peahens Not Prefer Peacocks with More Elaborate Trains?' *Animal Behaviour* 76 (2008): pp.e5–e9.

Lukes, Steven. *Individualism*. Colchester: ECPR, 2006. First published 1973.

Lundgren, David. 'Social Feedback and Self-Appraisals: Current Status of the Mead-Cooley Hypothesis.' *Symbolic Interaction* 27 (2004): pp.267–286.

Lyell, Charles. *Principles of Geology, or the Modern Changes of the Earth and Its Inhabitants Considered as Illustrative of Geology*. 9th ed. New York: Appleton, 1854. First published 1830.

Lyell, Charles. *Principles of Geology, Being an Attempt to Explain the Former Changes of the Earth's Surface, by Reference to Causes Now in Operation, Vol.2*. London: Murray, 1835. First published 1832.

Mackintosh, James. *Dissertation on the Progress of Ethical Philosophy, Chiefly During the Seventeenth and Eighteenth Centuries*. Edinburgh: Adam and Charles Black, 1862. First published 1830.

Madigan, Robert, Susan Johnson, and Patricia Linton. 'The Language of Psychology: APA Style as Epistemology.' *American Psychologist* 50 (1995): pp.428–436.

Mallett, James. 'Hybrid Speciation.' *Nature* 446 (2007): pp.279–283.

Malloch, Stephen, and Colwyn Trevarthen, eds. *Handbook of Communicative Musicality*. Oxford: Blackwell-Wiley, 2009.

Malthus, Thomas. *An Essay on the Principle of Population, or a View of Its Past and Present Effects on Human Happiness; with an Inquiry into Our Prospects Respecting the Future Removal or Mitigation of the Evils Which It Occasions*. 6th ed. London: Murray, 1826.

Mandelbaum, Maurice. 'A Note on "Anthropomorphism" in Psychology.' *Journal of Philosophy* 40 (1943): pp.246–248.

Mandeville, Bernard. *The Fable of the Bees; or, Private Vices, Public Benefits*. Vol. 1, Indianapolis: Online Liberty Fund, 2011. First published 1739. http://files.libertyfund.org/files/846/Mandeville_0014-01_EBk_v6.0.pdf.

Manier, Edward. *The Young Darwin and His Cultural Circle. A Study of Influences Which Helped Shape the Language and Logic of the First Drafts of the Theory of Natural Selection*. Dordrecht: Reidel, 1978.

Marcus Aurelius. *The Thoughts of the Emperor Marcus Aurelius Antonius*. London: Bell & Sons, 1880.

Maris, Kathryn. 'The Death of Empiricism.' *Poetry Review* 108 (2018): p.34.

Martin, Tovah. *Once Upon a Windowsill: A History of Indoor Plants*. Portland OR: Timber Press, 1988.

Maxwell, Scott, Michael Lau, and George Howard. 'Is Psychology Suffering from a Replication Crisis?: What Does "Failure to Replicate" Really Mean?' *American Psychologist* 70 (2015): pp.487–498.

Maynard-Smith, John. 'The Concept of Information in Biology.' *Philosophy of Science* 67 (2000): pp.177–194.

Maynard-Smith, John. 'Group Selection and Kin Selection.' *Nature* 201 (1964): pp.1145–1147.

Maynard-Smith, John. 'Theories of Sexual Selection.' *Trends in Ecology and Evolution* (1991): pp.146–151.

Mayr, Ernst. 'Darwin's Influence on Modern Thought.' *Scientific American* 283 (2000): pp.78–83.

McCalman, Iain. *Darwin's Armada: Four Voyages and the Battle for the Theory of Evolution*. New York: Norton, 2009.

McDonald, Roger. *Mr Darwin's Shooter*. Sydney: Random House, 1988.

McMahon, Blair, and Roger Evans. 'Foraging Strategies of American White Pelicans.' *Behaviour* 120 (1992): pp.69–89.

McNeill, David. *The Acquisition of Language: The Study of Developmental Psycholinguistics*. New York: Harper & Row, 1970.

Mead, George. *Mind, Self, and Society: From the Standpoint of a Social Behaviorist*. Chicago: University of Chicago Press, 1934.

Meehl, Paul. 'Why Summaries of Research on Psychological Theories Are Often Uninterpretable.' *Psychological Reports* 66 (1990): pp.195–244.

Menzies-Lyth, Isabel. 'The Functioning of Social Systems as a Defence against Anxiety.' *Human Relations* 13 (1959): pp.95–121.

Merleau-Ponty, Maurice *La Nature. Notes—Cours Au Collège De France*. Paris: Seuil, 1995.

Mesoudi, Alex. *Cultural Evolution: How Darwinian Theory Can Explain Human Culture and Synthesize the Social Sciences*. Chicago: University of Chicago Press, 2011.

Micheneau, Claire, Johnson, Steven, and Fay, Michael. 'Orchid Pollination: From Darwin to the Present Day.' *Botanical Journal of the Linnean Society* 161 (2009): pp.1–19.

Mill, John. *On Liberty*. London: Longman & Green, 1867. First published 1859.

Miller, Merle. *Plain Speaking: An Oral Biography of Harry S Truman*. London: Gollancz, 1974.

Montgomery, Georgina. 'Darwin and Gender.' In *The Cambridge Encyclopaedia of Darwin and Evolutionary Thought*, edited by Michael Ruse, pp.443–450. Cambridge: Cambridge University Press, 2013.

Montgomery, William. 'Charles Darwin's Thought on Expressive Mechanisms in Evolution.' In *The Development of Expressive Behaviour: Biology-Environment Interactions*, edited by Gail Zivin, pp.27–50. London: Academic Press, 1985.

Moore, James, and Adrian Desmond. 'Introduction.' In *Charles Darwin: The Descent of Man and Selection in Relation to Sex*, edited by James Moore and Adrian Desmond, pp.xi–lxvi. Harmondsworth: Penguin Books, 2004.

Morawski, Jill. 'Organizing Knowledge and Behaviour at Yale's Institute of Human Relations.' *Isis* 77 (1986): pp.219–242.

Morgan, Lloyd. 'Mind in Man and Brute.' *Nature* 39 (1889): pp.313–315.

Morss, John. *The Biologising of Childhood: Developmental Psychology and the Darwinian Myth.* Hove UK: Lawrence Erlbaum, 1990.

Moscovici, Serge. 'Notes toward a Description of Social Representations.' *European Journal of Social Psychology* 18 (1988): pp.211–250.

Moscovici, Serge. 'Preconditions for Explanation in Social Psychology.' *European Journal of Social Psychology* 19 (1989): pp.407–430.

Motzkau, Johanna. 'Exploring the Transdisciplinary Trajectory of Suggestibility.' *Subjectivity* 27 (2009): pp.172–194.

Muller, Gerd. 'Evo–Devo: Extending the Evolutionary Synthesis.' *Nature Reviews: Genetics* 8 (2007): pp.943–949.

Muller, Johannes. *Elements of Physiology.* London: Taylor and Walton, 1842.

Munafo, Marcus, and John Sutton. 'There's This Conspiracy of Silence around How Science Really Works.' *Psychologist* 30 (2017): pp.46–49.

Murray, John. 'Letter to Darwin, 1st July.' In *Correspondence of Charles Darwin* Cambridge: Darwin Correspondence Project, 1870. https://www.darwinproject.ac.uk/letter/?docId=letters/DCP-LETT-7259.xml.

Murray, John. 'Letter to Darwin, 10th October.' In *Correspondence of Charles Darwin* Cambridge: Darwin Correspondence Project, 1870. https://www.darwinproject.ac.uk/letter/?docId=letters/DCP-LETT-7339.xml.

Murray, John. 'Letter to Darwin, 28th September.' In *Correspondence of Charles Darwin* Cambridge: Darwin Correspondence Project, 1870. https://www.darwinproject.ac.uk/letter/?docId=letters/DCP-LETT-7329.xml.

Muthukrishna, Michael, Michael Doebeli, Maciej Chudek, and Joseph Henrich. 'The Cultural Brain Hypothesis: How Culture Drives Brain Expansion, Sociality, and Life History.' *PLoS Computational Biology* 14 (2018): 1–37. https://doi.org/10.1371/journal.pcbi.1006504.

Nadel, Jacqueline, and Hélène Tremblay-Leveau. 'Early Perception of Social Contingencies and Interpersonal Intentionality: Dyadic and Triadic Paradigms.' In *Early Social Cognition: Understanding Others in the First Months of Life*, edited by Phillipe Rochat, pp.189–212. Mahwah NJ: Lawrence Erlbaum, 1999.

Newson, John. 'The Growth of Shared Understandings between Infant and Caregiver.' In *Before Speech: The Beginnings of Human Communication*, edited by Margaret Bullowa, pp.207–222. Cambridge: Cambridge University Press, 1979.

Newton, Isaac. *Newton's Principia: The Mathematical Principles of Newton's Philosophy.* New York: Adee, 1846. First published 1687.

Nicholson, Daniel. 'Organisms ≠ Machines.' *Studies in History and Philosophy of Biological and Biomedical Sciences* 44 (2013): pp.669–678.

Nicholson, Daniel. 'The Return of the Organism as a Fundamental Explanatory Concept in Biology.' *Philosophy Compass* 9 (2014): pp.347–359.

Nicholson, Daniel, and John Dupré, eds. *Everything Flows: Towards a Processual Philosophy of Biology.* Oxford: Oxford University Press, 2018.

Nietzsche, Friedrich. *Thus Spoke Zarathustra.* Harmondsworth: Penguin Books, 1975. First published 1891.

Nowak, Martin. 'Evolving Cooperation.' *Journal of Theoretical Biology* 299 (2012): pp.1–8.

Nowak, Martin. 'Why We Help.' *Scientific American* 307 (2012): pp.34–39.

Nowak, Martin, Corina Tarnita, and Edward Wilson. 'The Evolution of Eusociality.' *Nature* 466 (2010): pp.1057–1062.

Nyhart, Lynn. 'Embryology and Morphology.' In *The Cambridge Companion to the Origin of Species*, edited by Michael Ruse and Robert Richards, pp.194–215. Cambridge: Cambridge University Press, 2009.

Nyhart, Lynn. 'Natural History and the "New" Biology.' In *Cultures of Natural History*, edited by Nick Jardine, James Secord and Emma Spary, pp.426–443. Cambridge: Cambridge University Press, 1996.

Nyhart, Lynn, and Scott Lidgard. 'Individuals at the Center of Biology: Rudolf Leuckart's "Polymorphismus Der Individuen" and the Ongoing Narrative of Parts and Wholes. With an Annotated Translation.' *Journal of the History of Biology* 44 (2011): pp.373–443.

O'Shea-Wheller, Thomas, Sendova-Franks, Ana, and Franks, Nigel. 'Differentiated Anti-Predation Responses in a Superorganism.' *PLoS ONE* 10 (2015): 1–10. e0141012. doi:10.1371/journal.pone.0141012.

Odling-Smee, John, Kevin Laland, and Marcus Feldman. *Niche Construction: The Neglected Process in Evolution*. Princeton NJ: Princeton University Press, 2003.

Ospovat, Dov. *The Development of Darwin's Theory: Natural History, Natural Theology, and Natural Selection, 1838-1859*. Cambridge: Cambridge University Press, 1981.

Owen, Wilfred. 'Insensibility.' In *The Collected Poems of Wilfred Owen*, edited by Cecil Day Lewis, pp.37–38. London: Chatto & Windus, 1963. Composed 1918.

Oyama, Susan. *Evolution's Eye: A Systems View of the Biology-Culture Divide*. Durham NC: Duke University Press, 2000.

Oyama, Susan, Griffiths, Paul, and Gray, Russell. *Cycles of Contingency: Developmental Systems and Evolution*. Cambridge, MA: MIT Press, 2001.

Paget, James. 'Letter to Darwin, 9th July.' In *Correspondence of Charles Darwin* Cambridge: Darwin Correspondence Project, 1867. https://www.darwinproject.ac.uk/letter/?docId=letters/DCP-LETT-5582.xml.

Paine, Thomas. *The Rights of Man: Being an Answer to Mr Burke's Attack on the French Revolution*. London: Sherwin, 1817.

Pandora, Katherine. *Rebels within the Ranks: Psychologists' Critique of Scientific Authority and Democratic Realities in New Deal America*. Cambridge: Cambridge University Press, 1997.

Panksepp, Jaak. 'Affective Consciousness: Core Emotional Feelings in Animals and Humans.' *Consciousness and Cognition* 14 (2005): pp.30–80.

Panksepp, Jaak, and Jules Panksepp. 'The Seven Sins of Evolutionary Psychology.' *Evolution and Cognition* 6 (2000): pp.108–131.

Papathedorou, Theodora. 'Being, Belonging and Becoming: Some Worldviews of Early Childhood in Contemporary Curricula.' *Forum on Public Policy Online* 2 (2010): pp.1–18.

Park, Katherine. 'Observation in the Margins, 500-1500.' In *Histories of Scientific Observation*, edited by Lorraine Daston and Elizabeth Lunbeck, pp.15–44. Chicago: Chicago University Press, 2011.

Parker, Alan, and Gerald Scarfe. 'Pink Floyd—The Wall.' Beverly Hills CA: Metro-Goldwyn-Mayer, 1982.

Parker, Geoff, and Tommaso Pizzari. 'Sexual Selection: The Logical Imperative.' In *Current Perspectives on Sexual Selection: What's Left after Darwin?*, edited by Thierry Hoquet, pp.119–163. Dordrecht: Springer, 2015.

Parker, Ian. *The Crisis in Modern Psychology and How to End It*. London: Taylor & Francis, 2013. First published 1989.

Patmore, Coventry. *The Angel in the House*. Boston: Ticknor and Fields, 1856.

Patricelli, Gail, Eileen Hebets, and Tamara Mendelson. 'Book Review of R.O. Prum, "The Evolution of Beauty".' *Evolution* 73 (2018): pp.115–124.

Paul, Diane. 'Darwinism, Social Darwinism and Eugenics.' In *The Cambridge Companion to Darwin*, edited by Jonathan Hodge and Gregory Radick, pp.214–239. Cambridge: Cambridge University Press, 2003.

Pearce, Trevor. '"A Great Complication of Circumstances"–Darwin and the Economy of Nature.' *Journal of the History of Biology* 43 (2010): pp.493–528.

Perry, George, Nathaniel Dominy, Katrina Claw, et al. 'Diet and the Evolution of Human Amylase Gene Copy Number Variation.' *Nature Genetics* 39 (2007): pp.1256–1260.

Peters, Anne, Ralf Kurvers, Melissa Roberts, and Kaspar Delhey. 'No Evidence for General Condition-Dependence of Structural Plumage Colour in Blue Tits: An Experiment.' *Journal of Evolutionary Biology* 24 (2011): pp.976–987.

Phillips, Adam. 'Against Self-Criticism.' *London Review of Books* 37 (2015): pp.13–16.

Phillips, John. 'Award of the Wollaston Medal and Donation Fund.' *Quarterly Journal of the Geological Society of London* 15 (1859): pp.xxiii–xxiv.

Piaget, Jean. *The Moral Judgement of the Child*. London: Routledge & Kegan Paul, 1932.

Piaget, Jean, and Jacques Vonèche. 'Reflections on Baldwin.' In *The Cognitive-Developmental Psychology of James Mark Baldwin: Current Theory and Research in Genetic Epistemology*, edited by John Broughton and John Freeman-Moir, pp.80–86. Norwood NJ: Ablex, 1982.

Pickering, Andrew. 'From Science as Knowledge to Science as Practice.' In *Science as Practice and Culture*, edited by Andrew Pickering, pp.1–28. Chicago: University of Chicago Press, 1992.

Pigliucci, Massimo. 'Do We Need an Extended Evolutionary Synthesis?' *Evolution* 61 (2007): pp.2743–2749.

Pigliucci, Massimo. 'Phenotypic Plasticity.' In *Evolution: The Extended Synthesis,* edited by Massimo Pigliucci and Gerd Muller Cambridge MA: MIT Press, 2010.

Pigliucci, Massimo, and Gerd Muller, eds. *Evolution: The Extended Synthesis*. Cambridge MA: MIT Press, 2010.

Pinker, Steven. 'The False Allure of Group Selection.' In *Handbook of Evolutionary Psychology*, edited by David Buss, pp.867–880. Hoboken NJ: Wiley, 2016.

Pléh, Csaba. 'Two Conceptions of the Crisis of Psychology: Vygotsky and Bühler.' In *Karl Bühler's Theory of Language*, edited by Achim Eschbach, pp.407–413. Amsterdam: John Benjamins, 1988.

Plomin, Robert. *Blueprint: How DNA Makes Us Who We Are*. London: Penguin, 2018.

Plotkin, Henry. *Evolutionary Thought in Psychology: A Brief History*. Oxford: Blackwell, 2004.

Popper, Karl. *Objective Knowledge: An Evolutionary Approach*. Oxford: Clarendon Press, 1972.

Porter, Duncan. 'Charles Darwin's Notes on Plants of the Beagle Voyage.' *Taxon* 31 (1982): pp.503–506.

Pradeu, Thomas. 'The Organism in Developmental Systems Theory.' *Biological Theory* 5 (2010): pp.216–222.

Priestley, John Boynton. 'The Linden Tree: A Play in Two Acts.' In *Time and the Conways and Other Plays*, pp.221–302. London: Penguin Books, 1969.

Prigogine, Ilya, and Isabelle Stengers. *Order out of Chaos: Man's New Dialogue with Nature*. London: Fontana Paperbacks, 1984.

Prodger, Phillip. 'Illustration as Strategy in Charles Darwin's "The Expression of the Emotions in Man and Animals".' In *Inscribing Science: Scientific Texts and the Materiality of Communication*, edited by Timothy Lenoir, pp.140–181. Stanford CA: Stanford University Press, 1998.

Prodger, Phillip. *Darwin's Camera: Art and Photography in the Theory of Evolution*. Oxford: Oxford University Press, 2009.

Proust, Marcel. *Against Sainte-Beuve and Other Essays*. Harmondsworth: Penguin Books, 1988. First published 1954.

Prum, Richard. *The Evolution of Beauty: How Darwin's Forgotten Theory of Mate Choice Shapes the Animal World--and Us.* New York: Doubleday, 2017.

Prum, Richard. 'The Lande–Kirkpatrick Mechanism Is the Null Model of Evolution by Intersexual Selection: Implications for Meaning, Honesty, and Design in Intersexual Signals.' *Evolution* 64 (2010): pp.3085–3100.

Rad, Mostafa, Alison Martingano, and Jeremy Ginges. 'Toward a Psychology of *Homo Sapiens*: Making Psychological Science More Representative of the Human Population.' *Proceedings of the National Academy of Sciences of the United States of America* 115 (2018): pp.11401–11405.

Radick, Gregory. 'Animal Agency in the Age of the Modern Synthesis: W.H. Thorpe's Example.' *British Journal for the History of Science: Themes* 2 (2017): pp.35–56.

Radick, Gregory. 'Darwin on Language and Selection.' *Selection* 1 (2002): pp.7–16.

Radick, Gregory. 'Darwin's Puzzling *Expression*.' *Comptes Rendus Biologies* 333 (2010): pp.181–187.

Radick, Gregory. 'Evidence-Based Evolutionism.' *Biological Theory* 5 (2010): pp.289–291.

Radick, Gregory. 'How and Why Darwin Got Emotional About Race: An Essay in Historical and Deep-Historical Reconstruction.' In *Historicizing Humans: Deep Time, Evolution, and Race in Nineteenth-Century British Sciences*, edited by Efram Sera-Shriar, pp.138–171. Pittsburgh PA: University of Pittsburgh Press, 2018.

Radick, Gregory. 'Is the Theory of Natural Selection Independent of Its History?' In *The Cambridge Companion to Darwin*, edited by Jonathon Hodge and Gregory Radick, pp.143–167. Cambridge: Cambridge University Press, 2003.

Radick, Gregory. *The Simian Tongue: The Long Debate About Animal Language.* Chicago: University of Chicago Press, 2007.

Ramsay, Edward. 'Letter to the Editor.' *Ibis* 9 (1867): pp.456–457.

Reddy, Vasu. 'Before the 'Third Element': Understanding Attention to Self.' In *Joint Attention: Communication and Other Minds*, edited by Naomi Eilan, Christoph Hoerl, Teresa McCormack and Johannes Roessler, pp.85–109. Oxford: Clarendon Press, 2005.

Reddy, Vasu. 'Infant Clowns: The Interpersonal Creation of Humour in Infancy.' *Enfance* 53 (2003): pp.247–256.

Reddy, Vasu. 'On Being the Object of Attention: Implications for Self–Other Consciousness.' *Trends in Cognitive Sciences* 7 (2003): pp.397–402.

Reed, Ed. 'Darwin's Earthworms: A Case Study in Evolutionary Psychology.' *Behaviourism* 10 (1982): pp.165–185.

Rennie, David, David Watson, and Michael Monteiro. 'The Rise of Qualitative Research in Psychology.' *Canadian Psychology* 43 (2002): pp.179–189.

Rheinberger, Hans-Jorg, and Peter McLaughlin. 'Darwin's Experimental Natural History.' *Journal of the History of Biology* 17 (1984): pp.345–368.

Richards, Evelleen. 'Darwin and the Descent of Woman.' In *The Wider Domain of Evolutionary Thought*, edited by David Oldroyd and Ian Langham, pp.57–111. Dordrecht: D. Reidel, 1983.

Richards, Evelleen. *Darwin and the Making of Sexual Selection.* Chicago: Chicago University Press, 2017.

Richards, Graham. *On Psychological Language and the Physiomorphic Basis of Human Nature.* London: Routledge, 1989.

Richards, Robert. *Darwin and the Emergence of Evolutionary Theories of Mind and Behaviour.* Chicago: University of Chicago Press, 1987.

Richards, Robert. *The Meaning of Evolution: The Morphological Construction and Ideological Reconstruction of Darwin's Theory.* Chicago: University of Chicago Press, 1992.

Richardson, Angelique. '"The Book of the Season": The Conception and Reception of Darwin's Expression.' In *After Darwin: Animals, Emotions, and the Mind*, edited by Angelique Richardson, pp.51–88. New York: Rodopi, 2013.

Richardson, Angelique. 'Darwin and Reductionisms: Victorian, Neo-Darwinian and Post-Genomic Biologies.' *19: Interdisciplinary Studies in the Long Nineteenth Century*, no. 11 (2010): pp.1–28.

Richerson, Peter, and Robert Boyd. *Not by Genes Alone: How Culture Transformed Human Evolution*. Chicago: University of Chicago Press, 2005.

Ricoeur, Paul. *Freud and Philosophy: An Essay on Interpretation*. New Haven CT: Yale University Press, 1970. First published 1965.

Ridley, Matt. *The Evolution of Everything*. New York: Harper, 2016.

Riley, Denise. *War in the Nursery: Theories of the Child and Mother*. London: Virago, 1983.

Rizvi, Jamila. *Why Women Do the Work and Don't Take the Credit*. Podcast audio. Big Ideas 2018. http://www.abc.net.au/radionational/programs/bigideas/why-women-do-the-work-and-dont-take-the-credit/10051596.

Robert, Jason, Brian Hall, and Wendy Olson. 'Bridging the Gap between Developmental Systems Theory and Evolutionary Developmental Biology.' *BioEssays* 23 (2001): pp.954–962.

Roberts, David. *The Social Conscience of Early Victorians*. Stanford CA: Stanford University Press, 2002.

Robinson, Peter. 'Literacy, Numeracy, and Economic Performance.' *New Political Economy* 3 (1998): pp.143–149.

Romanes, George. 'Charles Darwin, Part V. Psychology.' *Nature* 26, (1882): pp. 169–171.

Romanes, George. *Mental Evolution in Animals, with a Posthumous Essay on Instinct by Charles Darwin*. London: Kegan Paul, Trench & Co, 1883.

Romanes, George. *Mental Evolution in Man: Origin of Human Faculty*. London: Kegan Paul & Trench, 1888.

Romanes, George, and Ethel Romanes. *The Life and Letters of George John Romanes*. 2nd ed. London: Longmans, Green and Co., 1896.

Rose, Hilary, and Steven Rose, eds. *Alas, Poor Darwin: Arguments against Evolutionary Psychology*. London: Jonathan Cape, 2000.

Rose, Steven. *Lifelines: Biology, Freedom, Determinism*. Harmondsworth: Penguin Books, 1997.

Rose, Steven, Leon Kamin, and Richard Lewontin. *Not in Our Genes: Biology, Ideology and Human Nature*. Harmondsworth: Penguin Books, 1984.

Rosenthal, Robert, and Kermit Fode. 'The Effects of Experimenter Bias on the Performance of the Albino Rat.' *Behavioural Science* 8 (1963): pp.183–189.

Ross, Alex. *The Rest Is Noise: Listening to the Twentieth Century* London: Harper, 2009.

Ross, Sydney. 'Scientist: The Story of a Word.' *Annals of Science* 18 (1962): pp.65–85.

Rudwick, Martin. 'Charles Darwin in London: The Integration of Public and Private Science.' *Isis* 73 (1982): pp.186–206.

Rudwick, Martin. 'The Strategy of Lyell's Principles of Geology.' *Isis* 61 (1970): pp.4–33.

Ruskin, John. *Modern Painters*. 5 vols. Vol. 5. Boston: Estes and Lauriat, 1894. First published 1860.

Russell, James. *The Acquisition of Knowledge*. London: Macmillan, 1978.

Ryan, Joanna. 'Early Language Development: Towards a Communicational Analysis.' In *The Integration of a Child into a Social World*, edited by Martin Richards, pp.185–213. Cambridge: Cambridge University Press, 1974.

Rylance, Rick. *Victorian Psychology and British Culture, 1850–1880*. Oxford: Oxford University Press, 2000.

Sapp, Jan. 'The Nine Lives of Gregor Mendel.' In *Experimental Inquiries: Historical, Philosophical and Social Studies of Experimentation in Science*, edited by Homer Le Grand, pp.137–166. Dordrecht Netherlands: Kluwer Academic Publishers, 1990.

Sarton, May. *The House by the Sea: A Journal*. New York: Norton, 1981.

Schachter, Stanley, and Jerome Singer. 'Cognitive, Social, and Physiological Determinants of Emotional State.' *Psychological Review* 69 (1962): pp.379–399.

Schaffer, Rudolph. *The Growth of Sociability*. Harmondsworth: Penguin, 1971.

Schmidt, Karen, and Jeffrey Cohn. 'Human Facial Expressions as Adaptations: Evolutionary Questions in Facial Expression Research.' *Yearbook of Physical Anthropology* 44 (2001): pp.3–24.

Schmitt, Cannon. *Darwin and the Memory of the Human: Evolution, Savages, and South America*. Cambridge: Cambridge University Press, 2009.

Schwanz, Lisa. 'Parental Thermal Environment Alters Offspring Sex Ratio and Fitness in an Oviparous Lizard.' *Journal of Experimental Biology* 219 (2016): pp.2349–2357.

Schweber, Silvan. 'Darwin and the Political Economists: Divergence of Character.' *Journal of the History of Biology* 13 (1980): pp.195–289.

Scott, John. 'Letter to Darwin, 4th May.' In *Correspondence of Charles Darwin* Cambridge: Darwin Correspondence Project, 1868. https://www.darwinproject.ac.uk/letter/?docId=letters/DCP-LETT-6160.xml.

Secord, James. *Visions of Science: Books and Readers at the Dawn of the Victorian Age*. Chicago: University of Chicago Press, 2014.

Selby, Jane. 'Feminine Identity and Contradiction: Women Research Students at Cambridge University.' PhD thesis. Cambridge University, 1985.

Selby, Jane, and Ben Bradley. 'Infants in Groups: A Paradigm for the Study of Early Social Experience.' *Human Development* 46 (2003): pp.197–221.

Selby, Jane, Ben Bradley, Jennifer Sumsion, Matt Stapleton, and Linda Harrison. 'Is Infant Belonging Observable? A Path through the Maze.' *Contemporary Issues in Early Childhood* 19 (2018): pp.404–416.

Shakespeare, William. *Shakespeare's Hamlet*. New York: Henry Holt, 1914.

Shakespeare, William. *The Tempest,* edited by Jonathan Bate and Eric Rasmussen, Basingstoke, UK: Macmillan, 2005.

Shapin, Steven, and Barry Barnes. 'Darwin and Social Darwinism: Purity and History.' In *Natural Order: Historical Studies of Scientific Culture*, edited by Steven Shapin and Barry Barnes, pp.125–139. Beverly Hills CA: Sage, 1979.

Shapin, Steven, and Simon Schaffer. *Leviathan and the Air-Pump: Hobbes, Boyle, and the Experimental Life*. Princeton NJ: Princeton University Press, 1985.

Shapiro, Ian. *The Flight from Reality in the Human Sciences*. Princeton NJ: Princeton University Press, 2005.

Shotter, John. 'Agential Realism, Social Constructionism, and Our Living Relations to Our Surroundings: Sensing Similarities Rather Than Seeing Patterns.' *Theory & Psychology* 24 (2014): pp.305–325.

Shotter, John. *Images of Man in Psychological Research*. London: Methuen, 1975.

Shotter, John. 'The "Poetry" Prior to Science.' *New Ideas in Psychology* 11 (1993): pp.415–417.

Shotter, John, and Susan Gregory. 'On First Gaining the Idea of Oneself as a Person.' In *Life Sentences: Aspects of the Social Role of Language*, edited by Rom Harré, pp.3–9. New York: Wiley, 1976.

Sidgwick, Henry. *The Methods of Ethics*. 7th ed. London: Macmillan, 1907. First published 1874.

Simard, Suzanne, David Perry, Melanie Jones, David Myrold, Daniel Durall, and Randy Molina. 'Net Transfer of Carbon between Ectomycorrhizal Tree Species in the Field.' *Nature* 388 (1997): pp.579–582.

Simpson, George. 'The Baldwin Effect.' *Evolution* 7 (1953): pp.110–117.

Singer, Tania, and Olga Klimecki. 'Empathy and Compassion.' *Current Biology* 24, (2014): pp. R875–R878.

Skinner, Burrhus. *The Behaviour of Organisms: An Experimental Analysis.* New York: Appleton-Century-Crofts, 1938.

Skinner, Burrhus. *Beyond Freedom and Dignity.* London: Cape, 1972.

Slater, Peter. *A Field Guide to Australian Birds, Vol.2. Passerines.* Adelaide: Rigby, 1974.

Sloan, Phillip. 'Darwin, Vital Matter, and the Transformism of Species.' *Journal of the History of Biology* 19 (1986): pp.369–445.

Sloan, Phillip. 'Darwin's Invertebrate Program, 1826–1836: Preconditions for Transformism.' In *The Darwinian Heritage*, edited by David Kohn, pp.71–120. Princeton: Princeton University Press, 1985.

Sloan, Phillip. 'The Gaze of Natural History.' In *Inventing Human Science: Eighteenth Century Domains*, edited by Christopher Fox, Roy Porter, and Robert Wokler, pp.112–151. Berkeley CA: University of California Press, 1995.

Sloan, Phillip. 'John Locke, John Ray, and the Problem of the Natural System.' *Journal of the History of Biology* 5 (1972): pp.1–53.

Sloan, Phillip. 'The Making of a Philosophical Naturalist.' In *The Cambridge Companion to Darwin*, edited by Jonathon Hodge and Gregory Radick, pp.17–39. Cambridge: Cambridge University Press, 2003.

Sloan, Phillip. 'Natural History.' In *The Cambridge History of Eighteenth Century Philosophy*, edited by Knud Haakensson, pp.903–938. Cambridge: Cambridge University Press, 2006.

Smith, Adam. *The Theory of Moral Sentiments.* 6th ed. Sao Paulo: Soares, 2006. First published 1790.

Smith, Andrew. 'Letter to Darwin, 26th March.' In *Correspondence of Charles Darwin* Cambridge: Darwin Correspondence Project, 1867. https://www.darwinproject.ac.uk/letter/?docId=letters/DCP-LETT-5465.xml.

Smith, John. 'Message, Meaning, and Context in Ethology.' *American Naturalist* 99 (1965): pp.405–409.

Smith, Roger. *The Fontana History of the Human Sciences.* London: Fontana Books, 1997.

Smith, Roger. *Free Will and the Human Sciences in Britain 1870-1910.* Pittsburgh PA: University of Pittsburgh Press, 2016.

Smithson, Michael. *Confidence Intervals.* Thousand Oaks CA: Sage, 2003.

Smocovitis, Betty. 'Unifying Biology: The Evolutionary Synthesis and Evolutionary Biology.' *Journal of the History of Biology,* 25 (1992): pp.1–65.

Snow, Catherine. 'Mothers' Speech to Children Learning Language.' *Child Development* 43 (1972): pp.549–565.

Snow, Charles. 'The Two Cultures.' *Leonardo* 23 (1990): pp.169–173. First published 1959.

Snyder, Peter, Rebecca Kaufmann, John Harrison, and Paul Maruff. 'Charles Darwin's Emotional Expression "Experiment" and His Contribution to Modern Neuropharmacology.' *Journal of the History of the Neurosciences* 15 (2010): pp.158–170.

Sober, Elliott. 'Darwin on Natural Selection: A Philosophical Perspective.' In *The Darwinian Heritage*, edited by David Kohn, pp.867–900. Princeton NJ: Princeton University Press, 1985.

Sober, Elliott. *Evidence and Evolution: The Logic Behind the Science.* Cambridge: Cambridge University Press, 2008.

Sober, Elliott. 'Language and Psychological Reality: Some Reflections on Chomsky's *Rules and Representations*.' *Linguistics and Philosophy* 3 (1980): pp.395–405.

Sober, Elliott, and Wilson, David. *Unto Others: The Evolution and Psychology of Unselfish Behaviour*. Cambridge MA: Harvard University Press, 1998.

Spencer, Herbert. 'Bain on the Emotions and the Will.' In *Essays: Scientific, Political, and Speculative (Second Series)*, pp.120–142. London: Williams and Norgate, 1863.

Spencer, Herbert. *The Principles of Psychology*. London: Longman, Brown, Green, and Longmans, 1855.

Spencer, Herbert. *The Principles of Psychology*. 2nd ed. 2 vols. Vol. 1. London: Williams & Norgate, 1870.

Spencer, Herbert. *The Principles of Psychology*. 2nd ed. 2 vols. Vol. 2. London: Williams & Norgate, 1872.

Sperber, Dan. 'In Defense of Massive Modularity.' In *Language, Brain, and Cognitive Development: Essays in Honour of Jacques Mehler*, edited by Emmanuel Dupoux, pp.47–58. Cambridge MA: MIT Press, 2001.

Sponsel, Alistair. 'An Amphibious Being: How Maritime Surveying Reshaped Darwin's Approach to Natural History.' *Isis* 107 (2016): pp.254–281.

Sponsel, Alistair. *Darwin's Evolving Identity: Adventure, Ambition, and the Sin of Speculation*. Chicago: University of Chicago Press, 2018.

St. Mivart, George. '[Review of] The Descent of Man, and Selection in Relation to Sex.' *Quarterly Review* 131 (1871): pp.47–90.

Stam, Hank, ed. *The Body and Psychology*. London: Sage, 1998.

Stam, Hank. 'Unifying Psychology: Epistemological Act or Disciplinary Maneuver?' *Journal of Clinical Psychology* 60 (2004): pp.1259–1262.

Stam, Hank, and Tanya Kalmanovitch. 'E. L. Thorndike and the Origins of Animal Psychology: On the Nature of the Animal in Psychology'. *American Psychologist* 53 (1998): pp.1135–1144.

Stanislavski, Constantin. *An Actor Prepares*. London: Methuen, 1988. First published 1936.

Start, Edward. 'Is Politeness Innate?' *New Scientist* 71 (1976): pp.586–587.

Stauffer, Robert. 'Ecology in the Long Manuscript Version of Darwin's *Origin of Species* and Linnaeus' *Oeconomy of Nature*.' *Proceedings of the American Philosophical Society* 104 (1960): pp.235–241.

Stedman, Gesa. *Stemming the Torrent: Expression and Control in Victorian Discourses on Emotions, 1830-1872* Aldershot: Ashgate, 2002.

Steegmuller, Francis. *Cocteau: A Life*. London: Macmillan, 1970.

Stengers, Isabelle. *Another Science Is Possible: A Manifesto for Slow Science*. Cambridge: Polity Press, 2018. First published 2013.

Stengers, Isabelle. 'Who Is the Author?' In *Power and Invention: Situating Science*, pp.152–173. Minneapolis: University of Minnesota Press, 1997.

Stengers, Isabelle, and Ilya Prigogine. 'The Reenchantment of the World.' In *Power and Invention: Situating Science*, pp.33–60. Minneapolis: University of Minnesota Press, 1997.

Stenner, Paul. 'A.N. Whitehead and Subjectivity.' *Subjectivity* 22 (2008): pp.90–109.

Stenner, Paul. *Liminality and Experience: A Transdisciplinary Approach to the Psychosocial*. London: Palgrave Macmllan, 2017.

Stenner, Paul. 'On the Actualities and Possibilities of Constructionism: Towards Deep Empiricism.' *Human Affairs* 19 (2009): pp.194–210.

Stenner, Paul. 'Psychology in the Key of Life: Deep Empiricism and Process Ontology.' In *Theoretical Psychology: Global Transformations and Challenges*, edited by Paul

Stenner, John Cromby, Johanna Motzkau, Jeffrey Yen and Yu Haosheng, pp.48–58. Ontario: Captus, 2011.

Stetsenko, Anna. 'Darwin and Vygotsky on Development: An Exegesis on Human Nature.' In *Children, Development and Education: Cultural, Historical, Anthropological Perspectives*, edited by Michalis Kontopodis, Christoph Wulf, and Bernd Fichtner, pp.25–40. Dordrecht: Springer, 2011.

Stocking Jr, George. *Victorian Anthropology*. New York: Free Press, 1987.

Stoddart, Greta. 'Who's There?' *Poetry News*, no. Spring (2017): p.11.

Stokoe, Elizabeth. *Talk: The Science of Conversation*. London: Little Brown, 2018.

Stotz, Karola. 'Extended Evolutionary Psychology: The Importance of Transgenerational Developmental Plasticity.' *Frontiers in Psychology* 5 (2014): 1–14. https://www.frontiersin.org/articles/10.3389/fpsyg.2014.00908/full.

Strum, Shirley. 'Darwin's Monkey: Why Baboons Can't Become Human.' *Yearbook of Physical Anthropology* 55 (2012): pp.3–23.

Sulikowski, Danielle. 'Evolutionary Psychology: Fringe or Central to Psychological Science?' *Frontiers in Psychology* 7 (2016). https://www.frontiersin.org/articles/10.3389/fpsyg.2016.00777/full.

Sulikowski, Danielle, and Darren Burke. 'From the Lab to the World: The Paradigmatic Assumption and the Functional Cognition of Avian Foraging.' *Current Zoology* 61 (2015): pp.328–340.

Sulloway, Frank. *Freud, Biologist of the Mind: Beyond the Psychoanalytic Legend*. Cambridge MA: Harvard University Press, 1992.

Suls, Jerry, and Ralph Rosnow. 'Concerns About Artifacts in Psychological Experiments.' In *The Rise of Experimentation in American Psychology*, edited by Jill Morawski, pp.163–187. New Haven CT: Yale University Press, 1988.

Swinburne, Algernon. *Laus Veneris*. Portland ME: Thomas B Mosher, 1909. First published 1866.

Sylvester-Bradley, Peter. 'An Evolutionary Model for the Origin of Life.' In *Understanding the Earth*, edited by Ian Gass, Peter Smith and Richard Wilson, pp.123–141. Horsham: Open University Press, 1971.

Sylvester-Bradley, Peter. 'Evolution and the Destiny of Man.' *Syracuse University: Geology Contributions* 6 (1979): pp.1–23.

Sylvester-Bradley, Peter. 'Evolutionary Oscillation in Prebiology: Igneous Activity and the Origins of Life.' *Origins of Life* 7 (1976): pp.9–18.

Syme, Patrick. *Werner's Nomenclature of Colours: With Additions, Arranged So as to Render It Highly Useful to the Arts and Sciences, Particularly Zoology, Botany, Chemistry, Mineralogy, and Morbid Anatomy. Annexed to Which Are Examples Selected from Well-Known Objects in the Animal, Vegetable, and Mineral Kingdoms*. 2nd ed. Edinburgh: Blackwood, 1821.

Taine, Hippolyte. 'The Acquisition of Language by Children.' *Mind* 2 (1877): pp.252–259.

Tassy, Pascal. 'Trees before and after Darwin.' *Journal of Zoological Systematics and Evolutionary Research* 49 (2011): pp.89–101.

Taylor, Charles. *Sources of the Self: The Making of the Modern Identity*. Cambridge, MA: Harvard University Press, 1989.

Taylor, Eugene. 'William James on Darwin: An Evolutionary Theory of Consciousness.' *Annals of the New York Academy of Sciences* 602 (1990): pp.7–34.

Taylor, Gabriele, and Sybil Wolfram. 'The Self-Regarding and Other-Regarding Virtues.' *Philosophical Quarterly* 18 (1968): pp.238–248.

Terrall, Mary. *Catching Nature in the Act: Réaumur and the Practice of Natural History in the Eighteenth Century*. Chicago: Chicago University Press, 2014.

Thomas á Kempis. *The Imitation of Christ*. New York: Dover, 2003. First published c1423.

Thwaites, George. 'Letter to Darwin, 22nd July.' In *Correspondence of Charles Darwin* Cambridge: Darwin Correspondence Project, 1868. https://www.darwinproject.ac.uk/letter/?docId=letters/DCP-LETT-6285.xml.

Tinbergen, Niko. 'On Aims and Methods of Ethology.' *Animal Biology* 55 (2005): pp.297–321. First published 1963.

Tishkoff, Sarah, Floyd Reed, Alessia Ranciaro, et al. 'Convergent Adaptation of Human Lactase Persistence in Africa and Europe.' *Nature Genetics* 39 (2007): pp.31–40.

Tolman, Edward. 'The Determiners of Behaviour at a Choice Point.' *Psychological Review* 45 (1938): pp.1–41.

Tolstoy, Leo. *War and Peace*. London: The Reprint Society, 1960. First published 1867.

Tomlinson, Charles. 'A Meditation on John Constable.' In *The Penguin Book of Contemporary Verse*, edited by Kenneth Allott, pp.364–365. Harmondsworth: Penguin Books, 1962.

Tooby, John, and Leda Cosmides. 'The Evolutionary Psychology of the Emotions and Their Relationship to Internal Regulatory Variables.' In *Handbook of Emotions*, edited by Michael Lewis, pp.114–137. New York: Guilford Press, 2008.

Tooby, John, and Leda Cosmides. 'The Past Explains the Present.' *Ethology and Sociobiology* 11 (1990): pp.375–424.

Tooby, John, and Leda Cosmides. 'The Psychological Foundations of Culture.' In *The Adapted Mind: Evolutionary Psychology and the Generation of Culture*, edited by Jerome Barkow, Leda Cosmides and John Tooby, pp.19–136. New York: Oxford University Press, 1992.

Tooby, John, and Leda Cosmides. 'The Theoretical Foundations of Evolutionary Psychology.' In *The Handbook of Evolutionary Psychology*, edited by David Buss, pp.3–87. Hoboken NJ: Wiley, 2016.

Touraine, Alan. 'Is Sociology Still the Study of Society?' *Thesis Eleven* 23 (1989): pp.5–34.

Tourangeau, Roger, and Phoebe Ellsworth. 'The Role of Facial Response in the Experience of Emotion.' *Journal of Social and Personality Psychology* 37 (1979): pp.1519–1531.

Trevarthen, Colwyn. 'Communication and Cooperation in Early Infancy: A Description of Primary Intersubjectivity.' In *Before Speech: The Beginnings of Human Communication*, edited by Margaret Bullowa, pp.321–348. Cambridge: Cambridge University Press, 1979.

Trevarthen, Colwyn. 'Conversations with a Two-Month-Old.' *New Scientist* 62 (1974): pp.230–235.

Trevarthen, Colwyn. 'Descriptive Analyses of Infant Communicative Behaviour.' In *Studies in Mother-Infant Interaction*, edited by Rudolph Schaffer, pp.227–270. London: Academic Press, 1977.

Trevarthen, Colwyn. 'Two Mechanisms of Vision in Primates.' *Psychologische Forschung* 31 (1968): pp.299–337.

Trevarthen, Colwyn, and Kenneth Aitken. 'Infant Intersubjectivity: Research, Theory, and Clinical Applications.' *Journal of Child Psychology and Psychiatry* 42 (2001): pp.3–48.

Trivers, Robert. 'The Evolution of Reciprocal Altruism.' *Quarterly Review of Biology* 46 (1971): pp.35–57.

Trivers, Robert. 'Parental Investment and Sexual Selection.' In *Sexual Selection and the Descent of Man, 1871–1971*, edited by Bernard Campbell, pp.136–207. Chicago: Aldine Press, 1972.

Trollope, Anthony. *The Last Chronicle of Barset*. London: Smith & Elder, 1867. http://www.gutenberg.org/files/3045/3045-h/3045-h.htm.

Tylor, Edward. *Primitive Culture: Researches into the Development of Mythology, Philosophy, Religion, Language, Art, and Custom.* London: Murray, 1871.

Van Ommeren, Ron, and Thomas Whitham. 'Changes in Interactions between Juniper and Mistletoe Mediated by Shared Avian Frugivores: Parasitism to Potential Mutualism.' *Oecologia* 130 (2002): pp.281–288.

Vines, Sydney, and Dukinfield Scott. 'Reminiscences of German Botanical Laboratories in the 'Seventies and 'Eighties of the Last Century.' *New Phytologist* 24 (1925): pp.1–16.

Voloshinov, Vladimir. *Marxism and the Philosophy of Language.* New York: Seminar, 1973.

Voss, Julia. *Darwin's Pictures: Views of Evolutionary Theory, 1837–1874.* New Haven CT: Yale University Press, 2010. First published 2007.

Vygotsky, Lev. *Mind in Society: The Development of Higher Psychological Processes.* Cambridge MA: Harvard University Press, 1978.

Vygotsky, Lev. *Thought and Language.* Cambridge MA: MIT Press, 1962. First published 1934.

Wallace, Alfred. *Contributions to the Theory of Natural Selection.* London: Macmillan, 1870.

Wallace, Alfred. *Darwinism: An Exposition of the Theory of Natural Selection, with Some of Its Applications.* London: Macmillan, 1889.

Wallace, Alfred. 'Letter to Darwin, 29th May.' In *Correspondence of Charles Darwin* Cambridge: Darwin Correspondence Project, 1864. https://www.darwinproject.ac.uk/letter/?docId=letters/DCP-LETT-4514.xml.

Wallace, Alfred. 'Letter to Darwin, 1st May.' In *Correspondence of Charles Darwin* Cambridge: Darwin Correspondence Project, 1867. https://www.darwinproject.ac.uk/letter/?docId=letters/DCP-LETT-5522.xml.

Wallace, Alfred. 'Letter to Darwin, 24th February.' In *Correspondence of Charles Darwin* Cambridge: Darwin Correspondence Project, 1868. https://www.darwinproject.ac.uk/letter/?docId=letters/DCP-LETT-5922.xml.

Wallace, Alfred. 'Letter to Darwin, 1st March.' In *Correspondence of Charles Darwin* Cambridge: Darwin Correspondence Project, 1868. https://www.darwinproject.ac.uk/letter/?docId=letters/DCP-LETT-5966.xml.

Wallace, Alfred. 'Letter to Darwin, 15th March.' In *Correspondence of Charles Darwin* Cambridge: Darwin Correspondence Project, 1868. https://www.darwinproject.ac.uk/letter/?docId=letters/DCP-LETT-6012.xml.

Wallace, Alfred. 'The Origin of Human Races and the Antiquity of Man Deduced from the Theory of "Natural Selection".' *Journal of the Anthropological Society of London* 2 (1864): pp.clviii–clxxxvi.

Walsh, Denis. *Organisms, Agency, and Evolution.* Cambridge: Cambridge University Press, 2015.

Walsh, Denis. 'The Struggle for Life and the Conditions of Existence: Two Interpretations of Darwinian Evolution.' In *Evolution 2.0: Implications of Darwinism in Philosophy and the Social and Natural Sciences*, edited by Martin Brinkworth and Friedel Weinert, pp.191–209. Berlin: Springer-Verlag, 2012.

Walsh, Denis. 'Two Neo-Darwinisms.' *History and Philosophy of the Life Sciences* 32 (2010): pp.317–340.

Watson, David. 'Mistletoe: A Keystone Resource in Forests and Woodlands Worldwide.' *Annual Review of Ecology and Systematics* 32 (2001): pp.219–249.

Watson, John. *Behaviourism.* London: Kegan Paul & Co, 1925.

Watson, John. 'Psychology as the Behaviourist Views It.' *Psychological Review* 20 (1913): pp.158–177.

Watson, John, and Will Durant. 'Is Man a Machine? A Socratic Dialogue.' *The Forum* 32 (1929): pp.264–271.

Watson, Matthew. 'Performing Place in Nature Reserves.' *Sociological Review* 51 (2003): pp.145–160.

Watt-Smith, Tiffany. 'Darwin's Flinch: Sensation Theatre and Scientific Looking in 1872.' *Journal of Victorian Culture* 15 (2010): pp.101–118.

Wcislo, William. 'Behavioural Environments and Evolutionary Change.' *Annual Review of Ecology and Systematics* 20 (1989): pp.137–169.

Weiten, Wayne. *Psychology: Themes and Variations.* 9th ed. Belmont CA: Wadsworth, 2013.

West-Eberhard, Mary Jane. 'Darwin's Forgotten Idea: The Social Essence of Sexual Selection.' *Neuroscience and Biobehavioral Reviews* 46 (2014): pp.501–508.

West-Eberhard, Mary Jane. *Developmental Plasticity and Evolution.* New York: Oxford University Press, 2003.

West-Eberhard, Mary Jane. 'Mary Jane West-Eberhard.' *Evolution & Development* 11 (2009): pp.8–10.

West-Eberhard, Mary Jane. 'Toward a Modern Revival of Darwin's Theory of Evolutionary Novelty.' *Philosophy of Science* 75 (2008): pp.899–908.

Wetherell, Margaret. *Affect and Emotion: A New Social Science Understanding.* London: Sage, 2012.

Whewell, William. *The Philosophy of the Inductive Sciences.* 2 vols. Vol. 2. London: Parker, 1840.

Whippo, Craig, and Roger Hangarter. 'The "Sensational" Power of Movement in Plants: A Darwinian System for Studying the Evolution of Behaviour.' *American Journal of Botany* 96 (2009): pp.2115–2127.

White, Gilbert. *The Natural History of Selborne.* Oxford: Oxford University Press, 2013. First published as part of *The Natural History and Antiquities of Selborne* 1789.

White, Paul. 'Darwin Wept: Science and the Sentimental Subject.' *Journal of Victorian Culture* 16, no. 2 (2011): pp.195–213.

White, Paul. 'Darwin's Emotions: The Scientific Self and the Sentiment of Objectivity.' *Isis* 100 (2009): pp.811–826.

White, Paul. 'Darwin's Home of Science and the Nature of Domesticity.' In *Domesticity in the Making of Modern Science*, edited by Donald Opitz, Staffan Bergwik, and Brigitte Van Tiggelen, pp.61–83. Basingstoke UK: Palgrave Macmillan, 2016.

White, Paul. 'The Emotional Specimen: Darwin, Duchenne and the Science of Expression.' Seminar presentation, 24th May 2017.

White, Paul. 'The Face of Physiology.' *19: Interdisciplinary Studies in the Long Nineteenth Century*, no. 7 (2008): 1–22. http://doi.org/10.16995/ntn.487.

White, Paul. 'The Man of Science.' In *A Companion to the History of Science*, edited by Bernard Lightman, pp.153–163. London: Wiley, 2016.

White, Paul. 'Reading the Blush.' *Configurations* 24, no. 3 (2016): pp.281–301.

White, Paul. 'Sympathy under the Knife: Experimentation and Emotion in Late Victorian Medicine.' In *Medicine, Emotion and Disease, 1700–1950*, edited by Fay Alberti, pp.100–124. Basingstoke: Palgrave-Macmillan, 2006.

White, Paul. *Thomas Huxley: Making the 'Man of Science'* Cambridge: Cambridge University Press, 2003.

Whitehead, Alfred. *Nature and Life.* Cambridge: Cambridge University Press, 1934.

Whitehead, Alfred. *Process and Reality: An Essay in Cosmology.* New York: Free Press, 1978. First published 1929.

Wholleben, Peter. *The Hidden Life of Trees: What They Feel, How They Communicate.* Carlton, Vic.: Black Inc., 2016.

Wiener, Phillip. 'Peirce's Metaphysical Club and the Genesis of Pragmatism.' *Journal of the History of Ideas* 7 (1946): pp.218–233.

Wilde, Oscar. *The Decay of Lying.* 1889. http://cogweb.ucla.edu/Abstracts/Wilde_1889. html.

Wilde, Oscar. *The Picture of Dorian Gray.* London: Simpkin Marshall Harrison, 1920. First published 1891.

Williams, George. *Adaptation and Natural Selection.* Princeton NJ: Princeton University Press, 1966.

Williams, Raymond. 'On Dramatic Monologue and Dialogue (Particularly in Shakespeare).' In *Writing in Society*, pp.31–64. London: Verso, 1991.

Williams, Raymond. 'Social Darwinism.' In *Herbert Spencer: Critical Assessments*, edited by John Offer, pp.186–198. London: Routledge, 2000.

Williams, Tennessee. *Cat on a Hot Tin Roof.* New York: New Directions Books, 2004. First published 1954.

Wilson, David, and Edward Wilson. 'Rethinking the Foundation of Sociobiology.' *Quarterly Review of Biology* 82 (2007): pp.327–348.

Wilson, Edward. *Sociobiology: The New Synthesis.* Cambridge MA: Harvard University Press, 1975.

Wilson, Robert, and Matthew Barker. 'The Biological Notion of Individual.' In *Stanford Encyclopedia of Philosophy*, edited by Edward Zalta 2017. https://plato.stanford.edu/arch-ives/spr2017/entries/biology-individual/.

Wimsatt, William, and Monroe Beardsley. 'The Intentional Fallacy.' *Sewanee Review* 54 (1946): pp.468–488.

Winch, Peter. *The Idea of a Social Science and Its Relation to Philosophy.* London: Routeldge and Kegan Paul, 1958.

Winnicott, Donald. 'Mirror-Role of Mother and Family in Child Development.' In *Playing and Reality*, pp.149–159. Harmondsworth: Penguin Books, 1967.

Wispé, Lauren. 'The Distinction between Sympathy and Empathy: To Call Forth a Concept, a Word Is Needed.' *Journal of Personality and Social Psychology* 50, no. 2 (1986): pp.314–321.

Wittgenstein, Ludwig. *Zettel.* Berkeley CA: University of California Press, 1967.

Woolf, Virginia. 'Mr. Bennett and Mrs. Brown.' In *The Captain's Deathbed and Other Essays*, pp.94–119. San Diego: Harcourt Brace Jovanovich, 1978. First published 1924.

Woolf, Virginia. 'Professions for Women.' In *The Death of the Moth and Other Essays*, pp.235–241. New York: Harcourt, Brace & Company, 1942.

Wright, Chauncey. 'Review [of *Contributions to the Theory of Natural Selection. A Series of Essays* by Alfred Russell Wallace].' *North American Review* 111 (1870): pp.282–311.

Wundt, Wilhelm. *Principles of Physiological Psychology.* Translated by Edward Titchener. New York: Macmillan, 1910. First published 1874.

Wundt, Wilhelm. *Völkerpsychologie : Eine Untersuchung Der Entwicklungsgesetze Von Sprache, Mythus Und Sitte, Teil 1: Sprache.* Leipzig: Wilhelm Engelmann, 1904.

Wynne, Clive. 'What Are Animals? Why Anthropomorphism Is Still Not a Scientific Approach to Behaviour.' *Comparative Cognition & Behavior Reviews* 2 (2007): pp.125–135.

Young, Jacy. 'The Baldwin Effect and the Persistent Problem of Preformation versus Epigenesis.' *New Ideas in Psychology* 31 (2013): pp.355–362.

Young, Robert. 'Darwinism *Is* Social.' In *The Darwinian Heritage*, edited by David Kohn, pp.609–638. Princeton: Princeton University Press, 1985.

Young, Robert. *Darwin's Metaphor: Nature's Place in Victorian Culture.* Cambridge: Cambridge University Press, 1985.

Young, Robert. *Mind, Brain, and Adaptation in the Nineteenth Century: Cerebral Localization and Its Biological Context from Gall to Ferrier.* History of Neuroscience. New York: Oxford University Press, 1990. First published 1970.

Young, Robert. 'Scholarship and the History of the Behavioural Sciences.' *History of Science* 5 (1966): pp.1–51.

Zirkle, Conway. 'Natural Selection before the *Origin of Species.' Proceedings of the American Philosophical Society* 84 (1941): pp.71–123.

Index

Note: Tables and figures are indicated by *t* and *f* following the page number

For the benefit of digital users, indexed terms that span two pages (e.g., 52–53) may, on occasion, appear on only one of those pages.

thought (*cont.*)
   facial movements revealing, not
      expressing 106, 317–18
   as inherently conscientious, 255
   of others about us 163–64
   two types of 171
threat displays 196–97
Thwaites, George 125
Tinbergen, Niko 52–53, 149–50
toddlers *see* infants/children
tools 34*f see also* collecting equipment
   for dissection 33–34
   geological 33
   of macho violence 195
   microscope designed by Darwin 36–37
   primates' making of 240*t*, 263–64
   use of 240*t*, 263
Totem and Taboo (Freud) 258, 259
transdisciplinary study 329, 331–32, 348
transitional habits *see* habits
transmission (inheritance) 340, 342, 349
   of acquired habits 7
   cultural 89, 275, 281–82, 288, 343
   development and 85–86
   of DNA 344
   of gemmules 342
   of heritable material 2, 17, 85–86
   law of equal transmission 185–86, 199
   of memes 343
transmutation of species 44, 112–13, 181
travel, scientific advance and 28
trees 90
   asexual reproduction in 74–75
   cooperation between 90
trees of life 37, 39*f*, 80–81, 81*f*, 180
Trevarthen, Colwyn 300, 305*f*
trial and error 69, 70, 89, 98, 296, 297–98, 317, 347
triangulation *see* methods
tribal self 260
tribes 236–37 *see also* clan; groups; savages
   competition between primeval 246
   human ornamentation in *see* body adornment
   Native American 84–85
   proto-human 238
   sense of moral right and wrong 274–75
   standards of beauty 217–18

   women as cause of war in Australian tribes 213
Trivers, Robert 230
Tversky, Amos 171
twayblade orchids (*Listera ovata*) 102*f*
Tylor, Edward 217, 268, 281

unconscious 171 *see also* Bion, Wilfred; Freud, Sigmund; psychoanalysis
   actions 116–17, 133, 158
   dream-thoughts 323
   human agency 59
   selection 217–18
universal emotions 146–48
   blushing 125
unseen
   dreams and 281
   God 281
   human belief in unseen or spiritual agencies 59, 281
use and disuse of parts 59, 77–78, 84, 349

vanity
   animal 192, 213–14
   human 213–14
variability 80–83 *see also* variation
   origins of 85
   in secondary sexual characters 228–29
   sexual selection dependence on 188
The Variation of Animals and Plants under Domestication (Darwin) 85
variation, variations 77, 349 *see also* correlated variation
   development produces 316
   individual differences as variations 349
   theatre of agency generates 77–78, 83, 267–68, 271
   variation as process 349
Victorian Psychology and British Culture, 1850–1880 (Rylance) 318
Völkerpsychologie. Eine Untersuchung der Entwicklungsgesetze von Sprache, Mythus und Sitte (Wundt) 140–41
von Haller, Albrecht 66
Vygotsky, Lev 92, 300–1

Wallace, Alfred 11, 27–28, 200–1, 228, 239, 273, 285–86, 290